5-
5/23

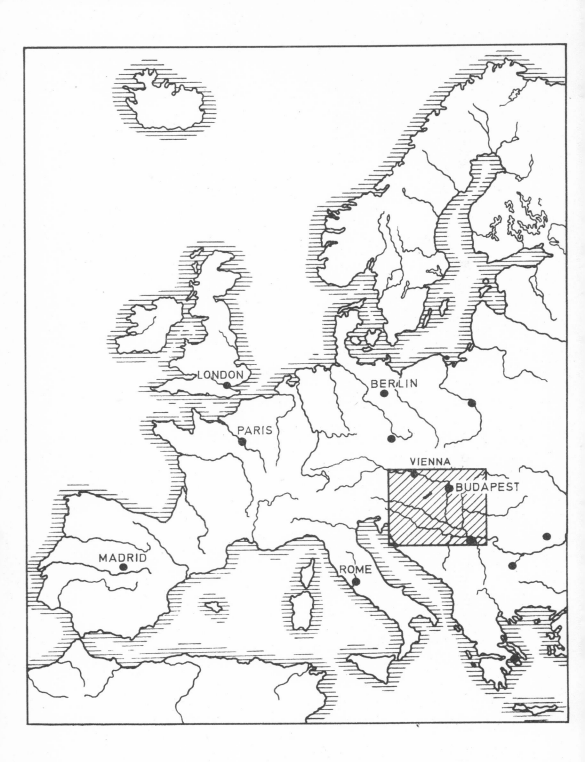

# THE ARCHAEOLOGY
# OF ROMAN
# PANNONIA

*Contributors*

László Barkóczi

Éva B. Bónis

György Duma

Jenő Fitz

Ferenc Fülep

Imre Jakabffy

Alfonz Lengyel

Imre Lengyel

Klára Póczy

G. T. B. Radan

Ágnes Salamon

Katalin Bíró-Sey

Sándor Soproni

Ágnes Cs. Sós

Edit B. Thomas

Ottó Trogmayer

# THE ARCHAEOLOGY OF ROMAN PANNONIA

A. Lengyel and G. T. B. Radan

*Editors*

**THE UNIVERSITY PRESS OF KENTUCKY**

**AKADÉMIAI KIADÓ, BUDAPEST**

ISBN 0-8131-1370-9 (University Press of Kentucky)

ISBN 963 05 1886 4 (Akadémiai Kiadó)

Library of Congress Catalog Card Number: 77-080463

A statewide cooperative scholarly publishing agency serving Berea College, Centre College of Kentucky, Eastern Kentucky University, The Filson Club, Georgetown College, Kentucky Historical Society, Kentucky State University, Morehead State University, Murray State University, Northern Kentucky University, Transylvania University, University of Kentucky, University of Louisville, and Western Kentucky University.

*Editorial and Sales Offices:* Lexington, Kentucky 40506

Joint publication by the University Press of Kentucky, Lexington, Kentucky and Akadémiai Kiadó, Budapest, Hungary

Printed in Hungary

# CONTENTS

# CONTRIBUTORS

*László Barkóczi*, head of department of ancient studies in the Archaeological Institute of the Hungarian Academy of Sciences, Budapest

*Éva B. Bónis*, chief researcher and deputy department chief in the Hungarian National Museum, Budapest

*György Duma*, professor of ceramic studies and researcher in the Archaeological Institute of the Hungarian Academy of Sciences, Budapest

*Jenő Fitz*, general director of the County Museum of Székesfehérvár

*Ferenc Fülep*, general director of the Hungarian National Museum, Budapest

*Imre Jakabffy*, department chief, Museum of Applied Arts, Budapest

*Alfonz Lengyel*, dean, Institute for Mediterranean Art and Archaeology, Cincinnati, Ohio

*Imre Lengyel*, researcher in the Archaeological Institute of the Hungarian Academy of Sciences, Budapest

*Klára Póczy*, field director of Aquincum Excavations and chief researcher in the Historical Museum, Budapest

*G. T. B. Radan*, professor of art and archaeology at Villanova University, Villanova, Pennsylvania

*Ágnes Salamon*, chief researcher in the Archaeological Institute of the Hungarian Academy of Sciences, Budapest

*Katalin Bíró-Sey*, chief researcher in the Hungarian National Museum, Budapest

*Sándor Soproni*, chief researcher and department chief in the Hungarian National Museum, Budapest

*Ágnes Cs. Sós*, chief researcher in the Hungarian National Museum, Budapest

*Edit B. Thomas*, chief researcher and department chief in the Hungarian National Museum, Budapest

*Ottó Trogmayer*, general director of the County Museum of Szeged

11

# ABBREVIATIONS

| | | |
|---|---|---|
| ActaA | = | Acta Archaeologica, Copenhagen |
| Acta Anat. | = | Acta Anatomica |
| ActaAntHung | = | Acta Antiqua Academiae Scientiarum Hungaricae |
| ActaArchHung | = | Acta Archaeologica Academiae Scientiarum Hungaricae |
| Acta Ethnogr. Hung. | = | Acta Ethnographica Academiae Scientiarium Hungaricae |
| Acta Physiol. Scand. | = | Acta Physiologica Scandinavica |
| AEpigr | = | Année Epigraphique |
| Amer. Jour. Phys. Anthrop. | = | American Journal of Physical Anthropology |
| Amm. Marc. | = | Ammianus Marcellinus |
| Annalen Physik. | = | Annalen der Physik |
| Ann. sci. nat. | = | Annales des sciences naturelles |
| AntCl | = | L'Antiquité Classique |
| Anthrop. Közl. | = | Anthropológiai Közlemények |
| AntTan | = | Antik Tanulmányok |
| ArchAnzeiger | = | Archäologischer Anzeiger |
| ArchAu | = | Archaeologia Austriaca |
| ArchÉrt | = | Archaeologiai Értesítő |
| ArchHung | = | Archaeologia Hungarica |
| ArchIug | = | Archaeologia Iugoslavica |
| Arch. exp. Pathol. Pharmakol. | = | Archiv für experimentelle Pathologie und Pharmakologie |
| Archiv. f. Österr. Geschichte | = | Archiv für österreichische Geschichte |
| ArchJ | = | Archaeological Journal |
| Arch. Suisses Anthrop. Gén. | = | Archives Suisses d'Anthropologie Générale |
| ArchKöz. | = | Archaeologiai Közlemények |
| Aur. Vict. Caes. | = | Aurelius Victor, Caesares |
| AV | = | Arheološki Vestnik |
| BayVb | = | Bayerische Vorgeschichtsblätter |
| Biochem. J. | = | Biochemical Journal |
| Bonner Jb. | = | Bonner Jahrbuch |
| BRGK | = | Bericht der Römisch-Germanischen Kommission |
| BSA | = | British School at Athens, Annual |

| | | |
|---|---|---|
| BudRég | = | Budapest Régiségei |
| Bull. et Mém., Soc. d'Anthrop. Paris | = | Bulletins et Mémoires de la Société d'Anthropologie |
| Bull. Soc. Chim. France | = | Bulletin de la Société Chimique de France |
| Bull. Soc. Préhist. Française | = | Bulletin de la Société Préhistorique Française |
| Chem. Rev. | = | Chemical Reviews |
| C.I.L. | = | Corpus Inscriptionum Latinarum |
| DissPann | = | Dissertationes Pannonicae |
| Diss. Phil. | = | Naturwiss. Fak. d. Univ. Basel |
| EpigrStud | = | Epigraphische Studien |
| ESA | = | Eurasia Septentrionalis Antiqua |
| Eutrop | = | Eutropius |
| FolArch | = | Folia Archaeologica |
| FontesArchHung | = | Fontes Archaeologici Hungariae |
| Hist. Aug. Aurel. | = | Historia Augusta Aurelianus |
| Hist. Aug. Comm. | = | Historia Augusta Commodus |
| Hist. Aug. Max. | = | Historia Augusta Maximianus |
| ILS | = | Inscriptiones Latinae Selectaes, Dessau |
| Ind. Eng. Chem., Anal. Ed. | = | Industrial and Engineering Chemistry, Analytical Edition |
| Intercisa | = | Intercisa I–II. Geschichte der Stadt in der Römerzeit. Budapest, 1954–1957 |
| It. Ant. | = | Itinerarium Antonini |
| Jahrb. f. Landeskunde v. Niederösterr | = | Jahrbuch für Landeskunde von Niederösterreich |
| J. Biol. Chem. | = | Journal of Biology and Chemistry |
| JfNGeldg | = | Jahrbuch für Numismatik und Geldgeschichte |
| JÖAI | = | Jahreshefte des Österreichischen Archäologischen Instituts |
| Jour. Anthrop. Soc. Japan | = | Journal of the Anthropological Society, Japan |
| Jour. Applied Chem. | = | Journal of Applied Chemistry |
| JPMuzÉv | = | Janus Pannonius Múzeum Évkönyve |
| JRGZ | = | Jahrbuch des Römisch-Germanischen Zentralmuseums |
| JRS | = | The Journal of Roman Studies |
| Klin. Wschr. | = | Klinische Wochenschrift |
| KölnerJb | = | Kölner Jahrbuch für Vor- und Frühgeschichte |
| Mitt. d. Num. Ges. in Wien | = | Mitteilungen der Numismatischen Gesellschaft in Wien |
| MTA BiolOK | = | A Magyar Tudományos Akadémia Biológiai Osztályának Közleményei |
| MTAK II | = | A Magyar Tudományos Akadémia II Osztályának Közleményei |

| | | |
|---|---|---|
| MTA TTOK | = | A Magyar Tudományos Akadémia Társadalmi-Történeti Tudományok Osztályának Közleményei |
| Not. Dig. | = | Notitia Dignitatum |
| NumKözl | = | Numizmatikai Közlöny |
| NumListy | = | Numismatické Listy |
| NZ | = | Numismatische Zeitschrift |
| Praehist. Ztschr. | = | Praehistorische Zeitschrift |
| Proc. Geol. Soc. London | = | Proceedings of the Geological Society, London |
| ReiCretActa | = | Rei Cretariae Romanae Fautorum Acta |
| R.E.Pauly–Wissowa | = | Realenzyklopedie der klassischen Altertumswissenschaft (Begr. Pauly u. Wissowa) |
| RéTan | = | Régészeti Tanulmányok |
| RIU | = | Die Römischen Inschriften in Ungarn |
| RLiÖ | = | Der Römische Limes in Österreich |
| RömMitt | = | Mitteilungen des Deutschen Archaeologischen Instituts, Römische Abteilung |
| RVillP | = | Thomas: Römische Villen in Pannonien |
| SA | = | Slovenska Archeologia |
| SowArch | = | Sowjetskaja Archaeologija |
| StudArch | = | Studia Archaeologica |
| SzekszárdiMúz.Évk. | = | Szekszárdi Múzeum Évkönyve |
| TheoSz | = | Theologiai Szemle |
| TIR | = | Tabula Imperii Romani |
| Wiss. Ztschr. d. Humboldt-Univ. zu Berlin | = | Wissenschaftliche Zeitschrift der Humboldt-Universität zu Berlin |
| WPZ | = | Wiener Prähistorische Zeitschrift |
| Zeitschr. f. anorg. u. allg. Chemie | = | Zeitschrift für anorganische und allgemeine Chemie |

# INTRODUCTION

Today the word *archaeology*, the study of man's past by means of the material relics left to posterity, is, of course, familiar. But the public still associates the concept of archaeology with that of the so-called classical world, for the public is familiar with the spectacular discoveries of the last century in Greece, Italy, Asia Minor, Egypt, and Mesopotamia. Relatively unknown, however, are more recent discoveries in the various provinces of the Roman Empire, where some of the most important large-scale excavations are currently taking place. The artifacts uncovered bring to light the picture of a fascinating culture, suggesting the ingenuity of the indigenous population. We catch a glimpse of life at the time of the Romans, before and during the early years of Christianity, not at the usual centers, but at the frontiers of Central Europe. When the power of the Roman Empire was on the wane, these provinces became as vital to the Romans as the Italian peninsula itself. The fate of the borders became more crucial than ever before.

Pannonia was one of these border provinces in Central Europe, lying in the Danube Valley, between the Alps and the Carpathian Mountains. Its borders were the Danube River in the north and east, the Alps to the west, and the Sava River in the south. The territory corresponds to present-day Hungary, the so-called trans-Danube area. Northwest Pannonia was the present-day Burgenland of modern Austria, but it stretched into the Vienna Valley as well. South Pannonia covered the area between the Drava and Sava rivers which belongs to modern Yugoslavia. In its time this area was one of the most important provinces of the Roman world, indelibly stamped with a kaleidoscopic variety of characteristics. In establishing its historic borders the *Tabula Peutingeriana*, the *Itinerarium Antonini*, the *Anonymus of Ravenna*, and the *Notitia Dignitatum* are helpful. When the name *Pannonia* first appeared in Polybius, its etymology was unclear. The meaning of the term is not completely definite even today. Sometimes it is used as a collective name for the geographical area, sometimes it refers to the Illyrian tribes populating the Sava Valley. Its acceptance as a geographical designation has gone beyond the chronological period dealt with in this book, since the kings of the Magyars proudly entitled themselves, in the Middle Ages, "Rex Pannoniae", hundreds of years after the fall of the Roman Empire in the West.

The work of Hungarian archaeologists within their own country is not entirely separated from those of colleagues who live in Austria, Yugoslavia, Czechoslovakia, and Romania because of international symposiums held frequently on Pannonia and related matters. The borders that concern archaeologists are iden-

17

tified with those of the Roman Empire, from Britain to the Pillars of Hercules, Africa, and across to Asia.

The idea of acquainting the world with this area as it had been in ancient times was reborn in 1968 when the Serbian Academy of Sciences and the Smithsonian Foreign Currency Program jointly launched the first large-scale expedition in the late Roman capital of Sirmium (under the leadership of Alfonz Lengyel). After a successful campaign, Lengyel left the expedition in 1969. Unfortunately, for a variety of reasons the expedition under new leadership was not a success, and the resulting publications, following in 1971 to 1973, shed little new light (Sirmium I–III, ed. E. Ochsenschlager and V. Popović). But as part of this cooperative effort, it was decided to try to bring forth the results of Pannonian research before the English-speaking public, which had looked upon such efforts before 1960 as entirely dependent on Roman Imperial history, rather than as specialized archaeological research. Although it is true that research in these areas cannot be separated from the history of the Roman Empire and that in a book tied to chronological subdivision and historical principles, the chronology of the empire will be dominant, it is also true that the two researches differ in details. While intensive research relating to the empire goes back hundreds of years, scientific research in the provinces, especially Pannonia, in terms of scientific methods, does not antedate the last one hundred years. But scholarship tied to the empire has passed its zenith, while research related to Pannonia intensified after the Second World War and is now reaching its peak.

Among the various provinces Pannonia played a unique role. It was located at the crossroads of East and West; the area lay in the heart of Europe, embraced by the Carpathian Mountains. It is not surprising, then, that it was a crossroad of invasions and suffered from many wars.

The remains of a rich cultural heritage lies buried under the earth. As early as Paleolithic times cultures started and developed, reaching relatively high levels. The result of enemy attacks and armies raking the territory was the eradication of these cultures, but generally survived some trace of their existence. On the ruins of the old, new life started. Levels of civilizations, sometimes poor, sometimes wealthy and well-developed, piled one above the other, provide us with an accurate record. To an archaeologist the rediscovery of the strata of bygone civilizations in Pannonia seems like the turning of the pages of a book.

Interest in the Pannonian past goes back centuries. But this interest had to do with the legends and associations attached to particular sites. Simon Kézai in his *Gesta Hungarorum*, dating from the Middle Ages, reports that one of the Béla kings of the Magyars erected his royal palace in the twelfth century on the spot where the ruins of the city of Attila, king of the Huns, were still visible near Sycambria.

King Matthias Corvinus of Hungary, whose houses at Buda were built "ad italicorum aedificiorum symmetriam" by Italian masons and decorated by Italian sculptors, was the only Renaissance ruler outside of Italy in the fifteenth century. He is known to have funded an archaeological collection in addition to establishing the well-known Corvinian Library. The historian Bonfini refers to this collection in his work, and part of the collection is thought to have been discovered during the recent excavations at Buda (L. Gerevich, *Excavations at the Burg of Buda* [Budapest, 1966] in Hungarian) and Szombathely—Savaria (E. B. Thomas, *Il primo Museo Ungharese* [Budapest, 1966] in Italian). J. Marsigli, the noted Italian military architect, in the seventeenth century made several sketches and sketch maps of ruins along the Danube River, noting the still visible ruins of the great Roman defense line, the *limes*. The English travelers J. Milles and R. Pococke traveled to Hungary in the eighteenth century to study the antiquities there and made illustrated diaries of their journey. Recent excavations have authenticated the accuracy of their observations. The travel fever of the Baroque period and the ambition for discovery continued during the reign of Queen Maria Theresia and Joseph II. The extensive surveys and road construction projects of this period resulted in further archaeological discoveries.

Count Ferenc Széchényi established the Hungarian National Museum in 1802, the year Lord Elgin sent his two hundred chests of marbles from Athens to England, where they became the heart of the collection of the British Museum. It was also at this time that Ennio Quirino Visconti was about to open his "l'antiquarium" in the summer home of Anne of Austria and the illustrious Louvre was still called the "Musée Napoléon". Thus the Hungarian National Museum became one of the earliest European national archaeological collections.

During the last century some of the first Hungarian archaeological publications appeared in Latin and Hungarian. F. Rómer was the first Hungarian archaeologist systematically to plan excavations in Hungary. He directed excavations and published reports on the excavated material. After him, two internationally known archaeologists, F. Pulszky and J. Hampel, started to systematize all findings, while one of the most basic systematic researches is identified with the name of B. Kuzsinszky, who founded the Museum of Aquincum. As was customary at that time, Hampel divided archaeology into two parts: prehistory and the medieval history of Hungary. The proceedings of the Eighth International Congress of Pre-History and Anthropology held in 1876 were published in French (in the international tradition of Hungarian archaeology, a tradition that has never changed). Since 1859 the *Archaeological Notes (Archaeologiai Közlemények)*, and since 1868 the *Archaeologiai Értesítő*, have been published in several languages. When A. Alföldi, who later was professor at the Institute of Advanced Study at Princeton, established the two series of monographs called *Dissertationes Pannonicae*, they were directed to the international public.

During the second half of the nineteenth century, in the wake of the great revival of antiquarian interests, archaeological societies were established on local levels which gave impetus to the founding of city museums and private collections. Around the end of the century the publication of archaeological monographs, mostly of regional importance, relating to counties and cities, began. These were written by specialists, though systematic excavation in the modern sense did not exist in the area of greater Hungary. Research was directed toward art-historical and aesthetic evaluations of single objects, rather than placement in an archaeological framework, even though several excellent researchers, well versed in archaeology, were among the historians doing this work. Still, at the threshold of the twentieth century, there were exceptions in the field of Pannonian research, a field that managed to advance slowly because of the few critical and balanced works from this period. Among these were K. Torma, B. Posta, and M. Wosinsky. This generation, however, was followed by B. Kuzsinszky, L. Márton, F. Tompa, L. Nagy, J. Banner, and A. Alföldi. Most of the contributors to this volume are the beneficiaries of this older generation of Hungarian archaeologists.

After World War II the number of archaeological periodicals multiplied; there appeared the *Acta Antiqua Academiae Scientiarum Hungaricae*, the *Acta Archaeologica Academiae Scientiarum Hungaricae* and *Budapest Régiségei* (the Antiquities of Budapest) and the *Folia Archaeologica*, to cite some of the best-known examples. Others are the *Alba Regia, Gorsium, Antik Tanulmányok*, and *Cumania*. In addition to these a dozen or so other journals of regional importance have appeared, such as the *Fejér Megye Története* (History of the County Fejér), *Arrabona, Antiquitas Hungarica*, and, of course, the outstanding museum publications such as the *Bulletin du Musée Hongrois des Beaux-Arts*. The postwar era established periodicals with great flexibility, capable of publishing new achievements and preliminary excavation results without long delays. A great many monographs have provided international forums for archaeologists.

The postwar era also brought new laws for the protection of ancient monuments, as many nations began to pay more attention to their cultural heritage. Museums and other institutions in Hungary started to build systematic registers to document and protect the Hungarian national heritage. After the war-torn museums and galleries were rebuilt, more excavations were undertaken.

On the several monographs published on Roman cities and settlements, and Pannonia in general, the first after the war was on Intercisa, in two volumes (1954, 1957). After 1956 Edit B. Thomas edited the *Archaeologische Funde in Ungarn*, and István Bóna edited *Hungarian Art till the Arrival of the Magyars* (in Hungarian). The early compilations of archaeological finds from prehistory to the medieval period were systematically published in these books. In 1962 the first volume of the *Archäologische Bibliographie des Mittel-Donaubeckens* was published by J. Banner and I. Jakabffy. It was a complete bibliographical work, organized according to subject matter, and included a complete international bibliography on Pannonia. The book appeared again in 1968 with a cumulative bibliography and will be republished periodically. In 1962 A. Mócsy published his monograph in the *Pauly-Wissowa Real Encyclopedia*, the first comprehensive work on Pannonia, written by an archaeologist who comes from that area. Articles of lesser value were published previously in the *Encyclopedia dell'Arte Antica* and *Fasti Archaeologici*, which suffered from a lack of expert knowledge. Among the newest of the comprehensive works is Mócsy's *Pannonia and Upper Moesia* (London, 1974), one of the series of *The Provinces of the Roman Empire*, which, with his "Die Bevölkerung von Pannonien bis zu den Markomannen-kriegen" (*JRS* 52 [1962]), constitute a substantial contribution.

The first textbook was published in 1963, written by L. Barkóczi, I. Bóna, and A. Mócsy, and edited by J. Harmatta. It introduced students of archaeology to prehistoric Roman Pannonia, and to the decline and fall of the Roman world in the province; it included a selected bibliography. It was soon followed by another text written by Mócsy in 1972, in Hungarian, which presented a well-balanced historical analysis, using ancient sources to describe the whole ancient pre-Roman Pannonian panorama. Mócsy dealt here, in depth, with controversial issues, by going to Greek sources for the interpretations of names. But many of the explanations remain hypothetical, without enlisting supporting archaeological evidence.

Meanwhile, Hungarian archaeology introduced new methods to fit situations not encountered before, with previously unavailable manpower. With special pride in the past and in their national treasures, the Hungarian government promoted excavations, frequently with the help of the army, and using new technology in archaeology. This new methodology was already published in 1954 in an excellent archaeological technical manual under the joint authorship of J. Banner, Gy. László, I. Méri, and A. Radnóti. These innovative techniques produced some revolutionary logistics, among them A. Salamon's punch-card system for the evaluation and organization of Avaric and fifth-century material. This was an inexpensive, computerized method, useful at any site, that eliminated expensive computer evaluation and delay that risks the loss or damage of material. ("The Fifth Century and Avaric Material Evaluated with the Punch-card System", in Hungarian *ArchÉrt* 93 [1966]: 284–90).

Among historians and archaeologists from outside Hungary who dealt with comprehensive works on Pannonia, P. Oliva stands out. His *Pannonia and the Onset of the Crisis in the Roman Empire*, published in English in 1962, was researched in a remarkably short period of time. In it, however, nineteenth- and twentieth-century sources are treated in a somewhat opinionated manner; missing is the well-balanced critical perspective found in Oliva's other historical works. In his foreword he projected some ideas on the Roman outposts in Slovakia and Moravia, but unfortunately failed to elaborate on these. Recently it was suggested that some buildings there were residences of Roman clients.

For students of the history of pre-Roman Pannonia it would seem that due to a lack of any cohesive force the area was subject to frequent power changes, with the population periodically absorbed by brutal, if temporary, superpowers. With the expansion of Roman power in the area, life in Pannonia stabilized and an economic upsurge resulted. The Romans, realizing the strategic importance of this border province, organized it as a military defense area, a type of buffer state. But as Roman brilliance in politics dimmed – the result of internal decay – political mistakes were committed. Instead of expanding their territory into the Carpathian Mountains, using these natural borders as a line of defense with the Danube area as a second line, the Romans gave up Dacia. It was an expensive mistake which was paid for later, during the period of military crises.

In order to present a complete picture to the reader, the authors of this volume have introduced a summary prehistory of Pannonia during the Neolithic, Mezolithic, and Copper ages. There a descriptive rather than analytic approach was taken, since this period is not directly related to the Roman period. The reader should observe that an east–west migration pattern was characteristic of the Paleolithic period of this area. This partially changed into a south–north migration, as in the "Linear Pottery culture", the well-known "Lengyel culture", of Neolithic or Eneolithic times. The Lengyel culture eventually moved into the

Copper Age and was replaced by the "Pécel" culture, which moved from the Balkans, south to north. During the Middle Bronze Age another migration pattern developed with the arrival of the "Tumulus" culture, which moved in from the west and northwest, through the Transdanubian area, toward the Balkan south. Much of the reconstruction of the four centuries before the Roman period in the area later to be called Pannonia is still based on ancient historical sources alone. The archaeological evidence to corroborate this data is still meager. For the time being, until scientific archaeological undertakings deliver more evidence, theories creating relatively simple images must satisfy, and historical events of the immediate pre-Roman period must be understood against the political background of Rome itself and the sources the Romans themselves provided.

Hellenistic scholarship gives us a knowledge of ancient pre-Roman Pannonia which is episodic and wrapped in a hazy mythology. The tribes mentioned by Herodotus and Hecataeus in the middle-Danube area and on the Great Hungarian Plain play no part in the later pre-Celtic history of Pannonia. Many things remain obscure despite the enormously advanced archaeological research in Hungary. There is, for instance, no possibility at the present time of being able to place the once troublesome tribe of the Triballi into their historical–geographical place, for the geographical names referred to by ancient historians are not always determinable. Mócsy, with the aid of linguistic research, divided the tribes of southeast Europe into three main language areas: Thracians, Illyrians, and Celts. He stated, however, that the tribal divisions of the Thracians and Illyrians are no real proof of their division into language families. But the confusion concerning the language and the dialects spoken by the Illyrians and Thracians is slowly clearing up. No longer is it believed, for instance, that the people of northern Dalmatia spoke the Illyrian language. Careful analysis, based on mapping place-names, produced the theory that the Thracians belonged to two distinct linguistic groups. There is, however, evidence of the distribution of Celtic name forms in Pannonia. The concentration of these name forms is around Aquincum, from Sopianae diagonally to the bend of the Danube, following the Amber Route from Poetovio to Savaria and Carnuntum. In southern and southeastern Pannonia, Illyrian names are prevalent.

Nevertheless, some tribes of the pre-Roman middle-Danube area remain the subject of controversy. Let's take, for example, the Dardanians, whose troublesome existence in the Ancient World is matched by the troubles they have created in modern research. We find them in the service of the Roman army as "alae" and "cohortes Dardanorum", and thus the information concerning their provenance should be adequate. Yet they are still believed by some historians to be Illyrians, while others strongly maintain that they were of Thracian descent. Such confusions, which stem from the frequent difficulty of recognizing ethnic and linguistic characteristics, may be expected to endure until further research and excavation reveal more substantial evidence.

Another puzzling problem still awaiting solution is that no Celtic influence can be detected among the descendants of the Scordisci in southeast Pannonia, even though these people lived under a lengthy Celtic domination, or were Celts themselves. The answer to this question probably lies in the thorough mixing of influences of the Illyrians and the Thracians, in addition to that of the Celts, creating a mélange where the once dominant linguistic characteristics have disappeared and ethnic and social change ensued. It will take many more years of excavations to uncover epigraphic and other evidence before specific demarcation lines and power relationships among the Illyrians, Thracians, and Celts in these areas can be determined.

The second century B.C. affords more than guesswork in many areas of Roman Pannonia, but the years between the Celtic expansion of the fourth century up to the second century are still in need of clarification. It was in the fourth century that the Celts reached the middle-Danube areas, inhabited originally by the Illyrians (about the same time that the Celtic invasion reached Italia).

The fluid state of the divisions between the Dardanians and the Celtic Scordisci was probably due to anarchy in tribal organization. We do know of their political alliances with Macedonia and Rome at the time of their rise in importance. During the third century the picture is very nebulous. One may suppose that the Scordisci established hegemony after eradicating the pre-Celtic civilizations at the confluence of the Sava and the Danube rivers. This period of Scordisci control, as shown by Mócsy, was to be known later as *Civitas Scordiscorum* and lasted as late as the first century A.D. The correct extent of their expansion and the whole power structure is yet to be determined. Also there is the problem of establishing political structures for the pre-Roman period. It remains to be seen whether the warrior classes were always the initiators of political and geographic change, or whether some of these changes were the results of peaceful integration.

As for physical and ethnic borders among the various pre-Roman Celtic tribes, no conclusive evidence has emerged so far, beyond indications that the Taurisci were dominant in southwestern Pannonia and the Boii in northwestern Pannonia. The migration of the Celts as described by Caesar, remains a question in the pages of history, just as their physical geography, as described by Strabo (VII 2,2), also has led to confusion. Similarly the question of the settlement of the Taurisci in Pannonia still heavily taxes scholarship. Most scholars agree that the dominant power structure was divided among the Celtic Boii, the Taurisci, and the Scordisci, but the borderline between the Dacians and the Anartii, i.e., the separating line between the Celts and the Dacians, has not yet been determined. According to Poseidonius and Strabo, the Celtic wave which appeared at the end of the second century B.C. was able to establish eventual political continuity in the Boii territory of the northern and southern sectors of the future Pannonia. The Boii capital has never been definitely located, though it is suspected to have been somewhere in the area of Bratislava (Pozsony).

It is interesting that the geographic distribution of Celtic names largely corresponds to the wagon burials on tombstones and tumuli burials. Since this type of burial is not characteristic of the Illyrian tribes, this particular phenomenon may help in further determination of the divisions just discussed. Other evidence comes from studies by É. B. Bónis (*Die spätkeltische Siedlung Gellérthegy–Tabán* [Budapest]), including her contribution to this volume (*ArchHung* 47 [Budapest, 1969]). She clarifies certain relationships through ceramic evidence. Her research, culminating in this volume, spans four decades (*DissPann* II, 20 [Budapest, 1942]).

Ultimately, it appears that the natural borders at the Drava and Sava rivers, an area which served as a bridgehead to and from the Italian peninsula and Dalmatia many times during history, was a first line of defense and a catalyst between East and West, but was never recognized as an important boundary by either the Illyriciani or the Celtic tribes. These tribes never became a political entity of any significance, in contrast to the Romans, who recognized early the area's geopolitical significance, and with their uncanny sense of strategy laid siege to Siscia as early as 156 and 119 B.C. in order to secure that land between the Drava and Sava rivers (G. Alföldy, *ArchÉrt* 89 [1962] pp. 146 ff.). Future research will have to treat the Pannonians in greater depth, although here the veil of time remains obscure. The relatively scanty information available on the Pannonians can be attributed to their primitive tribal organization, which did not develop into a centralized power. Some semblance of such centralization emerges only after the grip of neighboring powers was weakened in 88 B.C. when the Scordisci were defeated. This retarded development may be attributed to the thorough absorption of Celtic influences, making it difficult to isolate those characteristics that were typical of the Pannonians in pre-Celtic times. The only useful description of the geographical distribution comes from Appian (I, 11, 14).

One must agree with Mócsy, who feels that while the Late Iron Age of Pannonia requires further investigation, evidence for the time being must be derived from those characteristics of the La Tène culture, which, he says, "could be attributed to variations in the degree of Celtization" (*Pannonia and Upper Moesia*, p. 26). Interest in this area of archaeological research has increased among European scholars since the early 1950s. Gordon Childe has felt that the Central Danube area was of capital importance in the deciphering of European prehistory. As for future research in this area, archaeological discoveries must be regarded as the most important future source of information to broaden our perspective regarding such areas as ethnic characteristics and language.

It is characteristic of the changing Roman military strategy that certain provinces were occupied with a speed comparable only to a modern Blitzkrieg, while the annexation of other provinces was accomplished in several stages. The latter method was used in Pannonia. Some historians believe that when the Romans founded Aquileia in 121 B.C. they already had a plan in mind for advancing into the Danube region, and that this city, with its extensive ports, was to serve as a hinterland and supply bridgehead. By the end of the century Aquileia had developed into a great political, economic, and military center, located as it was at the head of the Amber Route, which led to the north.

During the step-by-step military conquest of Pannonia, Romanization was aided by a resettlement of the population. By regrouping ethnic units the Romans achieved a threefold goal: they accelerated the process of Romanization; they decreased the potential for uprising; and they helped create conditions for an economic upswing. In this volume we have emphasized this penetration policy of the Romans as an integral element of their foreign policy, and this became an important factor affecting the life of Pannonia. The conquerors soon realized, however, that they could not leave the defense of the province up to the "client states", and the idea of the Danubian defense line, the so-called *limes*, was developed in Pannonia. One of the leitmotivs developed in this book is that the archaeological evidence leads to the conclusion that the *limes* served first as a fortified base for raids into Barbaricum, rather than as a defensive establishment. In this early formative stage the *limes* was built in one line through the use of palisades and wooden material in general. Gradually there was a changeover to brick and stone in the second century, while in the third century the *limes* was solidified and transformed in depth into a Maginot Line. Recently S. Soproni summarized the historical and topographical importance of one of its most significant parts in Pannonia (Der Spätrömische Limes zwischen Esztergom und Szentendre. Das Verteidigungssystem der Provinz Valeria im 4. Jh. Budapest 1978).

Behind the *limes* the cities developed from *oppida* into *municipia*, urbanization being the normal process of Romanization. It is significant that they took the form of *municipia* rather than *civitates*. Eventually the new Roman cities received the rank of *colonia*. In Pannonia the colony had a higher rank than in Italia. The non-Italian settlers formed the *municipia* and after they obtained the language and cultural levels of Roman cities they rose to the rank of *coloniae*. (In Italia the *colonia* had a lower rank than *municipia* because the latter were formed by the original Italian inhabitants.)

Certain territories were attached to each city, and *coloniae* and *municipia* extended their administration to the adjacent territories. The most important urbanization processes took place under the Flavians, Hadrian, and the Severan emperors.

Much of the economy was geared to the needs of the military. Here the reader is introduced to the legionary fortresses, *canabae*, and the *vici* systems. Some settlements were established in proximity to military camps. Army veterans who inhabited these settlements became wealthy from their land while forming an integral part of military security. Their sons formed the next generation of military personnel, coming from a home environment that nurtured an army-oriented ideology and providing army recruits with a ready source of raw material. The examination of military diplomas shows that at first the military leadership was exclusively Roman, as might well be expected. Until the beginning of the second century only Italian settlers were trusted and drafted into military service. But from the second century on, native Pannonians served in the local legions and after twenty-five years of service received Roman citizenship.

After the Dacian wars, Trajan divided Pannonia into two sectors. By the time of the Tetrarchy it was divided again, this time into four provinces. Of these, Pannonia Secunda ranked highest politically because of its capital, Sirmium, which became one of four late capitals of the empire. Archaeological documentation demonstrates that the foundations of the cities in southern Pannonia are earlier in general than those in the north, where the territory was annexed more gradually. The city administration was in the hands of civilian personnel, but territorial administration was in the hands of the military, at least in the beginning. The Roman central government demonstrated great sensitivity in deciding what functions were important to safeguard the interests of the empire, and which might safely be relegated to local governments. Eventually the military administration extended its role over civil administrations as political conditions changed.

25

The nuances of such administrative changes throughout Roman times are dealt with in that chapter of the volume.

The *villa Romana* is an important episode in the history of settlement and the development of agricultural wealth in Pannonia. While this book gives the subject adequate coverage the reader who wants more detail is referred to E. B. Thomas's *Römische Villen in Pannonien*. These villas may be divided into five groups: those near Lake Balaton, those near Lake Fertő, those between the Drava and Sava rivers, those situated along the *limes*, and those scattered in the interior of Pannonia. As in other parts of the Roman Empire, the purpose of the villa system was primarily economic, but these villas later became of strategic importance. Though the system was organically related to the slave economy, it appears that the landlords did not cultivate their lands with slaves at the beginning. The impoverished original settlers were used as agricultural laborers and it seems that certain groups of small landholders created a cooperative type of villa. The survival of this type of economic institution has great importance. Recently we found evidence of similar systems in Italy, with a lifespan extending into the seventh or even the eighth century A.D. (A. Lengyel, G. Radan et al., "The Excavations at Castelliere di S. Fedele", *Bulletin of Mediterranean Archaeology* 1 [1975]: 9 ff.).

Pannonia never adapted itself completely to the Roman way of life. Its culture kept resembling that of a frontier settlement, a transitory kind of culture. Since they lived in a frontier province, one that was selected to perform primarily a military function, the inhabitants sought fulfillment in their military roles. This fits well with the survival of the native warlike character of the Pannonians, a spirit not conducive to art and literature. Illiteracy must have been extensive and pottery marks in the form of imitative letters, instead of signatures, may be indicative of that state of affairs. Documents bear further evidence of ill-understood Latin. Much of the epigraphy comes from foreign groups such as the Syrians, Jews, and latinized Greeks. Historians who wrote about the eastern military campaigns of the Pannonian legions were critical of the cultural levels of these troops. Aurelius Victor, in his book on the emperors, pointed out that those emperors who descended from Pannonia lacked humanity. But the word *humanity* may, in this case, have had a somewhat different connotation.

Many mythological scenes on stone monuments point to the fact that mythology was taught in schools, but in essence Pannonian literature consisted of verses for graves. The Pannonians regarded Sylvanus, an Italian import, as their chief deity, and he is depicted in different forms. The Capitoline Triad appeared occasionally, but never reached the degree of popularity it attained in Italia. The Danubian rider-god combined characteristics of Mithras Epone and Dioscuri and was popular among the Danubian military. When the Greek gods reached Pannonia they became Pannonized. The Syrian influence, mostly imported by soldiers, brought exotic gods. In Savaria and Poetovio, Isis sanctuaries were discovered along with the Serapis cult and the cult of Amon–Jupiter. But this versatile pantheon was, in essence, transitory. In the third and fourth centuries Mithraeums were built and Mithraism was the dominant Oriental cult. Christianity appeared in the middle of the third century and during the persecutions

Diocletian, several bishops were martyred. Irenaeus of Sirmium, Quirinus of Siscia, and Victorinus of Poetovio became victims of Diocletian's terror.

Burial, as might be expected, was strongly tied to religion. From the period of the early Roman penetration, the Pannonians buried their dead outside of settlement boundaries. From the second century on, family burials with monuments appeared inside the city limit. A wagon was often represented on tombstone monuments, a characteristic of Celtic burial customs. These tombstones are rich sources of information about Pannonian customs. Much can be learned about ethnic background, movements of military and ethnic groups, costume, jewelry, and religion. The reader is referred to the works of G. Erdélyi, who, before her untimely death, studied the ethnic characteristics of these monuments, and to the new series by L. Barkóczi, *RIU* (referred to earlier), which deals with several counties in northwest Hungarian Pannonia. Jews in Pannonia are dealt with by G. T. B. Radan ("Comments on the History of the Jews in Pannonia", *Acta Arch.* 25 [1963]) and Edit B. Thomas ("Eins ist der Gott!" Über jüdische Denkmäler der Römerzeit in Pannonien. *Magazin für europäische Zusammenarbeit* [Eisenstadt], V. Jahrgang, No. 3–4. 1977, pp. 21–25, Figs. 1–10) and are mentioned in Alice Burger's description of a late Roman cemetery.

Artistic influences were largely limited to the imports brought from Italia by the Imperial troops during the first century of the Christian era. Latest results were published in F. Fülep: *Roman Cemeteries on the Territory of Pécs (Sopianae)*, Budapest, 1977; Edit B. Thomas: *Savaria Christiana*, Szombathely, 1977, pp. 1–60, Figs. 1–50; Edit B. Thomas: Pannonische Reliquienaltäre (Vorbericht), *Arheološki Vestnik* 29, pp. 573–88, Figs. 1–20.

The chapter on ceramics deals with a pivotal point in Roman archaeology. By the second century of Roman rule imported wares were replaced, and local shops were able to supply the entire needs of the province, except for ornamental vessels. Here the previously mentioned studies by É. B. Bónis and E. B. Thomas are recommended. *Terra sigillata*, and its various moves across the continent, has been the subject of much research in the last two decades. Its appearance in Pannonia is an important part of the history of this ceramic. In addition to the famous work of H. Comfort, *De la Céramique Sigillée*, the reader is referred to the work of D. Gabler. Another extremely important work in this field is "Angaben zur Verbreitung der Sigillaten in Pannonia", *ArchÉrt* (1964), pp. 94–110. A recent work on the small, so-called lighthouses is G. Radan in the *Alba Regia* 4 (1972) and *Gorsium* I (1974).

Continued research on Pannonian glassware received impetus from the recent work of L. Barkóczi. The finer pieces came from the Rhineland and, of course, from Syria. However, domestic workshops were established and produced some forms little known in Western or Oriental centers. Here the reader is again referred to the earlier work of Fremersdorf and the newer works of Barkóczi.

For details on arms the reader is referred to Edit B. Thomas in H. Klumbach: *Spätrömische Gardehelme*, Munich, 1973, pp. 39–51, 103–111, Pls. 12–18, 45–57 and *Helme, Schilde, Dolche, Studien über römisch-pannonische Waffenfunde*, Amsterdam, 1971.

Coinage did not change during the first century of Roman occupation. The indigenous inhabitants continued using their own currency. Later, as more Roman coins came into circulation, mints were set up in Viminacium and later in Carnuntum. Constantine the Great founded the mint in Sirmium. After Valentinian the money supply ceased and the inhabitants had to make do with the available currency. Finally, north of the Drava, the circulation of coins ceased entirely after A.D. 375. This factor, according to some Hungarian researchers, is related to the loss of this part of the province. For varying opinions on this the reader should consult, in addition to the other sources, the new *Bibliography of Hungarian Numismatics* (1977) by M. F. Fejér and L. Huszár. For further research in numismatics see the following works: W. Hahn: Carnuntum, in *Die Fundmünzen der römischen Zeit in Österreich*, III. Vol. 1, Vienna, 1976; Katalin B. Sey: Coins from Identified Sites of Brigetio and the Question of Local Currency, *Régészeti Füzetek*, Ser. II, no. 18, Budapest 1977. Especially important is J. Fitz's *Der Geldumlauf der römischen Provinzen in Donaugebiet Mitte des 3. Jahrhunderts*, Vols 1–2, Budapest, Bonn, 1978.

By the beginning of the sixth century the province of Valeria was lost to the Huns as "foederati". A. Alföldi partially explained this event in his classic *Untergang der Römerherrschaft in Pannonien* (Berlin, 1924–1926). These theories were not seriously challenged until the 1960s. Mócsy analyzed these events and corrected some earlier errors in the *PWRE* supplement. But the issue of the change of guard in this area remains controversial especially in view of L. Várady's *Das letzte Jahrhundert Pannoniens* (1969), pp. 376–476; and T. Nagy's "The Last Century of Pannonia in the Judgement of a New Monograph", *ActaArchHung* 19 (1971): 299–345. To some of the new theorists it now appears that the Huns controlled, but did not populate, the province. According to these theories the Romanized population gradually left Valeria beginning in A.D. 380 and the Huns were active only in the non-Romanized Alansland.

The real Hun advancement must have occurred after the Ravenna treaty in A.D. 433, when the Romans ceded Pannonia Prima. Although the archaeological evidence is insufficient to reconstruct a picture of the social and cultural situation, it now appears that the traditional theory of "Untergang" is being challenged and that the takeover was not as sudden as once believed. Some survivals surely existed until the great migratory period. Recent works by L. Barkóczi and Á. Salamon are expected to shed some more light on this subject.

The reader will find an insightful and entirely unknown dimension when reading about I. Lengyel's paleoserological research and his results. Lengyel is a practicing physician and a professor at the University of Budapest. His novel method of blood-typing with fluorescent antibodies is little known in the Western world. This method, invented by Lengyel, provides a new approach to the determination of the ethnogenetic and biological developments of the past. It has been used with considerable success in thousands of cases on skeletal remains from the Copper Age to the Middle Ages. Recently the editors of this book used the method to determine genetic relationships in examining the findings of their excavation of monastic cemeteries in Italy.

Conventional methods of anthropology, normally based on morphological examinations, preclude the use of intact or almost intact skeletal remains and cannot determine changes in biological procedures at certain ages, as before puberty or during the aging process. Thus, since we are lacking the method of determining the age of certain groups, it is not suitable in many cases to determine demographic changes. Lengyel's complex serological bone-tissue examinations have overcome these deficiencies. No longer enlisting morphological methods, but rather, using the determination of blood groups, it sheds light on certain diseases based on hormonal changes, vitamin deficiencies, infections, and other factors.

Hungarian achievements in anthropology in the postwar years are reflected in the work of J. Nemeskéri. His "Fifteen Years of the Anthropological Department of the Hungarian National Museum" (*Yearbook of the Nat. Hist. Mus.* [1961], pp. 615–30) summarizes much material. The work of P. Lipták, especially his experience and study in the Soviet Union and his collective works on Hungarian anthropology, are commendable. Many other related works by Gy. Regöly Mérei, A. Kralovánszky, and K. Éry in such periodicals as the *Anthropological Notices (Anthropologiai Közlemények)*, *Acta Biologica Univ. Szegediensis*, and others are also to be recommended.

After finishing this book the reader should have a broad view of the new dimensions in Pannonian research since the Second World War, including the many well-documented and modern excavations such as that at Gorsium (Tác), which is presently the largest in Central Europe. However, imbalances and lacunae still exist in the process of establishing one of the finest archaeological schools on the Continent. A long-awaited synthesis has been written by Mócsy in the *PWRE* supplement and has been kept up to date (*Eirene* 4 [1965]: 133–55; *ActaArch* 21 [1969]: 340–75, 26 [1973]: 375–403). However, the approach still remains more historical and analytical, rather than entirely based on archaeological evidence. Although some monographs have appeared, the comprehensive analysis of the arts of Pannonia remains to be written. The same is true for the comprehensive study of Early Christianity in Pannonia. The director of the National Museum, F. Fülep, produced some spectacular results at Pécs in the area of the cathedral. In large-scale excavations it appeared that the area was a paleo-Christian cemetery, but the historical presentation in the context of paleo-Christian research and a corpus of Early Christian finds is still to be written.

The strategic situation of the Danube River offered a perfect solution to the problem of consolidating the Roman Empire in its formative years and of establishing frontiers which would assure lasting peace and safety in the southeastern flank of Europe. The Danube was not only a line of demarcation between Romans and Barbarians but also an integral part of a vast shipping network in Roman Europe. It connected the east with the west, the sea with the continent. A Roman river from its source to its delta, the Danube became a dynamic commercial artery, providing cheap and efficient routes for bulk cargoes. Several brief works have dealt with the historical and commercial implications and military importance of the river. The great, but speculative, works produced before modern excavations belong to the last century and were not firmly based on concrete factual information. These elaborate pieces of scholarship belong to J. Szentkláray, Zs. Fekete (later a legislator and a well-trained dilettante, whose works commanded special attention and were republished in Buenos Aires in 1976), and several engineers such as G. Téglás, Gy. Neudeck. None of these men were historians or archaeologists. These works detected the great accuracy of observations of L. Marsigli made some two centuries earlier. They also copied Imperial inscriptions at Orsova/Dierna at the Iron Gate where excavation is currently under way and traced the steps of the magnificent Imperial expeditions and the steps of the legio VII Claudia, Cohors II Hispanorum, and the invasion of Dacia.

In spite of the above, as in other provinces, many questions have not been dealt with. It is suspected, but not really known, that in the rapid consolidation of river transport between the eastern province and the empire, river ports and ferry posts of the defeated natives were incorporated into the newly unified network. Of the twenty-five cities of Pannonia in the third century, of which more than twenty can be correctly located, twelve had ports and landing facilities. Most of these landing facilities have not been located at all. Famous bridges, one of which was located by the editors of this book in the late 1960s at Sirmium, have rarely been surveyed and none have been excavated.

Although underwater archaeology has some limitations in murky river waters, such excavations have now been carried out at several places. Some efforts to consolidate the present situation as a base for further study were begun by E. B. Thomas, who plans to publish a volume dealing with the objects found in the Danube from Esztergom, Százhalombatta, Intercisa, and other spots. Mócsy pointed out the importance of this matter in his article in the *PWRE* and raised the question of the many installations which were needed for the accommodation of the classic Histrica, which also serviced Moesia Prima and Dacia Ripensis. He also identified some unique river-landing fortifications. The further study and excavation of these installations may shed light not only on ferry crossings, bridgeheads, and river-landings in Pannonia but also on other river ports in Europe during Roman times. The reader is referred to "Ein spätantiker Festungstyp am linken Donau Ufer", *Roman Frontier Studies* (1969, 1974), pp. 191–96; "Die spätrömische Schiffslande in contra Florentiam", *Folia Archaeologica* 10 (1958); "Il problema della condizioni del suolo attribuito alle unità militaria nelle province danubiana", *Accademia Nazionale dei Lincei* (1974). Nevertheless, large-scale research in this area, which includes commerce as well, is not yet planned.

The editors of this book spent their younger years in sight of many of the excavations mentioned in this volume. They felt that the aim of the book should be to recapitulate the investigations by Hungarian archaeologists in Pannonia, especially in the postwar period. Some chapters produce catalogs of catalogs, which are useful and should be of assistance to students of archaeology for years to come. For the student of the Roman provinces (whose numbers are rapidly increasing as available classical sites become exhausted or difficult to reach), this book is a "condicio sine qua non". These aspects of the life of the ancient world will replace classical sites as centers of continuing interest. For the casual reader this volume will be a revelation of a cultural heritage and an evaluation of technical and scientific material thus far generally unavailable. But to both amateur and scholar it will be a valuable visual archive of picture material, much of which is new, never before published.

The book is the result of cooperative efforts among a number of specialists. A work of this type, based on spectacular, incompletely publicized discoveries, has been long overdue in the Western world. The contributors were given a free hand to explain the main achievements in their various fields. Thus the reader may get some repetitions where certain fields overlap, but such repetitions are unavoidable and, in most cases, necessary. Nevertheless, the editors feel that it is a comprehensive work, planned and written over a nine-year period, and that it presents the best coverage to date of Pannonian research since the Second World War.

Alfonz Lengyel<br>
George T. Radan

# PANNONIA RESEARCH IN HUNGARY

FERENC FÜLEP

The first collection containing fragments of Roman monuments was made in Szombathely, where, according to recent research, the assembling of the stones found there occurred in the 1460s during the reign of King Matthias Corvinus (1458–1490).[1] The eminent historian Bonfini, who lived in the court of Matthias Corvinus, mentioned them in his work.[2] They were later described by W. Lazius, who visited Szombathely about the middle of the sixteenth century,[3] and by Marsigli, who wrote of these monuments and sketched them in his maps which appeared in 1726.[4] The same "lapidarium" was mentioned by Matthias Bél, the historian of the first half of the eighteenth century, in his important unpublished historical work. In addition, two English travelers, Richard Pococke and Jeremiah Milles, visited Hungary in 1733 and 1736 and mentioned the inscriptions of the Roman stones which they had seen in Szombathely.[5] However, antique monuments of Szombathely were first published in the work of Stephan Schoenwisner in 1791.[6]

The first methodical archaeological excavations in Hungary were started in 1780. As a result of them, a painted Early Christian tomb chamber was discovered at Pécs. A report of this discovery was published in Latin by J. Koller in 1804.[7]

For archaeological excavations in Hungary, the establishment of the Hungarian National Museum in 1802 was a milestone. It established the basis for scientific research which became the guideline for all Pannonian researchers in the nineteenth century and which eventually during the twentieth century gave the key to all important research on Pannonia. The real beginnings of Pannonian research — the methodical and systematic topographical determination of the sites, surveying them, after local inspection, and the determination of the field by sketches — is associated with the name of F. Rómer (1815–1889). He was also important as an organizer, working for decades on the salvaging and conservation of archaeological finds. As a result of his work numerous provincial museums were established in the second half of the nineteenth century. These played important roles in the history of Pannonian research, in the organization of excavations and in the foundings of publications. *Archaeologiai Közlemények* which first appeared in 1859 and ceased publication in 1899, and *Archaeologiai Értesítő*, which began publication in 1869 and still exists, were the most important forums for Pannonian research.

At the end of the nineteenth century and the beginning of our own, another great personality, J. Hampel (1849–1913), played an important role in Pannonian archaeological research and publication. In Budapest, the establishment of the museum of Aquincum and the beginning of systematic research is linked with the name of B. Kuzsinszky (1864–1938), who was a major figure in the research centered on Aquincum. The periodical *Budapest Régiségei*, which was founded in 1899, mostly published on the archaeological remains found in the area of Aquincum. The archaeological survey of the Lake Balaton area and the material published about it were also B. Kuzsinszky's work.[8] During the period between the two world wars, L. Nagy and his pupils pursued systematic excavations at Aquincum and published the results of their work. Historical research on Pannonia, except what was done for the important provincial publications, was concentrated at the University of Budapest. A. Alföldi methodically organized the collection of the archaeological material on Pannonia and its publication. As a result, two series of *Dissertationes Pannonicae* were published. The original idea — the examination of detailed questions and, eventually, a history of Pannonia based on these details — was unfortunately not accomplished; nevertheless, the two series of *Dissertationes Pannonicae* were among the most important undertakings in the elaboration of the archaeological material on Pannonia.

Among the publications of the prewar years, two comprehensive works are important. First is the article on the collection of Early Christian material in Pannonia by L. Nagy.[9] The second gives important data relating to the history of Christianity in Pannonia from the time of the great migration period. This article, by A. Alföldi, illuminates the crucial problem of the survival of the late Roman population of Pannonia.[10]

A few years later, the history of Budapest in the ancient world was published as a result of research done by L. Nagy and A. Alföldi.[11]

A few joint publications were issued after World War II. A. Mócsy and L. Barkóczi wrote on the ethnic makeup of Pannonia from the Roman occupation until the end of the third century.[12] Under the auspices of L. Barkóczi a history was published in the form of a university textbook,[13] while A. Mócsy contributed an article to Pauly–Wissowa on the history of Pannonia, compiling the results of all research.[14] Edit B. Thomas wrote a comprehensive book on Pannonian villas.[15] Also important is the work of L. Várady in which he describes Pannonia's crucial last century on the basis of the critical evaluation of sources.[16]

In more detailed research, two areas are important: first, historical research on the *limes*, and second, research on the interior cities of Pannonia. This study will attempt to describe the history of this research in chronological order, attaching a selected bibliography as an appendix. Obviously we cannot compress into a limited space all research on all archaeological sites, and so we will concentrate only on the most important ones.

A new periodical, *Acta Archaeologica*, was published first in foreign languages in 1951, later in 1958 the Archaeological Institute of the Hungarian Academy of Sciences was founded to promote Pannonia research.

## LIMES RESEARCH

### Brigetio (Szőny)

Brigetio was a city on the right bank of the Danube, opposite the estuary of the river Vág. Its name appeared on inscriptions, in *Itinerarium Antonini* (246.4;

263.2; 265.3), and in *Notitia Dignitatum* (Occ. 34.51). Ammianus Marcellinus mentioned it (17.12.21; 13.5.15), and the name also appeared in other places (Fig. 19).

Until about 214, Brigetio belonged to Pannonia Superior (Fig. 6). During frontier modification under Caracalla, it was attached to Pannonia Inferior. Marsigli mentioned it in 1726 (vol. 1, Table 5; vol. 2, Table 1, Figs. 3, 4). Brief information about it was published by the English traveler R. Pococke and by Samuel Mikovinyi, a colonel in the Austrian Imperial Army in the middle of the eighteenth century. As Mikovinyi worked out a plan for the draining of the marshy areas of Szőny, he described the still visible ruins of Brigetio. His diagrams were never published. (See also S. Takáts, "The Environment of Komárom and Brigetio about the Middle of the 18th Century", *Komárom Megyei Múzeum Egyesület Értesítője*, 1907, p. 26.) Topographical research in the area of the ruins was carried out by F. Rómer.

I. Paulovics, on behalf of the Hungarian National Museum, started regular excavations in 1927. He found the remains of an aqueduct and opened two cemeteries in that year. During 1928–1929, he excavated in the camp area and opened three more cemeteries. I. Paulovics and A. Radnóti excavated at Brigetio in the 1930s, and A. Radnóti and L. Barkóczi worked there in the early 1940s.

The first Roman *castra* in the territory of Brigetio (which may have been a camp for an *ala*) was built in the neighborhood of the second decade of the first century. Remains of this camp are to be found east of a legionary camp which can be seen today (Figs. 1 and 19).

Earlier than the legionary camp, however, still another first-century camp was built in the territory of Brigetio, the remains of which are in the northern forepart of the legionary camp between the Danube and the northern wall of the camp. The legionary camp, somewhat remote from the Danube, was built circa A.D. 100 by detachments of three legions: the XI Claudia, the XXX Ulpia Victrix, and the I Adiutrix. The last of these returned to Brigetio in A.D. 118 or 119 after a short absence, and from then until the end of the fourth century constituted the permanent local garrison of the camp. Brigetio also served as an important station for the Danube flotilla.

A military city was on the western and southwestern side of the camp, and west of this, in the village of Szőny, was the civilian town. The town obtained the rank of a *municipium* and later *colonia* in the third century.

A part of the cremation cemetery is to be found under the legionary camp. Later burials surrounded the civilian town, the military town, and the camp. At the end of the fourth century, burials were extended to the territory of the military town, which by then had been abandoned; later, at the end of the fourth and during the fifth century, there were burials in the territory of the camp. From one of the graves emerged a gold coin of Anastasius (A.D. 491–518).

Opposite the camp, on the left bank of the Danube, a counterfortification is to be found at Celamantia by Ptolemaios (Leányvár), where J. Tóth Kurucz excavated between 1906 and 1913. The name of the counterfortification does not appear in *Notitia Dignitatum*; thus Celamantia must have been abandoned by the end of the fourth century.

## Castra ad Herculem (Pilismarót)

The military camp Castra ad Herculem (*It. Ant.* 266) once stood on the right bank of the Danube at one of the important points of the *limes* area, now within the village of Pilismarót (Fig. 15). The camp was first mentioned in the last century by R. Fröhlich. About 3 kilometers west of this military camp, close to the Danube, S. Soproni excavated one of the smaller advanced military fortifications in 1959. The ground plan of this edifice measured about 30 by 45 meters. Its kernel was a square tower with strong walls $1^1/_2$ or more meters thick. The supposedly one- or two-storied tower was surrounded on three sides by a U-shaped court with a thick precinct wall. In one of the corners of the court was constructed the bath of the fortification. A second court was attached on the fourth side of the tower. The living quarters were situated in this narrow court. The defence system of the fortress was strengthened with a rampart (Pl. X. 2). S. Soproni found two furnaces containing rich deposits of late Roman pottery.

This small fortress, where thirty to forty soldiers were stationed, was built at the beginning of the 370s and was destroyed at the beginning of the 400s.

## Ulcisia Castra (Szentendre)

Ulcisia Castra was the most important fortification (Figs. 7 and 8; Pl. VII.2) in Pannonia Inferior north of Aquincum (*It. Ant.* 266). Excavations started here in the 1930s under T. Nagy. The first inhabitants of the place were the Eravisci, but because of its excellent strategic situation the Romans established a military camp there. The Roman settlement consisted of two parts: one was the actual military camp, while the other was the *canabae* or civil town. The ground plan of the camp was trapezoidal, 205 by 134 meters. The generally 1.2 meter-thick walls were reinforced at the corners and sides with towers. The *porta praetoria* and the *porta decumana* were excavated.

According to the latest research, the building of the camp occurred in the first half of the second century, but was altered several times. Its last rebuilding dates from the end of the fourth century.

The garrison changed several times. The cohorts Ulpia Pannoniorum and Milliaria Nova Surorum Sagittariorum and the Legio II Adiutrix exchanged details with each other. The last garrison consisted of units of the Equites Dalmatae. At this time, in the fourth century, the camp received the name Castra Constantia.

Houses decorated with wall paintings and supplied with hypocausts were excavated in the civil settlement. A small Early Christian burial chapel, originating from the last century of the Roman settlement, was also unearthed. One house in the *canabae* was converted into a Christian basilica. Among the findings, most outstanding was a bronze plate from a chest, decorated with symbols of pagan-Christian syncretism.

**Aquincum** (Budapest III)

The capital of the province of Pannonia was situated at an excellent strategic point — at the meeting of land routes and waterways, on the right bank of the Danube River (Fig. 8). It was mentioned in *Itinerarium Antonini* (245.7; 263; 264) and *Notitia Dignitatum* (Occ. 34.54); among ancient authors, Ptolemy (2.15.3), Ammianus Marcellinus (35.13), and Sidonius Apollinaris (5.107) mentioned it. The name also occurs in inscriptions on stone monuments (Mommsen *CIL*, III, 439) (Fig. 21).

The English traveler R. Pococke spoke of the ruins of Aquincum around 1730 during his sojourn in Hungary. In 1778, while a lime hole was being dug in Florian Square (house no. 3), two premises of *thermae maiores* were discovered. Eventually these baths were excavated by Stephan Schoenwisner. The *frigidarium* was found in the same way in 1930; after other details were found, it was conserved in 1932. Some additional parts of it were discovered in 1960 under a modern apartment house. The basement of the apartment house has been converted into a museum, and now the whole complex can be seen.

In 1860 more Roman baths were found by A. Sacken, a Viennese scholar, in the area of the Óbuda shipbuilding company. Between the years 1860 and 1880, graves of a large third- and fourth-century cemetery were found near the Danube in the so-called Raktár fields. In 1875 F. Rómer, T. Orvay, and G. Zsigmondy began excavations in Aquincum's civilian quarter at Papföld, which is the area situated around the museum of Aquincum.

In 1898 remains of contra-Aquincum fortifications were found at the Pest-side abutment of Elizabeth Bridge. The southwestern corner tower of the camp at Eskü square was found at the same time. L. Nagy went on with these excavations in 1932 and then, under V. Bertalan, the fan-shaped corner tower of the camp, lying under Belvárosi Church was excavated. Excavations at this site are currently in progress.

In 1925 B. Kuzsinszky and L. Nagy began excavations of the amphitheater of the military city (Pl. XVI.2) in the area within Nagyszombat, Ottó Korvin, and Szőlő streets. This amphitheater was believed by A. F. Marsigli, an Italian military engineer, to be a military fortification here at the end of the seventeenth century. The excavations were continued between 1935 and 1939 by J. Szilágyi and completed in 1940 by T. Nagy. Finally, the walls were preserved according to plans by L. Gerő.

Around the property of 171 Szentendrei Street, along the road leading from the military camp toward the southern gate of the civilian city and continued by the Roman road, graves of cremation burials were found in 1929 and above them, parts of a cemetery from the fourth century.

In 1934 L. Nagy discovered a large Roman villa on the south slope of Csúcshegy. But the complete excavation of the building, with its picturesque surroundings, was abandoned. Altogether, twelve rooms were excavated — rooms with central heating, frescoes, stucco decorations, and so forth.

In 1930 the *cella trichora* was excavated in Raktár Street. It was standing in the center of an Early Christian cemetery and, at the time of the excavation, was intact.

However, this building was badly damaged during World War II. It was restored in 1961 and now has many visitors.

In 1936 L. Nagy opened three hundred graves in the city's oldest cremation cemetery, under a house in Bécsi Street.

After World War II, research on Aquincum was intensified. The most spectacular find was made on Hajógyár Island (Shipbuilding Island). As long ago as 1851, while the foundations of the Óbuda shipbuilding works were being laid, G. Zsigmondy, the city engineer, had discovered rooms of a huge public building decorated with frescoes and mosaics. In 1941–1942, research was undertaken under the guidance of J. Szilágyi; between 1951 and 1956 a part of a huge palace (which Szilágyi thought to be the palace of the Roman Governor of Pannonia) was completely excavated (Fig. 25; Pl. XVII). To date, the excavated ground plan of the palace is 120 by 95 meters. On its eastern side is a round, domed reception room, with a diameter of 75 meters. To the south are rooms for servants and stores, while to the north are living rooms and the bathing quarters. The palace had a central heating system, and its walls were decorated with frescoes and stucco (Pl. XVII).

One of Pannonia's most beautiful mosaic floors, portraying Heracles and Deianeira, was found in 1958 when a modern building was being erected in Meggyfa Street (Pl. XVIII). Here in 1958–1959, under the leadership of I. Wellner, a decorative villa with a ground plan of 110 meters square was excavated. This villa goes back to the second or third century. In the cellar of the school that was erected above the site, a modern display room and a museum were established, dealing with local history relating to the findings.

Under the auspices of the leaders of the museum of Aquincum, topographical research has been carried on for the last twenty years. Between 1949 and 1951, at the abutment of Árpád Bridge in the neighborhood of Kerek and Tavasz streets, the walls and *fossae* of an early imperial legionary camp were discovered.

Between 1949 and 1951, Roman houses were unearthed along Kiscelli Road, while from 1950 to 1953 under the auspices of Klára Póczy, apartment houses were excavated above which a modern museum was built. At that time, the building periods of the Aquincum *canabae* were first certainly established.

The year 1958 saw the excavation of two Roman watchtowers in Lajos Street. In the same year, the problem of the source of the Aquincum water supply began to be explained.

The excavations of cemeteries are integrally connected with research on the military and civilian city of Aquincum. After World War II many cemeteries were discovered accidentally, when modern buildings were being constructed. In the year 1953, for instance, late Roman brick graves were found on Elek Fényes Street. In 1956 and 1957, excavations of the fourth-century cemetery in Bogdáni Street continued under Györgyi Parragi. The early imperial cemetery excavations, which began under L. Nagy and have been continued under Melinda Kaba, started near Aranyhegyi Patak in 1957. In 1958, a Roman necropolis dating from the late period was unearthed by Györgyi Parragi.

In addition to accidental discoveries, systematic excavations were undertaken to date the building period of the ruins in the fields surrounding the museum of

Aquincum. Before the large buildings of the old civic areas were permanently preserved (buildings such as the thermae, markets, private houses, mithraeums, etc., opened up in the last century), new excavations were undertaken to clarify certain chronological as well as stratigraphical questions. Large-scale excavations were initiated in 1966 by Klára Póczy and Gy. Hajnóczi opposite the area in front of the Aquincum museum, on the western side of the railroad tracks, in order to reach into the western section of the civilian quarter of the city. The successful excavations defined the full perimeter of the walls of the civilian city and the location of the watchtowers as well as of the city gates. The systematic investigation of the *insulae* within the town walls is being continued on a permanent basis.

In 1967 Melinda Kaba started systematic excavations in an urnfield in Elek Benedek Street which so far has yielded two hundred extraordinarily rich graves.

With the modernization of Miklós Square and the building of a new housing development, organized excavations aimed at determining the extent of the military camp and the building periods there.

## Albertfalva (Budapest)

The Roman camp south of Aquincum on the right bank of the Danube is now part of the eleventh district of Budapest, which previously was known as the village of Albertfalva. The camp itself was unknown until 1947. Since 1947 excavations have been conducted by T. Nagy. The center part of the settlement was the auxiliary camp which sealed off the entry to the Rózsa valley from the Jaziges. A garrison of five hundred equestrian troops guarded this strategic point. The early palisade camp stretched out in a 2.8-hectare area, surrounded by a double turf rampart. The gates were defended by wooden towers. In the interior of the camp, the commander's building and the main roads have been investigated and excavated. The palisade camp was destroyed about 92–93 during the wars of Domitian.

Later, under Trajan, a permanent stone camp was erected at the same place on a 3.6-hectare area, where a garrison of a thousand equestrians was stationed. Its walls were made of 140 centimeter-wide ashlar blocks, and the corners were rounded; the interior towers are missing. Three gates of the camp have been found. In the center, a U-shaped commander's building dominated the camp. The ashlar walls were surrounded by a 20 meter-wide double *vallum*. In the second century, this stone camp was destroyed during the Sarmatian wars of Marcus Aurelius, to be rebuilt later. In the third century, however, Jazigian and Roxolan attacks resulted in its final destruction, and it was never rebuilt. Its Latin name is not even known.

The *canabae* were on the northern, western, and southern sides of the camp.

## Campona (Budapest XXII, formerly Nagytétény)

South of Aquincum, on the right bank of the Danube, stretched the second Roman camp, as mentioned in *Itinerarium Antonini* (245.6). Unfortunately, its name does not occur in Roman inscriptions. Among Hungarian researchers,

Katancsich, Schoenwisner, and Salamon identified Campona with Nagytétény. On the second and seventh plates of his book, Marsigli mentioned oblong Roman ruins which may also be identified with Campona.

Many findings in the nineteenth century and at the beginning of the twentieth have pointed towards the existence of a camp at this location. Among these was a hoard of more than 10,000 coins discovered in 1887.

In 1932 the camp itself was found by G. Dáni, director of the local museum. The first excavations were undertaken by I. Járdányi-Paulovics. His excavations yielded the northeastern and southwestern corner towers and the *porta praetoria*; he excavated around the *porta principalis sinistra* as well. In 1934 I. Paulovics discovered a mithraeum on the hills north of the Roman camp.

From 1949 until 1957 and again in 1960 F. Fülep continued excavations in the camp itself, in the territory of the *canabae*, and in the Roman necropolis. His excavations yielded the complete *porta principalis sinistra*. In addition, he successfully recovered and excavated the *porta principalis dextra* and even remains of the *porta decumana*. The two missing corner towers and several other towers of the camp were also found. He succeeded in clearing away the system of earthworks surrounding the camp and he excavated various details of buildings within the camp.

The camp measured 178 by 200 meters. All four gate towers were fortified with square battlements; during Constantine's rule huge fanlike bastion towers were built before the four, square watchtowers on the corners.

The camp was built at the beginning of the second century A.D. Currency circulation may be demonstrated here until the end of the fourth century — even until the beginning of the fifth century. The troops stationed here were Ala I Tungrorum Frontoniana and Ala I Thracum Veterana Sagittariorum Civium Romanorum. At the end of the fourth century, according to *Notitia Dignitatum* (Occ. 32.35), the Equites Dalmatae were its garrison.

The *canabae* were on the eastern and northern sides of the camp. Also on the eastern side, along the highway leading towards Aquincum, was the Roman necropolis.

In 1961 F. Fülep, who was in charge of the excavations, unearthed a Roman sarcophagus north of the camp. On the hills north of the camp where I. Paulovics excavated the mithraeum, traces of a villa settlement exist.


**Vetus Salina** (Adony)

Vetus Salina, mentioned in *Itinerarium Antonini* (254.4), was near the Danube on the site of the modern village of Adony. Its distance from Aquincum, 36 miles, was reported on a milestone (*CIL* III, 3723, 10631).

The Roman camp at Adony was first mentioned by R. Fröhlich. The first excavations were carried on here by L. Barkóczi and Éva Bónis in 1949–1950.

The camp was built during Domitian's reign for the purpose of counterbalancing the Sarmatian pressure. It was surrounded by several systems of trenches. Three additional palisade camps and two stone camps have subsequently been excavated.

Cohorts II Batavorum, II Alpinorum, and then III Batavorum were stationed in the *castra*.

About two kilometers north of the camp traces of a village have been discovered. Archaeological material points towards the existence of considerable numbers of natives here, sometime between the period of the fourth palisade camp and the first stone camp. Traces of native characteristics in the archaeological material survived here much longer than in any other Romanized Pannonian center.

### Intercisa (Dunapentele–Dunaújváros)

The humanist historians of the fifteenth and sixteenth centuries, Bonfini, Lazius, and Sambucus, all mentioned ruins and inscriptions which emerged at Dunapentele at that time. In his 1726 work Marsigli mentioned Roman walls south of the village near Öreghegy. It was, however, Schoenwisner who in 1780 identified Intercisa with the village of Dunapentele. This identification was confirmed by Mommsen as well (*CIL* III, 4330; *It. Ant.* 245.3).

In the second part of the nineteenth century, and especially at the turn of the century, a multitude of archaeological findings turned up at Dunapentele. Under the auspices of E. Mahler and A. Hekler, regular excavations, which aimed primarily at opening the graves in the necropolis, were carried out at Öreghegy between 1906 and 1913. By the end of this period 846 graves of both cremation and inhumation burials had been excavated. The excavations of the next year yielded six interconnecting buildings. After World War I, Z. Oroszlán excavated another 66 graves and another building in this area. Research and topographical definition of the camp is associated with the name of I. Paulovics and his excavations in 1926. The excavations of 1932 yielded the *porta decumana* and the northwestern and northeastern corner towers of the camp.

After World War II, L. Barkóczi (Fig. 11), under the auspices of the Hungarian National Museum, carried on large-scale excavations here in connection with the building of the Danube Ironworks. These excavations cleared the walls, gates, and the trench and earthwork systems, correcting some assumptions of I. Paulovics and fixing the various building periods of the camp. During the same period, K. Sági succeeded in unearthing a cemetery section containing 99 graves. During the years in which various building activities in the city and the construction of the Danube embankment were being undertaken, Eszter Vágó, who died in 1970, carried on large-scale research in Intercisa, in the course of which 1,528 graves, various buildings in the *canabae*, and an Early Christian basilica were unearthed.

The ground plan of the Roman camp (Fig. 11), on the right bank of the Danube, measured 175 by 240 meters. The garrison consisted of Cohort I Alpinorum Equitata, Ala I Thracum, Cohort I Milliaria Hemesenorum, and, after reorganization under Diocletian, Equites Sagittarii (*Not. Dig.* Occ. 34, Dux Valeriae 25–26). Finally, near the end of the camp's existence, Cuneus Equitum Dalmatarum and then Cuneus Equitum Constantianorum were stationed here. The *canabae* were situated at the southern side of the camp prior to the Marcomannian wars, but eventually they surrounded the whole camp at almost all vacant places.

41

Cemeteries belonging to different periods were also found on all sides of the camp.

It is believed that a large number of Syrian military personnel and civilian inhabitants occupied the camp towards the end of the second and the beginning of the third century. Among Eastern emigrants settled here were Jews of the Diaspora: an inscription which refers to a synagogue was found here. In the fourth century, according to findings and buildings situated here, Intercisa had a large Christian community.

### Lugio (Dunaszekcső)

Lugio was an important camp on the *limes* of Pannonia Inferior, stretching along the right bank of the Danube. It was the terminus of important arteries leading through the area between the Danube and Tisza rivers which was occupied by the Sarmatians. It was mentioned in *Itinerarium Antonini* (244.2), by Ptolemy, in the *Tabula Peutingeriana*, in *Ravenna Anonymus*, and in the *Codex Justinianus* (9.20; 10.11). Curiously, however, it was not mentioned in *Notitia Dignitatum*. One may suppose that by the time that work was compiled, for reasons unknown the name of the camp had been changed to Florentia (*Not. Dig.* Occ. 33.43).

Archaeological traces have turned up since the end of the nineteenth century in Várhegy at Dunaszekcső: graves and numerous other findings have pointed to the existence of the camp in this neighborhood. In the last century M. Wosinszky and B. Pósta, and later A. Graf, in his work (which appeared in 1936), all placed Lugio at Várhegy, Dunaszekcső. On the left bank of the river, opposite Lugio, Roman ruins had been found, which were identified by R. Fröhlich as the Contra Florentiam *burgus* (blockhouse) and which were mentioned in the *Notitia*. M. Wosinszky, on the other hand, believed that the ruins belonged to a harbor installation. A. Mócsy reconstructed this harbor and ship station known as Contra Florentiam in a study that appeared in 1958.

According to the testimony of the *terrae sigillatae* from the Po Valley and from southern Gallia, excavated in the camp and dated between A.D. 60 and 90, the building of the camp was set at the end of the first century A.D. In the fourth century, the garrison was changed from the units of Cohort I Noricorum Equitata to Equites Dalmatae. According to R. Fröhlich (*Arch. Epigr. Mitteilungen* 14 [1891]: 51), the November 5, 293, decrees of Emperor Diocletian (*Cod. Iust.* 9.20.10,11) were issued at Lugio and naturally he must have been there at the time of the issue. In fact, A. Mócsy connects the building of the port with Diocletian's sojourn here.

One Roman necropolis was discovered in the area of the district called Püspökhegy. Among the numerous findings there was one gold-glass cup with a portrait of a man and a woman and with the inscription "Semper Gaudeatis in Nomine Dei". L. Nagy and, recently, F. Fülep have studied this piece. A decorated gold pendant with a Constantinian type of Chi Rho design in the center, probably had its provenance from the same Early Christian cemetery.

## Scarbantia (Sopron)

Scarbantia was established near Lake Fertő at the crossing of the roads to Vindobona and Carnuntum. In the first half of the first century, this city, situated near the Amber Route, was already populated by veterans who had been settled there and by merchants. It was mentioned in *Itinerarium Antonini* (233.6; 261; 262; and 266.4), in *Notitia Dignitatum* (Occ. 34.30), and in the *Tabula Peutingeriana* (5.2). Among ancient authors, Pliny (*H.N.* 3.146) and Ptolemy (2.14.4) mentioned it.

The Roman city was at the center of the part of the modern city that was surrounded later with a medieval city wall (Fig. 22; Pl. XX). Toward the end of the nineteenth century, Roman cemeteries here were excavated by L. Bella and, when the city hall was constructed, monumental sculptures of Jupiter, Juno, and Minerva were found. The ancient city's *capitolium* was situated there (Pl. XIII).

M. Storno excavated in the area of the amphitheater in 1925. Excavations after World War II were undertaken only in the 1950s when A. Radnóti and Mária R. Alföldi excavated in the Roman city area and in the necropolis.

Since 1959, Klára Póczy has conducted systematic excavations in the area of the Roman city. The most substantial results of these excavations have been the discovery of the Roman city walls, bonded into the medieval city walls. Since these had projecting battlements, Póczy succeeded in identifying the ancient city gates. She also excavated in the necropolis outside the city walls (Fig. 22).

The city, which was founded in the first century, presumably received municipal rank under Emperor Domitian. A few buildings decorated with frescoes and mosaics have been found. A short section of a water-pipe system was discovered in the last century. The city's amphitheater, close to a sanctuary of Nemesis, was situated outside the city walls. This has not yet been excavated. The city was surrounded by a stone wall towards the end of the third or the beginning of the fourth century. In the fourth century, the city had a sizable Christian community.

## Savaria (Szombathely)

Savaria, which stretched near one of the important junctions of the Amber Route (Pl. XII), was founded by the Emperor Claudius (A.D. 41–54) and obtained its colonial rank during his reign. Its foundation was linked to a settlement of veterans (*It. Ant.* 233.5; 261; and 262) (Fig. 20).

The collection of the city's monuments was started by Bishop János Szily in the second half of the eighteenth century. The inspiration came during the construction of an elaborate bishop's residence when traces of the Roman city came into view. Under orders from the bishop, Schoenwisner prepared a corpus of the stone monuments which had been discovered and summarized the history of the city from Celtic times until the end of the eighteenth century.

In the nineteenth century L. Bitnitz described his own collection in the periodical *Tudományos Gyűjtemény*. F. Rómer published the objects which had been collected by J. Varsányi and, in the process, determined the location of the cemetery of

Savaria and thus the basis for future topographical research. Topographical research was further developed by V. Lipp, a high school teacher at the Premonstratensian school. He determined the street network, traced the system of canals and drew attention to the need for a search for the amphitheater. When he was transferred to the city of Keszthely, research began to stagnate. Large-scale excavations began in 1939 and went on until 1941 under I. Járdányi-Paulovics. As a result of this work, the so-called Quirinus basilica and its surroundings were archaeologically clarified and the topography of the surrounding network of roads determined.

An increasing amount of archaeological material has come to light since large-scale building began in 1950. The Isis sanctuary excavated in 1955 is especially important; under the leadership of T. Szentléleky, it has been opened only after ten years of work. In 1963 the city of Szombathely displayed the famous sanctuary in the center of a park containing other ruined monuments. Topographical research has continued under Teréz P. Buócz, and these excavations are contributing to understanding of the ancient city's topographical situation. The fact that the so-called Qiuirinus basilica was a basilica at all has been recently questioned, and now it is bel eved by E. Tóth to have been a representative and impressive public building erected about the end of the fourth century.

There is a strong possibility that the cultural and economic center of the city, its forum, was in the cathedral area. The Capitoline triad — the remains of the monumental statues of Jupiter, Juno, and Minerva (Pl. CVII) — was discovered very close by, and thus the location of the ancient capitol was placed in this area. The city was actually the center of Pannonia Superior and of its imperial cult. An economic upswing in the city may be attributed to the appearance of the Syrians at the end of the second century and the beginning of the third, and to the development of the *latifundium* system.

With the spread of Christianity, a large Early Christian community must have developed. Our sources point to the execution of Bishop Quirinus here in 309. Martinus (Saint Martin), the bishop of Tours, was born here in 316.

Savaria had a splendid imperial palace in which several emperors spent part of their time during the fourth century. During Constantinian times, the city became the administrative center of Pannonia Prima, and it is possible that in the first half of the fourth century a bishopric was established there.

The last flowering of the city was connected with the name of Valentinian I. Ammianus Marcellinus stated that the city was in a state of general decline toward the end of the fourth century, although it was the only city that offered a winter residence to Valentinian. It was finally destroyed by an earthquake in 455.

### Gorsium (Tác)

South of Székesfehérvár, at the border of the municipality of Tác (Fig. 19), archaeological remains have been noted ever since the middle of the last century. At that time, vineyard cultivation destroyed many of the graves and stone relics. T he *Itinerarium Antonini* (264, 265) mentions Gorsium as being at the point where two main north-south roads (connecting Sopianae with Aquincum and Brigetio, respectively) converge and for a short distance follow the same route. Following Th. Mommsen's calculations of the mileage from Sopianae, scholars placed the town in the general vicinity of Székesfehérvár. More recent archaeological research has discovered some other roads which passed through Gorsium.

The first excavations in this area, begun under the direction of A. Marosi in 1934, were in the area of two Roman villas situated 400 meters apart. In 1936 and in 1939 research continued under the leadership of A. Radnóti. His excavations resulted in the opening of the northern third of one of the villas which was then preserved at the beginning of the 1940s. The results of these excavations were published by Edit B. Thomas in 1955 (Pls. XXI–XXIII).

In 1954 excavations were executed in the area of the second villa and the fourth- or fifth-century necropolis nearby. After these excavations, a systematic investigation of this area began under J. Fitz with the cooperation of a large team of archaeologists.

Between 1958 and 1967, excavations were centered on previously opened areas. The excavation of a larger villa was completed in 1961. Its ground plan measured 65 by 49 meters. This villa, which originated in the first century A.D., had several building periods, the latest in the fourth century. Some 60,000 objects were found around it.

On the eastern side of the villa what seems to have been an Early Christian basilica was discovered as a result of new excavations. The building was destroyed by intense fire, but several of its sections once had fine stucco decorations of the Eastern type. Several levels of the basilica go back to the second century. A large oval well attached to the eastern side had a diameter of 3.3 meters, a depth of about 10 meters.

Besides these discoveries, numerous streets and blocks became identifiable. Also recognizable was a city center, built in the third quarter of the second century. One result of the excavations of the last few years is a large municipal building in the forum.

Excavations, begun in the 1930s, have continued some five hundred meters to the south in the area of another villa. Here, close to a smaller villa abandoned in the last third of the fourth century, was a cemetery used by the inhabitants of Gorsium even in the fifth century A.D. The excavations resulted in the covering of 256 tombs (Pls. XXII and XXIII). Two of these contained sarcophagi; the rest were brick and earth graves. Under the cemetery are the remains of brick houses that once belonged to a first-century military camp. Excavations in the area of the two adjacent sites yielded a large number of stone relics and other findings. Altogether the yield consisted of some 400,000 pieces.

The first military unit in Gorsium, Ala I Scubulorum, was replaced by Cohort I Alpinorum. The earliest Roman traces are a plank camp and its *canabae*. The camp was dissolved under the reign of Domitian. From this primitive settlement, situated at an eminently strategic position where several important arteries met, a very wealthy city developed. It flourished for four centuries of Roman rule and weathered several wars.

After the great and disruptive Roxolan invasion in 260, life began again in the last decades of the third century. At that time, the name Gorsium was changed to Herculia in honor of Emperor Maximianus Herculius. Building activity can be traced until Valentinian I's reign. The city was not sacked during the wars of the fifth century. Beginning in the second third of the fifth century, however, the inhabitants slowly trickled away, and the vicinity of the large villa was eventually used as a cemetery by the Longobards.

### Keszthely—Fenékpuszta

Mátyás Bél first mentioned this fortified Roman settlement around 1730. In 1786, Korabinszky wrote that the owners, the Counts Festetics had unearthed fosses, buildings, and Roman tombs at the site. Finally, F. Rómer published an account of the buildings at Fenékpuszta. From 1899 to 1909, the Museum of Lake Balaton at Keszthely commissioned Á. Csák to excavate systematically the ruined area at Fenékpuszta. This work resulted in the opening of some sixteen buildings. Under the directorship of A. Radnóti the excavations continued, with some interruptions, from 1947 to 1959. After Radnóti, L. Barkóczi and K. Sági (the latter the director of the Museum of Lake Balaton at Keszthely) took over the leadership of the excavations. The original purpose was to define the extent of the fortifications. An Early Christian basilica (No. 2) and a larger horreum were opened. After 1959 one hundred graves were unearthed near the southern walls of the fortification, all dating from the migration period. In 1969, work continued on an Early Christian basilica (No. 1) (Figs. 40 and 41).

The earliest Roman settlement in this area was in what is now called Újmajor, a part of the modern town of Keszthely, close to the Balaton Museum. This settlement originated in the first century, and one may suppose that it was a settlement of veterans. The urnfield belonging to the settlement stretched along the road leading to Aquincum near the modern cemetery of Szentmiklós. An inscription on a tombstone indicates that the couple buried beneath were given their citizenship by Emperor Tiberius.

A somewhat smaller Roman settlement was built around the Festetics castle in the area of Toronydomb. Traces of Roman buildings have also been found in the passage of Sömögye and in the border area of Keszthely toward the village of Gyenesdiás. In the third century, an impoverished settlement existed in the territory that is now Fenékpuszta. In the second half of the third century and the beginning of the fourth century, the aforementioned Roman settlements were destroyed as a result of enemy attacks. The devastated area was populated again by Constantius II, and the settlement at Fenékpuszta dates from this period, in an area of about 400 by 400 meters. This fortified settlement was designed to defend a crossing point of Lake Balaton. Its first necropolis was in a border area called Halászrét. A small fifth-century Christian chapel was erected above one of the tombs in this cemetery. Life in the fortified city continued in the fifth and even in the sixth century.

**Tricciana** (Ságvár)

The first traces of this fortified Roman city (Fig. 6), which was built along the main artery connecting the cities of Sopianae (Pécs) and Arrabona (Győr), reach back to the second and third centuries (*It. Ant.* 267.7). Its fortified characteristics developed in the fourth century A.D. Its dimensions were 290 by 268 meters and its walls were between 2.30 and 2.60 meters thick. The excavations were started in 1937 by I. Járdányi-Paulovics of the Hungarian National Museum.

The Hungarian National Museum conducted excavations between 1937 and 1942 close to the city on Tömlöchegy, with A. Radnóti in charge. A description of late Roman and Early Christian material from the 342 graves was published by Alice Burger in 1966.

Our present knowledge would suggest that there must have been a *beneficiarius* station in the city. The *beneficiarii*, mentioned in inscriptions on the stone monuments of the second and third centuries, supervised traffic and road crossing points. The theory that the Emperor Gratian was in the city in 379 has also been supported. According to the evidence of the necropolis, the Romanized population survived until the beginning of the fifth century.

In the center of the Early Christian necropolis of Tömlöchegy there was a double-apsed funerary building containing three sarcophagi. West of this building were two tomb chambers also ending in apses. It is to be noted that the cemetery had several other vaulted tomb chambers as well. The tombs stretched around the central funerary building in a fanlike manner. Among the 342 graves, brick graves were most frequent. A great number of wooden coffins, stone sarcophagi, and simple earthen graves have also been found.

It appears to be a new feature that in the east-west oriented tombs without any grave goods, Jewish ritual can be demonstrated along with Christian. The cemetery's plentiful glassware came from the Rhine area, the East, and from northern Italia. Apart from the material with an Early Christian character, the tombs contain rich amounts of goods reflecting what is generally found in Pannonian cemeteries.

Judging from the evidence of the rich numismatic material, it appears that money circulation increased during the period of Constantius II. But no coins issued after 375 were found at Ságvár.

**Sopianae** (Pécs)

According to *Itinerarium Antonini* (232, 267), the city of Sopianae was situated in the territory between the Danube and Drava rivers at the intersection of several roads (Fig. 6). The first painted tomb chamber (No. 1) was discovered in 1780 near the southern wall of the cathedral when a medieval building was demolished. One may suppose, however, that this was not the first Roman finding in Pécs, since I. Szalágyi (Salagius) in the fourth volume of his work, written in that same year, placed the Roman city in the area of the modern inner city. The theory that the medieval city wall and the medieval cathedral were of Roman origin also stems from him.

The discovery of the painted tomb chamber in 1780 was described by J. Koller in 1804. Because his description was written in Latin, then an internationally known language, the extraordinary building in an Early Christian cemetery of Pécs became widely known.

During the restoration of the cathedral between 1805 and 1827, a considerable amount of archaeological material must have been discovered, since the cathedral of Pécs lies in the middle of the late Roman–Early Christian necropolis. The findings of this period were summarized in a work by M. Haas in 1845. He placed a Roman camp in the inner city and applied the name "Sopiana" to the Roman city.

During the last century, a subterranean corridor leading to the painted tomb chamber No. 1 was built. It produced a wrong impression in visitors, as if the tomb chambers at Pécs were catacombs. This subterranean corridor was mentioned in an important work dealing with Pécs written by I. Henszlmann in 1873. A very momentous remark in his work — that the chamber's walls rose above the lower parts (i.e., the *cubiculum*, which was used for burials) — indicates that he believed that the walls were part of a funerary chapel above the tomb chamber. These walls were demolished during the restoration of the tomb chamber in the last century.

Toward the end of the last century, L. Juhász was occupied with the collection of Roman material at Pécs. He published antiquities of the city and, in one of his works, dealt with the origin of the name "Sopianae".

Large buildings of the Roman city were first discovered in 1903–1904, when the present central post office building was erected. A 50-meter-long Roman building came into view at that time. It was divided into a multitude of 6-by-4- and 4-by-4-meter sections. All the buildings were destroyed as the post office was erected; they are known to us only from drawings executed in situ by F. Cséfay, city engineer. But further excavations were conducted in 1904 by P. Gerecze at this site and the results of his excavations were published in *Archaeologiai Érte-sítő*.

After 1910 O. Szőnyi opened sections in the area of the Roman cemetery. In 1913 he opened a number of brick graves, mostly in the vicinity of the cathedral. His most noted work was the investigation of the triple-apsed funeral chapel (*cella trichora*) which he accomplished with I. Möller, a professor at the Technical University of Budapest. This cemetery chapel, built in the fourth century, consisted of an oblong narthex (under the floor of which a tomb was found in 1922), a central area, and the three apses. Its walls were decorated with two layers of frescoes. The earlier fresco was contemporaneous with the building of the chapel, while scholars disagree on whether the second layer should be dated in the ninth or the eleventh century.

From the middle of the 1920s, large canalization projects were undertaken in the inner city of Pécs. These works were completed in 1927. They cut through many buildings of the ancient Roman town, which was situated under the western section of the inner city, and through the Roman necropolis under the northern section of the modern city. Gy. Fejes, director of the Pécs museum, was in charge of the project's archaeological salvage work. His work resulted in processing some 40 Roman houses and 170 Roman tombs.

When the church of the inner city (the old Turkish mosque) was enlarged in 1936, Gy. Török excavated under the church in Széchenyi Square and the vicinity. This work also resulted in the discovering of Roman buildings and tombs.

The square in front of the cathedral was opened up in 1938. Here Gy. Gosztonyi, city engineer and director of the project, discovered painted tomb chamber No. 2. Symbols of the Eucharist, jug and beaker as well as vine leaves, were on the frescoes, and Paradise was depicted on one side. Gy. Gosztonyi also opened the seven-apsed cemetery chapel (cella septichora) near the cathedral in 1939. It was, however, covered up again. His two books summarized all information about the Roman city known at that time.

From 1939 to 1941, Gy. Török directed excavations east of the cella septichora and uncovered late Roman buildings and tombs. During 1939–1940 he also excavated under a house in Rákóczi Street (39/c–d) — that is, in the area of the Roman town. On the basis of his findings, he placed the capitolium in this area of the city.

In 1954, following numerous accidental discoveries, J. Dombay and P. Lakatos performed salvage excavations in Hal Square.

During operations to preserve the cella trichora from water damage in 1955, renewed archaeological research started under the leadership of A. Radnóti and continued under F. Fülep. As a result of the research, the earlier ground plan of the cella trichora was modified: the presence of a second tomb under the floor of the narthex was discovered. In addition, Fülep published observations explaining the destruction of the cella trichora and the origin of the second series of frescoes, which he dated to the 11th–12th centuries A.D.

Fülep discovered some new Roman tomb chambers in the same area and definitely established the existence of an Early Christian chapel along the ascending walls above the cubiculum. To the west, he successfully excavated one main artery of the water system which supplied the Roman town.

Starting in 1958, excavations were conducted by F. Fülep in the court of the regional library (Eta Geisler Street 8), which falls within the area of the Early Christian cemetery. Here a painted double tomb, a large vaulted tomb chamber containing fourteen graves, and another decorated tomb chamber with two niches in its western wall were found. These chambers were surrounded with some thirty-five brick graves. West of here (István Square 12 and Eta Geisler Street 14), a heretofore unknown section of the Early Christian cemetery was discovered by F. Fülep in 1968–1969. The tombs in the cemetery were placed around the apsidal funerary chapel, in the interior of which four graves were sunk under the floor. Later on, at the very end of the fourth or at the beginning of the fifth century, an altar and a bench for priests were built in the apse of the chamber and it was converted into an Early Christian chapel. Children's burials were in a special section of the cemetery. In the western end of the court, another frescoed tomb was found. Outstanding among the findings are some ninety late Roman glass vessels. In Early Christian tombs, the ensemble of jug and beaker was often present, symbolizing the Eucharist.

# ASPECTS OF THE FUTURE

A regular network of archaeological research on Pannonia with an ever-increasing capacity has rendered it possible that important results may be expected within the next ten years. However, it is not only the matter of well-planned excavations that we should consider but also the necessity of preserving archaeological remains that emerge haphazardly in the course of building processes and other enterprises involving excavations.

Also numerous evaluations are under way which must necessarily be pursued. The publication of the new volumes in the series *Magyarországi Régészeti Topográfiája* (*Hungarian Archaeological Topography*) under the auspices of the Archaeological Institute of the Hungarian Academy of Sciences will be continued also in the coming years. During that time emphasis will be given to the study of the counties of western Hungary, so that a systematic survey of the Pannonian archaeological material and places of provenance will be completed and the results published. Two other comprehensive works will appear. A new edition of Pannonian inscriptions is also to be expected (RIU, Vol. I. 1972); five volumes of this monumental work will be published in the coming years. Publication of a corpus on a collection of stone relics without inscriptions is also being considered.

Archaeological research continues in Aquincum, the capital city of the province. The great building program in Budapest makes such research imperative, particularly the excavation of buildings in the legionary camp and military city (by Klára Póczy and others). Since an exact topographical survey of the legionary camp has been successfully accomplished and the successive periods determined, in the coming years efforts will be concentrated in the given area with the hope of achieving a precise definition of the buildings which existed in the camp and in the military city as well. Another important task involves the excavation of the sumptuous buildings in the villa district, north of the legionary camp. It is evident that the splendid villa decorated with mosaics in Meggyfa Street was by no means a unique specimen, since many similar ones are emerging from the adjoining area. In the coming decades new excavations will be made in the area of the civilian city, particularly because of the new road constructions. Also the western section of the civilian city will be excavated, and further excavations will be carried out in the cemeteries of this city. Another urgent task is the local conservation and display of important buildings which have come to light (the gates of the Roman camps, towers, etc.).

Great efforts are being exerted in further research on the Danube *limes* in Intercisa (under Z. Visy). New research is expected in the camp at Dunaszekcső (Lugio). Minor research is expected to be carried on in a southern camp at Altinum (Kölked) under F. Fülep.

On the northern section of the *limes* research will be continued (under S. Soproni) on a late Roman fortification, unearthed on the Sibrik-domb of Visegrád, where the camp will not only be reconstructed but also displayed. In recent years successful preliminary work has been carried out in late Roman advanced fortifications in Barbaricum, in Hatvan–Gombospuszta and at Felsőgöd. In the latter places the systematic excavation of the fortifications can be expected. The coming years will see the publication of a comprehensive monograph by S. Soproni on the history of the Roman *limes* at the Danube bend. The results of research carried on in the Roman camp of Almásfüzitő will soon be published, too (by F. Fülep and E. Biró).

The most significant step forward in the study of Roman cities in the interior of Pannonia has occurred at Gorsium, where J. Fitz intends that the entire city with the adjoining cemeteries will be unearthed. It is hoped that the results of the excavations will also be published in a monograph and that a monographic series will thus be launched.

After thorough excavations at Keszthely–Fenékpuszta (by K. Sági and others), emphasis there is on the preservation of the emerging relics and buildings, as well as the publication of the results.

Recent excavations under E. Tóth at Tricciana (Ságvár) have confirmed the topography of the fortified settlement. The next step will be the systematic excavation of the cemeteries. The topography, extension, and building periods of the fortified settlement of Heténypuszta have been established by S. Soproni. The next task (undertaken by Alice Burger) will be to unearth the cemeteries as well.

At Sopianae (Pécs) it is to be expected that the great building program there will contribute to the emergence of new parts of the Roman city; beyond this, we are planning to excavate new portions of the cemeteries, primarily that of *cella septichora*. A volume dealing with research to date will offer a full monograph on the city (by F. Fülep), throwing light also on the economic and social conditions of the late Roman population and offering valuable contributions to the history of Christianity in Pannonia.

At Savaria (Szombathely) the topographical study of the Roman city will be continued. It is of great importance to excavate the late Roman, Early Christian cemeteries (E. Tóth). The coming decade will see the publication of important excavation results, notably the findings at the Iseum by T. Szentléleky. It is also expected, that the results of earlier excavations carried on at Sopron (Scarbantia), primarily those in the city and city walls, will be published by Klára Póczy.

The systematic excavation of an identified Pannonian *municipium* (Municipium Sala) will go on at Zalalövő under the direction of A. Mócsy.

As far as comprehensive research is concerned, Pannonian glassware is being studied by L. Barkóczi who is uncovering important data on Pannonian glass manufacture, as well as the problems of glass import into Pannonia. Research into brick-etchings (by Edit B. Thomas) will contribute new data on the history of Pannonian Christianity through the study of a so-far neglected group of finds. The study of Roman cemeteries in the county of Baranya and in southeast Pannonia by Alice Burger provides data to help determine the composition of the Pannonian population and the demographic conditions of the province.

### NOTES

[1] E. B THOMAS, "Affreschi di Dorffmaister a Szombathely". *Acta Historiae Artium* 12 (1966): 113–54.

[2] A. BONFINIUS, *Historia Pannonica*, Lib. I (Leipzig, 1771), p. 26.

[3] W. Lazius, *Reipublicae Romanae in exteris provinciis, bello aquistis constitutae commentariorum libri XII* (Frankfurt a. M., 1598), pp. 965–68.

[4] A. F. Marsigli, *Danubius Pannonico Mysicus observationibus geographicis, astronomicis, hydrographicis, historicis, physicis . . .*, vol. 2 (The Hague, Amsterdam, 1726) Table 39, pp. 101–3.

[5] R. Pococke and J. Milles, *A Description of the East and Some Other Countries* (London, 1743–1745).

[6] S. Schoenwisner, *Antiquitatum et historiae Sabariensis ab origine usque ad praesens tempus libri IX* (Pestini, 1791).

[7] J. Koller, *Prolegomena in historiam episcopatus Quinqueecclesiarum* (Posonii, 1804).

[8] B. Kuzsinszky, *A Balaton környékének archaeológiája* (Budapest, 1920).

[9] L. Nagy, "Pannonia Sacra" in *Szent István emlékkönyv* (Budapest, 1938), pp. 31–148.

[10] A. Alföldi, "A kereszténység nyomai Pannóniában a népvándorlás korában", in *Szent István emlékkönyv*, pp. 151–70.

[11] A. Alföldi and L. Nagy, *Budapest Története I: Budapest az ókorban 1–2* (Budapest, 1942).

[12] A. Mócsy, *Die Bevölkerung von Pannonien bis zu den Markomannenkriegen* (Budapest, 1959); L. Barkóczi, "The population of Pannonia from Marcus Aurelius to Diocletian". *ActaArchHung* 16 (1964): 257–356.

[13] L. Barkóczi, I. Bóna and A. Mócsy, *Pannónia Története* (Budapest, 1963).

[14] A. Mócsy, "Pannonien". *R. E. Pauly-Wissowa*, Suppl. 9, pp. 515–776.

[15] E. B. Thomas, *Römische Villen in Pannonien* (Budapest, 1964).

[16] L. Várady, *Das letzte Jahrhundert Pannoniens (376–476)* (Budapest, 1969).

## LITERATURE

**Brigetio** (Szőny)

Barkóczi, L. *"Brigetio, I—II"*. Budapest, 1944–1951 *(DissPann* 2: 22).

Barkóczi, L. "Római díszsisak Szőnyből — Römischer Paradehelm von Brigetio". *FolArch* 6 (1954): 45–48, 200–201.

Barkóczi, L. "A New Military Diploma from Brigetio". *ActaArchHung* 9 (1958): 413–21.

Barkóczi, L. "New Data on the History of Late Roman Brigetio". *ActaAntHung* 13 (1965): 215–57.

Juhász, Gy. "A brigetioi terra sigilláták — Die Sigillaten von Brigetio". Budapest, 1936 *(DissPann* 2 : 3).

Láng, N. "Das Dolichenum von Brigetio." *Laureae Aquincenses* 2 (1941): 165–81 *(DissPann* 2 : 11).

Paulovics, I. "A szőnyi törvénytábla — La table de privilèges de Brigetio" Budapest, 1936 *(ArchHung* 20).

Paulovics, I. "Funde und Forschungen in Brigetio". *Laureae Aquincenses* 2 (1941): 118–64 *(DissPann* 2 : 11).

Radnóti, A. "Bronz Mithras-tábla Brigetioból — Le bas-relief mithriaque de bronze de Brigetio". *ArchÉrt* (1946–48), pp. 137–55.


**Pilismarót**

Fröhlich, R. "A pilismaróti római tábor" (The Roman Camp at Pilismarót). *ArchÉrt* (1893), pp. 38–47.

Soproni, S. "Római erőd Pilismaróton" (Roman Fortress at Pilismarót). *Dunakanyar Tájékoztató*, No. 2 (1966), pp. 51–53.


**Ulcisia Castra** (Szentendre)

Kuzsinszky, B. "Kiadatlan római kőemlékek Szentendrén — Unveröffentlichte römische Steindenkmäler in Szentendre". *ArchÉrt* (1929), pp. 45–57, 325–27.

NAGY, L. "Keresztény-római ládaveretek Szentendréről — Gli ornamenti di bronzo di una cassetta romana trovata a Szentendre". *Pannonia* (1936), pp. 3–21, 157.

NAGY, T. "Kutatások Ulcisia Castra területén — Indagini sul territorio di Ulcisia Castra". *ArchÉrt* (1942), pp. 261–85.

SOPRONI, S., BOROS, L., and SZOMBATHY, V. *Szentendre*. Budapest, 1961.


Aquincum (Óbuda)

ALFÖLDI, A. and NAGY, L. *Budapest története I: Budapest az ókorban 1–2* (History of Budapest I: The History of Budapest in Antiquity). Budapest, 1942.

BÓNIS, É. B. *Die spätkeltische Siedlung Gellérthegy—Tabán in Budapest*. Budapest, 1969.

BRELICH, A. "Aquincum vallásos élete" (La religiosità di Aquincum). *Laureae Aquincenses* 1 (1938): 20–142 (*DissPann* 2 : 10).

GERŐ, L. "Az óbudai amfiteátrum rekonstrukciója" (The Reconstruction of the Amphitheatre at Óbuda). *Magyar Mérnök- és Építész-Egylet Közlönye* (1941), pp. 316–23.

HAMPEL, J. "Aquincum, mint a pannoniai limes egyik védpontja" (Aquincum as One of the Defence Points of the Pannonian *Limes*). *Archaeologiai Közlemények* 8 (1871), pp. 169–78.

HAMPEL, J. "A papföldi közfürdő" (The Public Bath at Papföld). *BudRég* 2 (1890): 53–74.

HAMPEL, J. "Thrák vallásbeli emlékek Aquincumból" (Thracian Religious Monuments from Aquincum). *BudRég* 8 (1904): 5–47.

KABA, M. *Aquincum emlékei — Die einstige Römerstadt Aquincum*. Budapest, 1963.

KABA M. and PÉCSI, S. *Az aquincumi orgona* (The Organ at Aquincum). Budapest, 1965.

KORBULY, GY. Aquincum orvosi emlékei — Die ärztlichen Denkmäler von Aquincum. Budapest, 1934 (*DissPann* 1 : 3).

KUZSINSZKY, B. "Az aquincumi Mithraeum" (The Mithraeum at Aquincum). *ArchÉrt* 8 (1888). pp. 385–92.

KUZSINSZKY, B. "Az aquincumi ásatások 1882–1884 és 1889" (The Excavations at Aquincum 1882–1884 and 1889). *BudRég* 2 (1890): 77–160.

KUZSINSZKY, B. "Az aquincumi amphitheatrum" (The Amphitheater at Aquincum). *BudRég* 3 (1891): 82–139.

KUZSINSZKY, B. *Die Ausgrabungen zu Aquincum 1879–1891*. Budapest, 1892.

KUZSINSZKY, B. "A gázgyári római fazekastelep Aquincumban — Das große römische Töpferviertel in Aquincum". *BudRég* 11 (1932): 5–384.

KUZSINSZKY, B. *Aquincum, Ausgrabungen und Funde*. 7th ed. Budapest, 1934.

NAGY, L. *Az óbudai ókeresztény cella trichora a Raktár utcában — Altchristliche Cella Trichora in Óbuda*. Budapest, 1931.

NAGY, L. *Az aquincumi orgona — Die Orgel von Aquincum*. Budapest, 1934.

NAGY, L. "Az aquincumi múmia-temetkezések — Mumienb egräbnisse aus Aquincum". Budapest, 1935 (*DissPann* 1 : 4).

NAGY, L. "A csúcshegyi római villa Óbudán — Die römische Villa auf dem Csúcshegy in Óbuda". *BudRég* 12 (1937): 27–60, 301–302.

NAGY, L. "Gladiátor-ábrázolások az aquincumi múzeumban — Gladiatoren-Darstellungen im Museum von Aquincum". *BudRég* 12 (1937): 181–95, 308–9.

NAGY, L. "Dirke bűnhődése az aquincumi mozaikon — Il supplizio di Dirke su un mosaico di Aquincum". *BudRég* 13 (1943): 77–102, 498–99.

NAGY, T. "Il secondo amfiteatro romano in Aquincum." *Corvina* (1941), pp. 829–49.

NAGY, T. "Perióduskutatások az aquincumi polgárváros területén — Erforschung der Perioden im Zentralgebiete der Zivilstadt von Aquincum". *BudRég* 21 (1964): 9–54.

PÓCZY, K. Sz. "Aquincum a IV. században — Aquincum im 4. Jahrhundert". *BudRég* 21 (1964): 55–77.

PÓCZY, K. Sz. "Festett férfiportré egy aquincumi múmiasírból — Ein gemaltes Männerporträt aus einem Mumiengrab in Aquincum". *ArchÉrt* (1966), pp. 272–77.

SZILÁGYI, J. "Az aquincumi helytartói palota — Der Legatenpalast in Aquincum". *BudRég* 14 (1945): 31–153.

SZILÁGYI, J. "Új adatok Aquincum és Pannonia hadtörténetéhez — Neue Beiträge zur pannonischen Garnisonsgeschichte". *BudRég* 15 (1950): 513–34.

SZILÁGYI, J. *Aquincum*. Budapest, 1956.

SZILÁGYI, J. *Az aquincumi amfiteátrumok* (The Amphitheaters at Aquincum). Budapest, 1956.

SZILÁGYI, J. "Az aquincumi helytartói palota — Der Statthalterpalast von Aquincum". *BudRég* 18 (1958): 53–77.

SZILÁGYI, J. and HORLER, M. "A Flórián téri rómaikori fürdő" (Bath in Florian Square from the Roman Period). In *Budapest műemlékei*. Budapest, 1962, 2 : 335–38.

SZILÁGYI, J. and HORLER, M. "Az aquincumi helytartói palota romjai a Hajógyári szigeten" (The Ruins of Aquincum's Gubernatorial Palace on the Shipbuilding Factory Island). In *Budapest műemlékei*. Budapest, 1962, 2 : 352–60.

WELLNER, I. *Az aquincumi mozaikok — Die Mosaike von Aquincum*. Budapest, 1962.


**Albertfalva**

NAGY, T. "Az albertfalvai római telep" (The Roman Settlement at Albertfalva). *Antiquitas Hungarica*, (1948), pp. 92–114.

NAGY, T. "The Military Diploma of Albertfalva". *ActaArchHung* 7 (1956): 17–71.

NAGY, T. "Az albertfalvai római tábor feltárásáról" (The Excavation of the Roman Castra at Albertfalva). In *A Budapesti Történeti Múzeum tevékenysége a felszabadulástól napjainkig*. Budapest, 1965, pp. 47–53.


**Campona** (Nagytétény)

CSEREY, É. and FÜLEP, F. *Nagytétény műemlékei* (Monuments of Nagytétény). Budapest, 1957.

FÜLEP, F. "A nagytétényi római tábor" (The Roman Camp at Nagytétény). In *Budapest műemlékei*. Budapest, 1962, 2 : 643–52.

JÁRDÁNYI-PAULOVICS, I. *Nagytétényi kutatások* (Research at Nagytétény). Budapest, 1957.


**Vetus Salina** (Adony)

ALFÖLDI, A. "Egy batavus istennő oltárköve Adonyból" (The Altarstone of a Batavian Goddess from Adony). *Székesfehérvári Szemle* (1935), pp. 30–33.

BARKÓCZI, L. and BÓNIS, É. "Das frührömische Lager und die Wohnsiedlung von Adony (Vetus Salina)". *ActaArchHung* 4 (1954): 129–99.

MAROSI, A. "Az adonyi Bacchus-jelenetes bronz dombormű" (The Bronze Relief of Adony Representing a Bacchic Scene). *Székesfehérvári Szemle* (1939), pp. 41–42.


**Intercisa** (Dunaújváros)

FÜLEP, F. "New Remarks on the Question of the Jewish Synagogue at Intercisa". *ActaArchHung* 18 (1966): 93–98.

HAMPEL, J. "Intercisa emlékei" (Monuments of Intercisa). *ArchÉrt* (1906), pp. 221–74.

Intercisa: Geschichte der Stadt in der Römerzeit. 2 vols. Budapest, 1954, 1957 (*ArchHung* 33, 36).

PAULOVICS, I. *A dunapentelei római telep (Intercisa): A maradványok története, a kutatások irodalma, 1926 évi leletek — Die römische Ansiedlung von Dunapentele (Intercisa), Geschichte der Überreste, Bibliographie der Forschungen, Fundergebnisse von 1926*. Budapest, 1927 (*ArchHung* 2).


**Lugio** (Dunaszekcső)

FITZ, J. "Lugio." *R. E. Pauly-Wissowa*, Suppl. 9, pp. 391–96.

FÜLEP, F. "Early Christian Gold Glasses in the Hungarian National Museum". *ActaAntHung* 16 (1968): 401–12.

HAMPEL, J. "Dunaszekcsői római sírok" (Roman tombs at Dunaszekcső). *ArchÉrt* 17 (1897), p. 95.

NAGY, L. "A dunaszekcsői keresztény-római pohár" (The Christian Glass from Dunaszekcső). *Theológiai Szemle* (1929), pp. 158–63.

NAGY, L. "Római régiségek Dunaszekcsőről — Römische Altertümer aus Dunaszekcső". *ArchÉrt* (1931), pp. 267–71, 358.


## Scarbantia (Sopron)

ALFÖLDI, A. "Kapitóliumok Pannóniában" (Capitols in Pannonia). *ArchÉrt* (1920–1922), pp. 12–14.

BELLA, L. "Római sírok a soproni Deák-téren" (Roman Graves at the Deák Square in Sopron). *ArchÉrt* 11 (1891), pp. 287–88.

PÓCZY, K. Sz. *Sopron rómaikori emlékei — Die römischen Altertümer in Sopron.* Budapest, 1965.

RADNÓTI, A. "Sopron és környéke régészeti emlékei" (The Archaeological Monuments of Sopron and its Vicinity). In *Sopron és környéke műemlékei.* Budapest, 1956, pp. 23–38.

STORNO, M. "A római amfiteátrum és nemeseum Sopronban — Das römische Amphitheatrum und Nemeseum zu Sopron (Scarbantia)". *Soproni Szemle* (1941), pp. 201–16.


## Savaria (Szombathely)

BUÓCZ, T. P. *Savaria topográfiája* (The Topography of Savaria). Szombathely, 1967.

KÁDÁR, Z. "A szombathelyi bacchikus elefántcsont szobor — La statua bacchica d'avorio di Savaria". *FolArch* 14 (1962): 41–50.

KÁDÁR, Z. and BALLA, L. *Savaria.* Budapest, 1968.

LÁNG, N. "A savariai Dolichenus-csoportozat — Die marmorne Dolichenusgruppe von Savaria". *ArchÉrt* (1943), pp. 64–70.

MÓCSY, A. "Savaria utcarendszerének rekonstrukciójához — Zur Rekonstruktion des Straßensystems von Savaria". *ArchÉrt* (1965), pp. 27–36.

PAULOVICS, I. "Szent Quirinus savariai bazilikájának feltárása — Freilegung der Basilica des hl. Quirinus in Szombathely (Savaria)". *Vasi Szemle* (1938), pp. 138–52.

SCHOENWISNER, S. *Antiquitatum et historiae Sabariensis ab origine usque ad praesens tempus libri novem.* Pestini, 1791.

SZENTLÉLEKY, T. *A szombathelyi Isis szentély — Das Isis-Heiligtum von Szombathely.* Budapest, 1965.


## Gorsium (Tác)

FITZ, J. "Gorsium". *A táci római kori ásatások* (Gorsium: Excavations of the Roman Period at Tác). Székesfehérvár, 1964.

HORVÁTH, T. and MAROSI, A. "Jelentés az 1934 őszi tác-fövenypusztai ásatásról" (Report on the Tác-Fövenypuszta Excavations). *Székesfehérvári Szemle* (1935), pp. 35–38.

MAROSI, A. "Jelentés a tác-fövenypusztai ásatásról" (Report on the Excavations at Tác-Fövenypuszta). *Székesfehérvári Szemle* (1937), pp. 24–25.

THOMAS, E. B. "Die römerzeitliche Villa von Tác-Fövenypuszta". *ActaArchHung* 6 (1955): 79–152.


## Fenékpuszta

ALFÖLDI, A. *Der Untergang der Römerherrschaft in Pannonien.* Vol. 2. Berlin, 1926.

BARKÓCZI, L. "A Sixth Century Cemetery from Keszthely-Fenékpuszta". *ActaArchHung* 20 (1968): 275–311.

CSÁK, Á. "A keszthelyi római urnatemetőről (Roman urnfield at Keszthely). *ArchÉrt* (1912), pp. 374–75.

LIPP, V. "A fenéki sírmező" (The graveyard at Fenék). *Archaeologiai Közlemények* 14 (1886): 137–59.

SÁGI, K. *Fenékpuszta története* (History of Fenékpuszta). Veszprém, 1960.

SÁGI, K. "Die spätrömische Bevölkerung der Umgebung von Keszthely". *ActaArchHung* 12 (1960): 187–256.

SÁGI, K. "Die zweite altchristliche Basilika von Fenékpuszta". *ActaAntHung* 9 (1961): 397–459.

SIMONYI, D. "Fenékvár ókori neve" (Fenékvár's Name in Antiquity). *Antik Tanulmányok* 9 (1962): 13–30.

## Tricciana (Ságvár)

BURGER, A. Sz. "The Late Roman Cemetery at Ságvár". *ActaArchHung* 18 (1966): 99–234.

KÁDÁR, Z. "A ságvári későrómai szinkretisztikus ábrázolású ládaveret — La rapprezentazione sincretistiche dello scrigno tardoromano di Ságvár". *ArchÉrt* (1968), pp. 90–92.

RADNÓTI, A. "Római kutatások Ságváron — Römische Forschungen in Ságvár". *ArchÉrt* (1939), pp. 148–64, 268–76.

## Sopianae (Pécs)

DERCSÉNYI, D. and POGÁNY, F. *Pécs*. Budapest, 1956.

FÜLEP, F. "Neuere Ausgrabungen in der Cella trichora von Pécs (Fünfkirchen)". *ActaArchHung* 11 (1959): 399–417.

FÜLEP, F. *Pécs római kori emlékei — The Roman Monuments of Pécs*. Pécs, 1963.

FÜLEP, F. "Későrómai temető Pécs: Geisler Eta u. 8. sz. alatt — Early Christian Cemetery at Pécs: No. 8, Geisler Eta Street", *ArchÉrt* (1969), pp. 3–42.

GERECZE, P. "A pécsi postapalota alatt kiásott régi falak" (The Ancient Walls Excavated under the Post Office Building at Pécs). *ArchÉrt* (1904), pp. 322–25.

GERKE, F. "Die Wandmalereien der neugefundenen Grabkammer in Pécs (Fünfkirchen). In *Neue Beiträge zur Kunstgeschichte des I. Jahrtausends*. I. Halbband. Baden-Baden, 1952, pp. 115–37.

GERKE, F. "Die Wandmalereien der Petrus-Paulus-Katakombe in Pécs (Südungarn)". In *Neue Beiträge zur Kunstgeschichte des I. Jahrtausends*. II. Halbband. Baden-Baden, 1954, pp. 147–199.

GOSZTONYI, GY. *A pécsi Szent Péter székesegyház eredete* (The Origin of St. Peter Cathedral at Pécs). Pécs, 1939.

GOSZTONYI, GY. "A pécsi hétkarélyos ókeresztény temetői épület — Ein altchristliches Gebäude mit 7 Apsiden in Pécs". *ArchÉrt* (1940), pp. 56–61.

GOSZTONYI, GY. "A pécsi II. sz. ókeresztény festett sírkamra és sírkápolna — Die bemalte altchristliche Grabkammer und Grabkapelle No. II". *ArchÉrt* (1942), pp. 196–206.

GOSZTONYI, GY. *A pécsi ókeresztény temető* (The Early Christian Necropolis at Pécs). Pécs, 1943.

HENSZLMANN, I. "A pécsi ókeresztény sírkamra" (The Early Christian Tomb Chamber at Pécs). In *Magyarország régészeti emlékei* II. Budapest, 1873, 1 : 21–165.

RADNÓTI, A. "Sopianaeból kiinduló római utak — Die aus Sopianae ausgehenden römischen Wege". *Pécs Város Majorossy Imre Múzeumának Értesítője* (1939–1940), pp. 27–39.

SIMONYI, D. "Sull'origine del toponimo 'Quinque Ecclesiae' di Pécs". *ActaAntHung* 8 (1960): 165–84.

SZŐNYI, O. *A pécsi püspöki múzeum kőtára* (The Glyptothek of the Episcopal Museum at Pécs). Pécs, 1906.

SZŐNYI, O. *A pécsi ókeresztény sírkamra* (The Early Christian Tomb Chamber at Pécs). Budapest, 1907.

SZŐNYI, O. "Ásatások a pécsi székesegyház környékén 1922-ben — Grabungen in der Umgebung der Pécser (Fünfkirchner) Domkirche im Jahre 1922". *Országos Magyar Régészeti Társulat Évkönyve* (1923–1926), pp. 172–95, 384–86.

TÖRÖK, GY. "Rómaikori sírkamrák Pécs felsősétatéren — Grabkammern aus der Römerzeit an der oberen Promenade von Pécs". *ArchÉrt* (1942), pp. 207–15.

# GEOGRAPHY OF PANNONIA

SÁNDOR SOPRONI

A proper comprehensive description of the geography of Pannonia (Fig. 1) has never been attempted and source material is limited. Greek writers gave the earliest accounts of the area, but they had some false, romantic ideas about a faraway cold northern land. These faulty descriptions were even used by the Romans when they first became interested in the middle Danube area. These descriptions referred to the area as cold, deserted, and infertile. However, when Roman interest in Pannonia intensified and military men and traders visited the area, their knowledge of Pannonia became more accurate.[1]

Most of the existing geographical, hydrographical, and demographical data appear in the works of Strabo, Pliny, Appian, and Ptolemy. Some other writers also contributed bits of useful information to the general picture of Pannonia, but unfortunately their descriptions did not provide any overall view of this land. Most of their data referred to an earlier phase of Roman rule, depicting the southwestern and southern part of Pannonia only. Most of our current information about the central and northern part of Pannonia has come from itineraries and inscriptions. The *Notitia Dignitatum* also yielded extremely important geographical data on the locations of the camps and the frontier or *limes* of Pannonia.

The exact frontiers of Pannonia (Fig. 6), with the exception of the strip along the Danube, are not known.[2] Ptolemy mentioned Mons Cetius as the western border of Pannonia separating it from Noricum. The identification of this mountain is only hypothetical. It is generally accepted that the name referred to the eastern foothills of the Alps between Wienerwald and Poetovio. This boundary began at the military camp Cannabiaca (Klosterneuburg) by the Danube, a western outpost in Pannonia. From there it wound through the Wienerwald and the Schneeberg to the south. The exact course of the border is not known; it will be the task of scholars of the future to reconstruct it from the systematic examination of the inscriptions of stone relics and brick stamps. The border through the Wechsel mountains along the valley of the Strem reached the Raba River at Ad Vicesimum (Radkersburg). This point was documented in the *Tabula Peutingeriana* and by the Geographer of Ravenna in his description of the Amber Route. From there, the border went southwest (Fig. 20), and west of Poetovio crossed the Drava and the roads between Poetovio and Flavia Solva (Leibnitz); then it crossed the Poetovio-Celeia (Celje) road and, about five kilometers west of Neviodunum (Drnovo) at Brestanica, reached the Sava and the Neviodunum–Celeia road. From there, mainly in the northern part of the valley of the Sava, the line reached Atrans (Trojane). Over the summits of the mountains Kamniške

Planine and Karavanka, it continued west as far as Radovlica. From there, it turned to the south and reached the Julian Alps, where it ran between Ad Pirum (Hrušica-Valico del Pero) and Longaticum (Logatec), crossing the Amber Route. From this point, the line turned eastward, crossing Ocra Mons and the Kapella area. Between Sv. Peter na Mrežnici and Metulum (Čakovac), it intersected the road between Siscia (Sisak) and Senia (Senj). This road branched off at Romula from the Siscia–Neviodunum road. From an inscription we have learned that Topusko belonged to Pannonia, but next to Topusko, Ad Fines (Degoj) station was the border between Pannonia and Dalmatia. The *Tabula Peutingeriana* and the Geographer of Ravenna also referred to Ad Fines as a border station between Pannonia and Dalmatia. East of Siscia, the border was located 25 to 30 kilometers south of the Sava River at the edge of the valley of the Sava and the foothills of the highlands. The border followed the flow of the Sava into the Danube.

The Danube, which formed the northern and eastern borders of Pannonia from Cannabiaca to Taurunum, follows much the same course today, although the water level of the Danube was higher at that time. The islands of the Danube did not belong to Pannonia, but a few of the larger islands were under Roman surveillance. The Romans built bridgeheads or small fortresses on these islands, as for example on the Szentendre and Csepel islands.

The border described above was not a final or permanent arrangement in Roman times, especially not the western section. At the beginning of the first century, Savaria and its surroundings belonged to Noricum. Certain changes occurred during the second century, possibly after the Marcomannian wars when Emona was attached to Italia. From this time, the border moved south of Atrans and crossed the Sava line at Acervo (Št. Vid). During Diocletian's reforms, Poetovio and its neighborhood were annexed by Noricum.

At the beginning of the first century Pannonia was called Illyricum Inferior. In the second half of the first century, it became an independent province and was named Pannonia. The division of Pannonia into two provinces took place in A.D. 106–107 when Trajan, after the Dacian war, felt the need of reorganization. Ptolemy, the itineraries, and the inscriptions give information defining the border between the two Pannonias. The western part became *Pannonia Superior* and the eastern part, *Pannonia Inferior*. The northern end of the border between them ran north and south between Ulcisia Castra (Szentendre) and Cirpi (Dunabogdány). The garrison unit of Cirpi was mentioned on military diplomas from the second century onward among the troops of Pannonia Superior. The border ran through the Pilis and Vértes hills and along the land held by the Eraviscan and Azalian tribes as far as the Mór valley. It passed east of Lake Balaton, then turned toward its southern shore, where it went directly south, crossing the Drava River between Serena (Valjevo?) and Marinianis (Donji Miholjac). Between the Drava and the Sava, we are not certain of the border. Possibly it reached the Sava between Marsonia (Slavonski Brod) and Servitium (Bos. Gradiška), but Servitium may have belonged to Pannonia Inferior. In 213 under Caracalla, the Danube bend area up as far as Brigetio was attached to Pannonia Inferior.

Another general change occurred between 295 and 297, when Diocletian divided Pannonia into four parts: Pannonia Superior, from the Sava up to the northern

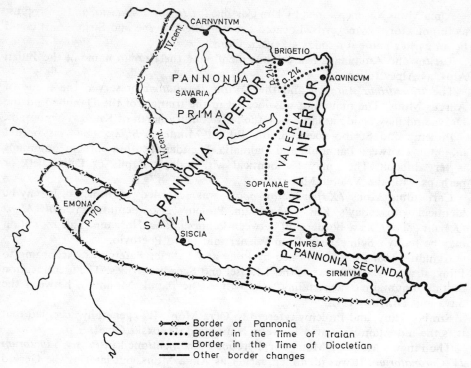

Fig. 1. Pannonia—Border of provinces

*limes*, was named Pannonia Prima; south of the Drava, Pannonia Superior became Savia; the northern part of Pannonia Inferior from the Drava River became Valeria; and the southern half became Pannonia Secunda. It is noteworthy that the border between the two parts was between Altinum (Kölked) and Ad Militare (Batina), which cut off a triangle between the Danube and the Drava. This area belonged to Pannonia Secunda.


## MOUNTAINS

Although Pannonia has many hills and mountains, only ten of their Roman names survived, and it is difficult to identify even these. Most of them relate to the foothills of the Alps, and they chiefly appear in the geographical descriptions of the Pannonian frontier. The names of mountains in the interior of Pannonia, particularly the Transdanubian parts, are hardly known.

Albius Mons (᾿Αλβανον ὄρος or ᾿Αλβιον ὄρος) was mentioned by Strabo and Ptolemy, as part of Alpes Iuliae and Ocra Mons in the territory of the Iapodes. It is possible that it can be identified with Velika Velebit of today.

Alma Mons was mentioned by Dio Cassius, the *Historia Augusta*, and Eutropius, as famous for its vineyards. Located between the Danube and Sirmium, it could be an earlier name for today's Fruška Gora.

Tacitus and Ammianus Marcellinus mentioned the Roman name of the Julian Alps, as Alpes Iuliae.

The *Itinerarium Antonini* and the *Notitia Dignitatum* preserved the name of Aureus Mons. The mountain was located in the triangle of the Danube and the Drava and may be identifiable with the hilly country north of Kneževi Vinogradi.

Ptolemy and Strabo speak of the Biblini Montes *(Βίβλια ὄρη)* as border mountains between Pannonia and Dalmatia. The identification of these mountains is very difficult. They may be identical with today's Papuk or Dilj (Djel), or perhaps with the Velebit Mountain.

Carusadius Mons *(Καρουσαδίον ὄρος)* was recorded by Ptolemy. It may be identical with today's Karst Mountain. Ptolemy also mentioned *Cetius Mons (Κέτιον ὄρος)* as a border hill between Pannonia and Noricum. This mountain may be today's Schneeberg, from Wienerwald up to Poetovio.

Claudius Mons was mentioned by Pliny and Velleius Paterculus. According to Pliny, it was located between the Taurisci and Scordisci tribes. The identification of this mountain is uncertain. Perhaps it is the Papuk Mountain between the Sava and Drava rivers.

Strabo, Pliny, and Ptolemy referred to Ocra Mons. It is generally accepted that it is the mountain pass area between Nanes and Trnovški Gozd.

The Pinguis Mons or Mons Porphyreticus was mentioned in *Passio Sanctorum IV Coronatorum*. It was identified earlier as Alma Mons, but later as the Geresd hills in County Baranya.


## RIVER NETWORK

The rivers of the province are better known than the mountains. The names of the three great rivers, the Danube (Danuvius), Drava (Dravus), and Sava (Savus) appeared frequently in ancient sources. Pliny speaks of the last two as tributaries of the Danube. The other rivers are little known, except those in the western and southern parts of Pannonia. Perhaps the names of several rivers were adopted from the Romans after the fall of Rome and preserved in a corrupt form. For example, today's Zala River was spelled Salle by the Romans. Several less important rivers which were referred to in ancient documents have new or altered modern names and their exact location is not identifiable.

The tributaries of the Danube and Sava are better known than those of the other rivers. Among the left bank tributaries of the Danube, five names remain from ancient times: the Marus (Morava); the Cusus, sometimes referred to as the Duria (perhaps the Vág); the Granua (Garam–Hron); and the Pathisus (Tisza–Tisa). These rivers reached the Danube along its Pannonian section, but because they were located on the left bank of the Danube, they were not considered Pannonian rivers.

Among the right bank tributaries, in addition to the Drava and Sava, only the Arrabo (Rába–Raab), which flows into the Danube at Arrabona (Győr), is known. The small stream Perint, which was called Sibaris in Roman times, was mentioned in the *Passio* of Saint Quirinus. It was one of the tributaries of the Arrabo River. *Itinerarium Antonini* refers to one of the stations of the Cariniana valley, which may preserve the Roman name of the Kapos River.

We are able to identify only one large tributary of the Drava: the Savarias River, mentioned by Ptolemy, is possibly identical with the Mura (Mur) River.

The names of most of the rivers which flow into the Sava are known, but their identification is still problematical. Pliny mentioned among the right bank rivers of the Sava the Nauportus (Ljubljanica), which flows into the Sava at Emona (Ljubljana). Strabo and the Geographer of Ravenna included the name of the Corcoras River, which is identical with Krka. Strabo, Dio Cassius, Appian, and Pliny made references to Colapis, which might be the Kulpa. Perhaps the stream Δράβος mentioned by Strabo is the small stream Odra, which flows into the Sava at Siscia. The other right bank tributaries are not certainly identifiable. Pliny referred to the Valdasus River, which could be the Una or the Ukrina. He also mentioned the Vrpanus, which could be identical with the Vrbas. Bathinus or Basante might be the Bosna River. Ptolemy, the *Tabula Peutingeriana*, and the Geographer of Ravenna recorded the name of Drinus, the present Drina River.

Among the smaller rivers of the area between the Drava and Sava rivers, Pliny and Ptolemy referred to the Bacuntius, which could be the Bosut. The ancient Ulca, a small stream which flows into the Danube at Vukovar, is now called the Vuka.

In addition to the rivers mentioned above, several others were referred to by ancient Pannonian sources (for example, by Jordanes) after the surrender of the province. Several attempts at identification of these streams have failed so far. All we know of the rivers Aqua Nigra, Scarniunga, Bolia, and Nedao is that they were supposed to be in Pannonia. Nor is it possible to be certain of the identity of Bustricius, mentioned by the Geographer of Ravenna.


## LAKES AND SWAMPS

Our knowledge today of ancient names of lakes and swamps is fragmentary. The Lacus Lugeon (Lugeae Paludes) might be identical with Ljubljansko Barje (Laibacher Moor) or with Lake Cerknica. Lake Balaton was mentioned in several important ancient documents: its Roman name was Lacus Pelso. The ninth-century *Conversio Bagoariorum et Carantanorum* mentions Lake Balaton as Lacus Pelissa Inferior. It is possible that the second large lake of Pannonia, Lake Fertő (Neusiedler See), was called Lacus Pelso Superior. This lake, which is next to the Deserta Boiorum was also called Lacus Pelso by Pliny. Thus the Romans called both large lakes Lacus Pelso.

The marshland of the Vuka River, south of Mursa (Osijek), was called Hiulca palus by the Romans. This name occurs in several ancient sources, but the area is referred to by Jordanes as Lacus Mursianus.

Pliny referred to two large islands, the Insula Metubarbis and Insula Segestica, at Siscia. The Insula Metubarbis could have been the tidal land between the Sava and Bosut rivers.

Various sources refer to marshes and to several swamps. During the reigns of Probus and Galerius, some swamps were drained. Under Galerius, the low area near Lake Balaton was drained in order to gain more land for agriculture.

## AGRICULTURE AND NATURAL RESOURCES

Even before the Romans came, the larger part of Pannonia was cultivated, because the climate was very favorable; during Roman times, cultivated land was increased, and great food surpluses resulted. Documents from the fourth century prove that wheat was exported to Italia. On the hillsides of Pannonia, the Romans planted large orchards and vineyards.

Animal husbandry was also highly developed. The cattle-breeding industry was most significant, but the bones of horses, pigs, goats, sheep, and poultry found in excavated settlements demonstrate the importance of these animals as well.

Pannonia was famous for its forests. In the hilly areas, the forests were either acorn-bearing (silvae glandiferae) or forest with grazing lands (silvae vulgaris pascuae). In the western part of Pannonia at the foot of the Alps, spruce, pine, and fir trees were common; in the central mountainous region, beech trees and oaks prevailed.

Literary sources and tombstones preserve evidence of a great hunting tradition in Pannonia (Pl. I. 1). Animal bones from excavations show the forests' wealth of game. Some of today's extinct species, such as the European aurochs (Bos primigenius), were frequent in those times. The most common species were deer, wild boar, wildcat, rabbit, and wolf. Their bones have been discovered frequently in Roman strata.

Pannonia was poor in minerals. There was some coal mining, primarily shallow surface mining. Iron metallurgy was practiced on the western frontier of Pannonia, and traces of goldwashing industry have been found near the Danube.

Quarry operations were very significant. Limestone, andesite, sandstone, and an inferior quality of marble were quarried throughout the whole of Pannonia in Roman times.

### NOTES

[1] On the ancient authors: I. BORZSÁK, "Die Kentnisse des Altertums über das Karpatenbecken" (Budapest, 1936) (*DissPann* 1:6). See also for general descriptions: J. O. THOMSON, *History of Ancient Geography* (Cambridge, 1948); E. H. BUNBURY, *A History of Ancient Geography among the Greeks and Romans* (London, 1959).

[2] For the frontiers of the province, see: A. GRAF, "Übersicht der antiken Geographie von Pannonien" (Budapest, 1936) (*DissPann* 1: 5); *Tabula Imperii Romani*, L 33 (Tergeste) and L 34 (Aquincum); A. MÓCSY, "Pannónia". *R. E. Pauly-Wissowa*, Suppl. 9, pp. 521 f.; M. PAVAN, "La provincia romana della Pannonia superior". *Atti dell'Accademia Nazionale dei Lincei* 6 (1955): 373–574; G. ALFÖLDY, "Eine römische Straßenbauinschrift aus Salona". *ActaArchHung* 16 (1964): 247–56; J. BOJANOVSKI,"Severiana Bosnensia". *Članci i Gradja*

*Zavičajnog Muzeja u Tuzli* 9 (1972): 37 f.; J. Šašel, "Emona". *R. E. Pauly-Wissowa*, Suppl. 11, pp. 540 f.

[3] For geographical names, see *Tabula Imperii Romani* L 33 and L 34. Some more recent articles on geography are: S. Petru, "Nekaj antičnih zemljepisnih pojmov o našik krajlik". *Arheološki Vestnik* 19 (1968): 375; M. Mirković, "Sirmium: Its History from the I. Century A.D. to 582 A.D.". *Sirmium* 1 (1971): 5 f; E. Tóth, "Adatok Savaria-Szombathely és környéke történeti földrajzához" (Contributions to the Historical Geography of Savaria-Szombathely and Its Neighborhood). *Vasi Szemle* (1972), p. 239 f.

# PANNONIA BEFORE THE ROMAN CONQUEST

OTTO TROGMAYER

Pannonia is in the heart of Central Europe, swept again and again during its history by a series of cultural trends from east, west, north, and south. The culture of this territory continually changed under the impact of the multiple migrations. The approximately square-shaped area is bordered in the north and east by the Danube, in the west by the Alps and Lake Fertő, and in the south by the foothills south of the Sava. In its center lies Balaton, the largest lake of Central Europe. Except for Kisalföld and Mezőség, which are flat, the terrain is generally hilly.

The history of Pannonia before the Roman occupation, a period spanning some fifteen thousand generations, is here summarized according to archaeological cultures.

## PALEOLITHIC AGE (Fig. 2)

### Buda Industry

In 1939 choppers were found in the territory of the Castle of Buda and were deposited in the Hungarian National Museum. It was not until 1964 that the character of the finds was described by L. Vértes, on the basis of the Vértesszőlős excavations.[1] The many-layered settlement was identified by M. Kretzoi as belonging to one of Mindel's interstadials of the upper Bihar period. The group of finds is characterized by choppers, chopper-tools, and Clacton splinters. Even a fireplace has been unearthed. Parallels of the excavation can be found at Choukoutien in Southeast Asia and in the Kafuan culture of Africa and perhaps in the West European Clactonien. The most important ensemble of finds comprises human teeth and part of a skull which in all probability belonged to a human being in the most ancient archaeological European culture. A. Thoma compares it with archanthropoid finds of China, Java, and East Africa. In the literature it has the name of Homo (erectus seu sapiens) paleohungaricus.[2]

### Moustier Culture

Even in the Paleolithic Age, the Danube was the frontier of our culture. Thus the Moustier cultures of Transdanubia must necessarily be distinguished from variants in North Hungary. Relying on published material in the two most important excavations we find two main sets of data in the Moustiers of Transdanubia.

65

*Tata* (GrN 3023–33600; GRO 2538–50000). The site lies in a limestone basin near hot springs, and comprehensive research places it in early Brörup interstadial. The archaeological material is typical Mousterian pebble, its main type being the bifacial scraping knife. Other sites are the cave of Kiskevély, Szelim, and the rock cavity at Csákvár. Its nearest parallel is Krapina, and L. Vértes believed that Tata derived from the Krapina culture or another identical one.[3]

*Érd* (Lower level of top layer: GrN 4444–44300). The site excavated by Vera Gábori lies in a small dip. Its oldest level is dated to the end of Riss-Würm, and above it another cultural level which can be divided into five levels has been unearthed. Its chronology is placed between the Amersfoort-Brörup and the first period of W 1. In the find cypress and the Ponthiniano-Charentien types are most characteristic. V. Gábori believes that Érd is more ancient than the find-ensemble of Tata, and that we must look for its origin across the Alps in the territories to the south.[4]

### Szeleta Culture

The Transdanubian version of the Szeleta culture can be fully distinguished from that of the Bükk Hills.[5] It is characterized by numerous bone instruments showing strong Levalloisien influence. The leaf-ends representing the most common type, unlike those at Bükk, have been produced with a firm technique.[6] The specific site is Lovas, where Europe's oldest mine (a hematic mine) has been discovered.[7] Special sites are: the Kiskevély cave, the Szelim cave,[8] the Csákvár rock cavity, the Bivak cave and a cave opening in Pilisszántó II. Genetic questions have remained unclarified, but the culture can probably be identified with the radiation of the local Mousterian. It goes back to W1–2.[9]

### Pilisszántó Culture

The name Pilisszántó was used by L. Vértes, who summarized under it the assemblage called earlier Magdalenian and later cave gravetti.[10] Its characteristic tools are the micro-gravette and the blunt-edged blade. It can be dated to the early W2, and its final impact can still be observed in postglacial times. Sites: Pilisszántó, the Jankovich cave (Pl. II.1), the Kiskevély cave and the youngest layer of the Szelim cave. M. Gábori believes it is of eastern origin.

### Eastern Gravettian Culture

The Eastern Gravettian culture emerged in South and Middle Russia. Its main site in Transdanubia is Ságvár (GroN 1959–18900, lower level; GroN 1783–17760, upper level).[11] It dates from W3. The finds are characterized by micro-gravettes, scrapers resembling ship's keels, blunt blades, and horn instruments. M. Gábori excavated the remains of a hut sunk into the earth from a typical reindeer hunter settlement, measuring 3.2 by 29 meters, and surrounded by post holes.[12]

In the Danube bend, another local variety of the eastern gravette is known. Its chief provenance is Pilismarót and it dates back to early W3.[13]

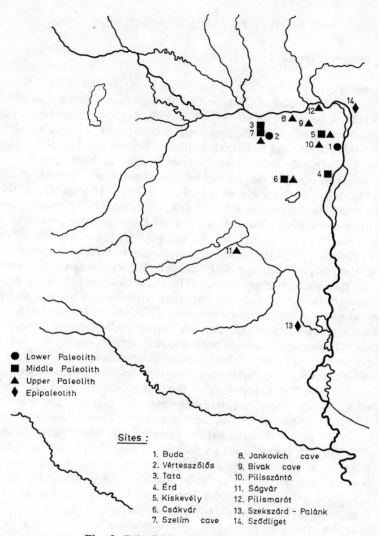

Sites:

1. Buda
2. Vértesszőlős
3. Tata
4. Érd
5. Kiskevély
6. Csákvár
7. Szelím cave

8. Jankovich cave
9. Bivak cave
10. Pilisszántó
11. Ságvár
12. Pilismarót
13. Szekszárd - Palánk
14. Sződliget

Fig. 2. Paleolithic sites of Transdanubia

The basic population of Mesolithic Transdanubia may have been identical with the people of the eastern gravette. Tools emerging at Szekszárd-Palánk (H-408-c-10350) display azilian reminiscences.[14] Yet the eastern gravetti element is more significant. Find-ensembles from the Mesolithic are difficult to explain.[15] Nevertheless, Mesolithic finds of a Transdanubian Microlithic type differ from the Macrolithic of the Bükk region. The finds excavated at Sződliget of a Tardenoise character relate to other Transdanubian materials.[16]

The basic population — or a part of it — of the Neolithic period may be found in the people of these Mesolithic materials. The impulse for neolithization arrived from a southern or southeastern direction.

# NEOLITHIC AND COPPER AGE

## Linear Pottery Culture

The Linear Pottery culture of Transdanubia is a part of a cultural complex covering the whole of Central Europe.[17] Although finds from the neighborhood of Budapest have been known since the last century, their comprehensive analysis is only now in progress. Extending over the whole territory uuder consideration, the area of this culture did not reach eastward beyond the Danube, while south of the Drava River only remains of the later period can be detected.

The settlements, on loess ridges, can be reconstructed on the basis of analogies found in Moravia and Bohemia as a cyclical system. Houses from the early stage were small, in later times larger, but in both periods resembled houses of the type found elsewhere in Europe. A cemetery from the later period, containing flexed skeletons has been discovered. Of the pottery, the globular forms are the most characteristic; they are decorated with deeply incised lines, sometimes with patterns in the early period, and with bundles of lines divided by incising in the later period; this is called Zseliz pottery (Pl. II.2). Not until the later period did painting appear; the space between two parallel lines was burned and covered with a porous red or white paint. Besides microliths, the tools found were shoe-last celts and awls made of bone. In the later period shell-bracelets appear. Several figurines seem to indicate relations with peoples farther south.

Although we have no exact data, theories of western origin and West Balkan origin make it seem most probable that the local Mesolithic population developed into a Neolithic one. The chronology of the culture can be fixed only indirectly. The earliest known pottery decorated with fingernail marks is not identical with the Linear Pottery culture of the southern part of the Great Plain. In spite of the generally accepted theory that the earlier phase of the linear pottery culture belonged to the Vinča A period, the fact that there was no pottery with fingernail decoration serves as proof of the contrary. Appearing on the left bank of the Danube and reaching northward as far as Kalocsa and southward to the Drava River, this culture did not enter into Transdanubia, although the proximity of the sites suggests a certain chronological parallelism.

The Linear Pottery with fingernail decorations south of the Drava River was discovered only in its later phase. Its youngest offspring, the Zseliz culture, is of the same age as the Szakálhát–Lebő group of the lowlands, which, according to thorough research, originated in the B period of Vinča. This period corresponds to the Otzaki I level in Thessaly. In the later phase of the Transdanubian Linear Pottery culture, strong Vinča influence becomes apparent as seen in examples at Bicske[18] in the same way as in the development of the Lengyel culture. On the basis of the determining values of the Tatárlaka (Tărtăria) tablets, the Vinča B period can be dated about 2800 B.C. Presumably this is also the date of one part of the Zseliz-type finds.

## Lengyel Culture

The Lengyel culture belongs to a widespread complex extending well beyond Transdanubia. Its western boundary is the Danube, but it extends beyond the river at Budapest and can be found also in Slovakia, Austria, and Moravia (Pl. II.3). Formerly, considered an offspring of the ribbon-ornamented culture, it is sometimes considered to have originated in the Balkans, presumably evolving from Zseliz, a local culture with ribbon decoration, under strong Vinča influence.

The settlements of the Lengyel culture are large and sometimes fortified. Their stratification attests their prolonged existence. The small houses, often sunk into the ground, are rectangular with tent roofs. Elsewhere, as at Zengővárkony, large dwellings have been unearthed.[19] The reconstruction of the large refuse pits by J. Dombay is in our opinion not acceptable. Indications of established burial rites can be found in burial places; contracted skeletons were placed in groups, their orientation always north-south or east-west, with an inventory of objects at their feet. Mutilated skeletons have also been found, sometimes with wild boar tusks on the skull.[20]

Characteristic pottery forms are large, hollow pedestal bowls, and mushroom-shaped pots, often with wart ornamentation applied in several ways. The painted decoration is usually red in the early period, but in the later period (which is less known in Hungary) white painting prevails, plastic decoration is more frequent, and incised patterns do not occur. Metal objects known in this complex are copper pearls and rings. Dentalium, silica, and obsidian have been found frequently, as well as trapezoidal and perforated stone axes, bone pins, stone battle-axes, antler hoes, and clay spoons.

The chronology can be defined on the basis of imported articles. The early Lengyel settlement of Aszód was located near the Luzianky group and has direct relations with the Tisza culture. The beginning of Lengyel culture could be placed at the end of the Vinča B-2-Larissa period, according to the figurines of enthroned gods unearthed at settlements of this culture. Its late period can be correlated, on the basis of the white encrusted painting, with the Vinča D, or Tiszapolgár culture. Thus in Hungary only these two periods can be ascertained; there are no periods of unpainted pottery known. For the time being it is impossible to reconstruct the chronology of the final phase; many elements of this phase were adopted by the Ludanice group in Slovakia and by the Balaton group in Transdanubia, but there is a hiatus which can only be filled by further research.

## The Balaton Group

Until recent years most scholars assumed that the Balaton group was a survival of the Lengyel culture. The separation of the former findings from the latter has been accomplished by N. Kalicz.[21]

The Balaton group, of the same age as the Bodrogkeresztur group in East Hungary and the Ludanice group in Slovakia, forms a part of the Middle Copper Age culture. This culture may have originated in a northward extension of some

Vinča-Pločnik elements. Influences from the foothills of the Alps near the Mond-see could also have contributed to its evolution. Its boundaries extend beyond Transdanubia both in the west and the south, but its northern and eastern boundaries are defined roughly by the Danube.

Settlements indicate a migratory life; numerous open-air settlements and traces of cave settlements have been found. There are remains of both rectangular houses, sunk in the ground, and lake dwellings of a village house type, while an unusual structure found in Fenékpuszta must have been used for public rites. In the settlement of Letenye evidence of a fortification has been found; on one side of a four-meter-wide ditch a palisadelike fence made of thick piles was built. Less well-known are the burial places. Generally one flexed skeleton and one bowl were found together in a single tomb. In the late phase of the group, cremation was also practiced.

The pottery of the culture is characterized by ribbed or fluted ornament types: bowls with pouring lips, biconical dishes, rough barrel-shaped pots. Encrusted ware with engraved or quilled ridges as decoration, jars with handles, and wide-necked pots are elements which originated in the Alps. Hollow bowls with pedestals seem to be of local origin. Metallurgy presumably was stimulated by the copper deposits in the Alps. Among the few metal objects found are fragments of copper spirals and rings. Fragments of crucibles and copper dross indicate local metallurgy. Pectorals are the characteristic gold jewels found at Csáford (Pl. III.1), while a horn made of a triton shell points to relations with the south.[22]

The earliest appearance of the group cannot be precisely dated; for a chronological indication of the final phase we consider the date of the Boleraz group which, in certain areas, put an end to the independent life of the Balaton group. It can, however, be assumed that the Balaton group was contemporaneous with the early period of the Baden culture in southwestern Transdanubia. Its western neighbor was the Kanzianberg–Villach culture, its southwestern neighbor the Lasinja culture. It is of the same age as the Bodrogkeresztur and Hunyadihalom groups of the Great Hungarian Plain.

In its initial phase this culture was greatly influenced by the intrusion of peoples from the south, and the new way of life must have been based upon agriculture. Some western materials coming along the routes of the copper trade indicate the penetration of a smaller people. The latest phase of the Balaton group had a strong influence on the independently developed pottery of western origin, ornamented with quilted ridges at home in southwestern Slovakia and near Budapest.

### The Boleraz Group

Researches of I. Torma have confirmed that the Boleraz group flourished in the whole territory of Transdanubia.[23] The chronology of the group can be ascertained on the basis of the corresponding levels in Moravia and Slovakia. Its origins have been examined from several points of view, the most detailed research work being done by V. Nemejcová-Pavúková. Her early theory that the group was a branch of the local Ludanice group, was later modified after she recognized a strong Eastern influence on the development of the local basic population.

Excavations on a large scale have been made at Andocs-Nagytoldipuszta. Here pits of irregular shape surround a central area. There has been an attempt to reconstruct several types of houses sunk in the ground. Results here are similar to those obtained in the excavations in Slovakia. Nineteen graves have been excavated at Pilismarót and Fonyód-Bézsenypuszta; in eight of them the bodies were covered by a flat stone, and in three the grave furniture, consisting of pottery and a single stone axe, was covered by a thick layer of pottery fragments. An *ustrina* has also been found. The funerary rites at the burial place at Pilismarót are similar to those of Andocs-Nagytoldipuszta. Cremation is characteristic of this group.

As to the pottery, dishes with fluted decoration on the inner side are rather common, their outer surface often ornamented with furrows around the shoulder. Bowls with biconical break-lines, rounded off, and pots with double or triple lines of fingertip prints under the brim are also frequent. The body of the pot is often decorated with a herringbone pattern. Occurring frequently among the cups are those of flattened globular shape, with a conical neck and with a brim protruding slightly, their bulges often decorated with fluted lines. Also characteristic are big, biconical short-necked bowls with handles on the bulge and with a herringbone pattern on the upper part.

The chronology of the group can be clearly fixed in spite of the absence of "classical" Baden material at many sites. The cemetery of Fonyód can presumably be classified as representing a later phase of the group. Because its geographical distribution coincides in Transdanubia with that of the Ludanice and Balaton groups, it cannot be of the same age as those. Consideration must be given to I. Torma's statement that populations of the Boleraz type appeared also outside the area of the groups which could be considered as antecedents. This underscores the importance of alien elements, independent of local ones. Thus the Boleraz group followed the later phase of the Balaton group and preceded the "classical" phase of the Pécel culture.

In discussing some chronological problems relating to the origin of the Boleraz group, we have already mentioned the assumption that it might be considered as the first wave of the migratory movement of the peoples in the southeast during the Late Copper Age. The new population assimilated those of the Balaton and Ludanice groups and preserved its own burial rite (cremation), hitherto unknown in Transdanubia. Because of the lack of comprehensive excavations, the changes in the way of life can only be surmised.

## Classical Period of the Pécel (Baden) Culture

Both the definition and the comprehensive analysis of the Pécel/Baden culture (Pl. III.2) has been achieved by J. Banner.[24] This culture formed one part of a huge culture complex in southeast Europe during the Early Bronze Age.[25] Its Balkan (i.e., in the last instance, Anatolian) origin emerges from pottery and metal objects and from what we know of its way of life. Recent finds from the northern part of Hungary have also drawn our attention to the relationship of the culture with its southern neighbors.

The sites go beyond the territories with which we are concerned, and are spread all over Pannonia, but only a few settlements have been excavated thus far. The strata are very thin. The population did not remain long at one place and the analysis of animal bones tells us that their life was based on hunting, probably of smaller ruminants, since those are the bones which are most numerous. The types of houses must have been similar to those of adjacent areas within the same cultural environment; from the later period some types with semicircular apses have been found.

The greatest cemetery excavated so far is that of Budakalász. At the beginning flexed skeletons preponderated, though in the later phase cremation was again practiced. Packing in stone occurs rarely. J. Csalog suggests that chariot burial was also practiced, for bones of cattle have been found among the grave furniture, and a small vessel in the shape of a wagon has also been found at Budakalász.

The following forms are characteristic of the pottery: tripartite bowls, fishing-boat-shaped vessels, dippers with high handles, stamped cups with fluted shoulders. Decorative elements include net-patterns, indented rows of points, flutes, and, in the latest phase, stamped patterns.[26] The diadem found at Vörs is one of the most outstanding remains of the culture, and the triton-horn found at Békás-megyer gives evidence of relations with the south.

The chronological position of the culture has been cleared up on the basis of the stratigraphics of the Hungarian Plain Tiszapolgár-Basatanya and Székely-Zöldtelek. N. Kalicz supposes, on the basis of very detailed comparative analysis, that the culture ended at the same time as Troy V. He fixes an absolute chronology between the years 2050 and 1900 B.C. The sites in the mountains in northern Hungary represent its latest phase.


# BRONZE AGE

## The Zók Culture

On the basis of some groups of findings in the southwest, F. Tompa has pointed out the occurrence of the Zók culture in Hungary. This culture had its roots in the Pécel culture, the sites of which were later occupied by a population of southern, presumably Anatolian, origin. Under the influence of the tribes coming from the south, the Vučedol pottery style took over the territory between the Alps and the Nyírség. The evolution of the independent Makó culture might be attributed to the influence from the northwest, from Retz, in the same epoch.[27]

The remains of the Zók culture (Pl. IV.1) are spread almost all over Hungary. According to the work of N. Kalicz, Transdanubia was invaded by the people known to us as the Makó group. In the triangle of Baranya the Vučedol group was also present. The settlements, established on ridges or hills, were occupied by shepherds who moved from camp to camp. The excavated houses were partly sunk into the ground, but the excavations made at Vučedol also produced some other types of houses and fortified settlements. The burial places show very few tombs but give evidence of cremation; ashes were placed in urns or piled up on the ground of the grave.

The principal pottery type is the bowl ornamented on the inner side, but there are also pots with an egg-shaped bulge and a cylindrical neck, sometimes with two handles on the shoulder, as well as slim jugs with one handle and a projecting rim. *Askoi* also occur.

The relative dating of this culture can be given exactly by stratigraphical data. According to some observations, it followed the Pécel culture, its independent life ending with the intrusion of the Somogyvár-Vinkovci group. Its absolute dating might be placed between 1900 and 1800 B.C.[28]

### The Somogyvár–Vinkovci Culture

I. Bóna and S. Dimitrijević[29] have pointed out the importance of the group of findings belonging to the Somogyvár–Vinkovci culture. According to I. Bóna,[30] it was contemporary with the decline of the Level V of Troy — that is, with the movement of Balkan peoples towards the southeast. This diffusion was considerable, covering the whole of Transdanubia, except the Mezőség, where we cannot find any trace of the characteristic pottery.

The settlements, situated on higher hills or wider ridges, show that the people stayed only a short time at any given place. We have little data about the buildings. The burial places are peculiar, generally hillock cemeteries. The skeletons were sometimes enclosed in stone cysts, laid on their backs, always, to our knowledge, in the east-west orientation. Both the settlements and the burial places indicate an aggressive and warlike people.

Characteristic of the pottery are amphorae with two handles under a cylindrical rim. There were also storage pots with almost funnel-shaped necks and with handles on their bulges; there were also jars with a single handle, a slender neck, and a marked bulge. Few metal objects have been found except for gold lock-rings, and, at adjacent sites, a bronze torque and a dagger.

The relative chronology can be established exactly. On the basis of recently excavated levels of Vinkovci, we can tell that this culture directly succeeded the Zók culture, and that during its short existence, it probably played an important role in the formation of the Nagyrév culture. The Vinkovci excavation also indicates the upper chronological limit of the Zók; it flourished in the first third of the nineteenth century B.C.

### The Bell-Beaker Culture

The easternmost region of this culture, called "Pan-European" by H. Thomas, was Transdanubia. Its origin appears unambiguous; the Hungarian branch arrived along the Danube from Central Europe and made islandlike block settlements along the Danube as far east as Budapest.[31]

The rather small settlements are known solely from some surface findings and some excavated refuse pits. Small villages, which were rather close to one another and which were occupied for only a short period, were found inside the blocks of settlements. The burial places exhibit flexed skeletons in small clanship cemeteries, although cremation occurred in the younger phase and was practiced even more later on. The rather peculiar pottery of the Bell–Beaker culture includes bell-shaped pots with ornamental hands, bowls with several legs, and slim cups with one handle (Pl. IV.2). In addition, bronze wrist-guards, triangular daggers, and pearls with V-shaped bore have been found.

The relative chronology of the Bell–Beaker culture can be determined on the basis of indirect data. It may have appeared shortly after the rise of the Somogyvár culture, later becoming one of the popular components of the Nagyrév culture. Its absolute date ought to be placed in the nineteenth century B.C. Its developed local variation was the Oka (Oggau)-Sarród group in the area of Lake Fertő.

### Problem of the "Litzenkeramik"

The chronology of the "Litzenkeramik" culture, although amply discussed, is still not exactly established in Hungary, but can be defined by the stratigraphy of Grosshöflein-Föllik in the Burgenland.[32] A tomb with stone packing, embedded in a Gáta stratum, has been found here. A flexed skeleton, cord-ornamented pottery and animal bones were its contents. Although the influence of this population of eastern origin has been traced in several sites in Transdanubia, no burial place of its own has been excavated so far.

### The Gáta Group

The Gáta group, influenced by the steppe cultures, developed as a local variety of the culture of the bell-shaped ceramic.[33] It extended throughout East and West Pannonia and through Burgenland (Wieselburg Gruppe). The houses were sunk into the ground, and the flexed type of burial was used. The types of ceramics so far discovered are two-handled amphoras and pot-bellied jugs with ribbon-shaped handles. Among the metal objects, the Cyprian pins, twist-headed pins, flat-headed pins, and triangular daggers give good chronological evidence of the period in which the group existed.[34]

### The Nagyrév Culture

The Nagyrév culture got its name from a settlement near the village of Nagyrév, on the left bank of the Tisza. This culture has been examined mainly by historians on the basis of its place in the Tószeg stratigraphy.[35]

In spite of some differing suggestions, it seems that the culture developed under an influence from the southwest, amalgamated with the civilization of the earlier inhabitants. Near Budapest the culture established prosperous contacts with, and was influenced by, the Bell–Beaker culture. The northern boundary of the culture was the bend of the Danube and the southern limit was Bölcske. In Transdanubia, east of Balaton and following the course of the Danube, the characteristic complexes of the four main groups have been found at Ökörhalom, in Sziget-szentmiklós and Kulcs.

The settlements were tell-like and built on riverside terraces. Only a few family houses have been excavated. Cremation was known; sometimes ashes were kept in an urn and sometimes they were scattered, the former practice becoming more widely used during the life of the culture. It was generally used from the beginning in the Kulcs group. The most common type of pottery was the jug with one handle in the middle of a cone-shaped neck, but pots with broom-stroke ornament and plastic ribs and flat bowls with projecting brims were also common. Characteristic decoration was vertical ribbing or moustache ornament. The culture did not have many metal objects, although some plates with twisted brim, lock-rings, and spirals have been found, as well as one golden object — an earring unearthed in the cemetery of Alsónémedi.[36]

The Ökörhalom group, partly of the same age as the Vinkovci–Somogyvár culture, had some connections with the Pitvaros group and with the early phase of the Hatvan culture. Its later groups might be paralleled with the early period of the Szőreg-Perjámos culture in the Maros region.

## The Tokod Group

Though the first evidence of the Tokod group was published by J. Hampel in 1876, the differentiation of the group was accomplished by I. Bóna in the 1950s. The Tokod group was one of the ethnic units of the Hatvan culture advancing westwards during its younger phase. It was located in a relatively small area, between Esztergom and Süttő, on the right bank of the Danube, and near the mouth of the Ipoly.[37]

The settlements were tell-like, often built on flat sites along river valleys. The small family houses were rectangular and placed in a ring. The Tokod group practiced cremation, usually placing the ashes in an urn, but sometimes scattering them on the ground. Their pottery was practically the same as that of the Hatvan culture. Big, egg-shaped urns with conical neck and a tunnel handle on their bulge are characteristic (Pl. IV.3). Plastic ribs and wart ornament were common on pots, as were textile-impression and broom-stroke decoration. Encrustation, which also appeared, might be attributed to the influence of the Kisapostag culture.

The Tokod group was practically contemporaneous with the late Hatvan culture, the beginnings of the Vatya culture, and with the culture of Encrusted Ware. According to G. Bándi, its absolute dating is from 1700 to 1600 B.C.

## The Kisapostag Culture

The Kisapostag culture was intensively studied by Amália Mozsolics in 1942, but now it seems that it should be considered as a phase of the Nagyrév culture.[38] According to Amália Mozsolics, its origins must be sought in the sphere of the "Litzenkeramik". Its immigration to Transdanubia is presumed. Again according to G. Bándi, the culture can be identified with a later phase of the Szigetszentmiklós group.[39]

The sites of the culture are found on a broad strip of land on the left bank of the Danube, and some also appear in Mid-Transdanubia. Its burial places are urn-fields, cremation having been a tradition of the indigenous population, belonging to the Zók-Nagyrév civilization. The few skeletons which have been found might be connected with a new population.

The most characteristic pieces of pottery are the urn with funnel-shaped neck, egg-shaped bulge, and two to four ribbon handles, and the cup with funnel-shaped neck, ball-shaped bulge, flattened bottom, and projecting brim. On the bulge of the urn a vertical or oblique bundle of lines appears as ornament, while encrusted decoration, often with zigzag pattern between parallel lines, is common on the neck. Among the metal objects that are found are plates with rolled edge and trapezoid shape, plates of a Pan's pipe shape, pins with shovel-shaped heads, lunulae, spiral tubes, and conical *tutuli*.

The relative chronology of the group can be defined on the basis of these metallic objects and can be put between the late phase of the Nagyrév culture and the early phase of the Vatya culture. Its influence on the formation of the Vatya culture can be traced beyond doubt. One of our most important tasks is to clarify the role the "Litzenkeramik" played generally in the history of the Early Bronze Age cultures in Transdanubia.

### Encrusted Pottery Culture of Pannonia

Encrusted pottery culture is a characteristic Transdanubian one already known in the nineteenth century. Its borders are well-defined, and it can be found practically all over Transdanubia with the exception of Mezőföld and Lake Fertő.[40]

There are two contradictory theories of its origin. According to I. Bóna,[41] it was a direct descendant of the Zók culture, with other smaller groups playing a role in its formation. According to G. Bándi,[42] the culture was established by the population of the "Litzenkeramik" coming from the west, but the Zók is its basic population.

The settlements are found mostly at waterside slopes, on gentler ridges. Some theories suggest that the houses, which were partly sunk in the ground, were somewhat larger in the early phases. The burial places show evidence of cremation, with the ashes sometimes in urns, sometimes scattered in the tomb. Objects have been found in the tombs, sometimes more than 30 vases in a single tomb. The pottery shows encrusted as well as incised decoration, with double warts on rough pottery (Pl. V.1). The urns are sack-shaped, with a funnel neck and a handle on the bulge. In household pottery, funnel neck and flattened ball-like body are common, although there is a great variety of forms. The perfectly executed pottery has fine, thin walls; it is lustrous, of yellowish, reddish, or greyish color. Metal articles include comb-shaped pendants, pins with spherical head, and solid torques.

The chronology can be defined from the imported articles. Although the dating of the beginning of the culture is still controversial, it ought to be placed at the beginning of the Reinecke-A period. It may be regarded as a uniform culture, the variants of which are local in northern and southern Pannonia. The more developed phase of the culture is of the same age as the Vatya-Szőreg-Füzes-abony cultures, and its early period is contemporaneous with the Hatvan culture. It ceased to be independent after the appearance of the tumulus builders, during the Reinecke-BB I period. Some groups went to the south and founded the assemblages of the Szeremle-Cîrna type.

## The Vatya Culture

The Vatya culture, one of the most characteristic cultures of the Middle Bronze Age in Hungary, has yielded many rich finds. Its comprehensive evaluation is due to the efforts of I. Bóna. The analysis of the material and the burial rites prove beyond doubt that the Kulcs group of the Nagyrév culture can be considered as the basis of the Vatya culture. According to I. Bóna, the transformation of the Nagyrév culture can be attributed to the Kisapostag culture which intruded from the northwest. G. Bándi, however, does not accept the theory of an independent Kisapostag culture.

The sites of the Vatya culture (Pl. V.2) can be found on both sides of the Danube, from Vác to Mohács. The central area of the culture was Transdanubian Mezőség, though it occupied the region of the Danube and the Tisza as well. The tell-like settlements are at Koszider, Kulcs, Bölcske. In the middle phase of the culture, earthworks were built, thus safeguarding the sites of the culture by a fortified line in the west. The fortified settlements were sometimes as large as 100 meters in width and 200 in length. Double lines of fortifications have also been found at the central settlements. The houses were rectangular pile dwellings with mud-walls, and there were a great number of beehive-shaped pits around the houses.

The very wide burial places were near the settlements, on ridges of hills. At Dunaújváros 1,600 tombs have been unearthed; they were placed in groups arranged in oval rings, nine to eighteen tombs per ring. In the early phase, only sepulchral urns were used, and pots and metals were placed first beside the urns, and then inside them. In the latest phase, burial places with skeletons were also common; this can be related to the Koszider group.

The most characteristic types of pottery are the urns with oval body and funnel-shaped neck. Their flat strap handles are joined in pairs on the bulge, sometimes running from the rim to the shoulder. The decoration of the urns consists of comb-shaped incised bundles of lines, and, in the later phase, reliefs. The bowls are strongly profiled; in the later phases they were ornamented with channeled warts. The most characteristic type of cup was spherical with a curved neck and with a strap handle from rim to shoulder.

Although little local metallurgy could be expected, there are relatively numerous metal objects from the long Vatya period. Chief among them are the following types: Cyprian pins, pins with willow-leaf head, pins with lamellar, twisted, shovel-shaped, ball, and ring heads, rectangular plate pendants, lunules, eyeglass spirals, heart-shaped plate pendants, diadems, *tutuli*, triangular daggers with two to four nail holes, axes of the Cófalva type, and shaft-hole axes. The only gold objects found from the whole period are a few buttons.

The chronology can be defined easily with the help of the rich findings. I. Bóna divided its inner chronology, mainly on the basis of the metal objects, into an early transitional period and three other phases. The Vatya culture may be partly of the same age as the Gáta culture, the encrusted pottery culture, the Szőreg-Perjámos culture, and the Tokod group. It lasted from the seventeenth to the fourteenth century B.C. In the fourteenth century it managed to resist for a while the invasions of the western tumulus builders, but eventually was destroyed by them.[43]

## Hoard Level of Koszider

Two studies published at almost the same time dealt with the hoard complexes which represented the end of the so-called Hungarian autochthonous Bronze Age.[44] The authors of both studies connected the hiding of the treasure with the migration of the people of the tumulus culture near the Danube. There is, however, a basic difference between their views. According to Amália Mozsolics, the hoards are remains of the indigenous populations, while I. Bóna, on the basis of more detailed analysis, connects them with an immigration of several waves of the tumulus builders.

The characteristic objects of the Koszider group are *tutuli*, ribbed lunulae, sickle-shaped pins, axes with shaft-hole, and arm guards, all of bronze. It is possible that one part of the finds, discovered under uncertain conditions, really belonged to the graves. The hoards of Koszider type do not indicate a sudden historical catastrophe, but are rather the remains of a longer period, undoubtedly connected with the immigration of the tumulus builders.

## The Tumulus Culture

Although our attention was drawn to the characteristic remains of the tumulus culture by V. G. Childe as early as the 1920s, it was treated in detail only in the 1950s. A scientific debate following the publication of the monographs by I. Bóna and Amália Mozsolics is still going on. The finds of this culture are distributed throughout the territory under consideration. In its strong advance towards the east, it crossed the Danube; its find-complexes can be found as far to the south as the Balkans. During its early phase it might have come to the Lowlands across Transdanubia, passing around the earthworks of the Vatya culture. In the middle period it inundated the whole territory of Transdanubia.

Its origins are twofold, one branch coming through Austrian territory, bringing several western types, and the other developing in Slovakia, also under western influence. The way of life of the tumulus people was different from that of the cultures of the local Middle Bronze Age they displaced; the newcomers were a livestock-raising people. The agrarian cultures of the region did have southeastern relations, but they were essentially locals; the tumulus culture, however, with its rather primitive remains, did not develop locally. Its slow development began during the middle period, interrupted in the late period by migration of other peoples. As the tumulus culture came to Transdanubia, it pushed toward the southeast the Pannonian encrusted pottery culture which, together with the Gerjen group on the right bank of the Danube, formed the Szeremle-Cîrna culture and presumably eventually absorbed the tumulus culture as well.

We know none of the settlements from the early and middle period, and have only some refuse pits from the late phase. The burial places of the early period were rather small, and we know only a few graves, most of them flat. Burial places consisting of several hundred small tumuli are also known (as at Farkasgyepű), but they are known so far only from the middle period. Tumulus graves occurred most frequently in the last period. In the early phase both cremation and burial are known, the latter disappearing entirely in the late phase.

Characteristic pottery includes amphoras with cylindrical neck, cups reminiscent of the Aunjetitz form, containing sharp-lined bulges and dishes with spouts. From the latest level there are bowls with faceted shoulders and urns with conical necks and sharp-lined shoulders. The metallurgy seems to be of foreign character. From the early phase we have found bronze objects of Koszider type: pins in spiked *tutuli*, sickle-shaped pins, arm rings, gold pendants, and boat-shaped lock-rings. These objects can be separated from the grave furniture of the middle period, which includes pins, heart-shaped pendants and lamellar belts. The latest level is characterized by pins with conical heads and by Peschiera bronzes.

On the basis of the metal objects the chronology of the tumulus culture can be placed between the phases Reinecke -BB and -BD. According to our hypothesis, the earliest complexes of finds appeared during the B period of the Bronze Age; it is contemporaneous with the Bogárzó horizon of the Great Hungarian Plain. The middle period was Bronze Age C, Tápé horizon of the Great Plain, while the latest remains can be fixed as of Bronze Age D, Csorva group in the Great Plain and Čaka group in Slovakia. Its absolute chronology cannot yet be given, but the culture may have flourished in Pannonia from the fourteenth to the twelfth century B.C.

### The Urnfield Culture

A comprehensive monograph on the Urnfield culture, previously considered as belonging to the Lausitz culture, has recently been written by E. Patek.[45] Two main theories exist about its origin.[46] Some archaeologists regard it as a local development, while others, such as E. Patek, suggest a great southeast migration of peoples at the end of the Bronze Age D period, a theory that gains credibility in view of the differences between the Lowland and Transdanubia.[47] The sites of culture, which reached as far east as the Danube, can be found throughout Transdanubia; what we know of their density depends on the extent of the excavations. Some related groups developed south of the Drava River and north of the Danube, the west Hungarian group forming a part of the great Middle European cultural complex. According to Patek, the culture can be divided into three groups: that of West-Transdanubia, of South-Transdanubia, and of the Val group.

As to settlements, the villages on the lower ridges lasted for a short time only; those built at higher places persisted for a longer time. These settlements, sometimes fortified, must have controlled some commercial routes. We have no data on the inner living areas of the fortifications. Pits, which some time ago were thought to be pit-dwellings, may have been refuse pits. At Pécsvárad, houses with mud walls strengthened by rectangular pillars have been excavated by J. Dombay. Excavations made by the author on Helemba Island have revealed round houses, partly sunk into the ground; the roofs were supported by a central pillar and the diameter of the clay floor was about 5 meters. Similar houses have been found at Velemszentvid.

The burial places show the exclusive use of cremation, with some of the ashes placed in urns, others scattered on the ground. In spite of the name of the culture, there are some tumuli dating from the beginning of the period, but these have not been thoroughly studied and may belong to the late tumuli culture.

Characteristic of the pottery were bellied urns with cone-shaped necks and two opposing handles on the shoulder; during the early period rims were faceted. The rim of the deep conical dishes protruded at an angle above the handle. Also characteristic were the cups with high handles and rough pots with ribbing around the shoulder.

The bronze industry was also highly developed. In the great workshops of Velemszentvid many commercial articles were produced, including fibulas of the violinbow, spectacles, bow, harp, and shield types; two-bladed and crescent-shaped one-bladed razors; vase-headed, spherical-headed, and onion-headed pins; ornamental discs; and bronze vessels. The great number of hoards may mark separate historical events.

The chronology falls within the periods Hallstatt A and B; exact dating of the culture's beginning is impossible until there is a detailed treatment of the culture of the late tumulus builders (Period D of the Bronze Age). The suggestion that the culture appeared in Transdanubia in the twelfth century B.C. seems acceptable.

## EARLY IRON AGE

It is unfortunate indeed that there is no comprehensive evaluation of the Early Iron Age period containing the legacy of the native population of Pannonia prior to the Roman era. At the beginning of the HB–HC period, elements of new ethnic groups appeared in east Hungary; the remains of the Mezőcsát group may be demonstrated at the peripheries of the eastern Transdanubian area (as at Pécs-Makárhegy and Kakasd) and give evidence of the first use of iron in this area. The openwork dagger sheaths and forms of bridles as well as the burial customs are indications of the group's eastern origin, traced by some to Cimmerian ethnics.[48]

A new wave of western people seems to have settled on the remains of the urnfield people of the north Transdanubian area. The aristocrats lived on heights surrounded by earthworks and buried their members under large tumuli, which have yielded painted and incised decorated ceramic pieces (Pl. VI.1). These objects show connections with the western Hallstatt civilization and suggest that this new group is a variant of the Kalendernberg culture. We know of the survival of the local people, who may be forerunners of the native Pannonian population. The area of this culture was north Transdanubia, north of Lake Balaton.

Along the Drava River, or, more accurately, in South Pannonia, a culture showing some affinity with the northwestern Balkan or the Veneto-Illyrian culture developed. Its burial rites differed from those of the groups in the areas north of the Lake Balaton.[49] Its extended burial tombs yielded numerous objects of Italian origin, such as the *certosa fibula*. The evidence of horse bits and the remains of some horse skeletons point to the possibility that the groups of this area played an intermediary role between the Scythae from the East, and the Ventic culturale groups.

## LATE IRON AGE

At the end of the Early Iron Age, Transdanubia (Dunántúl) was inhabited by Pannonians who spoke a language similar to Illyrian, and the Great Plain (Alföld) by peoples of steppe origin with a Scythian culture. However, the Late Iron Age introduced significant ethnic changes. In 400 B.C. a new tribe arrived in the Carpathian basin from the west. The Celts (Pl. VI.2), coming as conquerors from their native homes in Switzerland and the north of France, invaded the rest of Europe in several waves,[50] extending their dominion from Britain to Asia Minor. They devastated Rome and plundered Delphi.[51] We are not even familiar with the names of the first raiding tribes. In the second century B.C., the Romans applied the name *Taurisci* to the peoples of the eastern Alps.[52] Pompeius Trogus, a historian of Celtic origin, tells us that the Celts who invaded Pannonia had met with heavy resistance from the local Pannonians and could not overrun the southern part of Transdanubia.

The Celts who advanced as far as Delphi were defeated in 279 and were forced to withdraw to the northern part of the Balkan peninsula — to present Hungary. One tribe, the Scordisci, settled down in the eastern part of the territory between the Drava and Sava rivers. The Celts flourished in the Carpathian basin in the late third and second centuries.[53] Numerous Celtic cemeteries with rich finds prove conclusively that many settlements existed. In 114 B.C. the Germanic Cimbri invaded from the north, overrunning the mighty tribe of Boii in the northern part of Transdanubia.

Tribal leaders were named on Celtic coins and these included Biatec, Nonnos, and Evoiurix.[54] In Pannonia the minting of coins reached its height in the first century B.C., with Macedonian coins used as models at first, and Roman coins later.[55] Independent tribes had their own special coinage.

The Iron Age culture of the Celts is called La Tène, after the Swiss Celtic center. During the second century there arose Pannonian centers of the Celtic tribes, centers for agriculture, trade, industry, coinage, and even religious life. These centers were called oppida; they were in fact village settlements built on mountains and hills, surrounded by earthwork. In time of war, they also served as shelters. One of the important centers of the metal workers, the oppidum of Velemszentvid, was in the western Transdanubian region; and in the south was the earth fortification of Szalacska, in the valley of the Kapos River.[56] The present city of

Pest can be first distinguished in the middle of the first century B.C. It was in the territory of counties Fejér and Tolna, where the Eravisci tribe lived. Their significant *oppidum* was situated in the heart of present Budapest, on Gellért Hill and in the adjoining city of Tabán.[57] The potters' quarters of the settlement have been unearthed on the hill and near the Danube.

In the early first century A.D., the last free Celtic tribes were expelled in the north by the Germanic tribes and in the west and south by the Romans. In the Great Plain the settlement of the Jazyges denoted the end of Celtic rule. The highly developed material culture of the Celts in Hungary had left many ethnic traces and survived in the heritage of both the Romans and the Barbarian peoples.[58]

NOTES

[1] L. VÉRTES, *Az őskőkor és az átmeneti kőkor emlékei Magyarországon* (Prehistoric and Transitional Stone Age Relics in Hungary Budapest, 1965); L. VÉRTES, "Jelentés a vértesszőllősi őstelep újabb asatásairól" (Report on Recent Excavations at the Ancient Settlement of Vértesszőllős). *MTA II. Oszt. Közl.* 15 (1966): 111–14; L. VÉRTES, "Bilan des découvertes les plus importantes faites de 1963 à 1966 dans les fouilles du site paléolithique inférieur de Vértesszőllős (Hongrie)". *Revue Anthropologique*, 1967, p. 1–13.

[2] A. THOMA, "Az előember fogmaradványai a vértesszőllősi őstelepről" (Tooth Fragments of Prehistoric Man at Vértesszőllős Settlement). *MTA Biol. Oszt. Közl.* 9 (1966): 263–80; A. THOMA, "Az előember nyakszirtcsontja a vértesszőllősi őstelepről" (Occipital Bone of Prehistoric Man at Vértesszőllős). *MTA Biol. Oszt. Közl.* 10 (1967): 1–20.

[3] L. VÉRTES, *Tata, eine mittelpaläolitische Travertin-Siedlung in Ungarn.* Budapest, 1964 (*ArchHung* 43).

[4] V. GÁBORI-CSÁNK, "Vorläufiger Bericht über die paläolitische Fundstelle von Érd". *Móra F. Múzeum Évkönyve*, 1964–65/2. (1966), p. 5–9; V. GÁBORI-CSÁNK, *La station du paléolithique moyen d'Érd — Hongrie.* Budapest, 1968 (*Monumenta Historica Budapestinensis* 3); V. GÁBORI-CSÁNK, "C-14 Dates of the Hungarian Palaeolithic". *ActaArchHung* 22 (1970): 3–12.

[5] M. GÁBORI, "Regionale Verbreitung paläolithischer Kulturen Ungarns". *ActaArchHung* 21 (1969): 155–65.

[6] L. VÉRTES, "Gruppen des Aurignacien in Ungarn". *Archaeologia Austriaca* 19/20 (1956): 15–27.

[7] GY. MÉSZÁROS and L. VÉRTES, "A Paint Mine from the Early Upper Palaeolithic Age near Lovas". *ActaArchHung* 5 (1955): 1–34; GY. MÉSZÁROS and L. VÉRTES, "Őskőkori festékbánya Lovason" (A Paint Mine from the Palaeolithic near Lovas). *ArchÉrt* 82 (1955): 3–18.

[8] L. VÉRTES, "Die archäologische Funde der Szelim-Höhle". *ActaArchHung* 9 (1958): 5–17.

[9] L. VÉRTES, "Beiträge zur Abstammung des ungarischen Szeletien". *FolArch* 10 (1958): 3–15.

[10] VÉRTES, *Az őskőkor.*

[11] M. GÁBORI, "A ságvári paleolitikus telep újabb ásatásának eredményei—Les résultats des fouilles récemment effectuées dans la station paléolithique de Ságvár". *ArchÉrt* 86 (1959): 3–19; M. GÁBORI, *A késői paleolitikum Magyarországon* (Late Paleolithic in Hungary) (Budapest, 1964); M. GÁBORI and V. GÁBORI, "Etudes archéologiques et stratigraphiques dans les stations de loess paléolithiques de Hongrie". *ActaArchHung* 8 (1957): 3–117; V. GÁBORI-CSÁNK, "A ságvári telep abszolút kormeghatározása—La détermination de l'âge absolu de la station de Ságvár". *ArchÉrt* 87 (1960): 125–29.

[12] M. GÁBORI and V. GÁBORI, "Der erste paläolithische Hausgrundriss in Ungarn". *Acta ArchHung* 9 (1958): 19–34.

[13] V. GÁBORI, "Neue paläolithische Funde im Eipel-Tal". *Archeol. Rozhledy* 10 (1958): 55–61.

[14] L. Vértes, "Szekszárd—Palánk, Tata". ArchÉrt 88 (1961): 284; L. Vértes, A Szekszárd—palánki jégkorvégi őstelep (Szekszárd, 1963).

[15] V. Dobosi, "Mesolithische Fundorte in Ungarn", in Symposium über die aktuellen Fragen der Bandkeramik (Székesfehérvár, 1970); R. Pusztai, "Mezolitikus leletek Somogyból—Mesolithische Funde im Komitat Somogy". Janus Pannonius Múzeum Évk., 1957, pp. 96–105.

[16] M. Gábori, "Mezolitikus leletek Sződligetről—Mesolithische Funde von Sződliget". ArchÉrt 83 (1956): 14–16.

[17] N. Kalicz and J. Makkay, "Südliche Einflüsse im frühen und mittleren Neolithikum Transdanubiens", in [ed.], Symposium.

[18] É. F. Petres, "Újabb-kőkori sírok Bicskén—Neolithic Graves at Bicske". FolArch 6 (1954): 22–28, 197–98.

[19] J. Dombay, Die Siedlung und das Gräberfeld in Zengővárkony: Beiträge zur Kultur des Aeneolithikums in Ungarn. Budapest, 1960 (ArchHung 37).

[20] Zs. Zoffmann, "Data to the Burial Rites of the Lengyel Culture". Janus Pannonius Múzeum Évk., 1965, pp. 55–60.

[21] N. Kalicz, "A rézkori balatoni csoport Veszprém megyében—Die kupferzeitliche Balaton-Gruppe im Komitat Veszprém". Veszprém megyei Múzeumok Közl. 8 (1969): 83–90.

[22] J. Korek, "Die Goldscheiben von Csáford". FolArch 12 (1960): 27–33.

[23] I. Torma, "Adatok a bádeni (péceli) kultúra boleráz csoportjának magyarországi elterjedéséhez — Beiträge zur Verbreitung der Boleraz-Gruppe der Badener Kultur in Ungarn". Veszprém megyei Múzeumok Közl. 8 (1969): 91–108.

[24] J. Banner, Die Péceler Kultur. Budapest, 1956 (ArchHung 35).

[25] N. Kalicz, Die Péceler (Badener) Kultur und Anatolien. Budapest, 1963 (Studia Archaeologica 2).

[26] S. Mithay, "Rézkori sír Győrött — Kupferzeitliches Grab in Győr". Arrabona 10 (1968): 5–8.

[27] N. Kalicz, Die Frühbronzezeit in Nordost-Ungarn: Abriss der Geschichte des 19–16. Jahrhunderts v. u. Z. Budapest, 1968 (ArchHung 45).

[28] G. Bándi, "Remarks on the History of Research in the 'Vučedol Problem' ". Alba Regia 8–9 (1967): 23–33.

[29] S. Dimitrijević, "Četiri groba iz slavenske nekropole u Otoku kod Vinkovaca". Opuscule Archaeol. (Zagreb) 2 (1957): 21–38.

[30] I. Bóna, "The Peoples of Southern Origin of the Early Bronze Age in Hungary". Alba Regia 4–5 (1963–1964): 17–63.

[31] N. Kalicz, "Die frühbronzezeitlichen Brandbestattungen in der Umgebung der Gemeinde Alsónémedi". ActaArchHung 9 (1958): 195–209.

[32] G. Bándi, "Adatok a Dunántúl korabronzkori történetéhez—Contributions to the Early Bronze Age History of NE Transdanubia". Janus Pannonius Múzeum Évk. 1964, pp. 65–72.

[33] F. Kőszegi, "Az oroszvári bronzkori temető — The Bronze Age Cemetery at Oroszvár (Rusovca)". FolArch 10 (1958): 43–59.

[34] I. Bóna, "Az oroszvári 4. női sír melldísze: Adatok a közép-dunamedencei bronzkori viselethez—The Pectoral Ornament of the Female Grave 4 at Oroszvár: Data Illustrating the Costume of the Bronze Age in the Middle Danube Basin". ArchÉrt 87 (1960): 198–205.

[35] I. Bóna, "The Early Bronze Age Urn Cemetery at Kulcs and the Kulcs Group of the Nagyrév Culture". Alba Regia 1 (1960): 7–15.

[36] I. Bóna, "The Cemeteries of the Nagyrév Culture". Alba Regia 2–3 (1963): 11–23.

[37] G. Bándi, "Data to the Early and Middle Bronze Age of Northern Transdanubia and Southern Slovakia". Alba Regia 4–5 (1963–1964): 65–71.

[38] A. Mozsolics, A kisapostagi korabronzkori urnatemető — Der frühbronzezeitliche Urnenfriedhof von Kisapostag. Budapest, 1942 (ArchHung 26).

[39] G. Bándi, "The Cemetery of Ercsi-Sinatelep". Alba Regia 6–7 (1965–66): 11–25.

[40] G. Bándi, "Ursprung und innere Chronologie der Kultur der inkrustierten Keramik". Janus Pannonius Múzeum Évk., 1965, pp. 61–74; G. Bándi, A dél-dunántúli mészbetétes edények népe kultúrájának elterjedése és eredete (Pécs, 1967).

[41] I. Bóna, "The Peoples".

[42] G. BÁNDI and Zs. K. ZOFFMANN, "Középső bronzkori hamvasztásos temetők Baranyában—Brandgräberfelder der Mittelbronzezeit im Komitat Baranya". *Janus Pannonius Múzeum Évk.*, 1966, pp. 43–56.

[43] I. BÓNA, "Clay Models of Bronze Age Waggons and Wheels in the Middle Danube Basin". *ActaArchHung* 12 (1961): 83–111.

[44] I. BÓNA, "Chronologie der Hortfunde vom Koszider-Typus". *ActaArchHung* 9 (1958): 211–43; A. MOZSOLICS, "Archäologische Beiträge zur Geschichte der großen Wanderung". *ActaArchHung* 8 (1957): 119–56.

[45] E. PATEK, *Die Urnenfelderkultur in Transdanubien*. Budapest, 1968 (*ArchHung* 44).

[46] F. KŐSZEGI, "A piliscsabai koravaskori temető — Früheisenzeitliche Gräber von Piliscsaba". *FolArch* 8 (1956): 47–62; F. KŐSZEGI, "Funde in Érd aus der frühen Hallstattzeit". *ActaArchHung* 9 (1958): 289–98.

[47] S. MITHAY, "A koroncói koravaskori kunyhó — Die Hütte von Koroncó aus der jüngeren Eisenzeit". *Arrabona* 12 (1970): 5–16.

[48] S. GALLUS, *A soproni Burgstall alakos urnái — Die figuralverzierten Urnen vom Soproner Burgstall*. Budapest, 1935 (*ArchHung* 13).

[49] E. G. JEREM, "The Late Iron Age Cemetery of Szentlőrinc". *ActaArchHung* 20 (1968): 159–208.

[50] I. HUNYADY, "Kelták a Kárpátmedencében—Kelten im Karpatenbecken". Budapest, 1944 (*DissPann* 2:14).

[51] M. SZABÓ, *The Celtic Heritage in Hungary* (Budapest, 1971).

[52] J. TODOROVIĆ, "Kelti u Jugoistocnoj Evropi". Beograd, 1968 (*Diss. Muzej Grade Beograda* 8).

[53] J. FILIP, *Keltové ve středni Evropě* (Praha, 1956).

[54] V. ONDROUCH, *Keltské mince tipy Biatec z Bratislavy* (Bratislava, 1958).

[55] K. PINK, "Die Münzprägung der Ostkelten und ihrer Nachbarn". Budapest, 1939 (*DissPann* 2:15).

[56] É. BÓNIS, "Die Siedlungsverhältnisse der pannonischen Urbevölkerung und einige Fragen ihres Weiterlebens". *ActaArchHung* 23 (1971): 33–39.

[57] É. B. BÓNIS, *Die spätkeltische Siedlung Gellérthegy-Tabán in Budapest*. Budapest, 1969 (*ArchHung* 47).

[58] É. F. PETRES and M. SZABÓ, *A keleti kelta művészet—Eastern Celtic Art*. Székesfehérvár, 1974.

# HISTORY OF PANNONIA

## LÁSZLÓ BARKÓCZI

The former Roman province of Pannonia occupied parts of three present-day countries, i.e., the north of Yugoslavia (between the Drava and Sava rivers), the east of Austria (Burgenland and the Viennese basin), and the west of Hungary (Transdanubia).

Pannonia derives its name from the tribes of the Pannonii, a collective term applied to Illyrian tribes living in the Sava basin, a name which first occurs in the writings of Polybius.[1]

It is Pliny the Elder who among ancient authors gives a correct comprehensive picture of the geographical situation of the province and the tribes living in it. He writes as follows:

...Noricis iunguntur lacus Pelso, deserta Boiorum; iam tamen colonia divi Claudi Savaria et oppido Scarabantia Iulia habitantur. Inde glandifera Pannoniae, qua mitescentia Alpium iuga per medium Illyricum a septentrione ad meridiem versa molli in dextra ac laeva devexitate considunt. Quae pars ad mare Adriaticum spectat apellatur Dalmatia et Illyricum supradictum; ad septentriones Pannonia vergit: finitur inde Danuvio. In ea coloniae Aemona, Siscia. Omnes clari et navigabiles in Danuvium defluunt Dravus e Noricis violentior, Savus ex Alpibus Carnicis placidior CXVm.p. intervallo, Dravus per Serretes, Serapillos, Iasos, Andizetes, Savus per Colopianos Breucosque. Populorum haec capita; praeterea Arviates, Azali, Amantini, Belgites, Catari, Cornacates, Eravisci, Hercuniates, Latovici, Oseriates, Varciani. Mons Claudius, cuius in fronte Scordisci, in tergo Taurisci. Insula in Savo Metubarbis, amnicarum maxima. Praeterea amnes memorandi: Colapis in Savum influens iuxta Sisciam gemino alveo insulam ibi efficiens quae Segestica appellatur. Alter amnis Bacuntius in Savum Sirmio oppido influit, ubi civitas Sirmiensium et Amantinorum. Inde XLV Taurunum, ubi Danuvio miscetur Savus; supra influunt Valdasus, Urpanus et ipsi non ignobiles. *Hist. nat.* 3: 146–47.

This description, written about the middle of the first century A.D., pointed to a stable situation. Earlier, however, the territory had a long history of upheavals.

The emergence of the province of Pannonia was largely the result of the expansionist policy of Italia, its fate was significantly affected by the changing forces within the Carpathian basin, and its history was determined by population movements occurring outside the basin itself. The Roman occupation of the territory was largely prompted by the fact that the natural frontier of the Danube not only safeguarded the defense of the Roman provinces alongside the river but also prevented expansion of the Roman boundary. Every attempt to cross this river frontier had proved temporary, the real frontier remaining forever the Danube.

Strictly speaking, the history of Pannonia comprises two regions. Its essential part had always been the territory between the Drava and the Sava, since this passage supplied a direct route between Dalmatia and Italia and also contact between Italia and the east. In the beginning the part north of the Drava was of secondary importance; its final Roman occupation occurred later, when it was demanded by the strategic situation and the concept of a unified Danube frontier.

## HISTORY OF TERRITORY PRIOR TO ROMAN OCCUPATION

The history of the later Pannonian area encompasses the Celtic movements, the first wave of which reached the western part of Pannonia as early as the fourth century B.C. However, no information survives of the fights between the arriving Celts and the indigenous Illyrians. Nor do we know of new Celtic waves prior to the second decade of the third century. In the first two decades of the third century, Celtic tribes devastated the Balkans, invading even Greece. However, surviving sources mention no more than one date relating to Pannonia; namely, in 279 part of a Celtic tribe, returning home after ransacking Delphi, settled down near the confluence of the Danube and Sava rivers. Henceforth they called themselves Scordisci, a name which derived from Mons Scardus. The Scordisci played an important role in the later history of both Pannonia and the Balkans.

Although written sources make no mention of the settlement of other Celtic tribes in the area, archaeological evidence clearly shows that the Celtic invasion of the early third century B.C. marked the arrival of other Celtic tribes as well in the Transdanubian region. However, the names of these tribes have remained unknown, although it is likely that the Textosages tribe was among them.[2]

During the third century the Scordisci expanded their power over southern Pannonia and most of Serbia. In 179 Philip V attempted to persuade the Bastarnae in the lower Danube region to attack Italia, for the way to the Adriatic led through a land held by the Scordisci. And as early as 156, the Scordisci in alliance with the Dalmatians were already in conflict with the Romans. That same year the Romans first attempted to occupy Siscia, which lay in a key position at the confluence of the Kulpa and Sava rivers. In 119 the conflict between Rome and Dalmatia continued, but Rome was never able to occupy Siscia. The Scordisci fought again on the side of the Dalmatians. After several setbacks, the Romans were unable to deal their enemy a powerful blow until 88 B.C.[3]

In the south, power remained in the hands of the Scordisci; in the north, the Boiian state enjoyed supremacy. The Boii were Celts who had invaded Italia with other Celtic tribes in the early fourth century and, after their defeat in 191 B.C., had left the territory and settled down near the Danube. According to another theory, however, they first founded a state in the Bohemian mountains and moved to the Danube only later.[4]

According to archaeological evidence, too, a new wave of Celts that appeared at the end of the second century B.C. settled in the southern part of Slovakia and in neighboring northern Hungary, where, according to the testimony of Poseidonius, the Boii lived.[5] Strabo tells us that the eastern border of the land held by the Boii reached the Tisza (Tibiscus) River, but there were also other non-Celtic tribes under Boiian rule.[6] The seat of this Boiian settlement was north of the Danube somewhere around present-day Bratislava; their neighbors to the west were the Taurisci, who held also a small part of southwestern Pannonia. No matter where the Boii first settled after they left Italia, however, when they arrived at the Danube they had to fight the Dacians who held the entire territory — or at least part of it. Strabo tells us that later animosity between the Dacians and the Boii stemmed from the fact that the Dacians demanded the land from the latter which the Dacians pretended to have possessed earlier.[7]

In the early years of the first century B.C. the Dacians emerged under King Burebista as a dangerous power, holding an empire centered in Transylvania and reaching down in the south to the Danube, and occupying much of the Boiian imperium in the north. While their hold on the area between the Danube and the Tisza was only loose, the Dacians exerted considerable pressure in the Carpathian basin, in the lower Danube region and beyond, on the Greek cities of the Black Sea. Their first attack in the Carpathian basin was directed against the already weakened Scordisci, and then they made use of the defeated enemy as an ally. The next attack was launched by Burebista against the Boii and their allies, the Taurisci. According to Strabo, Burebista exterminated all his enemies.[8] The end came shortly after 44 B.C., when the Boii ceased to exist as a political entity. Upon the remains of their civilization the Romans later formed a *civitas* in northwestern Transdanubia.

Upon the defeat of the Scordisci, the Pannonian tribes were liberated in 88 B.C. From the plan of Mithridates, king of Pontus, to attack Italia we learn that the territory he was to cross already belonged to the Pannonians.[9]

After their defeat the Scordisci were located at the eastern part of the Srem, where they had retreated. With the dissolution of the Boiian hegemony and the ebb of Boii and Scordisci power, several local tribes appeared and regained strength as independent entities in the north.

When Burebista died, about the time of the assassination of Julius Caesar, the Dacian state disintegrated as well, but was still important as a political factor and was considered a potent enemy of the Roman Empire. We can be sure that a campaign against the Dacians would have been launched (as indeed Caesar himself had contemplated it) had not the civil war after his assassination prevented it. Thus they are mentioned merely as parts of various political schemes.

In 35 B.C. Octavian seems to have completed Caesar's plan in conquering a large part of the Sava valley in a campaign against the Iapodes, taking also Siscia, which was to play an important role as a base of operations in the campaign against the Dacians. In the following year, however, Octavian was already fighting in Dalmatia. Thus we can suppose that the occupation of Siscia and the Sava River valley was necessary not to prepare a Dacian campaign but rather to establish contact between Dalmatia and Italia.

According to Cassius Dio, Octavian's campaign in the Illyrian territories was due to the Iapodes' refusal to pay taxes to the Romans. The real reason, however, may have been that the geographical position of the Iapodes presented an obstacle to Roman expansion and to those closer contacts which the Romans intended to establish with Dalmatia through the Sava River valley.

In the spring of 35, Octavian attacked with two armies, one coming direct from Italia and the other crossing the sea from Dalmatia. He took the stronghold of Metulum, the center of the Iapodes, and moved toward Siscia. This town, which later played an important part in Pannonian history, held out through a month-long siege. The resistance which broke out during the following winter was quenched by the twenty cohorts left there by Octavian.[10] By then, the whole Sava valley had certainly bowed to Octavian who, according to Appian and Cassius Dio, subjugated all Pannonian tribes.[11]

With the defeat of the Scordisci and the Boii, the retreat of the Dacians, and the occupation of Siscia by the Romans, Pannonia emerged as an independent unit, well defined by the Danube. Strabo wrote about the later situation of this territory, when it was occupied by the Boii, Taurisci, Scordisci, and Pannonians who had been defeated by the Dacians.[12]

Later, only the Breuci, Amantini, and the Andizetes, western neighbors of the Scordisci, belonged to Pannonia, while the others lived in present-day Dalmatia, Herzegovina, and Bosnia.

Owing to the loose occupation of the Sava valley an attack was launched in 16 B.C. against the Istrian peninsula by the Pannonian and Norican tribes living in present-day Croatia and Bosnia. In the following year Tiberius, who later became emperor, defeated the Scordisci, who then became Rome's allies. In 14 B.C. another Pannonian revolt had to be subdued by the Romans. Augustus's son-in-law, M. Vipsanius Agrippa, was sent against Pannonia in 13. He died early in the following year. Three years of difficult campaigning by Tiberius, between 12 and 9, finally completed the occupation of the Pannonian area between the Sava and Drava rivers.[13] The Dalmatians were allied with the Pannonians during these revolts, which in itself points to the center of the revolts, with their organizers coming from the southern Pannonian groups. Their leader was a man called Bato, and the Breuci and Amantini tribes played a leading role.

Meanwhile the Dacians were not idle. In 29 B.C. Licinius Crassus drove back a Dacian attack directed southwards, and it was Cornelius Lentulus in the second decade B.C. who drove them out of the territory later called Moesia. They even invaded Pannonia in 10 B.C.[14] The occupied Pannonian territories were quiet at last, but peoples east and north of the Danube seemed to threaten the earlier occupations. It was to secure Pannonia that L. Domitius Ahenobarbus mounted a campaign in A.D.1 against the Germanic tribes extending as far as the Elbe.[15] Locally more important, however, was the campaign of M. Vinicius, against Celtic and Dacian tribes living east of the Danube bend.[16] With this campaign we see a tendency not only to organize the territory between the Drava and Sava but also to secure politically the areas north of the Drava.

Conditions around Pannonia had to be put in order also because the Marcomannians appearing in present-day Czechoslovakian territories, in the last decade B.C., were becoming a dangerous factor. Thus Augustus decided to check the Marcomannians in A.D.6 by attacking them from two sides; one army approached from the Rhine and the other one crossed the Danube at Carnuntum.[17]

The campaign was interrupted, however, by the outbreak of a revolt of the Pannonians and Dalmatians, who once again were attempting to shake themselves free of the Roman yoke. Rushing back from his nearly successful campaign against the Marcomannians, Tiberius was able to subdue the warlike Pannonian tribes only after more than three years of hard fighting.[18] The Romans estimated the insurgent army at 200,000; according to Suetonius, this was the most serious struggle since the Punic wars.[19] Velleius Paterculus who fought with Tiberius wrote that the rebels had conspired with the peoples living around them and that all Pannonia was in arms. According to Suetonius, Tiberius finally defeated all peoples between the Danube and the Adriatic.

At the time of the war only the peoples between the Drava and the Sava rivers were included under the name of Pannonians, and only the southernmost part of this area participated in the revolt.[20] The war otherwise was called "bellum Dalmaticum", and the second part of the war continued in Dalmatia after the defeat of the Breuci who had been in confederation with the Dalmatians during the revolt. The other tribes were either not very important or were in alliance with the Romans, as were the Scordisci. These tribes north of the Drava had no connections with the revolt.

Two tribes took leading roles in the revolt. One was the Breuci, under the leadership of Bato, and the other was the Daesidiates, under the leadership of Pinnes and another Bato. The insurgents attempted to attack from two directions, one army moving against the Dalmatian shores and thence toward Italia, the second moving against Sirmium and Macedonia. A third army remained to defend the homeland. After being defeated by a local military governor, the first army withdrew from the Dalmatian coasts and united with the second army in the siege of Sirmium. Caecina Severus, coming from Moesia, liberated Sirmium from the Pannonians, but he defeated them only the following year in the Fruska Gora Mountains, where they had reestablished themselves, while Caecina Severus was moving toward Siscia to join Tiberius's army. He was attacked by the Pannonians, but he fought back successfully and joined Tiberius. During the winter of 7−8 the war came to a halt. Tiberius's tactic was that of playing for time. The Pannonians destroyed the harvests left behind and finally got into hard circumstances and agreed to talks. On the third of August, A.D. 8 they acknowledged defeat and laid down their arms. Augustus went personally to Rimini to meet Tiberius.

The Breucus Bato delivered Pinnes to the Romans and became the vassal prince of his tribe. The Daesidiates, however, refused to surrender, and as the rebellion flared up once again, they continued to fight and even succeeded in capturing and executing Bato. This particular revolt was terminated by Plautius Silvanus. The leader of the Daesidiates pushed out from the territory between the Drava and Sava rivers and continued to fight in the mountains of Dalmatia. Finally, he surrendered to Tiberius in 9 and was exiled to Ravenna.

With the last great revolt, fighting calmed down around Pannonia. Illyricum (which included the Transdanubian area and the western part of Yugoslavia) was divided into two provinces in A.D. 8: Dalmatia, previously Illyricum Superius, and Pannonia, formerly Illyricum Inferius. The name Pannonia came into common usage only in the second part of the century.[21]

Eventually it became necessary to reestablish foreign relations. The Romans interfered in the hostilities between two kings: the Marcomannic Maroboduus and the Germanic Ariminius. In the conspiracy against Maroboduus, Rome supported its enemy Cataulda. When the Marcomannians expelled Cataulda, Drusus appointed Vannius, from the tribe of the Quadi, to lead the Marcomannians. In 50, Vannius, too, was expelled by the Marcomannians, and he and his followers were allowed to settle in Pannonia. His successors, Vangio and Sido, maintained good relationships with Rome.[22]

## THE ROMAN OCCUPATION OF PANNONIA AND THE FLAVIANS

The first step in the occupation of Pannonia was the foundation of Aquileia.[23] This town was the starting station of the Amber Route, which opened from the Adriatic, passed through the western border of today's Hungary, touched Savaria (Szombathely) and Scarbantia (Sopron), and led to the coast of the Baltic Sea. Aquileia thus became the starting point for Roman action toward the north and east. At the time of the foundation of Aquileia the Romans in order to establish a secure inland link between Italia and Dalmatia tried to occupy Siscia at the confluence of the Sava and Kulpa rivers. After two futile attempts in 156 and 119 B.C. the Romans relinquished their interest in occupying the Sava valley, turning instead toward Noricum (roughly, modern Austria), which had both iron and silver mines.

Following this period, the life of the peoples of the Carpathian basin changed, partly because of the defeat of the Scordisci in 88 by Scipio Asiagenus and partly because of the war between the Boii and the Dacians. While it is difficult to estimate the impact of these wars on the life of the Pannonian people, one obvious result was the thinning out of the population north of the Drava. With the final occupation of Siscia in 35 B.C. the whole Sava valley came under Roman rule, and the wars between 12 and 9 B.C. brought the whole territory of Pannonia under Roman rule as well. In his will called "Monumentum Ancyranum", Augustus referred to this period, declaring that as a result of the campaign of Tiberius, the borders of Illyricum were extended to the Danube.[24] It has been debated whether he meant the stretch of the Danube between the Drava and Sava or the whole territory of Pannonia. Scholars generally maintain that he meant the whole of Pannonia, but recent archaeological evidence proves that the Roman occupation of Pannonia was more differentiated.

The population north of the Drava River seemed to have had no part in the Pannonian-Dalmatian conflict. No data are available to indicate whether they participated in the wars or any previous revolts. Recent research has revealed that north of the Drava and east of the Scarbantia-Savaria line no monument exists which refers to any occupation in the Augustan period. Pannonia may have been occupied in gradual steps. What we know of the Pannonian territory north of the Drava and the beginning of the construction of the *limes* has been largely influenced by the well-known inscription from Aquincum. The reconstruction of this inscription, connected with the names of Drusus and C. Calpurnius Aviola, leads us to believe that a camp was already built at Aquincum, if not under Augustus, then under Tiberius. A later interpretation giving the names of Vespasian and C. Calpetanus would place the inscription and the associated camp building under the governorship of C. Calpetanus in Pannonia (73–76 or 73–78).[25]

All the Pannonian uprisings took place in southern Pannonia and involved the Dalmatian tribes. It was mainly the tribes of the Sava together with those of the Dalmatian valley that participated in the fights. During their occupation the Romans conducted continual expeditions attempting not only to secure the territory between the Drava and the Sava but also to expand north of the Drava. It is the Danube which makes Pannonia a single geographic unit. Knowing that Augustus occupied between 12 and 9 B.C. the territory between the Drava and the Sava and that the Romans penetrated Noricum and used the Amber Route at the western borders of Pannonia, it is hardly possible to believe that the Romans were not in control of the area north of the Drava and bound by the Danube, even though no evidence seems to corroborate this assumption. It cannot be accidental that during the reign of Augustus, Roman currency was imitated by a mint of the Eravisci at the bend of the Danube, a strategic point which was of great importance in the centuries to come.[26]

For some time it was suspected that the Azali, the western neighbors of the Eravisci, were forced to settle here from southern Pannonia during one of the Pannonian uprisings.[27] There is also a strong possibility that the population of the Regöly rampart fortification in the Augustan period received its new settlement from somewhere in south Pannonia. The Romans felt it necessary to repopulate the area north of the Drava, although politically they did not yet systematically extend their empire to the Danube. It is characteristic of the territory north of the Drava that in 14 the Roman legions were still stationed in south Pannonia; one, however, was moved to Carnuntum after a legionary revolt in the same year.[28] We know of some camps on the western frontier of Pannonia along the Amber Route such as Carnuntum and the areas near Gyalóka, Savaria, and Poetovio, which were few in number, thus could primarily have supervisory functions and were not defensive in character.

In the age of Augustus the Romans were in a kind of alliance with the peoples living in territories north of the Drava, observing and supervising them as well as repopulating the areas. The territory had strategic importance but no economic significance for the Romans. This territory had been quiet for a long time; the many veteran settlements in the age of Tiberius are evidence that it had been tied to Rome in the form of clientage already under Augustus.[29]

It was under Tiberius that new settlements of veterans were founded in the western part of the province, namely around Scarbantia, Savaria, in Emona, and near Lake Balaton. The territory of Savaria extended as far as the lake. Unknown in east Pannonia, it differed from the earlier Roman control which had been political and had emphasized surveillance. The settlement in Újmajor near Keszthely is probably also connected with the Tiberian period. The veteran settlements of the age of Tiberius moved toward that route (Sirmium–Sopianae–Keszthely–Fenékpuszta–Savaria–Scarbantia) which connected the east and west diagonally through Pannonia. This route was essential not only during the Roman period but even in the sixth century A.D.[30]

During the time of Augustus and Tiberius, the construction of the *limes* was still unnecessary, partly because of the clientage of the peoples on the left bank of the Danube and partly because the Romans had not yet penetrated to the Danube along its middle course. Such penetration came only in the period of Claudius, when the Romans occupied the territory north of the Drava; then there was a systematic integration into the Roman Empire and the appearance of settled Roman life. Claudius's presence could be felt in the interior of the province and along the *limes*,[31] even though the regular construction of the *limes* was not yet made. Permanent camps were built only at important fords, since concern at that time was not with defense but with the complete annexation of the territory.

Roman foreign policy further constructed the system of "client states". The Sarmatians–Iazyges were already situated between the Danube and the Tisza about A.D. 50, or possibly even earlier. They could have reached that area only with agreement from and alliance with the Romans. Obviously the Romans desired the establishment of a buffer state which would keep the dangerous Dacians separated from Pannonia and Moesia.[32] Possibly the arrival of the Iazyges, which may not have been a smooth one, prompted Claudius to occupy and to secure the most important crossing points of the *limes*. A similar "client" relationship or system tied the Quadi to the Romans, who, meanwhile, had strongly expanded into the territory of present-day Slovakia.[33] Thus the occupation of crossing points was necessary to maintain contact, if for no other reason. With the final settling of the Sarmatians–Iazyges, the overall stability of the peoples around the province was achieved, and the systematic unification of the territory north of the Drava and between the Drava and the Sava rivers could begin.

At first, Roman administration occupied itself primarily with the settlement of the situation outside the province. Circulation of money began in the Barbaricum under Nero — evidence of this emperor's active role. Also active during this time was Plautius Silvanus, who moved as many as 100,000 Barbarians from the left bank of the Danube to Moesia. Nero's governor in Pannonia, Tampius Flavianus, may have settled some 50,000 Barbarians in Pannonia, according to the reconstruction and interpretation of the Aquincum inscription.[34] Although more recent research refutes this theory, the significance of Tampius Flavianus's Pannonian governorship remains, for in his time began currency circulation in Barbaricum and, apparently, the first stages of the tendency to settle uniformly the *limes* of the central Danubian valley. One may suppose also that this governor brought about internal organization in the province itself.

The beginning of the systematic circulation of coins after the death of Nero, during Vespasian's reign, shows that Roman life was by then fully established in the territories north of the Drava. In any case, the new interpretation of the Aquincum inscription is in agreement with the known facts and illuminates Vespasian's role in the border defense of the province. From the comparison of the inscriptions of Calpetanus from Aquincum and from Carnuntum, it becomes evident that extended work on the *limes* is Vespasian's accomplishment. During

92

the Flavian period the Dacian state was again strengthened. It seems that Decebal succeeded in unifying it and from a position of strength it emerged as a dangerous enemy of Rome. In the winter of 85–86, the governor of Moesia fell victim to their first raid and in 87, near the lower Danube, they annihilated another Roman army. Although the Dacian army was dispersed in 88 by Domitian, a quick and favorable pact was made, since Pannonia itself was threatened by Suebian and Sarmatian tribes.

A war that was waged against the Germanic and Sarmatian tribes lasted nearly five years. Owing to scanty information we know next to nothing of the battles. In 88 the Marcomannians had scored a minor victory over the Romans which involved a visit by Domitian to the province. Fierce struggles followed when the Marcomannians and Quadians engaged in battle the Lugii living north of them. The latter received aid from Domitian. The Iaziges–Sarmatians who in 89 completely destroyed the Roman legion XXI Rapax, who had been sent from the west, also participated in this fight. Domitian returned to Pannonia and stayed for eight months until the end of the campaign.[35]

For the future of Pannonia the Suebian-Sarmatian wars were of determinative importance. Under Domitian the military situation centered in the Danube region; however, among the Danubian provinces the major role was allotted to Pannonia. Part of the fighting forces were brought into the province to defend the frontiers and from henceforth four legions and many auxiliaries were stationed here. Thus Domitian brought to a close the work begun by his father and established a stationary *limes*.[36]

After the death of Domitian, Nerva waged successful wars against the Germanic tribes living north of the Danube; however, no traces survive in the province of this expedition.[37]


## THE PERIOD OF TRAJAN, HADRIAN, AND ANTONINUS PIUS

Trajan could not tolerate the growing strength of the Dacians — chiefly due to the hasty peace pact of 89. Decebal executed a well-planned and organized expansion into the lower Danube area. Soon after Trajan's accession he provided for the concentration of troops there to make the province into a military base for his Dacian expedition.

Trajan launched two campaigns (101–102, 105–106) against the Dacians.[38] His wars did not directly affect Pannonia, but the new situation created by them basically changed its future history. With the wars over, the land of the Dacians was organized into a province, which was then annexed to the empire under the name of Dacia. The new situation involved also the reorganization of Moesia and Pannonia. When Dacia became a province the final settlement of the Sarmatians and the Quadians was completed in the foreground of the *limes*, and the subdivision of Pannonia, between 103 and 107, became necessary mainly to protect the frontiers. The new division demanded separate administrations for the Sarmatian and Quadian sections. The frontier divided the two provinces stretched from Ulcisia Castra southward through Pannonia, but it seems that the division did not disregard the tribal frontiers of the natives. Both sections had independent administrations, the seat of Upper Pannonia (Pannonia Superior) being Carnuntum and that of Lower Pannonia (Pannonia Inferior), Aquincum.

The transformation of Dacia into a Roman province brought great changes in the life of the Iazyges–Sarmatians as well.[39] The territory which they occupied stretched through the Banat to the Olt River; beyond it, along the Danube, lived the Roxolani tribe. Direct link with cognate tribes had been cut off by the Dacians, and after their defeat by Trajan, Sarmatian land had not been returned because it was this territory between the eastern edge of the Banat and the Olt River that formed a link between the new province and Moesia. Thus squeezed between Roman provinces on the south, east, and west of the Carpathian basin, the Sarmatians could expand only toward the north, which, however, was occupied by the Quadians. The situation was aggravated by the fact that the Romans tended to maintain connection between Dacia and Pannonia via routes through Sarmatian territory, along which they erected watchtowers that were tended by garrisons. We know about two such routes: one started at Aquincum, while the other started at Lugio (Dunaszekcső), crossed the Tisza River, and led into Dacia near the Maros River. According to an inscription, regular postal service was conducted on this latter route.[40]

The surrounding of the Sarmatians was the reason for their attack upon Pannonia in 107. The governor, Hadrian, later emperor, was stationed in Aquincum; he not only diverted this attack but also actually defeated the Sarmatians on their own soil.[41] The animosity continued, however, and the Iazyges decided on a large-scale expedition. After the death of Trajan in 117, they joined forces with the Roxolanians and attacked again; this attack endangered not only Pannonia but Moesia and Dacia as well, since Rome's forces were preoccupied with the Parthian wars. In 118 Hadrian appointed the experienced military leader Marcus Turbo as governor of Eastern Pannonia and Dacia in order to create a single command and a force capable of defeating the Iazyges from two sides. The aim was soon realized, and the Iazyges sued for peace. After returning from the east in 124, Hadrian increased the tribute to the Roxolanians, who were living in the lower Danube valley.[42] That the Quadians, who lived north of the Danube, and Iazyges had probably participiated in these attacks seems to be documented by an inscription referring to the "expeditio Suebica et Sarmatica".[43]

During the Dacian wars Pannonia's defensive and military system assumed its final form. The western troops concentrated there during the wars were placed into Pannonian camps. On the other hand, troops well acquainted with the local conditions were sent to Dacia, some never to return to Pannonia. The majority of the closely connected camps which Trajan established along the Danube were still only wooden. Their transformation into stone and the establishment of a watchtower chain were mainly the work of Hadrian. The date of this accomplishment and the final touches on the *limes* probably correspond to Hadrian's visit to Pannonia in 124.[44]

It was under Hadrian that the Pannonian military situation became consolidated.[45] Three legions were stationed on the Danube in Upper Pannonia, at Vindobona, Carnuntum, and Brigetio and were entrusted with guarding the frontier against the Quadians. In Lower Pannonia there was a legionary camp at Aquincum; south of this, along the Sarmatian border, only *limes* stations guarded mostly with auxiliary troops were to be found.[46]

In the last years of the reign of Hadrian we hear of restlessness among the Quadians. A Sarmatian uprising may also have had something to do with the temporary unification of the two Pannonian provinces under the command of L. Aelius Caesar in 136. Similarly, in 118, during the attack of the Iazyges and Roxolanians, the commands of East Pannonia and Dacia were united under one man. In 137 L. Aelius Caesar resided at one of the important sectors of the Quadian front, in Arrabona (Győr), but the campaign against the Quadians was brought to an end in 138 by the governor of Upper Pannonia, T. Haterius Nepos, who earned for his victory "ornamenta triumphalia". In 138 in Upper Pannonia and in 138 in Lower Pannonia, soldiers were retired: this indicates that the situation with the Quadians and the Iazyges had been settled.[47]

The two decades of Antoninus Pius's reign were quiet years, and various changes in Barbaricum had no substantial influence on Pannonia. On the reverse of some of Antoninus Pius's coins minted between 140 and 144 is the inscription "rex quadis datus" (king given to the Quadi), which refers not to the ending of a new campaign but to the reestablishment of the "foederatus" relationship.[48] It is characteristic of the consolidation that in 145, and again in 148, soldiers were discharged in the province, and in 149 three cavalry cohorts and two legionary detachments were ordered to Mauretania to deal with the Moorish uprising.

In the late forties and early fifties several coin hoards were concealed in the province, outside the province, and in Dacia as well. One may suppose that some of these hoards were buried during the time of Marcus Aurelius. The circulation of money could not be uniform over all the province, but certain geographic conditions in the Carpathian basin and the *limes* fortifications of Moesia under Antoninus Pius actually date the burial of the majority of these hoards to the time of this emperor. The movements of the Goths may have created disturbances in the provinces evidenced by the burial of the coins. From the Baltic Sea to the northwestern borders of the Carpathian Mountains, the Gothic migration pushed peoples through the passes into the Carpathian basin. While pressure weighed particularly on Dacia, it involved also the Quadians and the Sarmatians. The Gothic migrations on the outer perimeter of the Carpathians reached the Danube and made themselves felt along the Moesian *limes* as well. Although these attacks aroused some anxiety in the province, they could not have been overwhelming, and the actions against the Sarmatian border were easily repulsed by the defensive forces.[49] The movements in the Barbaricum were temporarily connected primarily with the settlement of various groups which drew the attention of the Iazyges and the newcomers for years. They exploded into destructive wars only during the period of Marcus Aurelius.

## THE WARS OF THE MARCOMANNIANS,
## QUADIANS, AND IAZYGES UNDER MARCUS AURELIUS

In Pannonia, as in other provinces, Roman border defense was based on two factors: on alliances with border nations and on a network of "client" states which formed the invisible frontier of the Roman Empire. Such an arrangement was established in the Pannonian region in the earliest time, known as the "regnum Vannianum". A similar contact existed with the Sarmatians, who were settled between the Danube and Tisza rivers, and with the Marcomannians and Quadians who dwelt north of the Danube. The Roxolanians who lived near the lower Danube received a yearly tribute from the Romans under Hadrian. Such arrangements probably existed with other tribes, too, in the Carpathian basin. The various inspection posts in the territories of the client state make us believe that Rome not only kept a watchful eye on these territories but actually influenced their politics. The reverse on coins "rex quadis datus" is just one example.[50]

The other system of defense was the line of the Danubian *limes*, which received its final structure and form under Trajan and Hadrian. This defense system, though not laid out in depth, served as a border check or as border police. They served their purpose as long as the system of client states existed. Because of the internal and external balance of the empire, it seemed unnecessary to change the system. The weaker points of this defense system did not soon become apparent, since, except for the Dacian wars, there were no great decisive wars in the Carpathian basin from the time of Claudius to Marcus Aurelius.[51] Not until the middle of the second century did the decisive change occur, when the appearance of the new peoples, driven by the Goths, upset the balance of the Carpathian basin.[52] This situation was rendered more serious by the occasional disturbances in the client states when troops were withdrawn from the territory because of wars in other provinces.

In spite of the changing balance of power in Barbaricum, troops were withdrawn because of the Parthian wars in the spring of 162, and Barbarian movements were becoming more dangerous. Although aware of the situation, the Roman high command tended to postpone a clash between the new and old neighbors until the termination of the eastern wars.[53] While this policy was successful, dangerous developments could not be halted permanently, and during the time of Marcus Aurelius a general Barbarian attack was directed against Pannonia.

During Marcus Aurelius's reign, the war was most harmful to Pannonia. Testimony of the destruction is supplied by thick charred layers in the camps and in the settlements in the interior of the province, dividing the whole of Pannonian history. The many hoards of coins discovered, the smaller number of inscriptions, the unsystematic burials, and troop concentrations are testimonials to this great war. According to scattered information, a coalition of eleven peoples under the leadership of the Quadians and the Marcomannians attacked the *limes* of Upper Pannonia.[54] That the situation grew more and more critical and that these peoples had exerted a pressure on the province ever since 164–165 are illustrated by the fact that the Romans began drafting in Italia for two legions in 165, and a year later, Iallius Bassus, a proved leader of the Parthian wars, was named governor of Upper Pannonia.[55]

The first serious invasion occurred in the winter of 166–167. Men from the Langobardi and Obii tribes who came from the Elbe area broke through to the right bank between Brigetio and Arrabona. This attack was repulsed by Macrinius Vindex, the commander of Ala I Contariorum in Arrabona, and an unnamed cohort commander.[56] The Marcomannians and Quadians did not participate in this attack, though they permitted the Langobardi and Obii to pass through their territory. According to Cassius Dio, after this defeat the Marcomannians and ten other nations petitioned the Romans for peace.[57] When Marcus Aurelius and Lucius Verus left for the Danubian provinces, because the troops had been deployed to the Parthian wars in 162, Pannonia was already inundated by Barbarian masses.[58] In 168 Furrius Victorianus suffered a defeat somewhere in Pannonia, and, at the same time, the Sarmatians started to move, turning against Dacia and Upper Moesia. At the end of 168 and the beginning of 169, the administrations of Dacia Apulensis and Upper Moesia were united under one man.

In 168, when the two emperors returned to Aquileia, the plague was at its peak in Italia as well as in the Danube provinces. When L. Verus died in February 169, the planned campaign was postponed to the end of that year, but apparently did not start until the spring of the following year. The Marcomannians and their allies unexpectedly broke through the Upper Pannonian border area, defeated Marcus Aurelius's army, and, crossing through the western part of Pannonia into Italia, besieged Aquileia and burned Opitergium (Oderzo). Only at the end of 170 and the beginning of 171, after heavy fighting and great sacrifices, did Claudius Pompeianus and the future emperor Helvius Pertinax succeed in repelling the Barbarians as they attempted to cross the Danube and succeeded in recapturing part of the spoils they had taken.

In the heavy fighting in Dacia, the governor of Dacia and Upper Moesia, Claudius Fronto, died in battle. United command over Dacia and Upper Moesia indicates that the Sarmatians had attempted to establish contact with their relations, the Roxolanians, who lived near the lower Danube. This contact had been severed when Dacia became a province. The year 171 was again a critical one; then the Romans were probably victorious over the invaders. Macrinius Vindex, praetorian prefect, died on the battlefield, and Marcus Aurelius, for the seventh time, was hailed as emperor.

In 172 and 173 the Roman armies against severe losses defeated the Marcomannians, the Quadians, and some other smaller tribes. In the second half of 173, the weight of the fight shifted to the Sarmatian front, and Marcus Aurelius moved his headquarters from Carnuntum to Sirmium. The Iazyges, anticipating the Roman preparations, broke through the south Pannonian front into the province in the winter of 173–174, only to be almost completely annihilated in a battle on the frozen Danube River. The second king of the Iazyges, Banadaspes, finally arrived at Sirmium with a peace mission, but the emperor refused to negotiate. Meanwhile the Quadians expelled their pro-Roman ruler and allied themselves with the Sarmatians; thus the Romans temporarily postponed a punitive expedition against the Sarmatians. The emperor himself led the campaign against the

Quadians, much of which took place in the Garam River valley. The Quadians were forced to surrender and were isolated from the Iazyges.

The Roman offensive against the Iazyges which began in 175 aimed not only at clearing the *limes* border area but also at reaching the Sarmatian heartland, situated in the Tisza River basin. The main body of the Roman army proceeded in the river valley from south to north, while two other armies, one from Pannonia and one from Dacia, outflanked the Barbarians. Although this three-pronged movement might have finished the Iazyges, the revolt of the Syrian governor, Avidius Cassius, in the same year, prevented the settlement of affairs in the Carpathian basin. A peace was made hurriedly on terms similar to those made with the Marcomannians and the Quadians. The Sarmatians were forced to return some 100,000 provincial inhabitants they had previously removed and were obliged to establish a force of 8,000 cavalry for the Romans. They were not allowed to settle closer than fifteen kilometers from the Danube and were permitted to barter with the Romans only on specified days and at specified places.

After the defeat of Avidius Cassius, Marcus Aurelius returned from Syria with his son Commodus in 176. In 177 the Sarmatian war broke out again, but was successfully checked by the legates of Upper and Lower Pannonia, Sex. Quintillius Maximus and his cousin Sex. Quintillius Condianus. After significant victories Marcus Aurelius was hailed for the ninth time and his son Commodus, for the second time, as emperors. In 178, however, the emperor himself had to be present in Pannonia again. Marcus Aurelius and Commodus left Rome at the beginning of August to begin a new campaign on the Sarmatian front. Bassaeus Rufus, praetorian prefect, managed to defeat the Sarmatians before the arrival of the emperor. The terms of peace must have been severe, since the Sarmatians sent delegates to Sirmium to ask the emperor for reduction. After the client relationship was reestablished, the Sarmatians may have helped the Romans fight the Quadians and Marcomannians; at the end of the year 179, Tarrutenius Paternus waged a decisive campaign against the Sarmatians, and Marcus Aurelius received the last of his tributes.

Marcus Aurelius saw clearly the situation in the Carpathian basin: both the position of the two provinces of Dacia and Pannonia and the dubious arrangement which granted security to these provinces. He realized that the peace with the Quadians, the Iazyges, and their neighbors was an uneasy one, but he also recognized that he had to keep the area, split up by the newly arrived tribes, in balance, a difficult task even under client conditions. In his last years, he began the occupation of the Barbarian territories and organized them into provinces under the names of Sarmatia and Marcomannia.

After the victory of 179, 20,000 Romans were stationed in the Quadian and the Marcomannian territories. Among others, a battalion from the Legio I Adiutrix was stationed in Laugaricio (Trenčín) under the command of the Pannonian-born M. Valerius Maximianus, whose *cursus honorum* (he was *praefectus gentium Marcomannorum, Naristarum et Quadorum*),[59] found on an inscription in Africa, indicates that the Barbarian tribes were under the supervision of Roman prefects. Also on enemy territory was stationed a *vexillatio* of Legio III Augusta from Africa. According to Cassius Dio, the military occupation preceding the organization of the province was at its height[60] when the death of the emperor halted these operations in 180. After Marcus Aurelius's death, Commodus continued the policies which his father had initiated, but discontinued the occupation of the Barbarian territories. He withdrew the troops, destroyed the fortifications, and eventually returned to the old frontier policy, reestablishing the client relationship of 175 with the Quadians and Marcomannians.

In essence, serious warfare, which had thinned the civilian population and decimated the armies, came to an end. However, the development which had started in Pannonia under the Antonine emperors had been grievously impeded. While Commodus started rebuilding with great vigor, the repercussions of the great wars could still be felt. Although the center of the enemy movements was the northern edge of Dacia, Pannonia received its share of Barbarian attacks. About 183, the praetorian prefect Perennis led a victorious expedition against a new Sarmatian movement, the originators of which may have been the Buri and the free Dacian groups, who had been pressed into the Carpathian basin. In 188, at the end of Commodus's reign, we hear about a campaign against the Germanic tribes which also brought a Roman victory. In any case, the watchtowers built during the reign of Commodus (184–185) along the entire lower Pannonian sector (*ripam omnem*) point to the fact that the balance of power on the left bank of the Danube had changed considerably.[61] Inscriptions on these towers refer clearly to their purpose: to check the intruding enemy from venturing into the *limes*.

## CIVITATES TOWN FOUNDATIONS, AND CITIZENSHIP FROM THE OCCUPATION OF PANNONIA UNTIL THE AGE OF MARCUS AURELIUS

After the conquest, the Pannonian tribes were organized into *civitates* with peregrine rights.[62] Some of the *civitates*, such as those of the Catarians and Tauriscians, lasted a very short time, while others, such as the *civitas Eraviscorum*, continued to exist until as late as the first half of the third century. The *civitates* maintained their original tribal borders, except that a single *civitas* united the Amantinians and Sirmienses, and a single one united the Boiians and Azalians. We know very little of the structure and internal life of the Pannonian *civitates*. The triple division of the *res publica Iasorum* was a special case. The nobles of the tribes went under the name of *principes*, though there were other leaders, too. We know of other *principes* in Pannonia of the Boii, Azali, Eravisci, and Scordisci tribes.

In general, the *civitates* came under military supervision, but there may have been exceptions. A centurion of a legion in the vicinity of a *civitas* became the commander (prefect) of the *civitas*.[63] We know about the prefect of the Colapiani who was the centurion of Legio XIII Gemina in Poetovio and the prefect of the *civitas Boiorum et Azaliorum* who simultaneously commanded that part of the *limes* which bordered on those *civitates*. After military supervision was ended, a few *civitates* remained as independent communities (*res publica, civitas*), but many evolved into *municipia* (e.g., those of the Latobici and the Varciani). The Calopiani, Serretes, Serapilli, Boii, Andizetes, Cornacates, and Amantini melted into or were coalesced with towns founded in these *territoria*. We are certain that the *civitas Eraviscorum* became part of Aquincum.[64]

Under Augustus, there was only military activity in Pannonia, but Tiberius elevated Emona (Ljubljana) to a colony and settled veterans in the neighborhood of Scarbantia, Savaria, and Lake Balaton and in the western territories of the province.[65] Important steps in Romanization were made north of the Drava River during Claudius's reign. The elevation of Savaria to the rank of colony opened the western border, and coins of Claudius have been found regularly on the *limes* border and in the interior of the province along the merchant route of the Kapos valley. What is more, we know of *terra sigillata* from the time of Claudius, which indicates mercantile activity or fixed camps.[66] Beside the "Savaria colonia" there were also veteran settlements in the western territories, and people were given Roman citizenship in both the western and eastern parts of the province.[67]

The Romanization policies of the Flavians were much broader than those of their predecessors. In West Pannonia three *municipia* and two colonies were established. Sirmium in the southeastern part of the province also became a colony. Although there were no cities in the eastern territory north of the Drava, a permanent *limes* was established and regular commerce was started. Intensive Romanization of the natives also began in this period. Since Pannonia became even more important after the occupation of Dacia, its Romanization became more intensive. There were extensions of citizenship at Poetovio, the only colony that Trajan founded in Pannonia, and more civil rights were granted in East Pannonia.

Extensive grants of citizenship went hand in hand with large-scale urbanization during the reign of Hadrian. During his rule, Carnuntum, Mogentiana, Aquincum, Mursa, and Cibalae became *municipia*. The activity of Hadrian marked the climax of the process which had started under Claudius and the Flavians. The citizenship drive started with Claudius and ended with the Flavians in the southern and western parts of the province, while in East Pannonia it began and ended during the reigns of Trajan and Hadrian.[68] Under Antoninus Pius, citizenship was granted on a smaller scale in Hadrian's towns. The extension of citizenship and of urbanization, which started in the western and southern parts of the province and proceeded gradually toward the east, demonstrates the trend of Romanization in this part of the empire. The final Romanization in the province occurred under Trajan and Hadrian, with the completion of the building of the *limes*.

## THE AGE OF THE SEVERANS

The situation in Pannonia changed considerably after the wars of Marcus Aurelius. As the balance of power shifted, the Quadians and the Iazyges were no longer Rome's main enemies in this area; the major threat now came from the movements of peoples from beyond the Carpathian basin. The arrival of these foreign groups into Pannonia and Dacia, which influenced also the movements of the Iazyges–Sarmatae and the Quadians, was already taking place under Marcus Aurelius.

The primary result of these repeated attacks was the destruction of the north Dacian *limes,* but their effect was felt in Pannonia, at the Danube bend. Moving along the range of the Carpathian Mountains, the invaders reached the lower Danubian area, where their activities almost paralleled those within the Carpathian basin. The history of the two areas cannot, therefore, be separated, especially since the Roxolanians, who lived in the lower Danubian area, were compelled in 260 to move to the area between the Danube and the Tisza rivers. Dacia, as a buffer state, repulsed and absorbed most of the Barbarian waves, but the province had eventually to be abandoned during the reign of Aurelian. Thus the situation in Pannonia became critical.[69]

The governor of Upper Pannonia, L. Septimius Severus, profited from the situation after the death of Helvius Pertinax, and, with the help of the Pannonian army, he assumed imperial power in 193. In a short while, he defeated his enemies, candidates from the Roman Senate and from the provinces, and marched into Rome. Historical sources do not give any account of military activities in Pannonia during the reign of Septimius Severus. On the basis of coin hoards and epigraphical data, however, we may conclude that the geographical positions of the various peoples shifted. The intensity of these shifts must have been felt primarily in the lower Danubian area, but it also affected the Carpathian basin, i.e., Pannonia.[70]

In the last years of the second century the migration of the Goths produced a shift in the position of the peoples to the east, north, and south of the Carpathian Mountains. Coin hoards, buried during the reign of Septimius Severus in Moldavia and in the Ukraine, are evidence of this. At the same time emerged the construction of the outer Olt *limes,* the *limes Transalutanus,* which the emperor was forced to build because of heavy pressure on the *limes* of the Olt River (the *Alutanus*). Owing to these shifts of population monetary exchange with the Baltic peoples was suspended after 196.

Portions and groups of the disturbed nations probably arrived to stay within the Carpathian basin. Among them was a Germanic group intermingled with Goths and a larger tribe of the Carpi. Breaking through the passes of the eastern Carpathian Mountains, these people primarily threatened the northern frontiers of Dacia; but they must also have excited pressure in the Great Plain, from where, to date, six coin hoards have been discovered. These waves did not leave Pannonia untouched, either. While the coin hoards in the Hungarian lowlands generally date from between 193 and 195, the relics which may be connected with these migrations in the province of Pannonia date from later periods. It is possible that the invaders did not reach the Pannonian borders until somewhat later, but it is also possible that Septimius Severus did not fight the invaders until after he had defeated Clodius Albinus.[71] The seriousness of the situation was shown by the fact that, after the defeat of Clodius Albinus, he sent Severus T. Claudius Claudianus, who had twice distinguished himself, to be governor of the endangered frontier of Lower Pannonia.

Before the summer of 197, the Romans rebuilt the presidium at Érd, a small fortification which had probably been the victim of an earlier attack. It is not inconceivable that a skirmish was in the background of a *rescriptum* made concerning the estate of a deceased centurion of Cohors II Alpinorum equitata.[72] The base of the statue erected by the garrison of Crumerum, the Cohors V Gallaecorum Lucensium, refers to such a fight. To this may be added that the repair of roads and the building of watchtowers (*burgi*), as well as restoration work was even continued in the following years. Similar reconstruction can be observed in Moesia and on the Olt *limes;* the outer *limes* of the Olt was built during these years, too. However, two finds of coins in Brigetio cannot be connected with these events, since the coins were buried some years later. Moreover, it is curious that these finds came from the neighborhood of Brigetio. Serious population movements took place east of the *limes* around 203. It was probably due to certain decrees issued west of the *limes* that people began to hide their money. Later, for unknown reasons, the money could no longer be retrieved.[73] According to extant data, the border area exposed to the Barbarians stretched from the Danube bend till about Brigetio. At about the same time, certain scattered tribes of Dacian origin settled north of Dacia and Brigetio. Later, they destroyed the northern Dacian *limes* and played such an important role in the district of Brigetio that the Legio I Adiutrix felt the need to engage a Dacian interpreter.[74] The fight with the Barbarians, the building of fortifications, and the restoration of the *limes* occurred sometime between 196 and 200, but further disturbances may have existed as late as 203. In any case, from the time of Septimius Severus's visit to Pannonia in 202 until his death, we know of no serious events corroborated by archaeological and other evidences.

The ascension of Septimius Severus to the throne brought important changes in Pannonia. He not only repaired the destruction caused by the Marcomannian wars and brought about an economic upswing but also tried systematically to consolidate the life of the provinces. A great upswing can be observed in the *limes* camps, in the military cities, and in civilian cities near the *limes* camps. Cities that proved their allegiance to the emperor were protected: in that way, Aquincum, Brigetio, Carnuntum, and Siscia increased their ranks. According to the evidence of inscriptions, many people from Lower Pannonia settled in the camps near the *limes* because of opportunities for work and a prosperity created by the military presence. The function of the cities close to the *limes* increased, receiving municipal rank primarily as military cities or camps; one may suppose that Brigetio received the municipal rank not as a regular city but as a military one.[75]

Under Septimius Severus, those relatively dormant movements in the Carpathian basin which had temporarily abated suddenly flared up into war in 212. The trouble affected not only the territory of Dacia but also extended to parts of Lower and Upper Pannonia, the so-called territory of the Danube bend. Hence the Quadians and the Iazyges were also involved. In a serious attack, the northern Dacian *limes* was overrun by Carpi and Vandal tribes; the camp of Porolissum had to be completely rebuilt as a result.[76] The inscriptions of Caracalla and the excavation results show us this reconstruction. Although we have no concrete information about the destruction of the Pannonian *limes*, we do know of building work in the *limes* during this period. Still there are evidences of Pannonian involvement in the fights. Evidence both of the movements during Septimius Severus's reign and of the fights under Caracalla was that the legion at Brigetio needed a Dacian interpreter. This makes us think that the Dacian people were close to the *limes*. It was presumably owing to the changed situation on the left bank of the Danube that Brigetio was annexed to Lower Pannonia in 214. In addition, the emperor had rebuilt the bridgeheads on the left bank along the Carnuntum–Aquincum stretch of the *limes* (i.e., Dévény became the counterfortification of Carnuntum, Leányvár that of Brigetio, and the fortification on present-day Március 15 Square in Pest became that of Aquincum). Imperial names given to the troops are further evidence of these movements from Brigetio to Intercisa. It is possible that the gravestone of the *librarius* of Legio I Adiutrix, who "decidit in expeditione Daisca", should be dated to this period as well. In any case, Caracalla's visit to Pannonia in 214 is definitely connected with these events.[77]

We have no information about outside enemy movements during the short rule of Elagabalus. The coin hoards in Pannonia around the Danube bend which date from the reign of Severus Alexander would not have much significance without taking into consideration the ancient writers and the similar hoards of coins beyond the Carpathians in Moldavia, Ukraine, Moesia, and southern Dacia under the same emperor. While excavations on the *limes* do not indicate that any serious action had occurred, names such as Severiana and Severiana Alexandriana, given to the troops in the Danube bend, and similar names given to some troops stationed in south Pannonia prove their participation, if not in full numbers, in some activity. The cause of this suspected movement must have been due to another contingent of the Gothic wave which, moving along the outer spine of the Carpathians, exerted a pressure on the interior of the Carpathian basin and, proceeding further toward the lower Danube area, came close to the Moesian and Olt *limes*. This phenomenon is similar to that which we observed under Septimius Severus; the Barbarians advanced and occupied territory under Septimius Severus and Severus Alexander, but the fully developed attack followed much later under Severus Alexander or Caracalla, and, as we will see, under Maximinus Thrax.[78]

The upswing which started during the reign of Septimius Severus continued under Caracalla and Severus Alexander. This was the most prosperous period in the third century in Pannonia. The privileges of the troops, the visit of Septimius Severus (202), Caracalla (214), and Elagabalus (219) gave an additional impetus to the rise of the province. Archaeological findings, sarcophagi, stelae, and altar stones demonstrate the kaleidoscopic wealth of life during this period.


## MILITARY EMPERORS

The movements beyond the provinces during Severus Alexander's reign exploded into wars during his successor's reign. Maximinus Thrax, who came to Pannonia in 236, remained in Sirmium until 238. The epithets *Dacicus*, *Germanicus* and *Sarmaticus* demonstrate that, during his short reign the whole Carpathian basin was in constant upheaval. According to the Maximiana epithets given to the troops, the Danube bend was endangered, and units from Lower Pannonia participated in the battles. Maximinus was unable to complete another expedition against the Sarmatians because of the revolt of the Senate and the elevation of Pupienus and Balbinus to the throne. There is little trace in Pannonia of this short rule, for he spent all his time in offensive wars against the Barbarians. One may suppose that the two coin hoards found in Poetovio were not due to the Barbarian threat, but were buried when the emperor withdrew to Italia.[79] It appears that Maximinus Thrax brought order to the Great Plain and to the area east of the Tisza River but had no time to conquer the lower Danube area.

Several coin hoards, terminating with coins from Gordian's reign, were found on the Olt *limes*; thus, the partial destruction of the *limes Transalutanus* may be dated to this time. The Goths had by then arrived at the lower Danube and at the Olt *limes* and were putting pressure also on Pannonia and Dacia. The movements in the Dacian and Pannonian passages are evidences of this. The milestones, the sculpture bases of the emperors' statues, and the epithet *Gordiana* which was given to the troops refer to these disturbances.[80]

Under Philippus, the Carps, a tribe of Dacian descent moved through the east Carpathian passes beyond Dacia and in 242 devastated the whole province. This raid affected indirectly the *limes* area between Brigetio and Aquincum. At this time, two hoards of coins were hidden behind this stretch of the *limes*; the *vexillatio* of the Britannic Legio II Augusta came to the rescue of the legion at Brigetio. These events are connected with two data: The military base in Brigetio, which was named for the honor of Philippus's wife Otacilia Severa, together with the city of Brigetio were raised to the rank of colony.[81] The violent attacks which started under Gordian in the lower Danubian area continued under Philippus, and in this period much of the *limes Alutanus*, the *limes* of Olt, was destroyed. The emperor then rebuilt the destroyed towns, elevating some of them to the rank of colony. The wars of Philippus came to an end in 247; his cognomen on his coins of 248 are *Carpicus* and *Germanicus*.[82]

Under Traianus Decius, serious enemy action shifted to Dacia again, but waves from these movements penetrated into the northern perimeters of Pannonia as well as into the area of the Danube bend. The enemy broke through the Olt *limes* again and into the valley of the Temes, intending to penetrate into Dacia through these passages, but Decius made great efforts to save Dacia and was later honored for his success with the names *Dacicus Maximus* and *restitutor Daciarum*.[83] We know of two hoards that were buried in the Danube bend under Trebonianus Gallus's reign. The military distinction *Galliana-Volusiana* may also refer to enemy attacks. We have no knowledge of other important events besides movements limited to this small territory.[84]

Although the Olt *limes* prevented the Roxolanians from moving into the Carpathian basin, they were greatly weakened under Philippus, and after Philippus they failed to hold back the enemy. About twenty-three different coin finds have come to light from the Lower Pannonian *limes* and the territory behind it, terminating with the period of Gallienus. Undoubtedly, a powerful Barbarian attack caused the hiding of these treasures at the beginning of 260; the finds are proof that the lower Danube area opened before the Roxolanians, whose movements pressed the Sarmatians against the *limes*. The Sarmatians then attacked the province with the Quadians, a raid which was considered the greatest catastrophe afflicting the territory of Pannonia and the *limes* in the third century. The extent of the destruction is known to us only in limited detail, but it can be observed at Aquincum. The abandonment of the camp at Albertfalva, south of Aquincum, and the *canabae* may also be dated to this period. This camp was not used again.[85]

The governor of Lower Pannonia, Ingenuus, defeated the Sarmatians with an army of peasants and urban inhabitants, and, at the news of the capture of Valerian, he was proclaimed emperor. As a result of these events, Gallienus came to Pannonia and defeated Ingenuus near Mursa with a cavalry led by Aureolus, Ingenuus himself dying in the battle. Following this defeat, the Moesian army and the Upper Pannonian armies backed Regalianus as counteremperor, while Lower Pannonia remained faithful to Gallienus. The Lower Pannonians also refused to back Macrian, who succeeded Valerian in 261.

We do not know the details of the outcome of the battles with the Sarmatians, but Regalianus still fought serious battles against them. It is likely that Gallienus renewed the federal relationship with the Roxolanians, thus forcing the Sarmatians to leave the area of the Lower Pannonian *limes*.[86] According to a source on this critical period, Gallienus, who married the daughter of the Marcomannian king Pipa, gave part of Upper Pannonia to the Marcomannians. It is highly probable, however, that this type of defense settlement occurred quite frequently in the history of Pannonia. The province remained quiet after 262; only toward the end of the rule of Gallienus, with whose name the foundation of the mint at Siscia is connected,[87] were coins buried again.

The great bath at Aquincum, in ruins for a long time, was restored during the reign of Claudius II in 268. There is no evidence of attacks on the Pannonian *limes* during his short reign. That he was occupied at the lower Danube is demonstrated by his epithet, Gothicus. It is very likely, therefore, that the Aquincum ruins are an inheritance from the rule of Gallienus, rather than the result of fresh enemy destruction.[88]

The attack of the Gothic nations of Dacian territory in 269–270, the simultaneous arrival of the Gepids at the upper Tisza region, the subsequent dispersal of the Vandals toward the west, and the invasion of Pannonia by the Vandals in alliance with the Suebians and Sarmatians created a new situation in the Carpathian basin. Aurelian succeeded in expelling the enemy from Pannonia; the Barbarians even delivered 2,000 horsemen to the Roman army. Nevertheless, Aurelian was compelled to abandon Dacia in 271, and its population had to be settled on the banks of the Danube in Moesia. With the loss of Dacia, the citadel which secured a buffer for Pannonia disappeared.[89]

We have scarcely any archaeological evidence to indicate that there was fighting alongside the *limes* in the two decades following Aurelian's rule; there are only a few hints. The Komin hoard was buried under Tacitus, and, according to recent research, a new layer of destruction can be traced in Aquincum after 275. Following the death of Probus, Carus scored a victory over the Sarmatians who repeatedly intruded into Pannonia. The latest coins found in Sirmium date to the time of Carinus, and not long afterwards Numerianus defeated the Quadians.[90]

We have no knowledge of new movements against Pannonia from the time of Aurelian until Diocletian. Perhaps the peoples who broke into the Carpathian basin, after occupying Dacia, had enough living space which stalled their move against the *limes*. They tried to secure their own position by allying themselves with the native population to stem the advance of new arrivals. A certain consolidation was observable in the province, with the emergence of a new, well-to-do eastern element among the population (they probably arrived with troops after the termination of Aurelian's eastern wars), but western import appears as well among the artifacts.[91]

## THE SITUATION IN THE PROVINCE IN THE PERIOD BETWEEN MARCUS AURELIUS AND DIOCLETIAN

During the wars of Marcus Aurelius, the composition of the population changed considerably in Pannonia.[92] The descendants of Italian families who had lived as veterans, civilians, or even as active soldiers before the time of Marcus Aurelius greatly diminished. Although the lines of descendants may sometimes be demonstrated, in other cases similar names refer to Africans or easterners, or can be traced to descent from other provinces. Taking into consideration the population possessing imperial *gentilicia,* the names Iulius, Claudius, and Flavius were more common later than they had been in the first two centuries. Only a few of these names can be linked with surviving groups; the majority emerged as new elements in Pannonia. The names Ulpius and Aelius, granted by civil right under Trajan and Hadrian, occur in the time after Marcus Aurelius as well. The extension of citizenship under Septimius Severus can best be seen in East Pannonia.

After Marcus Aurelius, Aurelius became the most common name. Citizenship was extended to the whole province, partly by Marcus Aurelius and partly as a result of the Constitutio Antoniana. But even this group of new Roman citizens was not homogeneous. The larger number of Aurelians came also from East Pannonia. In Carnuntum, Brigetio, and Aquincum, the population appeared to be mostly natives and easterners. In Aquincum, even a southern Thracian group was present. In Intercisa, there were mostly Syrians who received their citizenship from Marcus Aurelius. South of Aquincum, natives, Thracians, and many newly settled Cotini can be found. There was a considerable increase in the number of persons not possessing imperial *gentilicia* as compared with the first two centuries.

During the reign of Antoninus Pius, the troops who participated in the defeat of the revolt led by Maurus included a considerable number of north Africans. Another large contingent of north Africans came to Pannonia when the *vexillatio* of Legio III Augusta was integrated into the Pannonian legions to supplement lost ranks. The Cohors Maurorum arrived at the camp of Matrica (Százhalombatta) under Marcus Aurelius.

After the time of Marcus Aurelius, Pannonian troops who campaigned in the east brought many easterners back with them. Thus we find many easterners among the Julii, Claudii, and Flavii, fewer among the Ulpii, Aelii, and Septimii, but very many again among the Aurelii. Under Marcus Aurelius, the Cohors Hemesenorum arrived from the east to Intercisa as a closed unit. Another Syrian cohort was placed permanently at Ulcisia Castra. The troops who returned from the east through Thrace had also brought Thracians with them, most of whom settled in Aquincum but some appeared also in other parts of the *limes* and the province. With the soldiers also came the civilians. One must remember that Pannonian troops fought also in the west. *Vexillationes* from Pannonia were stationed even in north Italia during the century. Western and north Italian elements lived in Pannonia, but not in large numbers. A small number of people from the neighboring provinces of Noricum and Dalmatia lived in the western part of the province, as did Dacians, when legions from this province were sent to Pannonia. As seen from the above, military conditions decisively influenced the emergence of nationality groups in the province. But there was natural immigration too, mostly in the circles of merchants and craftsmen.

Romanization, the granting of citizenships, extended to the whole province of Pannonia after Marcus Aurelius; yet under Marcus Aurelius and Commodus, we do not hear of the founding of new cities. Urbanization in general and the strengthening of urban life occurred under Septimius Severus, while the cities emerged under Caracalla. As is evidenced by excavations, the military cities developing near the camps of the legions became two to three times larger than the regular cities. Those excavations have yielded rich finds and many inscriptions done by soldiers. The thatch and dried brick houses of the native population ceased to exist even in the auxiliary camps, giving way to luxurious stone houses. After Marcus Aurelius, preference in extending the title of city was given to military cities belonging to camps, or, better to say, it was this type of settlement that was first given this title.[93] Evidence of city life has also been found around the auxiliary camps toward the end of the second century. The rapid development of cities and their complete Romanization under Septimius Severus are known from their inscriptions and from the remaining monuments. With the native communities disintegrating and ceasing to exist, their role was taken over by the cities. Abandoning their former way of life, the native population became almost completely Romanized; their former identity is no longer recognizable in the surviving names and on monuments. This was also true among the groups who settled on the left bank of the Danube. In the vigorous urban life which developed, associations and *collegia* proliferated. Both prosperity and an extensive knowledge of reading and writing are indicated in many inscriptions, while religious dedications show a close relationship to and a feeling of solidarity with Rome. In the large-scale diffusion of Roman handicraft and manufactured items, and in the fully developed city life, easterners and western Pannonians played an important role. Additional inscriptions, found in the *limes* area rather than in the cities situated in the interior of the province, include many military inscriptions; this shows that the military was still very important though they were not the leading group in society in the first half of the third century. Noncommissioned ranks in the army were given to easterners and men of Italian descent, not to locals who were recently granted citizenship and were bearing the name of the emperor who granted it.[94]

The local population played no serious role in urban and military life until the middle of the third century.

Surviving inscriptions also indicate economic and social changes. Comparing them with topographical and archaeological observations, we can see that independent peasant farms and middle-sized estates had existed. Alongside Lake Balaton and between the Drava and Sava rivers in West Pannonia, villa cultivations had steadily developed. At the same time imperial estates must also have functioned.

It was particularly the inhabitants of foreign origin who wanted to possess land, while only a minority of the locals owned middle-sized estates; the majority of those, however, owned only small plots of land, near the *limes* in East Pannonia.

The first half of the third century, particularly the first four decades, are characterized by eastern influence, since the preponderance of soldiers were from the east, bringing with them eastern elements and cults. Eastern inhabitants formed the wealthiest sector of the population. Commercial and financial life was controlled by Syrian businessmen, so this group naturally possessed large fortunes. There are certain signs that *cives Suri* did not mingle with other groups among the inhabitants. After the death of Severus Alexander the inhabitants from the east lost their protector, and other citizens in the province turned against them. A striking manifestation of this spirit was the destruction of Dolichenus shrines (even near the Rhine) in 235 and some years after. These shrines were not repaired; so we may suppose that adherents of the sect had either escaped, decreased in number, or perished. Maximinus Thrax tacitly contributed to their destruction because he tried to cover his expenses for his wars from the capital accumulated by the Syrians. However, this move led to the destruction of an economically strong upper class, causing irreparable damage in the province.[95]

It is evident that in the second half of the century the province of Pannonia and the Danube valley as a whole suffered from various troubles, particularly owing to raids by the Barbarians. Various struggles, the fights of the counter-emperors, the annihilation of a wealthy leading class were responsible for serious decline in the economic and social life of the province. Earlier research speaks of catastrophic events in this period. Reality, however, has been much distorted because no thorough analysis of independent archaeological material has been carried out. The gloomy view concerning the period was also supported by the fact that only few remains with inscriptions could be dated back to the second half of the century. However, research in the past few years seems to establish that stonemasons had been regularly at work during the period.

No research has been done so far to produce a full picture concerning money circulation in the period and continuity in burial rites. Putting money into graves in funeral rites ceased with Trajanus Decius. This, however, does not mean that funerals themselves were suspended. The study of the second half of the century has been rendered difficult because of the arbitrary view that the great number of coin hoards hidden during Gallienus's reign was due to the fact that the province of Pannonia was greatly disturbed with the arrival of the Roxolanians.

Events in the Danube area naturally did not leave the life of the province unaffected, but it should not be forgotten either that the majority of the hidden hoards did not emerge from the vicinity of the *limes* but from the interior of the province. Hence concealment was not always due to external circumstances but to internal events as well. It could have been caused by military action directed toward Dacia, or even by tax collections. All this may have contributed to the worsening of economic life and a scarcity in stone monuments and grave goods; it is even possible that fewer coins had been deposited in the graves. Nevertheless, in general, there were no serious changes in the lives of the settlements. Research did not reveal such destruction near the *limes* or inside the province which could account for the large number of hidden coins under Gallienus, though, to a lesser degree, the archaeological material of the period does not lack in either eastern or western imports.[96]

New eastern elements that emerged in the last third of the century probably came with Aurelian's troops arriving from Syria. This small group lent interesting color to the life of the province. Their archaeological material is known mainly from Aquincum, Brigetio, and Intercisa. Rich textile products and other wealthy finds have been discovered in the graves.[97]

Side by side with the local military, the *vexillationes* of western legions also appeared from Britannia and Germania. At the same time, jewelry of a western style can also be found in the graves, and there is indication that the population lived on the left bank of the Danube.

## DIOCLETIAN AND THE TETRARCHY PERIOD

In the second half of the third century, owing to incessant warfare, the empire could not deal from sufficient strength with the external enemy. Pannonia too suffered considerably from both internal strife and external wars.[98] A new administration imposed by Diocletian brought about an end to the anarchistic situation in the province; his newly organized army kept the Germanic tribes and the Sarmatians clear of the borders of Pannonia.

Since 294 Emperor Galerius exercised actual power in the provinces of Illyricum and he was responsible not only for ending successfully the Sarmatian wars but also for carrying into practice the military, financial, and economic reform of Diocletian in Pannonia.[99]

Under the new system, Pannonia, Dalmatia, and Noricum were administered as one unit with the capital at Sirmium (Sremska Mitrovica). Pannonia itself was divided into four provinces: the previous two, Upper and Lower Pannonia, were subdivided into parts north and south of the Drava River. The part of Upper Pannonia north of the Drava became Pannonia Prima, and the part south of it became Pannonia Savia. The southern sector of eastern Pannonia was named Pannonia Secunda, while the sector north of the river was named Valeria after Galerius's wife, the daughter of Diocletian.[100] Civil power in the province was concentrated in the hands of the *praeses*, while military power was in the hands of the *dux*.

The modernization of the forts along the *limes* was followed by the reorganization of the territories behind the *limes*. It was in the framework of this activity that the remnants of the Carpi, who had been defeated near the lower Danube in 295, were settled in the Pécs sector (Sopianae) by Diocletian. The new settlements not only supplemented manpower but also enlarged areas with agricultural potential in order to increase the economic power of the empire. Settlements like this are characteristic of Pannonia in the period.[101] In spite of Diocletian's consolidation, his administrative, military, and economic reforms laid only the foundations of prosperity under the Constantine dynasty.

Prior to 291 the western Goths in Transylvania and the Romanian Taifals, situated south of them, fought bitter wars with the Vandals and Gepids living in East Hungary. As a result of the imbalance created by these wars, the Sarmatians were pushed against the *limes*. However, the Sarmatians were not the only ones who attacked Pannonia; we know that between 286 and 293, two expeditions were led against the invaders, the first probably in 290, the second in 292.

Diocletian spent the second half of 290 in Sirmium, and several months also in 293, when the counterfortification of Lugio (Dunaszekcső) was erected. This bridgehead was significant because an important military road against the Transylvanian Goths led from there to the delta of the Maros River; already in the second century it was one of the arteries connecting Dacia and Pannonia. In 294 the counterfortification of Bononia was built in the south; the counterfortification of Aquincum in the north at the Sarmatian frontier was built previously between 289 and 293 when the emperor had visited Aquincum.

In 294, using the strategic significance of the new bridgeheads, Galerius led a victorious expedition against the Barbarians, the third victory over the Sarmatians. In 295 we hear about his battles against the Carpi and Bastarni at the Danube delta. As a result of these expeditions, some Pannonian legions could be used against the Persians by 297. In 299 Galerius led his second expedition against the Sarmatians, but at the same time there were movements of the Marcomannians and Quadians. Between 299 and 311 he scored three more victories over the Sarmatians and three times acquired the name of *Sarmaticus*.[102]

The building of bridgeheads and counterfortifications as well as the series of expeditions indicate not so much defense at the *limes* as an aggressive new policy under Diocletian and Galerius — both near the *limes* and outside the province. This implies a new management of affairs in Barbaricum. We may well believe that the construction of a new outer *limes* line (a trench and plank system connecting with the Lower Pannonian *limes*) which has been attributed to Constantine was actually begun at this point. Its importance was only recognized recently. It was destined to protect the original inhabitants of the Hungarian lowlands, i.e., the allies of the Romans, against outside enemies and to offer a defense for Pannonia against constant population movements.


## THE PERIOD OF THE CONSTANTINIAN DYNASTY


At the conference of emperors at Carnuntum in 307, the new Augustus, Licinius, obtained the Illyrian provinces, as well as Pannonia. (In the autumn of 312 Constantine I succeeded in defeating Maxentius with a contingent of his Gallic army, although in 307 neither Severus II nor Galerius Augustus had been able to defeat Maxentius.) In 313 a new war was apparently fought against the Sarmatians, for by 314 both Licinius and Constantine bore the name *Sarmaticus*. In October of 314, near the Pannonian Cibalae (Vinkovci), the Gallic army under Constantine attacked troops from Illyricum under Licinius and defeated them, thus destroying the glory of the Illyrian army. Licinius escaped to Sirmium, and when this was also occupied by Constantine, he went to the eastern Balkans. Thus, by the end of 314 Pannonia and the Balkan provinces, except the Thracian dioceses, were attached to Constantine's empire.[103]

Constantine closely watched events near the middle Danube, because western Gothic raids into Moesia had to be stopped and life in the provinces needed to be settled. In the first fifteen years of his reign, he visited Pannonia almost yearly, keeping a permanent residence at Sirmium.[104] Since 320, events in the central and lower *limes* area assumed greater importance for the empire, and Constantine had to organize a larger expeditionary army, which in 322 successfully turned back the invaders.

About that time the Sarmatian king, Rausimodus, crossed the Danube, ransacking the province and encircling the camp of Campona (Nagytétény), south of Aquincum. Constantine appeared at the rear flank of the marauders, liberated the camp, and pursued the beaten enemy deep into Barbaricum. Rausimodus himself died while retreating. The emperor crossed the Danube in the vicinity of Bononia with numerous captives. The battle was colorfully told by Ammianus Marcellinus. The whole campaign took less than a month and by the end of July Constantine was again in Sirmium. A series of gold coins minted after the victory had *Sarmatia devicta* engraved on the reverse.[105] Although the Sarmatians themselves ceased their attacks on Pannonia after 322, they aided and abetted the army of Licinius, allied with the Goths, in his attack on Constantine in 324. That same year the latter emerged victorious from the wars, and after 325 for forty years, the Roman Empire was united once more under one leader, a fact which had important results for border defense.

In order to secure the new capital of Constantinopolis and the area surrounding it, the Romans had to defeat the Goths, which Constantine had already planned. Within the framework of the offensive Daphne plan, Constantine rebuilt the lower and middle Danube *limes* areas, fortifying them with new camps. While the evidence of archaeological excavations demonstrated building in both areas, more extensive work was carried on in the lower Danube area.[106] In connection with the building and fortifying of the Danubian *limes*, mention must be made of the Oltenian fence and palisade system, which started out from the Danube around Drobeta (Turnu Severin), cut diagonally through Oltenia, and returned to the Danube again south of Braila. Called now the Brazda lui Novac, this system was about 700 kilometers long; it encircled the area of the Danube and protected it north of the *limes*.

According to available information, Constantine started the fortification of the *limes* after the defeat of Licinius, rebuilding the camp of Constantiana Daphne and rebuilding Drobeta and other camps. According to research, the building of the Dobrudjan, Moldavian, and Bessarabian fences occurred in the fourth century. Further research in the last few years attempted to determine the extent of earthwork in Hungary. While results may not yet be taken as conclusive, it seems that earthwork and the palisade system started out from the Danube bend and embraced the whole of the Great Plain, constituting a unified system encircling an accurately determinable area, where, in fact, the Sarmatians were living.

With the loss of Dacia the road to the Pannonian *limes* was open to the Goths and other peoples, and Italia itself was vulnerable. Sarmatian as well as Roman interests demanded the creation of a new defense system which would repulse and disperse attacks coming from the east.[107] The raid of the Goths into Dacia in 269–270 and the abandonment of that province had created a new balance of power in the Carpathian basin. The Gepids, appearing in the upper Tisza region in 290, added to the confusion by pushing the Vandals westward. In 290 the Gepids, together with the Vandals, attacked the Goths, who allied themselves with the Taifals against the attack. At the same time, the Sarmatians broke into Pannonia, and Diocletian fought hard against them.[108]

In 332 the Goths, who had been expelled from the lower Danube area, attacked the Roxolanians and the Sarmatians. Under a previously arranged alliance, the latter called in the Romans, and together they repulsed the Goths. Some of the Goths were then recruited to replenish Constantine's depleted ranks. The Sarmatian slaves (*servi Sarmatae*) had been armed by their lords (*Argaragantes, Sarmatae liberi*) against this Gothic attack, but in the successive internal troubles the slaves expelled their masters, who escaped to the area inhabited by the Victovali and became their vassals. This was an area near the Quadians and, according to topographical evidence, outside the aforementioned fence-palisade system. The Goths and the escaping Sarmatians entered the Pannonian territory; evidence that the camp must have fallen to the attackers is provided by a coin hoard buried at that time. It was discovered near Campona and consisted of more than 11,000 coins.[109] Sometime during 338 – 340 Constantine scored a victory over the Sarmatians; two smaller coin finds probably refer to this event.[110]

Constantius II, while in Rome in 356, received the news that in alliance with the Quadians, the Sarmatians were devastating Pannonia and Upper Moesia. Coming to Sirmium in the winter of 357, the emperor received envoys from the Quadians and Marcomannians. In his successful offensive, which began in 358, Constantine crossed the Danube at Bononia, the port of Sirmium, moved along the left bank, and pursued the enemy toward the Danube bend. The account of Ammianus Marcellinus tells us that the defeated Sarmatians were those who had been expelled by their slaves (*servi Sarmatae*) from the Banat in 334. The Sarmatians were compelled to return the prisoners they had captured and to provide hostages, while their king, Zizais, became the vassal of Rome. On the other hand, Constantine had liberated them from the rule of the Victovalians. After the victory, he returned to the right bank and moved toward Brigetio along the *limes*; he crossed the Danube again and encountered the Quadians somewhat to the west. After some skirmishes the Quadians, under the leadership of Vitrodorus, sued for peace. Once again Constantine was successful.

Subsequently, the Limigantes (*servi Sarmatae*) who lived inside the boundary formed by the earthworks in the Banat, crossed into Moesia. Attacked from the

112

south by the Romans, from the north by the vassal king Zizais, and from the east by the Taifals, they were forced to give up their settlements. Their new habitations were outside the earthworks at the angle of the Szamos and upper Tisza rivers, and their place was taken over by their previously expelled masters. During the winter of 358−359 the emperor, who was at Sirmium, received the news that the Limigantes had left their new reservations and were wandering along the borders of Valeria. After making preparations, Constantius II came to Valeria in 359 and, at the request of the Sarmatians, received their envoy. Ammianus Marcellinus had vividly described how the Sarmatians, under the pretext of this mission, had plotted against the Romans. The Roman army mercilessly overran the perfidious Sarmatians, after which Constantine again secured Pannonia's borders and moved to Constantinople to prepare against the Persians.[111]

Meanwhile, on the Hungarian plain, bordered by the trenches, considerable ethnological changes occurred. In 332 parts of the Sarmatian settlements were destroyed or became depopulated due to Gothic attacks and internal strife. According to sources which may be exaggerated, some 300,000 people were settled in various regions of the Roman Empire, in the Balkans, and in Italia. This can be registered in the archaeological material, as well. The movement of the Gepids from the upper Tisza region to the area between the Maros, Tisza, and Körös rivers began in 340.[112] Discovered in the process of archaeological research, fragments of an inner trench system emerged which started out at the delta of the Tisza and Körös rivers and could be followed some distance further. It is believed to have been the border between the Sarmatians, who moved within the trench system, and the Gepids. According to another view the Taifals, possessing the Banat, together with the Gepids, appearing as allies of the Romans, took over from the Sarmatians the defense of the trench system south of the Körös River. If we study the movements of the peoples in relation to the outer defense line, we can see clearly the defensive character of the whole trench system. Between 322 and 374 an outer and inner trench system was built. The outer defense was built in 322 and the inner line sometime after 358.[113] The Sarmatians who had settled within the trench system did not attack Pannonia; only those who lived outside this defensive system made the attack in 357–358.

## THE GENERAL SITUATION IN PANNONIA DURING THE CONSTANTINIAN DYNASTY

Beginning with Diocletian, but particularly under Constantine, the tactical system of the *limes* and the purpose of the military camps changed. The change in the defense system was connected with the withdrawal of a central cavalry army of elite troops from the camps bordering the *limes*. Constantine repaired and modernized the fortifications; in addition, he built new fortifications for a better defense of the borders. In the large cemeteries built around the camps rich material has been found indicating the existence of local manufacture and intensified trade with other provinces. After the death of Constantine, during the rule of Constantius II, no basic changes occurred on the *limes*; no destruction is known to us, either.

For a long time, practically until the last decade, research considered Pannonia in the fourth century to be in a period of decline. This was due primarily to a one-sided treatment of the archaeological and the written sources. Later, when it became generally accepted that the province had been hit in the second half of the third century by serious disasters, scholars did not even attempt to systematize fourth-century findings according to their merit. Research was strongly affected by the fact that only a small number of relics with inscriptions and stonework were dated from the period, from which it was wrongly conjectured that life as a whole had deteriorated.

According to research done in recent years, however, we can say that stone-masonry can be regularly traced back at least to the age of Constantine and that it continued to exist even though it broke with traditionally conceived forms. The fact that the number of stone relics was smaller than in the previous centuries can be explained by a smaller production and by the spread of Christianity, since altars had disappeared. It is also known that later stone monuments were not architectural types, or only in rare cases, so that they could easily have gotten lost. Above all, it should not be forgotten that extensive large buildings were erected in the period, well represented by the capitols at Brigetio and Almásfüzitő (Azaum).[114]

In Pannonia, during the fourth century, large-scale imperial undertakings were initiated. Among these should be mentioned canalization near Lake Balaton under Galerius, deforestation, moorland reclamation east of Sirmium, and planting of vineyards. An item recorded in 383 seems significant, according to which grain was exported from Pannonia to Italia. The emergence of the latifundia had involved exports, so Pannonia exported wine, livestock, and cereals. Villas built previously were used continuously and even new ones were erected and luxuriously furnished. The extension of these villas bears witness to material wealth. Many of them belonged to imperial estates and were richly decorated. One of these was Villa Parndorf near Carnuntum (Murocincta), where Emperor Valentinian stayed in 375. Ammianus Marcellinus speaks of another imperial villa at Poetovio and of the palace of Savaria.[115] Recent research established that the imperial palace of Sirmium was a fine building. Owing to the situation and importance of Sirmium, the palace occupied a special place among the residences of the emperor.

From the wealth of the graves we may surmise that there was general prosperity in the period, which can be felt as long as two decades following the death of Valentinian. A new chapter began in the production of ceramics, and glass was already in general use. Owing to strengthening ties with the east, many easterners came to the province, both civilians and military population.

We come across two striking phenomena in fourth-century Pannonia, one being the emergence of large estates, latifundia, and the spread of *villae rusticae* throughout the province. Another important fact was that side by side with towns along the *limes* and inside Pannonia, many new settlements, towns, were established. There were, in fact, fortifications surrounded by thick walls and turrets, exclusively military strongholds, as established by research. The first period of their erection coincided with Constantine's building of the *limes*. The second phase must have occurred in the last third of the century. However, they were not built all at once, but step by step, continuously. Best known among them today is the Keszthely-Fenékpuszta fortification (Valcum); however, the ground plan does not seem to prove that it was a military establishment per se. It is also possible that they had originally been imperial estates supplying a sort of refuge for the local population, with their numbers gradually growing so that they developed into cities. The graveyards of these cities are characterized by rich finds. The function of these cities, however, changed after the death of Valentinian and it is possible that several ones in the interior of the province were formed into *foederati* settlements.[116]

## THE AGE OF VALENTINIAN

After the death of Constantius II, peace prevailed for a few years in the area east of the *limes*. Apparently Julian visited Pannonia, but we have no evidence of attacks on the *limes* during his reign. We do have evidence that, in 365, the Quadians and the Sarmatians were carrying out destructive raids in Pannonia. It is believed that their raids were avenged, since sometime prior to 370 Sarmatians were moved to Gallia.[117]

Enemies from beyond the Carpathians were kept away from the *limes* by an outer defense line of trenches stretching through the Hungarian Great Plain (Alföld), as well as by Sarmatian *foederati*. It appears from repairs on the *limes*, however, that its defensive character was changed from Valentinian onwards. The repairs seem to have become necessary because the trench line, as a buffer, ceased to exist.[118] In addition to repairing the *limes*, Valentinian established new camps. Numismatic evidence dates this large-scale construction, the remains of which are detectable in the Pannonian camps almost everywhere, to about 366–370. Building started in 366; in 367 monetary circulation increased considerably and building reached its peak. After the camps a chain of watchtowers was built between 371 and 373, complementing the line of fortifications. The inscription on the watchtower of Esztergom (Solva) called Commercium dates from 371, that of Visegrád from 372. Most of the watchtowers were found in the area along the Aquincum–Brigetio *limes*.[119] It cannot be denied that research has been quite intensive here, but it can also be supposed that Valentinian concentrated more care on the Aquincum–Brigetio *limes* than elsewhere.

In spite of all these precautions Pannonia became the victim of new devastations in 374, caused by the objections of the Quadians to the construction of fortifications in their territory. Valentinian continued the construction in spite of the opposition of Flavius Aequitius, commander of the Roman forces in Illyria. Marcellianus, the young *dux* of Valeria, decided on a radical solution to this problem; he invited Gabinius, Quadian king to Aquincum, and when he was about to leave he was assassinated. The allied Quadians and Sarmatians then invaded Pannonia; they massacred the population, engaged in harvesting, and carried off many of the survivors.[120] While several of the marauding groups got as far as the gates of Italia, the Sarmatians from the Banat broke into Moesia, but were defeated by Theodosius the younger, the Moesian *dux*, who later became emperor. The large coin hoards buried at the time of the invasion are evidence of the disaster.[121]

In the spring of 375, Valentinian started out from today's Trier to restore order in the central Danube area. Setting up his headquarters at Carnuntum, he spent the summer preparing an expedition against the Barbarians and, late in the summer, began his two-pronged attack against the Quadians. Merobaudes crossed the enemy border at Carnuntum with an infantry force, and the emperor himself attacked the Quadians from the bridgehead north of Aquincum at present-day Nógrádverőce. After ravaging their land, Valentinian returned to Pannonia at the same crossing sojourning first at Aquincum, and then establishing winter quarters at Savaria. However, he soon left to prepare another expedition, and while strengthening and fortifying the *limes* between Arrabona and Brigetio, he set up new headquarters at Brigetio. Here he was so angered when receiving a Quadian delegation that he suffered a stroke and died on November 17, 375.[122] After Valentinian's death, Merobaudes was recalled from enemy territory and the expedition was abandoned. On November 22 Valentinian II, who was residing in an imperial villa (Murocincta), was acclaimed emperor in Aquincum or Brigetio. The young emperor probably remained in Sirmium until 378.

Although the situation may have been anxious after 375 in the area east of the *limes*, nevertheless troops were sent from Pannonia to the Balkans at the end of 377. In 378 Gratian visited Pannonia. On January 19, 379, Theodosius was acclaimed emperor in Sirmium. It appears that while in Pannonia, the new emperor also visited Tricciana.[123]

## PANNONIA AFTER VALENTINIAN

We do not hear of new construction after Valentinian's death, and for the time being it is impossible to determine the architectural style of this period. Regular monetary circulation in the province ceased with Valentinian and Gratian, but the coins of the two emperors and also of earlier ones remained in use for a while. While new money rarely found its way to the province, the traces of continuous circulation of money can be found sporadically even in the early fifth century. This cessation of the regular circulation of money has greatly influenced theories concerning the history of post-Valentinian Pannonia and the dating of finds, graves, and buildings in general. The cessation of money economy together with the occurrence of Hunnish type of material in Roman settlements and in the province in general led to the opinion that the eastern part of the province (Valeria) was delivered to the Huns in 409 and that Pannonia existed in a state of stagnation after 375.[124]

More recent research pointed out that new types appeared in the Roman archaeological material after 375; new types of bottles and Hunnish octaeder earrings emerge here under Roman conditions. These new types of Roman material indicate that life was still active within the *limes*, but the emergence of Hunnish finds on the *limes* and in the interior of the province points toward the changes affecting future Pannonian life. This change explains both the cessation of monetary circulation and the transformation of defense with a waning importance of the *limes* in the last decades of the fourth century.[125] These archaeological observations receive support from the theory which connects the material of a Hunnish character with that segment of Huns which appeared in Pannonia at the end of the fourth century, rather than with the great Hunnish invasion of a later date. Material closely related to the Hunnish findings, discovered in various parts of the *limes* and in the interior of the province, point toward organized settlement of people with the purpose of general defense at the end of the fourth century. Certain ethnic elements, though perhaps not of common origin, who must previously have coexisted for a long time and mixed freely, appeared at this time. During this period, there was only one bigger settlement under Gratian (379), when the Hunnish, Gothic, and Alani settlers were brought to Pannonia as *foederati*.

New historical research insists that the Huns who separated from the larger tribe, either some two hundred years before or only in the fourth century and appeared within and outside the Roman imperial borders, are not identifiable with the Huns of Attila. Vithimir, leader of a group of eastern Goths, became allied with the Huns who had separated from their tribe and were now living in the southwest corner of the Alani settlement area. After the death of Vithimir, the regents during the minority of Vitheric, the Ostrogoth, Alatheus, and the Alan Saphrac retired to the west, the latter becoming leader of the Hunnish-Alani troops. At the border of the Roman Empire they appeared as a unified Ostrogoth–Alani–Hunnish group; they had already appeared united in Fritigern's raid against Constantinople, and now they participated actively in the battle of Hadrianopolis. Research had connected Alatheus's marauding with a passage in Ambrosius referring to Barbarian and heretic plunderings in the Danube provinces, including Valeria. But the beginnings of deterioration in Pannonia, the dissipation of the large fortunes of the province, and the general decline in the economic situation preceded the attack. The harsh Valentinian tax system, as well as the tax extortions of praetorian prefect Probus, were partly responsible for the critical situation in the province.[126]

Comes Vitalianus did not succeed against the tripartite group in their move against southeastern Pannonia. After his failure, Gratian was obliged to assume a federated relationship, permitting them to settle in Pannonia Secunda and in the southern strip of the Sava River. Economically weakened Pannonia was not really capable of providing federation benefits. The church sent Amantius, bishop of Iovia, as a missionary to the new settlers, who had difficulty in assimilating and were in a constant state of unrest. Their attacks and maraudings were checked by Theodosius in the summer of 379, but not without heavy devastation in several Pannonian towns, among them Mursa and Stridon.[127]

During the usurpation of Magnus Maximus, Bauto employed the Alani-Hunnish auxiliary troops, partially against the Juthungi. The *foederati*, however, could not agree with the Frankish military leaders, and, revolting against their rule in 385, they, together with the invading Sarmatians, devastated the province, leaving Pannonia Secunda and Savia virtually in a state of siege, while agriculture was completely paralyzed.[128] In 387 Theodosius used the Pannonian units of *foederati* who harbored open animosity against Maximus. The armies of Maximus then moved toward Pannonia Prima and Pannonia Secunda. In the battle of Siscia the Hunnish *foederati* played the most important role, while in the battle of Poetovio, Theodosius relied on the Ostrogoths and on Alani elements. In these battles the *foederati* no longer participated in individual groups but adopted the Roman pattern which was made necessary also by the Roman general supplies. It is because of this type of participation that cemeteries do not yield exclusively local types of graves with grave goods of local character.

Around 390 the western Goths first sided with Maximus, then escaped into the Macedonian marshes, and finally were included in the Roman *numeri*. Then they began to mingle with hordes of various nationalities under Hunnish leadership. Attacking Macedonia first, this group proceeded toward Italia in 392, but were halted by the Pannonian *foederati* and eventually forced them to return their captives.[129] These Pannonian *foederati* again played an important role fighting the usurper Eugenius in the last year of the reign of Theodosius. In this campaign, the commander of the Pannonian cavalry was the same general Saul who participated in the battle of Pollentia in 402. After the battle of Frigidus, the *foederati* troops, owing to enormous losses, revolted. A very difficult situation was created in the province when, under Arcadius and Honorius, pay was withheld between 395 and 399.[130]

According to current historical research, the Hunnish–Gothic–Alani *foederati* were first settled in southern Pannonia. Their violent history made it necessary to separate them; the Alani were placed in Valeria and the Goths in Pannonia Prima, while the Huns remained between the two rivers. The identification of ethnic groups is very difficult as yet. While there were uniformities in the various groups of the province, in some ways, they were also different. It is difficult to separate these ethnic groups on the basis of archaeological evidences, as these groups had shared a common habitation, and the settlements with defensive purposes are liable to change and obliterate ethnic characteristics. In addition, these people mixed with Romans and were integrated into Roman military units. With investigations advancing, it is possible perhaps to find purer ethnic patches in the middle of the province, but not in centers or in the *limes* which had a mixed population for defense. We have had opportunity to observe this phenomenon recently.[131]

From the end of the fourth century the intensity of Roman life seems to have gradually weakened. We have no evidence of further construction on the *limes*, neither do we hear of intensive destruction. Perhaps future research will uncover some signs, but the available evidence does not permit us to speak of a general collapse of the *limes*. The military population of the *limes* naturally became more and more barbarized, but archaeological indications show that *foederati* troops were used only in limited numbers for filling in the ranks. While the disruptive behavior of the *foederati* exerted harmful influence in Romanization, it eventually caused many people to flee from Pannonia owing to the restlessness of the province and the ever-changing loyalties. As regards the defense of the province, the *foede-*

*rati*, with all their instability, represented a powerful force.[132] Since the Hunnish-Alani relationship existed both within the province and outside it, the presence of the *foederati* served to pacify the province's neighbors. Finally, they maintained the administrative framework of the province in the face of declining Romanism.

On the other hand, the invasion of Alarich in 401, which passed through the western fringe of Pannonia, could have occurred only with the consent of the *foederati*. The settling of Alarich's tribe in southern Pannonia in 402 eventually hastened the disintegration of the Roman system. Alarich invaded Italia for a second time in 405. Of even greater intensity was Radagaisus's attack on Italia in 406. This invasion passed through the southern fringe of the province; it also could not have happened without the consent of the Hunnish *foederati*. According to historical sources, many people fled from the province at that time.[133] After 410 we have no knowledge of any large-scale action affecting Pannonia. This may indicate a more peaceful period.

Acceptance of the theory that the province of Valeria was ceded to the Huns in the first decade of the fifth century impeded more subtle research into this period. Research was also affected by an erroneous dating of historical periods, tombs, and cemetery portions and by the cessation of regular monetary circulation in the fourth century. No method has been developed for the right evaluation of later Roman graves. The topographical position of cemeteries was ignored, graves not containing objects were neglected, and no attempt was made to compare the supposedly Hunnish findings with Roman material. Recent research, however, tells us that Valeria was not delivered to the Huns; the recognition of the significance of the groups led by Alatheus and Saphrac greatly confirmed this view. New archaeological research already explains that certain types of Hunnish material may appear together with Roman findings; furthermore, the comparison of material from *foederati* settlements with that originating from other places permits speculation on organized Roman life which can be traced even into the first part of the fifth century. Archaeological research has taken only the first steps in this direction, but its initial success points to the possibility of describing the history of Pannonia in the early fifth century.[134] Active Roman life had existed in the province and evidence for it is furnished by the retirement of Dalmatius, governor of Gallia Lugdunensis Tertia, to his property in Pannonia at the beginning of the fifth century. Even the concept of the Danube as a river frontier contradicts the supposition that the province of Pannonia would have been torn away from other Danubian provinces and delivered to the enemy.

Complying with Rua's request in 425–427, the *foederati* status of the Huns, Alani, and Goths was stopped by the Romans. In 433 Rua made a new agreement on *foederation* with the Romans and this was when the "great" Huns entered Pannonia.[135] It is almost certain that side by side with the Hunnish material, a later Hunnish group can also be identified, but on the basis of the evidence it cannot be stated yet with certainty that they were identical with Rua's *foederati*. This type of Hunnish material can be found on the *limes* as well as in the interior of the province. In any case, its occurrence on the *limes* implies that the *limes* still existed. Archaeological research has not yet succeeded in revealing the fate of Pannonia under Attila, but it may be due to the fact that certain groups of findings have been wrongly dated. It is beyond doubt that, except for the *foederati* who were maintaining the old administrative organization, the original inhabitants vanished: they fled, disappeared, and owing to the lack of replacement their material culture vanished.

The province, however, was still acknowledged by the Romans even after the death of Attila. This is supported by the appearance of Emperor Avitus in Pannonia in 455 and arrangements concerning a *foedus* with the eastern Goths, both of which point to the fact that the Romans wanted to oversee the administrative framework of the province.[136] Just when, if at all, Pannonia was finally given up or abandoned cannot be determined at the present. Future discussion and scholarly contribution may clarify the debate.

It is beyond doubt that the remains of Roman natives could still be discovered near Lake Balaton as late as the sixth century and that the Byzantine Empire still maintained interest in the former province of Pannonia. The Byzantine occupation of Sirmium, between the Drava and Sava rivers and along the Sirmium–Sopianae–Keszthely-Fenékpuszta–Savaria–Scarbantia road connecting the east with the west, all point to this. The Byzantine Empire allied with the Longobards, who settled in the province east of this road, which shows that even at this late date, they attempted to maintain the administration of the province bordered by the Danube, by means of a *foederati* relationship.[137]

## NOTES

[1] I. BORZSÁK, "Die Kenntnisse des Altertums über das Karpatenbecken" (Budapest, 1936), (*DissPann* 1:6); Polyb. frg. 64.122, ed. Büttner–Wobst 4:523; A. MÓCSY, "Pannónia", *R. E. Pauly-Wissowa*, Suppl. 9, pp. 516 ff.

[2] I. HUNYADY, "Kelták a Kárpátmedencében—Die Kelten im Karpatenbecken I–II" (Budapest, 1942–1944) (*DissPann* 2:18); É. B. BÓNIS, "Die spätkeltische Siedlung Gellért-hegy—Tabán in Budapest" (Budapest, 1969), (*ArchHung* 47); K. PINK, "Die Münzprägung der Ostkelten und ihrer Nachbarn" (Budapest, 1939) (*DissPann* 2:15); G. ZIPPEL, *Die römische Herrschaft in Illyrien bis auf Augustus* (Leipzig, 1877); O. HIRSCHFELD, *Kleine Schriften* (Berlin, 1913), p. 9; I. BORZSÁK, K. VINSKI-GASPARINI, "Keltski ratnički grob iz Batine—Ein keltisches Kriegergrab aus Batina". *Arheološki Radovi a Rasprave* 1 (1959): 297; L. BARKÓCZI, I. BÓNA, and A. MÓCSY, *Pannónia története* (History of Pannonia) (Budapest, 1963), pp. 13 ff.; MÓCSY, p. 526.

[3] ZIPPEL, p. 133; A. ALFÖLDI and L. NAGY, *Budapest története I. Budapest az ókorban* (The History of Budapest, Budapest in the Antiquity) (Budapest, 1942), pp. 138 ff.; G. ALFÖLDY, "A szkordiszkuszok szállásterülete—Das Siedlungsgebiet der Skordisker". *ArchÉrt* (1962), pp. 147–59; MÓCSY. pp. 527 ff.

[4] ALFÖLDI and NAGY, pp. 142 ff.; A. ALFÖLDI, *Zur Geschichte des Karpatenbeckens im I. Jahrhundert v. Christi* (Budapest, 1942); MÓCSY, p. 529.

[5] Posidonius in Strab. 7.2.2, 3.2; ALFÖLDI; MÓCSY.

[6] ALFÖLDI.

[7] Strab. 7.5.2; ALFÖLDI and NAGY, p. 144; E. SWOBODA, *Carnuntum: Seine Geschichte und seine Denkmäler*, 3d ed. (Graz, 1958), pp. 19 f.

[8] MÓCSY, p. 531; C. DAICOVICIU, "Dacia capta (Zur Frage der Eroberung und ursprünglichen Organisation Dakiens)". *Klio* 38 (1960): 174–84.

[9] App. 3.5; Plut. *Pomp.* 41; App. *Mithr.* 109, 119; C. PATSCH, *Beiträge zur Völkerkunde von Südosteuropa V. 1. Teil: Bis zur Festsetzung der Römer in Transdanuvien* (Wien, 1932), pp. 35 ff.; ALFÖLDI and NAGY, pp. 144 ff.; MÓCSY, p. 530.

[10] App. *III.* 15–28, Cass. Dio 49.34–38; E. SWOBODA, *Octavian und Illyricum* (Wien, 1932);

120

F. MILTNER, "Augustus' Kamp um die Donaugrenze", *Klio* (1937), pp. 200–226; ALFÖLDI, p. 145; MÓCSY, pp. 538 f.

[11] App. III. 16; Cass. Dio 49. 37.6; E. SWOBODA, *Carnuntum*, p. 25.

[12] BORZSÁK, pp. 23 f.; ALFÖLDI, pp. 147 f.; SWOBODA, *Carnuntum*, pp. 28 f.

[13] Cass. Dio 54.20.2, 24.3, 28.1; E. RITTERLING, "Legio", *R. E. Pauly-Wissowa* 12:227; PATSCH, p. 98; ALFÖLDI, p. 152; MÓCSY, pp. 540 f.

[14] Cass. Dio 51.23.2, 54.36.2; Tac. *Ann.* 4.44; ZIPPEL, pp. 238 f.; PATSCH, pp. 101 f.; ALFÖLDI, p. 146.

[15] Cass. Dio 55.11a.2; Tac. *Ann.* 4.44; E. GROAG and A. STEIN, *Prosopographia Imperii Romani* (Berlin, 1943), p. 34, No. 128; PATSCH, p. 110; J. KLOSE, *Roms Klientel-Randstaaten am Rhein und an der Donau* (Breslau, 1934), pp. 62 ff.

[16] *ILS*, 8965; A. PREMERSTEIN, "Der Daker- und Germanensieger M. Vinicius und sein Enkel". *Jahreshefte d. Öst. Arch. Inst.* 28 (1933), 29 (1934); ALFÖLDI, pp. 153 ff.; B. SARIA, "Noricum und Pannonien", *Historia* 1 (1950): 443.

[17] L. SCHMIDT, *Geschichte der deutschen Stämme bis zum Ausgange der Völkerwanderung. II. Die Geschichte der Westgermanen* (München, 1938), pp. 153 ff.; ALFÖLDI, pp. 154 ff.; MÓCSY, p. 548.

[18] Cass. Dio 55 and 56; Vell. Pat. 2. 110–26; ALFÖLDI, p. 155; W. REIDINGER, *Die Statthalter des ungeteilten Pannonien und Oberpannonien von Augustus bis Diokletian* (Bonn, 1956), pp. 14 ff.; E. KOESTERMANN, "Der pannonisch–dalmatische Krieg 6–9 n. Chr.". *Hermes* 81 (1953), pp. 345–78.

[19] Sueton, *Tib.* 16.

[20] REIDINGER, pp. 12 f,; MÓCSY, pp. 519 ff.

[21] RITTERLING; REIDINGER, pp. 16 f.

[22] SCHMIDT; ALFÖLDI, pp. 179 f.

[23] ALFÖLDI, p. 138.

[24] MÓCSY, pp. 527–30; *Res Gestae Divi Aug.* 30: "Pannoniorum gentes, quas ante me principem populi Romani exercitus nunquam adit, devictas per Ti. Neronem qui tum erat privignus et legatus meus imperio populi Romani subiect, protulique fines Illyrici ad ripam fluminis Danuvi".

[25] Dunapentele: *Geschichte der Stadt in der Römerzeit in Intercisa* (Budapest, 1957), 2:499; E. TÓTH and G. VÉKONY, "Vespasianus-kori építési felirat Aquincumban — A Building Inscription in Aquincum from the Vespasian Period". *ArchÉrt* 97 (1970): 109–15.

[26] ALFÖLDI, p. 147; A. MÓCSY in *NumKözl* 60–61 (1961–1962): 15 ff.

[27] Hypothesis of A. MÓCSY.

[28] Tac. *Ann.* 1. 16–29; H. H. SCHMITT, "Der pannonische Aufstand des Jahres 14 n. Chr., und der Regierungsantritt des Tiberius". *Historia* 8 (1958): 378–83.

[29] A. MÓCSY, *Die Bevölkerung von Pannonien bis zu den Markomannenkriegen* (Budapest, 1959), pp. 45, 129; J. FITZ, "A Military History of Pannonia from the Marcomann Wars to the Death of Alexander Severus". *ActaArchHung* 14 (1962): 25–112.

[30] L. BARKÓCZI, "A Sixth Century Cemetery from Keszthely-Fenékpuszta". *ActaArchHung* 20 (1968): 309 ff.

[31] *Intercisa* 2: 497 ff.

[32] KLOSE; ALFÖLDI, p. 181; J. HARMATTA, *Studies on the History of the Sarmatians* (Budapest, 1950), p. 45; A. MÓCSY, "Zur Periodisierung der frühen Sarmatenzeit in Ungarn". *ActaArchHung* 4 (1954): 115–28.

[33] SCHMIDT, pp. 153 ff.; ALFÖLDI, p. 183.

[34] A. MÓCSY, "Tampius Flavianus Pannóniában—Tampius Flavius in Pannonien". *ArchÉrt* 93 (1966): 203–7.

[35] C. PATSCH, *Beiträge zur Völkerkunde von Südosteuropa V/2. Der Kampf um den Donauraum unter Domitian und Trajan* (Wien, 1937); SCHMIDT, pp. 159 ff.; ALFÖLDI, pp. 183 ff.; RITTERLING, pp. 1275 ff.; R. SYME, in *Cambridge Anc. Hist.* 11:175 f.; T. NAGY, in *ArchÉrt* 85 (1958): 203.

[36] L. BARKÓCZI and É. BÓNIS, "Das frührömische Lager und die Wohnsiedlung von Adony (Vetus Salina)". *ActaArchHung* 4 (1954): 183; TÓTH and VÉKONY.

[37] SCHMIDT, p. 161; ALFÖLDI, p. 188.

[38] PATSCH; DAICOVICIU.

[39] ALFÖLDI, p. 188; A. MÓCSY, "Zur Periodisierung". *ActaArchHung* 4 (1954): 125.

[40] ALFÖLDI, p. 190 ff.

[41] *Hist. Aug. Hadr.* 3.9; PATSCH, pp. 128 ff.; ALFÖLDI, p. 219.

[42] PATSCH, p. 189; ALFÖLDI, pp. 189 ff.; *Intercisa* 2:504 ff.; T. NAGY, "Buda régészeti emlékei II. Római kor" (The Archaeological Monuments of Buda. II. Roman Times), in *Budapest műemlékei* (Budapest, 1962), pp. 37 f.; E. STEIN, "Die Reichsbeamten von Dazien" (Budapest, 1944), pp. 14 ff.; *DissPann* 1:12.

[43] *CIL*, III, 6818; J. DOBIÁŠ, "A propos de l'Expeditio suebica et sarmatica de l'empereur Hadrian en l'an 118", in *Omagiu C. Daicoviciu* (Bucureşti, 1960), pp. 147–53.

[44] A. RADNÓTI and L. BARKÓCZI, "The Distribution of Troops in Pannonia Inferior during the Second Century A.D.". *ActaArchHung* 1 (1951): 191–230; T. NAGY, "The Military Diploma of Albertfalva". *ActaArchHung* 7 (1956): 17–71; BARKÓCZI and BÓNIS, pp. 129 f.; T. NAGY, "Buda régészeti emlékei", p. 38; *Intercisa*, 2:504.

[45] *Intercisa*, 2:504.

[46] RADNÓTI and BARKÓCZI, pp. 211 ff.

[47] ALFÖLDI, p. 190; L. BARKÓCZI, in *NumKözl* 56–57 (1957–1958): 17 f.; *Intercisa*, 2:506; T. NAGY, "Buda régészeti emlékei", p. 40.

[48] SCHMIDT, p. 162; E. SWOBODA, "Rex quadis datus". *Carnuntum-Jahrbuch* (1956), pp. 5–12; R. NOLL, in *ArchAustriaca* 14 (1954); 43 f.; FITZ, pp. 28 f.; T. NAGY, "Buda régészeti emlékei", pp. 40 f.

[49] NOLL; *Intercisa*, 2:506 ff.

[50] KLOSE: FITZ, pp. 28 ff.

[51] FITZ, pp. 26 ff.

[52] *Intercisa*, 2:508 ff.

[53] SCHMIDT, p. 164; ALFÖLDI, p. 191; T. NAGY, "Buda régészeti emlékei", pp. 41 ff.

[54] *Hist. Aug. Vita Marci*, 22.1; W. ZWIKKER, *Studien zur Markussäule* (Amsterdam, 1941); SCHMIDT, pp. 163 ff.; ALFÖLDI, p. 192; FITZ, pp. 32 ff.; NAGY, "Buda régészeti emlékei", p. 42.

[55] SCHMIDT, p. 164; FITZ, p. 24.

[56] Cass. Dio 71.3.1. (*Boissevain* 2:250 f.); SCHMIDT, p. 165; ALFÖLDI, p. 192; FITZ, p. 33; NAGY, "Buda régészeti emlékei", p. 42.

[57] *Hist. Aug. Ver.* 9.9; SCHMIDT, p. 165; NAGY, "Buda régészeti emlékei", p. 42.

[58] SCHMIDT, p. 166 ff.; NAGY, "Buda régészeti emlékei", p. 42.

[59] ZWIKKER; SCHMIDT, pp. 59 ff.; ALFÖLDI, pp. 193 ff.; NAGY, "Buda régészeti emlékei", pp. 41 ff.; H. G. PFLAUM, "Deux carrières équestres de Lambèse et de Zana (Diana Veteranorum), 2. M. Valerius Maximianus". *Libyca*, 3 (1955): 145; L. BARKÓCZI, "Die Naristen zur Zeit der Markomannenkriege". *FolArch* 9 (1957): 95.

[60] Cass. Dio 71.20; SCHMIDT, p. 175; NAGY, "Buda régészeti emlékei", p. 46.

[61] *Hist. Aug. Comm.* 6.1; Cass. Dio 72.81; F. STEIN in *R. E. Pauly-Wissowa*, 6: 955 f.; KLOSE; SCHMIDT, p. 176; *Intercisa* 2:515; J. FITZ, "Massnahmen zur militärischen Sicherheit von Pannonia Inferior unter Commodus". *Klio* 39 (1961): 199–214.

[62] A. MÓCSY, *Die Bevölkerung von Pannonien bis zu den Markomannenkriegen* (Budapest, 1959).

[63] MÓCSY, p. 108.

[64] MÓCSY, p. 110.

[65] MÓCSY, p. 129.

[66] *Intercisa*, 2:500.

[67] MÓCSY, p. 129.

[68] MÓCSY, p. 131 ff.

[69] *Intercisa*, 2 : 515 ff.

[70] SCHMIDT, p. 179; *Intercisa*, 2:516.

[71] J. FITZ in *NumKözl* 58–59 (1960): 7 ff.; FITZ, "A Military History of Pannonia", pp. 92 ff.

[72] *Cod. Just.* 2.50.1.

[73] L. BARKÓCZI and K. BIRÓ-SEY, "Brigetioi aranylelet—Trouvaille d'or à Brigetio". *NumKözl* 62–63 (1963–1964): 3–8, 109.

[74] L. BARKÓCZI, "Kiadatlan feliratos kövek Brigetióból—Iscrizioni inedite a Brigetio". *ArchÉrt* (1944–1945), p. 180; *Intercisa*, 2:518.

[75] L. BARKÓCZI, "The Population of Pannonia from Marcus Aurelius to Diocletian". *ActaArchHung* 16 (1964): 296.

[76] L. SCHMIDT, *Die Ostgermanen*, 2d ed. (München, 1941), p. 201; *Intercisa*, 2:516 ff.

[77] L. BARKÓCZI, 2:518 ff.; J. FITZ, "Il soggiorno di Caracalla in Pannonia nel 214". *Accademia d'Ungheria in Roma, Quaderni di Documentazione* 2 (1961): 5–23.

[78] *Intercisa*, 2:521 ff.; A. RADNÓTI, in *NumKözl* 34–35 (1935–1936): 2 ff.

[79] *Hist. Aug. Max.* 13.3; Herodian 7.4; B. SARIA, "Poetovio". *R. E. Pauly-Wissowa* 21:1174; SCHMIDT, p. 201; *Intercisa*, 2:523.

[80] SCHMIDT, *Die Ostgermanen*, p. 204; T. NAGY, "A cohors I. Noricorum equitata újabb feliratos emléke—Eine neue Inschrift der Cohors I. Noricorum Equitata". *ArchÉrt* (1940): 48–55.

[81] SCHMIDT, *Die Ostgermanen*, p. 205; *Intercisa*, 2:524.

[82] SCHMIDT; ibid.

[83] SCHMIDT, *Die Ostgermanen*, pp. 207 ff.; *Intercisa*, 2:526.

[84] SCHMIDT, *Die Ostgermanen*, pp. 208 ff.; A. RADNÓTI, "Trebonianus Gallus ezüstlemez mellképe—Silver Bust of Trebonianus Gallus from Brigetio". *FolArch* 6 (1954): 49–61; *Intercisa*, 2:526.

[85] SCHMIDT, *Die Ostgermanen*, pp. 209 ff.; K. BIRÓ-SEY, "A kistormási éremlelet—Le trésor de monnaies romaines de Kistormás". *FolArch* 15 (1963): 55–68; NAGY, "Buda régészeti emlékei" p. 55.

[86] SCHMIDT, *Die Ostgermanen*, pp. 212 ff.; NAGY, "Buda régészeti emlékei", p. 55; J. ŠAŠEL, "Bellum Serdicense". *Situla*, 4 (1961): 3–30.

[87] *Aur. Vict. Caes.* 33.1; SCHMIDT, *Die Ostgermanen*, p. 179; A. ALFÖLDI, in *NumKözl* 26–27 (1927–1928): 49 ff.

[88] A. ALFÖLDI, in *NumKözl* 26–27 (1927–1928): 49 f.; SCHMIDT, *Die Ostgermanen*, p. 213; NAGY, "Buda régészeti emlékei", p. 55.

[89] SCHMIDT, *Die Ostgermanen*, pp. 221 ff.; A. ALFÖLDI, "A gót mozgalom és Dácia feladása" (The Movement of the Goths and the Abandonment of Dacia). *Egyetemes Philologiai Közlöny* (1929), pp. 161–88; A. ALFÖLDI, "Über die Juthungeneinfälle unter Aurelian". *Serta Kazaroviana* (1950) pp. 21–24; DAICOVICIU.

[90] SCHMIDT, *Die Ostgermanen*, pp. 180, 222 ff.; K. SZ. PÓCZY, "Római épületek Óbudán a Kiscelli u. 10. sz. alatt—Römische Gebäude von Óbuda (Kiscelli-Strasse Nr. 10)". *BudRég* 16 (1955): 54.

[91] L. BARKÓCZI, "New Data on the History of Late Roman Brigetio". *ActaAntHung* 13 (1965): 215–57.

[92] L. BARKÓCZI, "The Population of Pannonia from Marcus Aurelius to Diocletian". *ActaArchHung* 16 (1964): 257–356.

[93] BARKÓCZI, "The Population", p. 296.

[94] BARKÓCZI, "New Data", pp. 250 ff.

[95] I. TÓTH, "Destruction of the Sanctuaries of Iuppiter Dolichenus at the Rhine and in the Danube Region, 235–238". *ActaArchHung* 25 (1973): 107–14.

[96] L. BARKÓCZI, "Beiträge zur Steinbearbeitung in Pannonien am Ende des 3. und zu Beginn des 4. Jahrhunderts". *FolArch* 24 (1973): 108 ff.

[97] From Brigetio, see note 91; from Aquincum, see K. SZ. PÓCZY, "Újabb aquincumi múmiasír—Ein neues Mumiengrab in Aquincum". *ArchÉrt* 91 (1964): 176–91.

[98] MÓCSY, p. 567; NAGY, "Buda régészeti emlékei". (Archeological relics of Buda), p. 55.

[99] D. VAN BERCHEM, *L'armée de Dioclétien et la réforme constantinienne*" (Paris, 1952); *Aur. Vict. Caes.* 39.40; Eutrop. 9.25.

[100] Amm. Marc. 19.11.4; A. H. JONES, in *JRS* 44 (1954): 21 ff.; ENSSLIN, in *R. E. Pauly-Wissowa*, 14:2521; A. ALFÖLDI, "Epigraphica IV". *ArchÉrt* (1941): 54; NAGY, "Buda régészeti emlékei", p. 53.

[101] Amm. Marc. 28.1.5; ALFÖLDI and NAGY, p. 674; NAGY, "Buda régészeti emlékei", p. 57.

[102] MÓCSY, pp. 570 ff.; NAGY, "Buda régészeti emlékei", p. 57.

[103] J. MOREAU, "Zur Datierung des Kaisertreffens von Carnuntum". *Carnuntum-Jahrbuch* (1960), pp. 7–15; Eutrop. 10.5; *Aur. Vict. Caes.* 41.5; Zosimus. 2.19; Anon. *Vales*, 16.18; E. STEIN, *Geschichte des spätrömischen Reiches* (Wien, 1928), 1:144–45; L. SCHMIDT, *Die Ostgermanen*, pp. 225 ff.; NAGY, "Buda régészeti emlékei", p. 58.

123

[104] Mócsy, p. 571.

[105] Zosimus 2.21; Optatianus Porfyrius *Carm.* 6.14; C. Patsch, *Beiträge zur Völkerkunde von Südosteuropa III. Die Völkerbewegung an der unteren Donau in der Zeit von Diokletian bis Heraklius* (Wien, 1928), pp. 16 ff.; E. Stein, in *R. E. Pauly-Wissowa* 2:21; A. Alföldi, "Epigraphica IV". *ArchÉrt* (1941): 676; *Intercisa*, 2:536 ff.; Nagy, "Buda régészeti emlékei", p. 59.

[106] *Aur. Vict. Caes.* 41.14; Schmidt, *Die Ostgermanen*, p. 226; *Intercisa*, 2:536; D. Tudor, *Oltenia romana* (Bucureşti, 1958), p. 356.

[107] S. Soproni, "Limes Sarmatiae", *ArchÉrt* 96 (1969): 43–53.

[108] Alföldi, "A gót mozgalom", p. 161 ff.; I. Bóna, "Az újhartyáni germán lovassír— Das germanische Reitergrab von Újhartyán". *ArchÉrt* 88 (1961): 206; Nagy, "Buda régészeti emlékei", p. 54.

[109] Patsch, pp. 28 ff.; Schmidt, *Die Ostgermanen*, p. 228; Alföldi and Nagy, p. 677; Bóna.

[110] *Intercisa*, 2:537; K. Sági, "Die spätrömische Bevölkerung der Umgebung von Keszthely". *ActaArchHung* 12 (1960): 187–256.

[111] Amm. Marc. 16.10.20, 17.12; Patsch, p. 37; Schmidt, *Die Ostgermanen*, pp. 181 ff.; Alföldi, p. 679; Soproni, p. 47.

[112] L. Barkóczi, "Transplantations of Sarmatians and Roxolans in the Danube Basin". *ActaAntHung* 7 (1959): 443–53; Soproni, p. 47.

[113] Soproni.

[114] Barkóczi, "Beiträge zur Steinbearbeitung", pp. 68 ff.

[115] A. Mócsy, "Pannónia". *R. E. Pauly-Wissowa*, Suppl. 9, p. 673.

[116] Mócsy, p. 673; A. Radnóti, "Pannóniai városok élete a korai feudalizmusban" (The Life of Pannonian Cities in Early Feudalism). *MTA TT. OK* 5 (1954): 494 ff.; Mócsy, p. 700.

[117] Auson. 5.8–9; Alföldi, p. 683; *Intercisa*, 2:443 ff.

[118] Soproni, p. 53.

[119] S. Soproni, "Über dem Münzumlauf in Pannonien am Ende des 4. Jahrhunderts". *FolArch* 20 (1959): 69–78.

[120] Amm. Marc. 29.6.1–16; C. Patsch, *Beiträge zur Völkerkunde von Südosteuropa. IV. Die quadisch-jazygische Kriegsgemeinschaft im Jahre 374/75* (Wien, Leipzig, 1929), p. 209; Alföldi, p. 684; A. Nagy, in *R. E. Pauly-Wissowa*, 7:2184 f.

[121] M. R. Alföldi, in *Antiquitas Hungarica* 3 (1949): 86 ff.; Sági, pp. 195 ff.

[122] Amm. Marc. 30.5.3, 5–6; Patsch; Alföldi, p. 685 ff.

[123] Amm. Marc. 10.10.5, 31.11.6; Alföldi, p. 686.

[124] Alföldi, pp. 686 f., 728; A. Alföldi, *Der Untergang der Römerherrschaft in Pannonien* (Berlin, 1924–1926), vols. 1–2.

[125] L. Várady, *Das letzte Jahrhundert Pannoniens, 376–476* (Budapest, 1969); A. Salamon and L. Barkóczi, "Bestattungen von Csákvár aus dem Ende des 4. und dem Anfang des 5. Jahrhunderts". *Alba Regia*, 11 (1971): 35–80; L. Barkóczi and A. Salamon, "IV. század végi és V. század eleji üvegleletek Magyarországról— Glasfunde vom Ende des 4. und Anfang des 5. Jhs. in Ungarn". *ArchÉrt* 95 (1968): 29–39.

[126] Várady, pp. 22 f., 34 ff.; L. Balla, "Savaria invalida: Megjegyzések a pannóniai városok Valentinianus-kori történetéhez—Savaria invalida: Notes to the History of Pannonian Towns in the Time of Valentinian". *ArchÉrt* (1963): 75–80.

[127] Várady, p. 37.

[128] Várady, pp. 44 ff.

[129] Várady, pp. 58 ff.

[130] Várady, 78 f., 82 ff.

[131] Várady, p. 388; Salamon and Barkóczi, pp. 35–80.

[132] Várady, pp. 82 ff.

[133] Várady, pp. 389 ff.

[134] Salamon and Barkóczi.

[135] Várady, p. 397.

[136] Schmidt, *Die Ostgermanen*, pp. 269 ff.

[137] Barkóczi, "A Sixth Century Cemetery", pp. 275–311.

124

# ADMINISTRATION AND ARMY

JENŐ FITZ

Roman power had taken long decades to establish itself finally in Pannonia. During the reign of Augustus, however, the conquest of the southern province took only a few years. For a long time the Roman legions were encamped south of the Drava River and their advance toward the Danube only occurred near the frontier of Noricum. From Claudius in the northern portion of the new province only a few outpost troops were stationed on strategically important points, road crossings, and crossing stations along the Danube, and the occupation did not interfere at all with the life of the local population.[1] The conquest, however, did not proceed in the same way in all parts of the province. Basing our estimates on the presence of the army, we can surmise that it was the territory south of the Drava and the western frontier region which first were incorporated into the empire. In the northeastern part of Pannonia this did not occur before the third phase of the development in the second half of the first century. Even prior to the Roman conquest there must have been differences between the western and eastern parts of the province, on one hand, and, on the other hand, the southern peoples living closer to Italia and the Greek territories. These differences were enhanced by the gradual expansion of the occupation which resulted in a step-by-step Romanization as well. The differences between the above-mentioned parts of Pannonia were remarkable for a long time.

In the first decades, but particularly after the great Pannonian-Dalmatian uprising the Roman army was stationed in Pannonia in an occupational role. However, even later, when the conquered no longer felt distrust toward the conquerors and the province obtained an administration similar to the other provinces, administration was run almost exclusively by the army. This meant, in the beginning, that the conquered were controlled by the Romans. In the course of later events, however, when the army stationed in the province was recruited locally and contained mainly Pannonians, the relationship between administrators and the administered on a higher level of government had completely changed. Even at this later stage, when the local population was eligible for service, in the legions and higher military ranks and when their financial status accorded for them the equestrian and senatorial ranks, the Pannonians could fill only minor posts in the management of the affairs of the province. Government, as in all the other provinces, was the privilege of the empire. Not only the swiftly succeeding governors but also all officers up to the rank of *centurion*, were representatives of the great imperial administration in the province. The constant exchange of senators, equestrians, and professional officers responsible for the interior affairs

125

of the province organized an administration independent from the local population and even aloof from it.

At the head of the province was a governor, elected from the members of the Senate who, since Pannonia was an imperial province, governed in the name of the emperor. His title was *legatus Augusti pro praetore*. Before Pannonia was divided, since the number of its legions was never less than two, it was governed by senators elected from among former consuls. During the decades following the occupation, the army in Pannonia was comparatively small (two legions and auxiliary troops not exceeding the strength of a legion) and its governors were less than outstanding. They, nevertheless, excelled from the time of Domitian onward, when during constant warfare the army rapidly expanded and the governors took an active role in the wars against the Dacians and Suebians. Among the well-known and successful war leaders, L. Funisulanus Vettonianus, Cn. Pinarius Aemilius Cicatricula Pompeius Longinus, L. Iulius Ursus Servianus, and others governed Pannonia.[2]

Owing to the altered military situation at the end of the Dacian wars, Emperor Trajan divided Pannonia into two parts. The larger western section under the name of Pannonia Superior with the three legions of the Marcomannian–Quadian front (Vindobona, Carnuntum, Brigetio) continued under the leadership of the consular legate. The long and narrow Pannonia Inferior controlled the Sarmatian front with one legion (Aquincum), and consequently its governor was chosen from among former praetors. Despite the military and administrative division of the province, the two parts continued to be in close contact. In more serious military situations, if need demanded, the two Pannonias were temporarily governed by one person. This happened in the last years of Emperor Hadrian's reign, when during the strong Barbarian threat, Aelius Caesar united the forces of the two provinces. It is even possible that during the wars with the Marcomannians it became necessary to unite the forces in a similar way. However, there also were other contacts between the two provinces. Beginning with Emperor Hadrian, the election of the governors followed a regular pattern. The Pannonian front, which at times was severely tried, became one of the most critical military points in the empire. This made it necessary to choose governors from *viri militares* with adequate military experience. Since local experience was equally indispensable, the legates were generally elected from those senators who had served in an earlier period of their careers in one of the Pannonias. Thus the governors of Pannonia Inferior generally succeeded to their posts from the head of a legion in Pannonia Superior. Previously they had been consuls and had served in different imperial offices in Rome. While this practice survived, the two provinces were governed by several outstanding soldiers whose competence proved them able to fend off the inroads of the Barbarians and also to settle local skirmishes. During the critical period of the century, in the Marcomannic wars, the best commanders headed the Pannonian armies, such as Ti. Claudius Pompeianus, the great military expert of the time who had been chosen by Marcus Aurelius to become his son-in-law and his adviser in matters of war.[3]

The gradually developed army had made Pannonia into one of the strategic centers of the empire. This had an ever-increasing effect not only externally but also internally regarding the power situation within the empire. Pannonia emerged first as a factor affecting public life after the murder of Commodus and became victorious in the struggle of the eastern-western military concentration. Septimius Severus was the first Pannonian governor to be raised to the imperial throne. In 214 Caracalla modified the division of the armies of the two provinces, and Brigetio became part of Pannonia Inferior. From that time on, both provinces were governed by consuls. This change brought an end to the earlier system of promotions. Even the number of outstanding soldiers diminished after the death of Caracalla. During the great crises of the mid-third century, when the empire was compelled to meet enemy attacks simultaneously in the east, along the Danube, and along the Rhine, the provinces were not strong enough to defend themselves. This emergency produced under Philippus the concentration of the Danube provinces: the forces of the two Pannoniae, the two Moesiae, and Dacia were united to form a great military unit under a military leader with exceptional ability. This concentration foreshadowed a later confederation which actually was realized in the time of Diocletian. Some commanders stationed at Sirmium, taking advantage of the difficulties of the empire, emerged as usurpers — such as Pacatianus, Decius, Aemilianus, Ingenuus, and Regalianus. Gallienus succeeded in restoring the unity and peace of the empire.[4]

After the edict of Gallienus the senators were ousted from the military leadership. For another decade the provinces were headed by legates whose power dwindled considerably. The military leadership was in the hands of the equestrians; the governor was replaced by the *agens vices praesidis* and in 270 was succeeded by the *praeses* from the equestrian order. The reform became final during the tetrarchy, when the provinces were newly divided. Pannonia Superior north of the Drava gave rise to Pannonia Prima and in the southern portion, to Savia. The northern part of Pannonia Inferior became Valeria, the southern, Pannonia Secunda. The civil administration of the new provinces was conducted by the *praesides* of equestrian rank, as *vir perfectissimus*, while military power was in the hand of the *dux* who was also *vir perfectissimus*. The position and administration of the four Pannonian provinces were not the same. The highest rank was allotted to Pannonia Secunda with Sirmium at its center — an imperial seat not only of the new province but also of the diocese of Pannonia and between 357 and 361 the residence of the Illyrian praetorian prefects. At a later stage, the *consularis* was at the head of the province, while military leadership was in the hands of a *dux* stationed at Sirmium. The *praeses* of Pannonia Prima was stationed at Savaria. At first, Savia was governed from Siscia by the *corrector* of praetorian rank. Nothing is known of the civil administration of Valeria. It had been governed by the *praeses*, who had probably been residing in Herculia.[5] To protect the Pannonia Prima frontier, it became united with Noricum, with the *dux*'s headquarters at Lauriacum.

The governor was supported in his work by the *officium consulare*, the governor's office. In Pannonia Superior and after 214 in Pannonia Inferior also, it had a staff of 200 persons, not counting the governor's guard, the *equites*, the *pedites singulares*, and the *stratores*. In Pannonia Inferior which had one legion (between 106 and 214) the *officium* consisted only of 100 persons. The administrative office, *officium corniculariorum* was headed by the *cornicularii*, officials of the highest rank. In Pannonia Superior their number was three. Three *commentarienses* dealt with legal matters. Criminal cases were investigated by the *speculatores*, numbering thirty in Pannonia Superior and ten in Pannonia Inferior. The *quaestiones* also dealt with legal matters. Among the less significant offices the function of the interpreters is also worth mentioning, necessitated by contact with the Barbarians.

In the governor's office in Pannonia Inferior, we know of one *interprex S(armatorum)*, one *interprex Ge(rmanorum)*, and one *interprex Dacorum*. Other officers were the *haruspex* (soothsayer) and the *victimarius*. Security services in the province were supplied by the *beneficiarii consulares* in the province. Inscriptions prove that they had a network of stations in the whole province, primarily at road junctions and at markets opened for the Barbarians. In consular offices they numbered sixty and under praetorian governors only thirty. Secret security services were filled by the *frumentarii* inside the *officium* and they also acted as supervisors of the post. Messengers (*principales*) and attendants (*immunes*) working in the *officium* were selected from the armed forces of the province, on the basis of ability and reliability. Under the Severi, many Syrians from Intercisa who showed devotion to the imperial house were recruited to serve in the governor's office.[6]

The postal service was also partly under the supervision of the governor: a diploma was issued by him for those who were eligible to use the postal service. Only few data survive of the organization of the Pannonian *cursus publicus*. We know of a single *praefectus vehiculationis* who managed the postal service of the two Pannoniae, Noricum and Moesia Superior in the rank of a *sexagenarius*.[7] However, in Virunum and Mediolanum another *praefectus vehiculorum* is known to have acted in Noricum, as well as a *f(isci) a(dvocatus) ad vehicula per Transpadanum et partem Norici*. It is possible, as in Italia, that some main roads and adjoining secondary roads were controlled by the *praefectus vehiculorum* in the provinces. It is possible that the postal service operated from Noricum to Moesia Superior on the imperial road from east to west. We have only a few data of the numerous officials employed by the postal service, with the exception of the *frumentarii* and *stationarii* who were in charge of the stations. *Itineraria* survive which furnish data on the imperial main roads and the stations on the adjoining secondary roads. There were postal stations within a day's distance which also offered accommodation for the night; *mansiones* were built throughout the province in smaller towns and settlements. The *mutationes* were stations for changing horses. The distance between two *mansiones* was generally from eight to ten kilometers. We know of several *mutationes* on the road between Aquincum and Brigetio, on the Csúcshegy road, and between Testvérhegy and Purbach.[8]

The extent of the postal service also proves that its activity was not limited within the boundaries of the provinces. The same may be observed for the majority of offices of an economic character. With this organization the imperial policy seemed to have tried to restrict the special power of the governors. This distinction was also expressed in choosing the place as the economic center of the province. The legate in Pannonia Superior had his seat in Carnuntum and the religious center of the province was Savaria; the offices of various procurators, on the other hand, were at Poetovio. The control of the military administration differed from the economic affairs also in a social sense: military affairs were in the hands of the legates who were recruited from the Senate, economic matters from the second half of the first century were concentrated in the hands of the equestrians.

Among the procurators who managed the economic affairs of the two provinces, the most important were the finance procurators. Until the division of the province, a *ducenarius* was responsible for the finances of Illyricum: this was the *Procurator Provinciae Pannoniae et Dalmatiae* who presided at Poetovio. After 106 this organization came to an end, and each independent province was headed by a procurator of the rank of *centenarius*. The procurator of Pannonia Superior remained at Poetovio, while that of Pannonia Inferior was located either at Mursa or Aquincum. The procurators were responsible for the financial affairs of the province under their charge. Their activities extended to the auditing of revenues, yet their main function was managing the expenditure, particularly disbursing soldiers' payments and financing state building plans. Their numerous jobs demanded the support of a smaller office (*officium*) and officials. The affairs of the *officium* were in the hands of the *cornicularius*, while local matters were dealt with by the *beneficiarii*. We know about *beneficiarii procuratoris* acting in Pannonia Superior in the towns of Siscia, Vindobona, and Carnuntum and in Pannonia Inferior in Aquincum and Mursa. Income and expenditure, i.e., the provincial cash box, was in the hands of the *dispensatores* who came from the ranks of imperial slaves. The central archive was also controlled by the procurator, as was the *tabularium provinciae* where the land register and the tax rolls were kept. Tax collectors gathered the taxes and on that basis, prepared the census. The *tabularium* of Pannonia at Poetovio had a large staff, headed by a *princeps tabularius*. The staff consisted almost exclusively of freed men. According to surviving data, the procurator had often succeeded to the office of a deceased or absent governor. We have no data to confirm this in Pannonia, but it seems likely, at least in Pannonia Superior, that the senior *legatus legionis* who temporarily acted in matters of jurisdiction also acted on behalf of the *consularis*.[9]

The seat of the *procurator hereditatium* who collected inheritance taxes was at Poetovio; he was in charge of both Pannonias. No procurator's name has come down to us; we know only a few officials. There was a 5 percent inheritance tax.

The sexagenarius *procurator famil(iarum) glad(iatoriarum)* whose office was extended to several provinces, including Aemilia, Transpadana, the two Pannoniae, and Dalmatia, controlled the supervision and recruitment of the gladiators. Caesius Anthianus held office in the second quarter of the third century.[10]

The mines had a special organization. From ancient times when the mines were managed by tenants (*conductores*) we have no data. Probably, in keeping with the general practice, the supervision of the *conductores* was arranged by means of the procurators of the rank of a *sexagenarius* since the time of the Flavians. The only name known among them, L. Crepereius Paulus, had supervised the mines as early as the second third of the second century, with the title of *proc(urator) Aug. argenta[riorum P]annonicoru[m]*. The mines went into state ownership during the reign of Antoninus Pius when the management came into the hands

129

of a procurator of the rank of *centenarius*, with the same procuratorial title as earlier. The next change came about in the management of the mines under Marcus Aurelius when the mines of Pannonia and Dalmatia were handled by the *procurator argentariarum* (or *metallorum*) *Pannonicarum et Dalmaticarum*. The stone quarries and the relatively small metal mines of the Sava basin were jointly managed with the prosperous Dalmatian ones. The procurator managed the mines from Dalmatia (Domavia); however, this organization probably supplied the requirements of the Pannonian army.[11]

Among the procurators of Pannonia Superior in the second century, we know about two freed men. Naturally, they did not occupy the position kept open for equestrians but they were in all probability supervisors of imperial estates as *procuratores patrimonii*.[12]

The customs duty organization was considerably larger than the unit of a province. The customs office of Illyricum comprised the two Pannoniae, Dalmatia, later also Noricum, both Moesiae, Dacia, Thracia, and part of northern Italia. In the early imperial period customs were handled as a hired office by *conductores* who were acting in a semiprivate and semiofficial capacity. The lease did not extend to the whole customs area. The money paid as customs duty went to the *fiscus* which was supervised by the *procurator provinciae*. Marcus Aurelius stopped the lease-system and introduced state control headed by a procurator. Both the *conductores* and, later, the procurators had their seat in Poetovio; it is possible, therefore, that it was the center of the whole customs area. Customs were payable not only on the frontiers of the empire and on the interior lines dividing the provinces, on transit and interprovincial trade, but also extended to roads, bridges, and crossing stations. Consequently, the customs stations were established near the great commercial roads, estuaries, crossing points, and bridges, even inside the provinces. Stations on the outward frontier were at Brigetio, Aquincum, Intercisa, Altinum and Statio Confluentes; on the Noricum border at Poetovio and Savaria; on the Dalmatian-Moesian border at Sirmium; inside the province near the roads leading to Dalmatia and Siscia. The officials of the customs stations were almost exclusively slaves or freed men. During the lease system, the customs stations were headed by the *vilicus* and his deputy, the *vicarius vilici*. Other officials were the *contrascriptor* (comptroller), the *contrascriptor ex vicario* (the former's deputy), the *arcarius* (money collector), the *vicarius arcarii*, the *scrutator* who searched the packages, and the *tabularius* doing office work. After the introduction of state control, the customs offices were headed by *praepositi*, whereas the names of the other officials remained unchanged. The slaves of the *conductores* were taken over by imperial slaves who, like their predecessors, were mainly Orientals, especially from the late second century on. Cosmius, the *praepositus* of the station at Intercisa, was at the same time cantor of the synagogue, *spondilla synag(ogae) Iud(a)eor(um)*.[13]

The imperial economic structure shows that the central government wanted to direct the regional economy independently from the territorial boundaries of the provinces, but most of the time the Pannonian city of Poetovio was the imperial economic center of the designated Middle Danubian provinces. This economic concentration resulted in adequate supplies for the Pannonian army and a closer collaboration among the economically interrelated Roman provinces. Even after the division of Pannonia into four political organizations the whole of Pannonia stayed in an economic unity.

In Pannonia the power of the state was preponderantly represented side by side with the imperial administrative staff and the imperial economic structure by the army. During the reign of Augustus the army kept its eyes more on the hardly defeated Pannonian tribes than on the Barbarians who were on the other side of the Danube River. Owing to this situation, the army was stationed south of the Drava, in the camps of Emona (Legio XV Apollinaris), Poetovio (Legio VIII Augusta) and probably Siscia (Legio IX Hispana). After A.D. 9 Legio XV Apollinaris marched alongside the Danube and built its camp at Carnuntum. This was necessitated by the growing strength of the Marcomannian Maroboduus and the defense of the pacified Noricum. After the suppression of the great Pannonian-Dalmatian uprising, the Pannonian tribes made no further attempt to restore their independence, and it became more and more apparent that they were gradually assimilated into the Roman life-style, making it unnecessary to keep them in check. Between 43 and 45, Legio IX Hispania left Pannonia for good, and for many decades to come the Roman army consisted of two legions and some advanced auxiliary troops. The legions, however, were frequently exchanged. Legio VIII Augusta was succeeded by XIII Gemina, XV Apollinaris between 62 and 71 took part in the eastern wars and was replaced by X Gemina (62–68), VII Gemina (68–69), and XXII Primigenia (69–71). Owing to the Dacian and Suebian wars of Domitian, it was necessary to withdraw and restation several legions. Legio XXI Rapax appeared in Pannonia about 89 or 90 and pitched camp at Mursella, near Mursa. During 92–93, when the Sarmatians destroyed this unity, Legio XIV Gemina took its place. Under Domitian even Aquincum received a legion, first II Adiutrix, then IV Flavia. In wartime, the troops were often exchanged and the final build-up along the Danube did not take place before Trajan. From Poetovio, Legio XIII Gemina moved to Vindobona, and around 100 construction on the Brigetio camp was begun. The long period of war ended with the triumphant conclusion of the second Dacian war, after which the Pannonian army received its final arrangement. In the camp of Vindobona was stationed Legio X Gemina, Carnuntum was defended by XIV Gemina, Brigetio by I Adiutrix, and Aquincum by II Adiutrix.[14]

The strategic situation changed basically with the stationing of the legions near the Danube, but it modified the function and duty of the army as well. The legions stationed south of the Drava River necessarily followed mobile tactics, forming a scattered frontier defense along the Danube. In the new situation the concept of the camp also underwent changes: it was originally the legions' quarters, providing protection against sudden attack; military action in normal cases always occurred outside the camp. However, the camps along the Danube served as fortifications and in the following centuries they were again reinforced. Under Domitian and Trajan the *limes,* with its smaller and larger fortifications as an uninterrupted chain sealed the Pannonian part of the Danubian frontier of the empire. This strong defensive line was intended to hold back the enemy attacks and secure the peace and prosperity of the inhabitants in this important frontier province.

The first line of this enormous defense system was the large river, the Danube itself. The legionary camps were built with special consideration to the Barbarians beyond the river and the more important strategic points. Behind the most dangerous Marcomannian front two legions were stationed at Vindobona[15] and Carnuntum.[16] Opposite the Quadians was built Brigetio[17] and opposite the Sarmatians, Aquincum.[18] Between these large camps, on a twenty-kilometer section, the Danube defense was conducted by auxiliary units in small camps. But in order to keep under observation the military road connecting the shore with the camps, a network of watchtowers was built, not further apart than two to four kilometers, so that they could signal each other at the slightest move from the enemy. It is possible that from the early occupational period a Danubian fleet was dispatched to patrol the left bank of the Danube, since Tacitus mentioned it in A.D. 50.

The final organization must have occurred under the Flavii, since it was called *classis Flavia Pannonica*. Ports existed at Carnuntum, Brigetio, Mursa, Novi Banovci, and Taurunum. The defense of the *limes* extended also to the control of the opposite bank of the Danube. Large military camps which controlled the crossing points along the roads that led into Barbaricum were defended by bridgehead fortifications, built on the opposite bank of the Danube. Such a bridgehead was the *Celemantia* across the Danube at Brigetio. Aquincum was defended through two counterfortifications. One was built close to the present-day Rákosbrook and another near present-day Elizabeth Bridge in Budapest. The bridgehead of Lugio was Contra Florentiam, and for Bononia it was the Castellum Onagrinum. In Barbaricum other fortifications were also built, particularly along the important trade routes. Finally, the defense system was also supported by a network of client-states which constituted the first line of defense in free Germania against the pressure of migrations beyond the Carpathians. The creation of the state of Vannius was the first significant example for such a buffer state between the empire and the Marcomannians. Later too we hear of the client relationship between the Sarmatians and Quadians of the expulsion of kings, executions, the installation of new kings, and the buying of their allegiance with gold.[19]

The army of the province around 106, after a large-scale development was over, comprised some 37,500 — 39,500 men, from which 22,300 were in the camp of Pannonia Superior and 15,000 — 17,000 were within the borders of Pannonia Inferior. In the second century, before the Marcomannic wars, some new camps were built with new formations. The postwar arrangement increased the strength of the army, primarily by means of exchanges, when troops of 500 men were replaced by *cohorts* of 1,000 men. This slow increase in numbers was responsible for the fact that in the early third century, some 23,000–25,000 men were stationed at the Marcomannian-Quadian front, while the army of the Sarmatian front numbered 19,000–20,500 men: the joint force of the two provinces must have reached 41,000–45,000 soldiers. According to the enumeration of *Notitia Dignitatum*, the Late-Roman armed forces stationed on the frontiers were distributed as follows: some 21,000 in Valeria; in Pannonia Prima and Noricum Ripense with which it formed one part from a military point of view, some 16,000; in Pannonia Secunda and Savia, 19,000 men. The full number, including the forces in Noricum (formerly 10,000 soldiers), was 56,000 men. A decrease could only have been illusory, expressing the situation after the reform initiated by Diocletian. During the reorganization two new legions appeared in the Pannonian provinces. Legio V Iovia and VI Herculia were sent to Pannonia Secunda which was left without any legions, the first in Bononia and Burgenae, the second in the camps of Aureus Mons and Teutoburgium. Constantine's reform created some innovations in the defense policy in Pannonia by placing highly mobile troops inside the province. This change in the defense strategy was necessary due to the fact of the demoralization of the army at the frontiers. In the western Danube provinces this defense, deployed in depth, was supplied by the newly organized Legio III Herculia and IV Iovia. The main reserve force, which was comprised of Legio I, II, III, and the Julia Alpina, was stationed in the Julian Alps.[20] The legions constituted the main units of the provincial armies. According to Vegetius a legion contained 6,100 foot soldiers and 726 mounted soldiers, divided into ten foot cohorts (cohors) and mounted soldiers. A cohort was divided into six centuriae of which three formed a manipulus. During the second century, until Caracalla's reforms, the army of Pannonia Superior, in addition to the three legions, contained 4 to 5 alae and 5 to 7 cohorts which were organized on the pattern of the legions. On the long Lower Pannonian front, where only one legion was stationed in the northernmost corner, the number of the auxiliary troops was larger. Military diplomas prove that there had been 4 to 5 alae and the number of cohorts went up to thirteen. The auxiliary troops, especially the equestrians, were used as independent tactical units under the direct command of the governor. In order to secure their better cooperation with the legions, military districts were organized which had territorial jurisdiction over every military unit stationed within the precinct. In Pannonia Superior, Vindobona, Carnuntum, Arrabona, and Brigetio could be regarded as centers of their respective districts. In Pannonia Inferior, the numbers and functions of auxiliary units changed often, resulting in changes in the extent of jurisdiction of the military districts. This fact is documented by late second-century military diplomas and the section designation of early third-century road repairs issued by the military districts.

The majority of the auxiliary troops contained 500 men. However, one equestrian ala comprising 1,000 men was stationed in each Pannonian province. These two equestrian units had prime importance not only in the defense of the frontier but also in the punitive raid against the Barbarians. Arrabona's garrison, Ala I Ulpia Contariorum Milliaria performed several important missions during the Marcomannian wars, especially under the command of two excellent officers by the name of M. Macrinius Avitus Catonius Vindex[21] and M. Valerius Maximianus.[22]

One part of the cohort consisted only of foot soldiers (peditata), while the second part also had mounted units (equitata): against 380 pedites there were 120 equites. In certain formations there were even specially armed soldiers, such as the sagittarii and catafractarii. These formations were generally equipped to meet the tactics used by the Barbarians on the opposite bank of the Danube. Thus some thousand archers were brought from Hemesa to fight here during the early reign of Commodus.[23]

133

The legions were led by a senator who had been appointed by the emperor generally chosen from the *praetorii*. In Pannonia Inferior where only a single legion was stationed, the governor also commanded the legion. The commander of an independent legion was given the title of *Legatus legionis*.

When speaking of the governors' careers, we have already mentioned that the governors of the two Pannoniae, in the majority of cases, were elected from the legates of the legion of Pannonia Superior, in order to better exploit the geographical conditions and the interests of the army stationed there. However, even those legionary commanders had made military careers outside Pannonia. They were eligible to become governors of the equally important two Germaniae and Moesiae; the most successful ones could even become governors of Britannia.[24]

The officers of the legions were *tribunes, prefects*, and *centurions*. The senior among the six *tribunes* was at the same time the legate's deputy as *tribunus laticlavius*, the only officer in the legion with the rank of senator. In the early career of a senator this was the first military assignment, before the age of twenty-three. The governor was generally responsible for the selection of the *tribunus laticlavius*, so it was often made with personal considerations in mind. During the governorship of his father in Pannonia Superior, L. Minicius Natalis Quadronius Verus[25] served in XIV Gemina; in Pannonia Inferior, Haterius Latronianus served during the governorship of his father between 161 and 164.[26] M. Caecilius Rufinus Marianus was the tribune of Legio IV Flavia while this unit was stationed temporarily at Aquincum, during the governorship of his parental relative.[27] Other tribunes of the legions were equestrians and with the exception of the *tribunus sexmestris* who commanded the mounted division, belonged to the staff of the legate. This rank was achieved in the equestrian career as the second stage of the compulsory *tres militia*, designated by *ab epistulis* imperial office, at the suggestion of the governor or the legion's legate. In this way the filling of the equestrian offices also occurred, in the majority of cases, by considering the recommendation of friends.[28]

The military command of the legions was primarily the duty of professional soldiers. These officers represented various stages in the *centurion*'s career. From their rank emerged the *praefectus castrorum*, the camp commander who performed first of all, technical and administrative functions. He was in charge of the military hospital, of the veterinary surgeon, of the *architectus, mensor, venatores*, of the military prison and clock (*horologiarius*), of the water-organ (*hydraularius*), etc. The duties of the company officers were performed by the *centurions*. Some of them obtained their rank by promotion, even if they started as ordinary soldiers (*ex caliga*), but former *centurions*, the sons of noncommissioned officers, accomplished young men, or those with good recommendations could also be appointed without any previous military training. In the first and second centuries, the centurions were recruited from Italici or fully Romanized local inhabitants who had the proper social background and culture. In the time of the Severi these strict rules were suspended and the Italici were replaced by local or Oriental inhabitants. To the latter belonged N. Aurelius Heraclitus, of Emesan origin, the son of a centurion from the Syrian cohort at Intercisa[29] and P. Aelius Proculinus, who came from the same background.[30] The *centurions* as well as the *tribunes* and other high-ranking commanders of the army, when being promoted, also went to other batallions, without regard to provincial frontiers. This practice was particularly beneficial to the unity of the imperial armies, but the officers too remained independent of local interests.

The noncommissioned officers of the legions came partly from the other legions and partly from the auxiliary units. The composition of the army, at an earlier stage came from Italian (Italici), and later from the native population. The noncommissioned staff formed two groups: the lower ranks were the *immunes*, subjected to the tactical ranks (*tesserarius-optio-signifer*) and corresponded to the noncommissioned officers of our present armies. The tactical ranks were called collectively *principales*, among which were the senior ensign, the *aquilifer*, and the sergeant, the *optio spei*. Members of the noncommissioned staff were the officials of the *officium* of the *legatus legionis* of the *tribunus laticlavius*, of the *praefectus castrorum*, of the *tribuni angusticlavii*, and of the *tribunus sexmestris*.[31]

There had always been basic differences between the legions and the auxiliary troops, as far as training, social and material esteem, and the chances of promotion were concerned. As far as command was concerned, there was strict discipline from top to bottom. Foot and mounted auxiliary troops were headed by officers from the equestrian order who began their military careers by commanding a *cohort* of 500. From here they were promoted generally to the staff of a legion where they received the rank of *tribunus angusticlavius*. (The more talented equestrians were given, instead of service in the legion, the tribunate of a *cohort* numbering 1,000.) The equestrians promoted from the legion went back to auxiliary troops and became *prefect* of an *ala* of 500 men; this could be followed by a procurature. From the middle of the second century the equestrian military career was supplemented by another function: command of an *ala* of 1,000 men. For the position of *quarta militia* the best officers were selected, from which position they could further advance their brilliant careers. Among the prefects of the Pannonian army the name of M. Macrinius Avitus Catonius Vindex deserves mention: in 167 he repulsed the attack of the Lombards and subsequently made a quick rise in his career, even being admitted into the senatorial order. He became governor of Moesia Superior and later of Moesia Inferior and died at the age of forty-two.[32] Another outstanding officer was M. Valerius Maximianus, who was believed to have killed the king of the Narists during the Marcomannian wars; he also became a senator and commanded a great number of legions in the Danube region, including Pannonia.[33] There were plenty of opportunities, even on a lower level, to transfer from the legions into auxiliary troops, or vice versa. This was originally possible only in auxiliary troops formed from Roman citizens; however, this restriction was later lifted.

Legions and auxiliary troops differed primarily in their composition. The Roman citizen was enlisted into the legion, whereas the peregrins, i.e., the freeman who was not granted civil rights, was enlisted into the *ala* and *cohort*. In the beginning the distinction referred to legions and local auxiliary troops, the former composed of Italici, the latter of the Romanized population of other provinces. At a later stage when the army was supplemented mainly from local personnel, the legions recruited their soldiers from the territories of the cities, while the auxiliaries relied on the small farming peasantry and the *civitas*. The legions recruited foreign elements only for wars that were fought far from the province. The auxiliary troops, however, continued in part to recruit where the division had been formed. Finally, fifty years after it began to be stationed in Pannonia, Cohors III Batavorum, the garrison of Vetus Salina was completed by Batavi. The garrison of neighboring Intercisa, the Cohors I Milliaria Hemsenorum, had retained its Syrian character up to the middle of the third century, owing to recruiting from the Syrian Emesa and its vicinity, where only a small number of Pannonian natives were enlisted. The *alas*, despite their national character, had recruited many Thracians. The recruitment from distant parts of the empire contributed to the hostilities which from time to time broke out among the legions and auxiliary units of different ethnic origins which had been drafted primarily from among local inhabitants. In the time of Severus Alexander, when the Pannonian legions showed more and more animosity to the ruling dynasty, the Syrians of Intercisa continued faithfully to support the Syrian emperor.

Young men began their military service between the ages of seventeen and twenty. During the time of Augustus, these men served sixteen years, plus an additional four; later it was raised to twenty plus five years. Strict military discipline was observed to maintain the spirit of comradeship; severe punishment was inflicted on those who infringed the regulations and discipline. All this and regular training imposed on the army confidence and courage which helped it achieve victory even when facing the attacks of superior Barbarian forces. This discipline also imposed a ban on the marriage of soldiers, with the exception of the *centurions*. Women who lived in the camp, as well as their children, were not recognized by the military government, nor were they recognized at the time of transfers or promotions. In this respect it did not matter whether the soldier was a Roman citizen or a *peregrinus*. The only possible solution was *concubinatus* (or *contubernium* if the woman was a slave). This strict rule came to an end at the time of the Severan Emperors. From the time of Septimius Severus, soldiers with Roman citizen rights, legionaries and auxiliaries both, could be legally married while in military service.

As a further step, the soldiers could rent out land on the territories of the legions. Under Severus Alexander the practice developed further and the soldiers' land could be inherited by their sons, provided they also entered the service. The legionaries who not so long before had observed strict discipline in the camp lived now with their wives and children outside the camp and only assembled occasionally for military service. When the soldier finished his term of service — *missus honesta missione* — he received a grant of land or money. In the time of Augustus this amounted to 12,000 *sesterces*; later it rose in proportion with the pay. Auxiliary soldiers, together with their families, received Roman citizenship. A discharged soldier with his allotment of money and land found his place in the top strata of the local society. Foreigners discharged from the army seldom returned to their native countries; the love of the army and of comrades, the atmosphere of the *canabae* during a long service made them strangers at home, whereas here they were citizens of the province.

The army stationed along the border fulfilled an excellent defensive function: to fend off the inroads of the Barbarians and expel the enemy entering the province. Neighboring peoples living in a client relationship had a considerably less developed economy, with less favorable economic conditions, than Roman citizens on the other bank of the Danube. However, the troubles of the empire and temporary weakening of the defense structure often made them forget their vows. Crossing the Danube, they plundered the province, drove away men and animals, and returned with their spoils to their land beyond the Danube. If the empire was not able to expel the enemy by swiftly summoned *vexillationes* and force it to return the spoils, peace was threatened. During these enemy raids uncertainty prevailed, so that coins were concealed, often buried in hoards. A great many of these coins were excavated recently. These numismatical findings, along with military diplomas, inscriptions commemorating victories, altars erected after major victories, and the burnt layers of the settlements are all excellent documentations for the fact that even during the peaceful periods of the second century, raids on both sides belonged to the daily life of the inhabitants of the *limes* provinces. These often-recurring skirmishes forced the Pannonian army to be on the alert constantly.

The empire also made use of the troops stationed along the *limes* in their wars on other fronts because these troops were combat-ready. The Pannonian army took part in most of the greater wars of the imperial period, even forming the nucleus of the western army from Septimius Severus to the eastern wars of Gordianus III. Recourse to the troops and the assembling of the expeditionary army occurred in several ways. The Pannonian forces played a major role in the Dacian wars, supplying full legions, *vexillationes*, and auxiliary troops. In the event of distant wars, however, the composition of the expeditionary army was arranged so as not to endanger the security of the *limes;* the strike force was left at home. From among the four legions, not more than one participated fully in the majority of cases. In the eastern wars of Trajan, Hadrian, L. Verus, Caracalla, and Severus Alexander it was Legio II Adiutrix that took part. Provision was made to supply a replacement legion in an emergency. In this way, Legio IV Flavia was sent several times to Aquincum from Moesia Superior, either in part or in toto. A legion sent on an expedition was augmented by the *vexillatio* of another legion in order to reach the required number. Up to the time of the Marcomannian wars, even auxiliary troops were added to the Pannonian army. Thus Ala I Flavia Britannica milliaria and a Batavian *cohort* took part in Trajan's eastern wars,[34] and a *cohort* made up of *vexillationes* also participated in Hadrian's wars in the Iudaea (*milites vexillationis cohortium Pannoniae Superioris*),[35] in the Mauretanian war of Antoninus Pius (Ala I Ulpia Contariorum, I Cannanefatium, I Hispanorum Aravacorum, I Augusta Ityraeorum, I Thracum veterana sag., III Augusta Thracum, and also Cohors XVIII Voluntariorum).[36] At the accession of Septimius Severus a strong army formed from four Pannonian legions made up a large part of the army sent to Italia, but the army sent against Pescennius Niger and later against Clodius Albinus was also dependent on the Pannonian forces. Ti. Claudius Candidus commanded the operating army as *dux*; he was also commander-in-chief of Severus's army.[37]

It was at the time of Septimius Severus's accession that it became apparent that among the three great military forces of the empire — the western army in Britannia and Germania, the legions of the Danube provinces, and the eastern legions — that of Illyricum was the strongest.[38] The strength and prestige of the Pannonian army reached its apex in the third century, especially in the serious military situation in the middle of the third century. In the person of Maximinus Thrax, the Pannonian army for the second time provided an emperor; later on, however, more Pannonian soldiers ascended the throne.

Side by side with strong military functions, the army carried out systematic building schemes along the *limes*, both inside the province and in Barbaricum. The camps were constantly kept on the alert and repairs, renovations, and rebuilding schemes were continuously under way. The earth camps of earlier times did not last any time. The first camp of Vetus Salina was probably built under Domitian, and, not much later, the second camp was built south of the first. The third camp was ready for use sometime between 118 and 120. A fourth earth camp was built before permanent stone camps were erected during Hadrian's reign, at which time the earth camps were rebuilt in stone. The walls, however, were not thicker than a palisade. After the Marcomannic wars, the renovation of the whole *limes* became necessary. The construction of the *limes* in the late second and third centuries demanded careful planning and execution in several stages. We have the greatest number of data on the construction immediately following the Marcomannic wars, of a replanning scheme from the same time, and of the *limes* building under Caracalla. After the Marcomannic wars, considerable regroupings occurred in the army of Pannonia Inferior which necessitated the rebuilding of the disarranged *limes*. The troops that had been stationed here for a long time were partly replaced by ones brought over mainly from the east or Africa (Cohors Hemesenorum, I Maurorum, I Numidarum). Syrian auxiliary units probably moved into Intercisa in 181 and this new formation began to rebuild the camp. The building inscriptions date from 182–183 and from 184. After the reconstruction of the camp — in Intercisa only demolished walls were restored and the interior of the camp renewed — watchtowers had been raised along the Danube between the individual camps. This took place in the Aquincum-Intercisa section between 184 and 185. It is very probable that the Sarmatian war had prevented the continuation of the building of the watchtowers south of Intercisa, nor was the *limes* road reconstructed. The *limes* construction under Caracalla seems to indicate that road repairs were planned to be executed in the later, perhaps even final, phase of the building. The road building under Caracalla is marked everywhere by Macrinus milestones; at two points Elagabalus ones indicate the conclusion of the building. Road repairs under Macrinus, Severus Alexander and Maximinus Thrax prove that reconstruction work had been started according to the military districts, and within each district the repairs were performed serially. Outside the building, along the *limes*, and even inside the province, several buildings fulfilled military functions, even after the interior camps were liquidated. In order to protect the interior roads, guard posts were organized at various points, as in St. Margarethen, in the Mecsek hills, in the Balaton area, and outside Aquincum.[39]

138

In order to supply the military construction schemes, great brick-baking kilns were established. Certain military units such as the Cohors I Aelia Caesariensis supplied the whole stretch of the *limes* with bricks produced in large quantities. When construction of the military buildings was stopped temporarily the brick-kilns supplied the civilian jobs with building material and contributed also to state, though nonmilitary, constructions.

Larger building schemes were generally produced with the help of the *vexillationes*, which worked under the command of a *centurion*. Tacitus speaks of working *vexillationes* in connection with the events of the year A.D. 14: "interea manipuli ante coeptam seditionem Nauportum missi ob itinera et pontes et alios usus, postquam turbatum in castris accepere, vexilla convellunt... retinentis centuriones... insectantur, praecipua in Aufidianum Rufum praefectum castrorum ira." Inscriptions and brick-stamps also confirm the existence of *vexillationes* at work. In urgent cases the local inhabitants were also recruited: on an early gravestone, we can see a peasant sitting on an ox-cart, watched over by a soldier.[40]

NOTES

[1] J. FITZ, "Römische Lager in Gorsium". *Actes du IX$^e$ Congrès International d'Études sur les frontières romaines* (Bucureşti, Köln, Wien, 1974), pp. 190–91.

[2] W. REIDINGER, *Die Statthalter des ungeteilten Pannonien und Oberpannoniens von Augustus bis Diokletian* (Bonn, 1956) (Antiquitas I/2).

[3] J. FITZ, "Über die Laufbahn der pannonischen Legaten". *Helikon* 3 (1963): 373–87.

[4] J. FITZ, "Die Vereinigung der Donauprovinzen in der Mitte des 3. Jahrhunderts", in *Studien zu den Militärgrenzen Roms* (Köln, Graz, 1967), pp. 113–21; J. FITZ, *Ingenuus et Régalien* (Bruxelles, 1966) (*Coll. Latomus* 81).

[5] J. FITZ, "Herculia im 4. Jahrhundert." *ActaAntHung* 24 (1976): 383–89.

[6] J. FITZ, *Les Syriens à Intercisa* (Bruxelles, 1972) (*Coll. Latomus* 122).

[7] *CIL*, III, 6075 = DESSAU, *ILS*, 1366.

[8] J. SZILÁGYI, "Kutatások Aquincumból—Les fouilles d'Aquincum". *ArchÉrt* 76 (1949): 67–71.

[9] H. G. PFLAUM, *Les procurateurs équestres sous le Haut-Empire romain* (Paris, 1950).

[10] *AEpigr*, 1908, 206 = DESSAU, *ILS*, 9014.

[11] J. FITZ, "Verwaltung der pannonischen Bergwerke". *Klio* 54 (1972): 213–25.

[12] *CIL*, XIV, 176 = DESSAU, *ILS*, 1484; *CIL*, VI, 796 = DESSAU, *ILS*, 391.

[13] Á. DOBÓ, "Publicum portorium Illyrici" (Budapest, 1940) (*DissPann* 2:16); F. FÜLEP, "New Remarks on the Questions of the Jewish Synagogue at Intercisa". *ActaArchHung* 18 (1966): 93–98.

[14] E. RITTERLING, "Legio", *R. E. Pauly-Wissowa*, 12:1211–1829; G. ALFÖLDY, "Die Truppenverteilung der Donaulegionen am Ende des I. Jahrhunderts". *ActaArchHung* 11 (1959): 113–41.

[15] A. NEUMANN, "Forschungen in Vindobona". *RLiÖ* 23–24 (1967–1968).

[16] E. SWOBODA, *Carnuntum: Seine Geschichte und seine Denkmäler*, 4th ed. (Graz, Köln, 1964).

[17] L. BARKÓCZI, "Brigetio" (Budapest, 1951) (*DissPann* 2:22).

[18] T. NAGY, "Római kor" (Roman Era) in *Budapest története az őskortól az Árpád-kor végéig* (Budapest, 1973).

[19] J. FITZ, "Pannonien und die Klientel-Staaten an der Donau". *Alba Regia* 4–5 (1963–1964): 73–85.

[20] J. SZILÁGYI, "Les variations des centres de prépondérance militaire dans les provinces frontières de l'empire romain". *ActaArchHung* 2 (1954): 117–223.

[21] J. Fitz, "Osservazioni prosopografiche alla carriera di M. Macrinius Avitus Catonius Vindex". *Epigraphica* 28 (1966): 50–94.

[22] H. G. Pflaum, *Les carrières procuratoriennes équestres sous le Haut-Empire romain* (Paris, 1960), pp. 476–94, Nr. 181 bis.

[23] Fitz, "Les Syriens".

[24] J. Fitz, *Die Laufbahn der Statthalter in der römischen Provinz Moesia Inferior* (Weimar, 1966).

[25] *CIL*, II, 4509 = 6145 = Dessau, *ILS*, 1029.

[26] *CIL*, III, 3473; *BudRég* 16 (1955): 407.

[27] *CIL*, III, 3463 = Dessau, *ILS*, 3638.

[28] J. Fitz, "Epigraphica II". *Alba Regia* 8–9 (1967–1968): 292–93.

[29] J. Fitz, "Die Domus Heraclitiana in Intercisa". *Klio* 50 (1968): 159–69.

[30] J. Fitz, "La carrière de P. Aelius Proculinus". *Latomus* 34 (1965): 565–75.

[31] A. Domaszewski, *Die Rangordnung des römischen Heeres*, 2nd ed. (Köln, Graz, 1967).

[32] Fitz, "Osservazioni".

[33] Pflaum, *Les carrières*.

[34] *CIL*, III, 3676.

[35] *AEpigr*, 1909, 235 = *AEpigr*, 1938, 13.

[36] *CIL*, XVI, 99; *CIL*, III, 4379; *AEpigr*, 1905, 240; Ritterling, pp. 1397, 1449, 1685, 1741.

[37] J. Fitz, "Réflexions sur la carrière de Tib. Claudius Candidus". *Latomus* 25 (1966): 831–46.

[38] J. Fitz, "Die Personalpolitik des Septimius Severus im Bürgerkrieg von 193–197". *Alba Regia* 10 (1969): 69–86.

[39] J. Fitz, "A Military History of Pannonia from the Marcomann Wars to the Death of Alexander Severus, 180–235". *ActaArchHung* 14 (1962): 25–112.

[40] R. Saxer, "Untersuchungen zu den Vexillationen des römischen Kaiserheeres von Augustus bis Diokletian". *Epigr. Stud.* 1 (1967): 87.

# POPULATION

JENŐ FITZ

## THE INDIGENOUS POPULATION

The indigenous population of Pannonia, found here at the time of the Roman conquest, was rather mixed because of the migratory movements of earlier centuries. Different Celtic waves did not reach the country everywhere with the same intensity, and the upper strata of the La Tène culture also amalgamated in certain places with the huge masses of the Illyrian–Pannonian population. When the Romans appeared in the province, more than fifty years after the last Celtic wave, we can account for an Illyrian–Pannonian population and four larger Celtic tribes within the territory of the province. The four tribes were the Boians, who occupied the northwestern part; south of them the Taurisci in the southwestern corner of Pannonia; the Eravisci in the northeast; and finally the Pannonianized Scordisci in the vicinity of Sirmium. Roman historical sources speak in addition of various native tribes. However, these names reflect the situation in Roman times, when the earlier power groups had been dissolved for security reasons, individual tribes had formed independent *civitates*, and the large tribal territories were broken up by artificially created frontiers. The tribal communities referred to in the sources were the following: the Catari near Emona; the Latobici at Neviodunum; the Varciani near Andautonia; the Colapiani near Siscia; the Oseriates and Seretes at Poetovio; the Serapilli in the district of Iovia; the Iasi near Aquae Balizae; the Boii, Arabiates and Azali at Brigetio; the Eravisci, Hercuniates, and Andizetes near Mursa; the Cornacates at Cornacum; the Amantini in the vicinity of Sirmium; the Scordisci and Breuci near Cibalae. Among those mentioned, besides the Boii, Eravisci, and Scordisci, we find some of the peoples referred to in earlier sources, such as the Andizetes, Breuci, Amantini, and Latobici. The place-names of other *civitates* yielded the names of the Varciani, Colapiani, Cornacetes and Azali tribes. We have as yet few data to trace ethnic differences everywhere. It is hard to decide whether the various *civitates* which emerged after the Roman Conquest were artificially created territorial units or had adjusted to earlier tribal or intertribal frontiers. Among the great and united peoples, however, there was no *civitas* to unite them into one: these peoples had been divided into districts.[1]

We have comparatively few data on the living conditions of the native population in the period before and immediately after the conquest. Among the few known or conjectured *oppida*, only the predecessors of Siscia and Sirmium had a municipal character. The majority of the inhabitants lived a peasant life in smaller settlements and villages. On Gellért hill, in the outskirts of Aquincum, excavators uncovered pits dug in the earth that served for living quarters in an Eraviscan settlement. Scanty finds furnish only an approximate picture of living conditions, clothing, religion, and property relations. There seems no doubt, however, that we cannot speak of aborigines, in general terms, in all parts of the province. Between the southern and northern tribes there had been distinct differences in culture too. Moreover, there were even different traits in behavior. In the north and in the west no traces remain of the pugnaciousness of the southern Pannonians, of their fierce resistance to conquest.

Most of our data concerning the native population refer to the middle or even late first century, when we cannot leave Roman influences out of consideration. Nevertheless, if we consider that the Roman occupation dates from the same period, there is no need to believe that in the earliest relics, in the coach scenes, in the appearance of women's clothing, there necessarily emerged either a revival of old traditions or new customs and ideas instigated by new influences. In the gravestones with coach scenes, it was not the decoration which was new but the habit of erecting gravestones. The characteristic clothing of Eraviscan women − their raised kerchiefs covered by veils, their dresses held together by *fibulae*, their torques and bracelets − although it naturally underwent changes in the course of time, nevertheless preserved and continued old traditions. Clothing fashions and gravestones with coach scenes, however, do not point to frontiers demarcated by the Romans but to real folk and ethnic boundaries. The same is true of the Lajta region or of southwest Pannonia, where the gravestones and the cemeteries of the native population preserved similar, though somewhat diverging, national costumes.[2]

For security reasons, the *civitas* was in the beginning under military supervision. The *praefectus civitatis* was generally the commander of the auxiliary troops stationed in the territory of the *civitas*, or if none were stationed nearby, a *centurio* of the legion to which it belonged. Such a *praefectus* was among the few whose name is known − L. Volcavius Primus, the commander of Cohors I Noricorum and *praefectus ripae Danuvii et civitatium Boiorum et Azaliorum* in early Flavian times. The *praefectus* thus even controlled two *civitates*, and besides, the involved stretch of the Danube was under his auspices. The state, however, did not interfere in the social and economic life of the native population beyond splitting up the larger political units and exercising organized control. Moreover, an ancient social and political organization had also survived. The gradual expansion of the empire inevitably infringed upon the rights of the local population, primarily in the economic field.[3]

## THE EMERGENCE OF TOWNS

The conquest of the province did not involve merely the presence of the occupying forces. Accompanying the army, and also independently, new inhabitants appeared and settled down, and within a few decades, they considerably changed the composition of the province's population. Next to the legionary and auxiliary camps, tradesmen, merchants, and the dependents of the soldiers settled down and thus produced spontaneously emerging settlements. In the beginning they had a transitional character, with barracks, living pits, and mud huts, all on a primitive

142

level of civilized life. As far as the future of Pannonia was concerned, the emergence of immigrant towns was of greater importance, since they became the bearers of ancient forms of life. The first towns emerged at the initiation of the state, following planned settlements; the first being Emona in A.D. 15. On the site of the legion camps, abandoned in the previous year, the veterans of Legio XV Apollinaris and landless Italici had been settled. The settlement was entirely apart from the local population. The next town, Savaria, was founded by Claudius; it was also intended to settle the veterans of Legio XV Apollinaris. Under the Julio–Claudian dynasty auxiliary soldiers were settled near Scarbantia and in the Lake Balaton area.[4]

After these humble beginnings urbanization made more headway under the Flavii. It was then that two new colonies were established, Siscia and Sirmium. It seems probable that both had been founded for demobilized soldiers of the fleet during A.D. 71. But the local population also had a part in these foundations, although the cities infringed upon the interests of the indigenous population. The earlier *civitas* became the *territorium* of the new city, while the local Celtic–Pannonian population was excluded from the ownership of land.[5]

Nevertheless, during the second half of the century, the *civitates* also went through a considerable development. The lack of confidence following the great Pannonian–Dalmatian uprising disappeared completely during fifty years of peace. The post of the military *praefectus* was handed over in certain *civitates* under the Flavii to the native *praefectus*. Such a *civitas* became from then on an autonomous territory, with a tribal council whose members, the *principes*, elected the *praefectus* from their ranks. All this went hand in glove with the granting of wider civil rights. During the reign of the Julio–Claudian dynasty, all this had occurred sporadically, but the gradual Romanization of the native aristocracy provided an opportunity for a swift rise in the numbers of a new civilian population. In the southern portion of the province where, owing to the establishment of cities, the long stay of the legions, and the proximity of Italia, the Roman life-style had been assimilated earlier than in the other parts of the province, two native *civitates* from the Varciani and Latobici organization changed into a *municipia*—the first under the name of Andautonia; the second, of Municipium Latobicorum-Neviodunum. In these the nucleus of the city was not produced by a new settlement, but rather civic rights had been granted to the community of the local population. These two native settlements became so much Romanized during the second half of the first century that they were able to assume city rank in the Roman sense too. Under the Flavii a third settlement too assumed the rank of a *municipium*; this was Scarbantia. However, in this case the process rather reminds us of the founding of a *colonia*. The establishment of Scarbantia had occurred under Tiberius, mainly through the settlement of auxiliary veterans. Here the raising to the rank of *municipium*, unlike that of the two other *municipia*, was the recognition of evolution. However, the settlement, owing to its veteran character, achieved under Tiberius some sort of autonomy under the name of Oppidum Scarbantia Iulia. Here we can find a *conventus civium Romanorum*, which also had its *decuriones*. The *territorium* of the *municipium* had been cut out from the land of the great Boian *civitas* in the same way as that of Savaria.[6]

143

The last city establishment connected with veteran settlements coincided with the rule of Trajan—this being Colonia Ulpia Traiana Poetovio. It was established by the veterans of Legio I and II Adiutrix, and Flavia IV, with double *missio agria* and single *missio nummaria* on the site of the abandoned legionary camp. The establishment of the city of Poetovio with a *deductio* resulted also in a grant of civil rights to the Serretes, who were local but not yet Romanized people.[7]

With the settlement of veterans, merchants, and tradesmen from North Italia, a considerable number of Italian inhabitants settled down in Pannonia. They constituted the majority of the city, population in Emona, Siscia, Savaria, Scarbantia, and perhaps also in Sirmium; they occupied a leading role in Carnuntum's *canabae* too. Moreover, their existence can be proved in Aquincum's *canabae*, which emerged under Domitian. In this new population, which had a higher level of culture, we can see the founders of the Roman life-style in Pannonia. In the early period, which was shorter in the west and south than in the east, a considerable social difference became evident between the newly settled population, which was granted land, and the natives, who were just as much deprived by the expropriation of their land as the native population within the city *territorium* had been in its legal status. The Italic masses could not obtain Roman civic rights, and their *peregrinus* status only emphasized the social differences. Only a narrow section of the native aristocracy could keep pace with the new citizens, and their self-conscious efforts obviously hastened their Romanization.[8]

Cities emerging owing to *deductio* bore the marks of a well-planned policy. A square-shaped network of streets has been found by archaeologists in Emona and Savaria, and evidently the same must have existed in other cities too. In Savaria, which was the center of the imperial cult, in one part of the city where the forum, the *Capitolium*, and the representative buildings and palaces were standing, we can find an arrangement differing from the *insula* system of streets.[9]

The building of the *limes* and the end of *deductio* city foundations denoted the close of a great period in the history of the Pannonian population. In subsequent decades only a flourishing economic life attracted new settlers in comparatively large numbers. As far as the composition of the population was concerned, early Barbarian settlements did not have any great significance. One group of the Cotini settled during the reign of Marcus Aurelius somewhere in the territory of Pannonia Inferior; this was followed by other settlements during the first century—the adherents of Vannius coming around A.D. 50 and another group during the governorship of Tampius Flavianus. New groups of merchants arrived from the Rhine region from the second century onward. Along with soldiers recruited into the garrisons of the *limes* camps and their dependents, many newcomers from other provinces had settled down during the second century, primarily in smaller and larger settlements emerging near the camps, in the so-called *canabae*. Moreover, the development of the town and the increasing Romanization had brought essential changes.

From the middle of the first century, soldiers had been recruited from the territories of the colonies into the legions stationed in the province and also from the Italic settlers. The Romanized inhabitants of the *municipia* appeared for the first

144

time in the legions during the early second century. Thus the composition of the legions did not change much in the beginning: the majority of the soldiers were of Italian origin, although they had come from the Pannonian cities. In Pannonia Inferior, where the process of Romanization had proceeded more slowly—its only colony founded by *deductio* being Sirmium—the Romanized population was not large enough to supply the Aquincum legion. Therefore, soldiers were recruited into this legion from other provinces as well. Recruitment from the native population, which from the time of the Pannonian–Dalmatian uprising had claimed a considerable portion of men capable of fighting for the various segments of the Roman army—particularly in other provinces—became in the course of time the strongest instrument of the process of Romanization. It is noteworthy that even in this respect the southern part of the province preceded by one phase other parts of the province. Under the Flavii, recruitment had been going on only from the South Pannonian tribes and *civitates* Varciani, Iasi, Sisciani, Colapiani, Breuci, Cornacates; that is, acceptance into the army was connected with a certain type of Romanization. The recruitment of Boian youths began somewhat later, but it was only extended to the whole province under Trajan, when the male population of South Pannonia had been accepted into the legions. Twenty-five years of military service in the framework of strong Roman discipline gave the soldiers, when they left, not only citizenship, but also a changed loyalty. It made them citizens of the empire, divesting them of the former tribal adherence. The auxiliary veterans, returning to their native land, became propagators of Romanization, and their relative wealth obtained them places in the new leading strata of the local society.

The quickening and expansion of the Romanization process can be traced in many other fields as well from the early second century onward. Whereas in the first century only fully Romanized natives could enter into the legions, under Trajan natives who were admitted received their rights of citizenship as they entered the legion. This reflects a loosening of the former rigid rule, but it is certain that the soldiers recruited in this way had met or approached the admission requirements of the legions. Under Trajan even the granting of citizens' rights became more generous. Until the end of the first century, in keeping with the measure of Romanization, natives received citizens' rights in larger numbers only in the west and south. In the early second century a great number of new citizens could adopt the name of Ulpius in all parts of the province. Whether this was the result of a self-conscious policy, in opposition to the reserved attitude of the first century, or the recording of the actual situation, as it had been before, is difficult to decide today.[10]

The process of urbanization had ended in the first period of Hadrian's rule. By that time in a new group of the native *civitates* Romanization had reached the stage which made it suitable to adopt a higher form of administration. Settlements in the area of many *civitates* achieved an urban character during a hundred years of development, and owing to returned veterans, settled tradesmen, merchants, and some other inhabitants, they became altogether Romanized. Next to the legionary camps established under Trajan, *canabae*, civilian settlements developed one or two kilometers away. Governors promoted the development of new camptowns with the resettlement of native settlers and the inhabitants of the *oppida* nearby.

It was in this way that the inhabitants of the Eraviscan *oppidum* of Gellért hill in Buda moved into the *canabae* of Aquincum. Since some of the inhabitants of the *canabae* were foreigners, urbanization also proceeded more quickly. During his stay in Pannonia in A.D. 124, Hadrian organized the internal conditions of the province, and then every *civitas* where it was possible was raised to the rank of *municipium*. Thus in the territory inhabited by the Boii emerged Carnuntum's civilian city; Mursella was somewhat to the southeast; perhaps in Arviates territory was Mogentianae; the civilian city of Aquincum was taken from the territory inhabited by the Eravisci; from the *civitas Andizetes* came Mursa and perhaps also Mursella; from the *civitas Iasorum* came Municipium Iasorum; from the center of the Breuci came Cibalae; in the territory of the Scordisci, Bassiana emerged. The establishment of towns occurred partly in the west, partly in the east and southeast, but mostly near the Danube—which shows the large role played by the army. Among the civilian cities which emerged near the legionary camps only Carnuntum and Aquincum obtained the rank of *municipium*. In the case of Carnuntum the obvious recognition of a lengthy development was expressed in this rank; in the case of Aquincum nothing more happened than allotting it the seat of the new province. Later, the towns mentioned above all became *municipia* with the exception of Mursa, which received the title of *colonia* without being a veteran settlement.[11]

It is possible that the list of towns from the Hadrianic period should be supplemented by two more sites where archaeological excavations corroborate city development; however, the municipal rank has not been conclusively proven without the presence of inscriptions. One of these towns was Sopianae, which rose alongside the main east–west route and in the course of time became the seat of the province of Valeria; it had large-scale buildings as early as the second century. The second is Gorsium, where we know of a *decurio*, although the term *municipium* never occurs; it was a municipal settlement on the site of an early military camp, with a forum and an enormous *area sacra* and other representative buildings from the time of Trajan. Here the majority of the new citizens were given the name Ulpius—other citizens belonged to the Papiria tribe—all this pointing to the fact that the conjectured higher rank may have been allotted at an earlier time.[12]

The *municipia* and as far as the layout of the city is concerned, Mursa was one of them, developed from earlier native settlements. Thus these cities did not have the strict square-shaped system of streets of the colonies founded by *deductio*. The *municipia* which emerged in the vicinity of military camps also had an irregular network of streets. As far as the layout of other cities is concerned, we have only an approximate idea of Gorsium. The roads which meet from various directions in the *cardo* form a rectangle in the *decumanus*, but in the external part of the city, which lay outside the onetime military camp, the streets follow no regular arrangement.[13]

Side by side with the towns which achieved municipal status, we know of a whole group of settlements with a similar character, one part of which, as with Vindobona and Brigetio, rose at a later stage to the rank of towns. Among these settlements were the *canabae* of the military camps which developed spontaneously alongside the camps and during the second century obtained the semblance of towns, with stone houses, streets, gutters, temples, and public buildings. Their homes, as far as luxury was concerned, did not lag far behind those of the *municipia*. The smaller *vici* which developed next to the auxiliary camps did not differ from the *canabae* of the legionary camps; to a lesser degree they were affected by provincialism. In these, until the Marcomannic wars, the houses were built of mud; but the majority of the population, following old local customs, still lived in pits. These settlements differed in particular from the *municipia* in lack of autonomy. The inhabitants, not having civic rights, necessarily counted as *peregrini*. However, development tended even here toward simpler forms of autonomy: they had a *quasi municipale* organization, an imitation of the *conventus civium Romanorum*. The inhabitants were *consistentes*, in Brigetio governed in the beginning by a *curator civium Romanorum*, later by *magistri*. The inscriptions in Aquincum also give evidence of *magistri* and *decuriones*. We can conjecture that similar organizations existed in the *canabae* of the auxiliary camps: *cives Romani consistentes* had been known in Vetus Salina too.[14]

## MUNICIPAL LIFE

Municipal life in the autonomous cities, although conditions were similar, appeared to be different. In the smaller cities the *magistratus* was often represented by the *scribae*. It was so in the unidentified Municipium Faustinianum, where the *scriba* was the *augustalis* of Siscia. It seems that in these places — Mursella is another example — the members of the *ordo* had insufficient erudition to deal with the affairs of the city, and so management was in the hands of a permanent *scriba*. In larger places the *magistratus* consisted, in keeping with the general practice of the empire, of *duumviri* elected by the body of *decuriones*. In certain places, as in Carnuntum and Emona, which differed from the general practice, the municipal organization was headed by IIII *viri jure dicundo*. As in other spheres of life, in the development of the cities, in the role of the forum and *capitolium* and the function of the *magistratus* Rome had been the paragon to be followed. Outside the two leading officials, we find the names of *quaestores* and *aediles* on the inscriptions, and the local municipal career or *cursus honorum* was just as binding as a Roman one. External forms corresponded to the customary: from Poetovio we know of the presentation of a *sella curulis*, the election of magistrates, and the leaders of certain bodies are referred to as *candidati*. In order to obtain a magistracy, a certain donation had to be offered to the city — such are mentioned on several inscriptions. The donations were moderate, restricted to the restoration or erection of a temple, or the building of *porticus* of a *nympheum;* and the inscription proclaimed the generosity of the donor.[15]

The composition of the municipal *ordo* was different in each city. In the colonies founded by *deductio* of early establishment, the local population's leading strata had hardly any function as compared with that of the Italian settlers. In western cities of later founding, as in Poetovio, Carnuntum, and Mogetianae, many examples point to this difference in status. In Pannonia Inferior, on the other hand, the situation was quite different. The *ordo* of Aquincum had been composed by Hadrian's order of the members of native Eraviscan aristocracy, most of whom were named Aelius. However, we have scanty data concerning the municipal aristocracy of several East Pannonian cities. The wealth of the *decuriones* was determined by the land they owned; however, many of them also took part in other ways in the economic life, for example a *decurio* from Mogentianae was the proprietor of a brick kiln.[16]

Another order of the municipal society was formed by the *seviri* and *augustales*. On the social scale, the *seviri* occupied a high rank, but their position in economic life could not compare with that of the *augustales*, who were mostly freedmen. In certain towns, as in Brigetio, there is no evidence of *seviri*; in other towns the two bodies had temporarily formed one. This union occurred in Emona in the second half of the first century, and in Scarbantia later on, the same thing occurred. Such transitional changes always expressed alterations in the economic strength of the second order. In Scarbantia the contraction can be explained by the poverty of the city. The number and function of the *augustales* were related to the wealth of each city. This was the most active stratum of the society in the economic field and at the same time it was the most given to sacrifices. Members of this group were responsible for organizing the economic life of the province; their far-reaching activities always coincided with a flourishing period. In early times the greatest number of *augustales* are known to have existed in Emona; some of them were freedmen from Aquileia. At the turn of the second and third centuries, when the cities along the *limes* began to flourish, we hear about many *augustales* in Brigetio and Aquincum. Their names have been preserved in numerous donations.[17]

The third order was composed of the *collegia*. We know them to have existed in many larger cities, but naturally not in all. There were *collegia* of different kinds: *collegium centonariorum, fabrum, fabrum et centonariorum, [con]veteranorum, dendrophorum, magnum Larum et imaginum dominorum n. Cae., Cereris, cultorum, naviculariorum, negotiantium, opificum, Victorianorum, montanorum, genii provinciae Pannoniae Superioris, scaenicorum, augustalium, iuventutis,* etc. Among other functions they performed was that of undertaking. It was particularly the Aquincum *collegia* which erected many gravestones during the second century. We know of many *collegium* members from there. From other cities the names of college *praefecti* and *patroni* have survived. The majority of the members were evidently not very rich, and the *collegium* provided only for the funeral — the gravestone usually had to be paid for by a member of his dependents. The members of Aquincum *collegia* were generally immigrants; natives did not take part in corporate life at all. The prefects and patrons of the *collegia* generally came from the *ordo decurionum*. Besides these we know of *magistri, quaestores* and *vexillarii* who were elected from the ranks of the members. *Collegia* with many members were divided into *decuriae*, with a *decurio* at the head of each. A *collegium* had its own house and meeting place. In Aquincum, a *praefectus* of the *collegium centonariorum* donated an organ to the house. In inscriptions we read of common feasts and excursions in which *vexillarii* and *tribuni* participated.[18]

Slaves and freedmen did not necessarily belong to the lowest strata of society. However, inscriptions name more *liberti* than slaves. In cities — in contrast with the usual landed estate—the number of slaves, as compared with the situation in other provinces, was not very high. Industry preferred to employ *liberti* and free wage earners. Most slaves were kept in the cities by *liberti*, with the purpose of teaching them a trade.[19]

## THE *TERRITORIUM* OF THE CITIES

The confines of the *territorium* outside the city can be defined only in exceptional cases by the stone monuments of city magistrates, decuriones, and by means of milestones and peculiarities of local populations. Between Emona and Neviodunum, Savaria and Scarbantia, Scarbantia and Carnuntum, Carnuntum and Vindobona we do know the boundaries. The *territorium* of Aquincum was closed by the frontier of the province, in the north and east; however, municipal monuments southward do not proceed beyond the Aquincum–Poetovio road. As far as we know, a *territorium* did not exceed a fifty-kilometer diameter. On this exceedingly large territory were situated settlements of different size and development. According to the customary division the *territorium* was subdivided into *pagi* and these into *vici*, such as pede Sirmense pago Martio vico Budalia.[20] Vindonianus, near Aquincum, was a municipal *vicus*; Vicus Fortunae was a district in Poetovio; Vicus Voleucionis was near Savaria; and Budalia was situated in the territory of Sirmium. We know next to nothing about the forms of settlement and extension of the *vicus*. Boii settlements—based on surface traces—are considered by scholars to have been small villages. Similar ones are also known in Eraviscan territories, in the area of Lake Velence. At Pogánytelek groups of small houses can be observed. The *Itineraria* speak of *mansiones*, which were postal stations in larger territories. Finds from various stations show different degrees of Romanization. Floriana (Csákvár), on the junction of the Savaria–Aquincum and Sopianae–Brigetio roads, was probably a large Eraviscan village. There Roman import commodities hardly occurred among the material objects, and the stone monuments preserve Celtic names and show women wearing native garments. On the other hand, Iasulones (Baracska), off the Poetovio–Aquincum road, presents the picture of a Romanized settlement without a native population. There are differences between individual localities in the Balaton area too: cremation cemeteries of different character suggest settlements with different populations.[21]

Another characteristic form of provincial settlement was provided by the *villae*, including smaller and larger farms and the centers of provincial–municipal middle-sized and large estates. Various villas have been fully excavated. They are of different sizes, sometimes very large, furnished according to financial ability with more or less luxury, with various outhouses attached to them, even a *horreum* at the very largest. Early villas date from the first century and reflect in general the growing claims of the local aristocracy. Among these we find villas of veterans — as extant remains of buildings at Pogánytelek, Eisenstadt, and Donnerskirchen — as well as the Parndorf and Winden am See villas which can be attributed to the centers of latifundiae. In the east the first buildings emerged somewhat later. The extensive Baláca villa surrounded by a large estate is the survival of early estate concentration in the Balaton area. The villas north of Aquincum may belong to the settlements of Trajan's period.[22]

The building of the *limes* and the emergence of the towns contributed to the prosperity of economic life. During the second century members of the municipal aristocracy had extended their landed properties and by clever management had accumulated relative wealth. This became manifest in a rise in their social status. The earliest Pannonian knights are known from the early second century in Neviodunum, Poetovio, Savaria, Carnuntum, and Mursa. Knights in the south and west of the province became the first great landowners. Among them, as in Carnuntum, were also the leading members of the native aristocracy. Until the middle of the century land concentration in the eastern section of the province did not reach that stage; however, the accumulation of wealth can be traced here also. While at the beginning of the century the ideas of the Eravisci concerning the life to come only appeared in coach scenes carved into the gravestones, from the middle of the century several coach burials point to the increased material welfare of the local aristocracy.[23]

## THE *CONCILIUM PROVINCIAE*

The municipal organizations of the province had been held in a loose unit by the provincial assembly. This *concilium provinciae*, as in the other provinces, could not have emerged in Pannonia much later than at the time of its organization at the latest when the *civitates* had developed. The yearly assembly primarily served the imperial cult and was directed in the presence of the delegates of the *municipia* by the high priest of the province. The *sacerdos provinciae* was elected from the worthy members of the municipal aristocracy. In Pannonia Superior, as we know, there was a relatively large number of high priests, practically all the known *sacerdotes provinciae* had been *duumviri*, with many knights among them. This office was regarded as the peak of the municipal career. While Pannonia was still undivided, the *ara Augusti* stood in Savaria, and Savaria remained the seat of the provincial council even after the division of Pannonia Superior.

The religious center of Pannonia Inferior was Gorsium. This fact seems to be confirmed by the altar stone raised by *totius provinciae sacerdote[s]*. This is also corroborated by the large number of cultic buildings which have been excavated in recent years from the *area sacra* of Gorsium. In a large western hall there was a dais, and a place for sacrifices appeared in front of it; this must have been the scene of the festivals of the imperial cult. The building itself must have been built under Trajan; that is, not much after the organization of the new province.

The provincial assembly safeguarded, on a lesser scale than with the imperial cult, the interests of the inhabitants. There is no evidence that in Pannonia, as in other provinces, the assembly should have protested in Rome against the excesses of its governors.[24]

# AFTER THE MARCOMANNIC WARS

The first great *caesura* in the history of Pannonia was drawn by the Marcomannic wars. The society of the province, which became welded together at the turn of the first and second centuries, had started on the road of healthy development in the successive decades. Romanization in economic and cultural life suffered an irreparable blow in the wars. Even though it may be an exaggeration to say that, with the signing of peace, the Sarmatians had to send home some 100,000 prisoners of war, it is indisputable that warfare had decimated the population of the province, and had destroyed towns, villa settlements, and villages; and the renewal inroads of the Barbarians had caused tremendous damage. The decades following the war, particularly the period of the Severi, repaired most of the damages and the province flourished in many respects more than before. Nonetheless, Pannonia was changed from what it had been before the war; it was led into another direction than it would have been, had the war spared the province.

The change can be measured best in the transformation in the composition of the population. The municipal population of Italic origin had partly disappeared after the Marcomannic wars which, from the point of view of Romanization, was a basic loss. New settlers arrived in their stead. Under the Severi surprisingly many Thracians appeared in Pannonia Inferior. Many of them had served as soldiers in various auxiliary troops and also in Legio II Adiutrix, where they existed even under Philippus. The number of Africans had increased also. Some of them had fought during the war in the *vexillatio* of Legio II Augusta, which had not been sent back to Lambaesis but was used to repair the losses of Legio II Adiutrix. The demobilized Africans remained in Aquincum after the war.[25]

Nevertheless, the most important event in the social history of the province was the appearance of the Orientals. One portion, as with the Thracians and Africans, arrived with the new military formations sent to the province. The first and most significant among them was Cohors I milliaria Hemesenorum, a thousand-strong auxiliary force recruited in the Syrian town of Emesa. During the reorganization of the *limes* this cohort had occupied the camp of Intercisa. Up to the middle of the third century the formation was reinforced from Emesa, so that the Syrian element formed the majority up to the end. Demobilized soldiers settled in Intercisa's *vicus*, where at the end of the century they constituted the majority of the population. An analysis of the inscriptions makes us suppose that during the whole duration of the Syrian diaspora, they kept aloof from the rest of the population and intermarried exclusively with Emesans, or Syrians, and that they kept no contact with other Orientals in the larger towns of the province. The peak of the diaspora coincided with the rule of the Severi, when emperors and empresses of Emesan origin had granted special favors to the auxiliary troops coming from their native land. During that period Intercisa not only differed from other Pannonian military camps, with its buildings and the wealth of its citizens, but also was a unique phenomenon in the long line of the empire's *limes*.

During the reign of the Severi another Syrian auxiliary unit had been sent to Ulcisia Castra, but since this occurred in the last years of the Severan period, there was little chance to produce such an independent management as in Intercisa.

The Orientals at Aquincum and Brigetio had come to Pannonia somewhat later than the Syrians of Emesa. Some of them had been recruited by Septimius Severus and Caracalla during the Eastern wars into Legio I and Adiutrix II; most of them were civilians. These were attracted by the economic boom of the Severan period. Among the Orientals the Syrians were in a majority, but many came from Asia Minor too. A considerable number of Jews also settled in the same period in certain cities. The Greeks who arrived at the same time were also Orientals. From the very beginnings the Orientals played an important role in the life of the towns, particularly in Aquincum and Brigetio. In the latter city, which obtained at that time the rank of a *municipium*, the majority of the *ordo* was made up of Orientals. Even in the western and southern cities the presence of the Orientals can be proved, although there they did not play a decisive role in city life.[26]

In the peregrine territories the population losses from the war were first of all made up by Barbarian settlements. Marcus Aurelius received into the empire large numbers of Barbarian masses, some of them receiving land and homes in Pannonia. Historians know only about the Celtic Cotini who, after being defeated by Tarrutenius Paternus, were allotted domiciles between Mursa and Cibalae. Under Commodus occurred the resettlement of the Osi, perhaps in the vicinity of Osones. But a considerable portion of Germanic names appearing in Pannonia may also be proof of repatriations during the reign of Marcus Aurelius.[27]

With the emergence of a new population, the earlier folk characteristics of the province mingled completely in one type of Romanization. Characteristic national clothing in Boiian territory, in South Pannonia, and among the Eravisci disappeared in the same way as the Celtic names and funeral rites. A far-reaching donation of civil rights served the same leveling tendency which accompanied the disappearance of local customs. However, new allotments of city rank did not really mean an advance in Romanization. Some of these changes in rank obviously were of a political character. Septimius Severus, who owed his power in particular to the cities of the Pannonian legions, raised Carnuntum and Aquincum to the rank of *colonia* which, nevertheless, only meant a change in title, as was the case with Cibalae and Bassiana. Siscia, which had been a *colonia*, became a *colonia Septimia*. In the Severan period two other civilian cities near legionary camps, Vindobona and Brigetio, also received municipal status.

These ranks, besides revealing a political tendency, also reflected a certain development: the Pannonian cities had reached in those decades their full flourishing, when in other parts of the empire the cities had declined, or at least had left behind their finest hours. This was true particularly concerning East Pannonia, which up to the wars could not catch up in culture and Romanism with the western and southern cities. Differences had been waning in this respect too, even though they did not fully disappear. At the turn of the century stone houses had greatly increased and had partly replaced earlier living pits and mud houses in the smaller settlements and the *canabae* of the auxiliary camps. Intercisa's *vicus* increased considerably because of the doubling of its garrisons, and richly ornamented buildings and temples were rising there. Although the Syrians here occupied a privileged position, we know of stone buildings in other military camps too.[28]

However, great building projects had been completed in this period in other Pannonian cities too. The first important group of these projects was completed about A.D. 202, when after several years in the East Septimius Severus returned with his family and his western troops to Rome and on the way visited the Danube *limes* and Pannonia. Along the emperor's route he was welcomed by scores of inscriptions which greeted him on his triumphal march through Pannonia. Intercisa, which on every occasion emphasized its relation with the imperial family, dedicated two temples: one to Deus Sol Elagabalus, the patron god of Emesa and the *cohors,* and the other to Diana Tifatina, who was venerated as the *paredros* of *Baal* of Emesa. At the same time, the *area sacra* of Gorsium, rebuilt at the emperor's expense, was consecrated, possibly in the presence of the emperor and his retinue. Inscriptions elsewhere also preserved the memory of imperial foundations, the most important being the *Kalendarium Septimianum* in Savaria.[29]

The prosperity during the Severian period is also expressed by a great number of inscriptions and stone monuments; altars and statues which celebrate the emperors date from this time. The greatest number of temples were built at that time by private individuals, and others were restored; *nymphea,* fountains, and ornamental buildings were erected. Municipal comfort also reached its highest provincial level. Interior decoration, wall paintings, stuccoes, and mosaics were widely used, even in the provinces. In larger settlements running water was installed everywhere, good drinking water being brought even from great distances, as to Brigetio from Tata and to Sirmium from the Fruška Gora.

At this high level of development the cities were surrounded with walls. The walls of Emona and Aquincum, built at an earlier time, and the fortifications of Scarbantia have been discovered by archaeological research; the walls of Savaria and Sirmium are referred to by several authors; the foundations of Cibalae have been recorded by Marsigli and by aerial photography. In general, the cities did not extend far. Aquincum was situated on an area of 50 hectares, Savaria on 42, Scarbantia (25 hectares), Emona (23 hectares), and Bassiana (19 hectares) were even smaller. In Gorsium where no walls have been found, the photographs show the city as much larger.

In the center of each city there was a *forum* and a *capitolium.* Among the *capitolia* large fragments of stone show that Savaria and Scarbantia had the largest; in the latter town even the site of the temple is known. An *area sacra* of enormous size has emerged at the excavations at Gorsium and from here derives the fragment of a statue of Minerva. In Savaria and Sirmium imperial palaces existed; in Carnuntum and Aquincum the governors had large residences. The towns were also adorned with many temples, public baths, palaestras, theaters, and amphitheaters. In Aquincum we know of a *macellum;* an inscription refers to Mursa's *tabernae* and the memory of basilicas and *horrea* also survives. Triumphal arches were raised after the Pannonian–Dalmatian uprising was put down, perhaps in Siscia and Sirmium. We know of later ones from inscriptions in Aquincum and Savaria. There were many statues, inscriptions, and reliefs everywhere. The appearance of the town, though in a simpler and more rustic form, followed that of Italian and western cities.[30]

At the same time that the towns were flourishing after the war, the municipal estates also were enjoying decades of prosperity. The numbers of villas increased — especially in the eastern part of the province, in the territory of Aquincum. These were larger in scale, and as far as luxury is concerned they far surpassed the earlier ones. The numbers of knights increased considerably in the east too, and the majority of estates were in the hands of senators. From such a senatorial family was born Emperor Dacius, near Sirmium. Earlier senators, such as an unknown one under Marcus Aurelius, and M. Valerius Maximianus, the famous general of the Marcomannic wars, had risen from the knighthood and were made members of *ordo amplissimus* in recognition of their merits. Similarly, though under less exceptional circumstances, Triccianus, the governor of Pannonia Inferior, became *consularis* in the time of Macrinus, and he governed his native land for a short time.[31]

Slave labor had been in general use in husbandry in the territories of the colonies but for the sake of greater efficiency it gradually gave way to wage labor. In Pannonia Inferior slavery reached its peak under the Severans; however, it was not used in agriculture but in municipal industry. Slaves also worked as house servants in luxurious homes.

The general boom of the Severan period gave hints of the diverging tendencies of development in the future. A rise in the living standard meant at the same time a certain leveling and a certain impoverishment on the larger social scale. Objects of daily use became simpler and mass-produced commodities replaced expensive articles. *Terra sigillata* and other imported ceramics, which arrived in the province during the second century, first became rare and then disappeared altogether. A tendency for simplification can be observed in bronze vessels. The quantitative increase in production had brought with it a certain uniformity.[32]

## THE LAST CENTURY

The period following the Marcomannic wars had lasted some hundred years when, during the middle of the third century, a more severe crisis was inflicted on the province. The elemental force of a Roxolani attack in A.D. 260 ravaged the greatest part of Pannonia and the decrepit *limes* could not fend off Barbarian attacks for plunder and arson. Only late in the century, under Diocletian, was the defense of the *limes* reorganized and life in the province restored.

It was no longer possible to replace the population which had perished or fled the province by drawing on the manpower of other provinces of the empire, since most had suffered badly from the wars inflicted upon the empire from all sides and from an internal political and economic crisis. Thus missing manpower in agriculture was primarily supplemented from Barbaricum. Settlements had been systematically established from the late third century and throughout the fourth century, but sources confirm only two of these. The first happened in A.D. 295, when Carpi, after their defeat, settled near Sopianae. The second under Gratian introduced the peoples of Alatheus and Saphrac into the province. The Constantine dynasty had repeatedly settled Sarmatians in various parts of the province; whether they were admitted into Pannonia is uncertain. Resettlement of the Vandals under Constantine is also debatable. Later settlements greatly increased, not only to expand tilling of the land but also because peoples moving westward under pressure from the Huns had been admitted to improve defense of the frontiers. Thus the Goths and Alani had also crossed the Danube.

Owing to these settlements the ethnic composition of the province was basically altered and the dwindling Roman population was inevitably mingled with various Barbarian tribes throughout the province. Their archaeological relics hardly help in separating them from the rest. The new settlers produced articles of local trade and commerce, but Roman industry also began to accommodate the taste and requirements of the new inhabitants. Barbarian preponderance was significant particularly in the zone of the *limes*, where great masses of Barbarians had settled to defend the frontier. The finds discovered in the northwest include many with Germanic characteristics, and excavations in Brigetio and Intercisa have unearthed many objects of Barbarian character. Finds from the end of the fourth century in the camps are entirely Barbarian in character. In the equipment of an arsenal of that period in Carnuntum, Asian types of arrows were found.[33]

The earlier population of the province withdrew into the towns in greater numbers. However, reforms introduced during the tetrarchy had inflicted their severest blows in these towns. In Pannonia this change was preceded by the great ravages of the second half of the third century; therefore the new trend that was unfavorable for the towns seemed more disastrous here than elsewhere. Only in a few places do archaeological excavations throw a clear light on the ruin of towns during the third century. For example, in Gorsium not only had the monumental buildings of the *area sacra* collapsed, its forum lying under the ruins, but also the whole town had become the victim of flames and destruction. The inhabitants continued to live in mud huts and in ramshackle houses built from the debris. Elsewhere, though the catastrophe was not so complete, the majority of the houses became uninhabitable. In the new period tremendous efforts were made to restore the province, which was needed both for defense of the frontiers and for its manpower. However, the towns were rebuilt according to different principles, and many of the once-flourishing cities were left to their fate. From Carnuntum we do not know of a single house which could have been built in the fourth century. Aquincum, which was also devastated in the third century was in an even worse position, and many of the ruined houses were never rebuilt. The situation was more or less the same in the *canabae*, where the population continued to live in ruined houses, drainpipes were not used, and fireplaces were built into the houses with hypocausts. Ammianus Marcellinus commented sadly on the situation in Pannonia: Valentinian could find suitable winter quarters only in Savaria, all other towns being impoverished and completely neglected. But the city gate even in Savaria was *locus adgestis ruderibus neglectus*. No provision was made by the state to repopulate these cities. The inhabited parts within the walls were cramped and cemeteries were made in the abandoned parts of Poetovio, Carnuntum, Vindobona, and Brigetio. In Aquincum the population withdrew into the camps, and the *colonia* was left derelict. The same process occurred in Brigetio and Carnuntum as well.[34]

155

The decline of the cities left to their fate presented only one side of the situation in the fourth century. At the same time in other cities the ravages of the war were cleared away and a new boom set in, though it was quite different from the former one. Among these cities was Sirmium, an imperial seat, where in 374 even a theater was built. Probably also among them was Sopianae, its excavated early Christian vaults are proof of a flourishing life. Above the ruins of Gorsium, which in the time of Diocletian received the name of Herculia in honor of Emperor Maximian Herculius, a new luxurious city was built, with enormous palaces of *villa urbana* type, basilicas, commercial buildings, and a huge industrial quarter. The great extension of the city, the accouterments of the still uncounted number of its buildings, and the tremendous growth of its citizenry do not remind one in any way the decline of the other cities. Until the reign of Valentinian, building had continued uninterruptedly and the money turnover of the city had reached its peak at the same time.

The restoration of Gorsium-Herculia shows many features in common with a new specific type of Constantinian settlement, which featured interior fortified settlements on a very large scale — although at present we do not know about the Gorsium fortifications. The fortified settlements were made with a common ground plan. They were more or less square, surrounded by a thick wall of 2.30 by 2.60 meters and fortified outside by round turrets. North of the Drava River we know of five such settlements: in Mursella, Környe, Valcum, Tricciana, and on the site of Alsóheténypuszta. Valcum was 395 by 448 meters; Tricciana 292 by 268 meters. The settlements were situated alongside the road leading from Sopianae. Each settlement had existed earlier, Mursella having been one of the old municipal cities. The large cemeteries inside the walls indicate cities with numerous inhabitants. Their foundations may date from the time of Constantin II. Inside the settlements no trace survives of the earlier principles for town planning; the houses were scattered, with Christian basilicas among them. South of the Drava River we know of a single fortified settlement, the one at Neviodunum. This settlement, however, cannot be listed with the above: it had uneven oval walls and measured about 440 by 150 meters.[35]

The new era, as demonstrated by the composition of the declining cities, did not promote the development of municipal life. At this time the state directly interfered in the affairs of the cities, autonomy surviving only in name. The complete cessation of municipal inscriptions clearly reflects the decline of the offices and not general poverty. The landowners' stratum, which had till then controlled the cities, reached its peak in the fourth century. A considerable portion of the villas known in the province were built then — among them the largest ones were the centers of the *latifundia*, which had grown to enormous dimensions. The Parndorf villa was built about A.D. 300 with mosaics and a rich interior. Among its auxiliary buildings there was an enormous *horreum* (56 by 26 meters). Here as elsewhere the villa, with its adjoining buildings, was surrounded by a fortified wall. Both the Sümegcsehi and the Regelsbrunn villas were fortified. In 375 Valentinian II stayed with his mother in a villa at Murocincta.[36]

In spite of a diverse development in the various parts of the province, and despite a growing Barbarian influence, the standard of living remained comparatively

156

high. This, however, was not due to a profitable local production but to state subsidies. Glass vessels, which had been formerly very expensive, did not mean luxury, nor did the onion-headed *fibulae* which were produced with great technical skill. The fourth-century cemeteries of Intercisa and the multiplicity of fine objects found there are proof of a surprising prosperity. Poverty could be seen in those spheres of life where the requirements sunk to a Barbarian level, as in architecture, in the equipment of homes, and in the lack of inscriptions.[37]

The adaptation of Barbarian customs made it easier for the old inhabitants to accept the new conditions and to coexist with the newcomers. Basic differences had ceased to exist between Romans and Barbarians in culture, in attitude to life, and in living standards, by contrast with the second and third centuries when the dividing line had stretched as between two different worlds. This transformation rendered it possible for the population of the Roman period to remain in the province for many decades after its loss, partly withdrawn to itself, partly mixing with the new inhabitants.

## NOTES

[1] A. Mócsy, *Die Bevölkerung von Pannonien bis zu den Markomannenkriegen* (Budapest, 1959), pp. 16 ff.; A. Mócsy, "Pannónia", *R. E. Pauly-Wissowa*, Suppl. 9, pp. 604 ff.; A. Mócsy, "Zur Geschichte der peregrinen Gemeinden in Pannonien". *Historia* 6 (1957): 488–98.

[2] Appian. Ill. 22–24; Strab. 7.5.2; Plin. N. H. 3.148; É. B. Bónis, *Die spätkeltische Siedlung Gellérthegy–Tabán in Budapest* (Budapest, 1969) *(ArchHung* 47); K. Sági, "Kocsiábrázolások Pannónia császárkori szepulchrális vonatkozású kőemlékein—Rappresentazioni del carro sui monumenti sepolcrali della Pannonia imperiale". *ArchÉrt* (1944–1945): 214–48; J. Fitz, "Az eraviszkusz női viselet—Die Tracht der Eraviskerinnen". *ArchÉrt* 84 (1957): 133–54; J. Garbsch, *Die norisch–pannonische Frauentracht im 1. und 2. Jahrhundert* (München, 1965); I. Čremošnik, in *Hommages à Marcel Renard* (Bruxelles, 1969) (Coll. Latomus), pp. 154 ff.

[3] *CIL*, V. 5363 = Dessau, *ILS*, 2737; *CIL*, V. 5364.

[4] Mócsy, "Pannónia", p. 596; J. Šašel, in *R. E. Pauly-Wissowa*, Suppl. 9, pp. 540 ff.

[5] *CIL*, XVI. 14; Mócsy, *Die Bevölkerung*, pp. 24 ff., 76 f.

[6] T. Flavius Proculus *pr(inceps) praef(ectus) Scord(iscorum)*; Mócsy, "Zur Geschichte", p. 488; Plin. N. H. 3.147; G. Alföldy, "Studia pannonica 1". *ArchÉrt* 88 (1961): 23–30; E. Swoboda, *Carnuntum: Seine Geschichte und seine Denkmäler*, 4th ed. (Graz, 1964), p. 242.

[7] B. Saria, "Poetovio", *R. E. Pauly-Wissowa*, 21:1167; Mócsy, "Pannónia", p. 598.

[8] Mócsy, *Die Bevölkerung*.

[9] J. Šašel, *Vodnik po Emoni—Guide d'Emona* (Ljubljana, 1955), pp. 44 ff.; J. Šašel, in *R. E. Pauly-Wissowa*, Suppl. 9, pp. 540 ff.; T. Buócz, *Savaria topográfiája (The Topography of Savaria)* (Szombathely, 1967), p. 148; Mócsy, "Pannónia", pp. 710 ff.

[10] Mócsy, "Pannónia", pp. 710 ff.; Ptol. 2.14.2; K. Kraft, *Zur Rekrutierung der Alen und Cohorten am Rhein und Donau* (Bern, 1951); G. Forni, *Il reclutamento delle legioni da Augusto a Diocleziano* (Roma, 1953); Mócsy, *Die Bevölkerung*, pp. 117 ff.

[11] Mócsy, "Pannónia", pp. 598 ff.; J. Szilágyi, in *R. E. Pauly-Wissowa*, Suppl. 9, p. 61; U. Laffi, *Adtributio e contributio* (Pisa, 1966), pp. 67 ff.; G. Alföldy, "Revidierte und neue römische Inschriften aus Nordwestungarn". *Epigraphica* 26 (1965): 95.

[12] In *AEpigr* (1965): 12.

[13] J. Szilágyi, Aquincum (Budapest, 1956); J. Szilágyi, *R. E. Pauly-Wissowa*, Suppl. 11, pp. 61 ff.; Swoboda.

[14] Mócsy, *Die Bevölkerung*, pp. 49 f., 69 f.; A. Mócsy, "Das Territorium Legionis und die Canabae in Pannonien". *ActaArchHung* 3 (1953): 179–200; Mócsy, "Pannónia", pp. 610 ff.; *CIL*, III, 10305.

[15] Mócsy, "Pannónia", pp. 601 ff.; A. Mócsy, "Scribák a pannóniai kisvárosokban—Scribae in den Kleinstädten Pannoniens". ArchÉrt 91 (1964): 16–17; D. Gabler, "Munera pannonica". ArchÉrt 93 (1966): 20–35.

[16] M. Pavan, La provincia romana della Pannonia superior (Roma, 1955), pp. 505 ff.; A. Radnóti, "Római tábor és feliratos kövek Környéről—Le camp romain et les monuments épigraphiques de Környe". Laureae Aquincenses 2 (1941): 89 (DissPann 2:11).

[17] G. Alföldy, "Augustalen- und Sevirkörperschaften in Pannonien". ActaAntHung 6 (1958): 433–59; Mócsy, "Pannónia", pp. 602 ff.

[18] Swoboda, pp. 116 ff.; Mócsy, Die Bevölkerung, p. 71; Mócsy, "Pannónia", pp. 603 ff.

[19] A. Mócsy, "Die Entwicklung der Sklavenwirtschaft in Pannonien zur Zeit der Prinzipates". ActaAntHung 4 (1956): 221–50.

[20] Sextus Aurelius Victor in Epitome de Caesaribus and Liber de Caesaribus mentioned Buballa as the birthplace of Decius, but Eutropius in his Breviarium ab Urbe Condita called it Budalia. Both sources located it near Sirmium. Alfonz Lengyel, "Excavation in Roman Sirmium", in Etruscans 2 (1970–1972): 6.

[21] Mócsy, "Pannónia", p. 601; J. Fitz, in Fejér megye története, vol. 1 (Székesfehérvár, 1970), chapter 4.

[22] E. B. Thomas, Römische Villen in Pannonien (Budapest, 1964).

[23] Mócsy, "Pannónia", p. 713; A. Alföldi, "Les chars funéraires bacchiques dans les provinces occidentales". AntCl (1939): 347–59; K. Sági, "Adatok a pannóniai császárkori kocsitemetkezések ethnikumának kérdéséhez—Sur le caractère ethnique des enterrements à l'époque impériale". ArchÉrt (1951): 73–78; Sági, "Kocsiábrázolások", pp. 214–48.

[24] J. Deininger, Die Provinziallandtage der römischen Kaiserzeit von Augustus bis zum Ende des dritten Jahrhunderts n. Chr. (München, 1965), pp. 116 ff.

[25] P. Oliva, Pannonia and the Onset of Crisis in the Roman Empire (Praha, 1962); L. Barkóczi, "The Population of Pannonia from Marcus Aurelius to Diocletian". ActaArchHung 16 (1964): 257–356; A. Alföldi, Zu den Schicksalen Siebenbürgens im Altertum (Budapest, 1944), pp. 39 ff.

[26] Intercisa (Budapest, 1954), vol. 1 (ArchHung 33); Intercisa (Budapest, 1957), vol. 2 (ArchHung 36).

[27] Cass. Dio 71.19.1; 71.11.4; CIL, VI. 32542, 32544; O. Fiebiger, "Inschriftensammlung zur Geschichte der Ostgermanen", Denkschrift d. Akad. Wien 60, No. 2 (1914): 302–3.

[28] Mócsy, "Pannónia", pp. 599 f., 699 ff.; T. Nagy, "Az albertfalvai római telep" (The Roman Settlement of Albertfalva). Antiquitas Hungarica 2 (1948): 92–114; T. Nagy, "Az albertfalvai típusú lakóházak egy újabb példája Rhaetiából—The Latest Example of the Albertfalva Type of Houses from Rhaetia". Antiquitas Hungarica 3 (1949): 135–37; T. Nagy, "Buda régészeti emlékei. II. Rómaikor", in Budapest műemlékei (Budapest, 1962), pp. 519 ff.; A. Mócsy, "A Százhalombatta–dunafüredi római tábor és település—Roman Camp and Settlement at Százhalombatta". ArchÉrt 82 (1955): 59–69; L. Barkóczi and É. Bónis, "Das frührömische Lager und die Wohnsiedlung von Adony (Vetus Salina)". ActaArchHung 4 (1954): 168 ff.

[29] J. Fitz, "Der Besuch des Septimius Severus in Pannonien im Jahre 202 u. Z.". ActaArchHung 11 (1959): 237–63; J. Fitz, "M. Campanius Marcellus". ActaAntHung 16 (1968): 313–23.

[30] Mócsy, "Pannónia", pp. 703 ff.; A. Mócsy, "Savaria utcarendszerének rekonstrukciójához —Zur Rekonstruktion des Strassensystems von Savaria". ArchÉrt 92 (1965): 27–36; A. Alföldi, "Kapitóliumok Pannóniában" (Capitols in Pannonia). ArchÉrt (1920–1922): 12–14; Cass. Dio, 56.17.1; E. Koestermann, "Der pannonisch–dalmatische Krieg 6–9 n. Chr.". Hermes 81 (1953): 377; CIL, III. 10917 (Savaria); in BudRég 12 (1937): 135 ff. (Aquincum); Swoboda, pp. 152 ff. (Carnuntum).

[31] H.-G. Pflaum, "Deux carrières équestres de Lambèse et de Zana (Diana Veteranorum)". Libyca 3 (1955): 134–55; H.-G. Pflaum, Les carrières procuratioriennes équestres sous le Haut-Empire Romain (Paris, 1960–1961), pp. 476 f.; J. Fitz, "Legati Augusti pro praetore Pannoniae Inferioris". ActaAntHung 11 (1963): 289 ff.

[32] Mócsy, "Die Entwicklung", pp. 229, 245.

[33] J. Fitz, Ingenuus et Régalien (Bruxelles, 1966) (Coll. Latomus 81); T. Pekáry, "Késő-

római sírok Fenékpusztán—Spätrömische Gräber in Fenékpuszta". *ArchÉrt* 82 (1955): 19–29; Eutrop. 9.25.2; Amm. Marc. 17.12.18; Euseb. Const. 4.6; L. BARKÓCZI, "Ethnische Zusammensetzung der pannonischen Bevölkerung am Ende des II. und in der ersten Hälfte des III. Jahrhunderts". *ActaAntHung* 7 (1959): 143 ff.; J. HARMATTA, *Studies on the History of the Sarmatians* (Budapest, 1950), pp. 62 ff.

[34] J. FITZ, *Gorsium* 3d ed. (Székesfehérvár, 1970), p. 71.

[35] Amm. Marc. 3.10.2; 29.6.11; F. FÜLEP, "Neuere Ausgrabung in der Cella trichora von Pécs (Fünfkirchen)". *ActaArchHung* 11 (1959): 399–417; F. FÜLEP, "Későrómai temető Pécs, Geisler Eta u. 8. sz. alatt—Early Christian Cemetery at Pécs, No. 8 Geisler Eta Street". *ArchÉrt* 96 (1969): 3–42; J. FITZ, in *Acta Arch. Carpathica* 10 (1968): 299 ff.; K. SÁGI, "Die spätrömische Bevölkerung der Umgebung von Keszthely". *ActaArchHung* 12 (1960): 254 f.; MÓCSY, "Pannónia", pp. 700 ff.

[36] SWOBODA, p. 126.

[37] MÓCSY, "Pannónia", pp. 707 ff.

# THE WAY OF LIFE

JENŐ FITZ

A striking difference between the culture of the indigenous population of Pannonia and the new settlers survived for a very long time. Romanization began with the assimilation of external features. The indigenous aristocracy quickly accommodated to the new conditions as far as objects of daily use are concerned — architecture and the circumstances of economic life. However, it continued to conduct burial rites according to the beliefs of the natives as late as the second half of the second century. In the wider strata of the population folk traditions did not disappear until the third century, especially in regions far from the towns and the military camps.

It is very likely that Celtic and Pannonian dialects were widely used in the *peregrine* territories and villages. No written records survive, however, because the inhabitants of the territory could neither read nor write. This means, at the same time, that literacy in Pannonia was always equivalent to the knowledge of Latin. When the custom of placing gravestones was adopted by the natives, even those who were not familiar with the language of the conquerors put Latin inscriptions on the stones. However, even prior to the conquest, the use of Latin reached a preliminary stage, at least in commercial and diplomatic relations, to a greater extent in the south than in the north and east. The words of Velleius on the knowledge of the Pannonians cannot be related to the whole of Pannonia but only to the leading strata of the population in the Sava basin: "in omnibus autem Pannoniis non disciplinae tantummodo, sed linguae quoque notitia Romanae, plerisque etiam litterarum usus et familiaris animorum erat exercitatio".[1]

Inscriptions on gravestones and altars began to appear with the advance of Romanization in native territories; however, the use of stone monuments was adopted sooner than writing. Illiteracy can be clearly seen on vessel stamps: here wording is imitated. In Northeast Pannonia the first stones with inscriptions show that those who carved the inscriptions were not identical with those who ordered them, because the artisans did not know Latin. Knowledge of Latin was disseminated through military service, the return or settlement of veterans, and new settlers and merchants throughout the second century. But in areas remote from urban and military centers the use of Latin in this period is far from probable.

Comparatively few persons were familiar with the Greek language. Those who did know it were generally immigrants. A *praeceptor Gr[aecus]* was active in Neviodunum; yet the only traces of the Greek tongue that survived in this area were in Bononia — an inscription incised into brick, a few stone relics which are bilingual, and the inscriptions of magic carpets, amulets, and other objects. Most of the Greek relics, however, emerged from Sirmium, but a few also survived from Siscia, Poetovio, Carnuntum, Aquincum, Gorsium, and Intercisa. Those who ordered inscriptions were all Orientals — Syrians, Jews, or Thracians — from Asia Minor and the eastern Balkans. During the third century, when the number of Orientals increased considerably, Greek speech in the cities became more frequent.[2]

About the culture of the Pannonians, with the exception of Velleius, the authors give us no favorable information. Fronto says about the eastern campaign of Lucius Verus: "non incuriose per militum contubernia transire, sed forte temere Syrorum munditas Pannoniorum inscitas introspicere, de cultu cuiusque ingenium arbitari".[3] Since Legio II Adiutrix participated in the war in full numbers, it is possible that the reference to the lack of culture referred to soldiers in Aquincum. Victor Aurelius recorded the lack of humanity of Pannonian emperors: "his sane omnibus Illyricum patria fuit; qui quanquam humanitatis parum, ruris tamen as militiae miseriis inbuti satis optimi rei publicae fuere."[4] *Historia Augusta* recorded that Apollonius of Tyana addressed the Emperor Aurelian in Latin: "haec latine, ut homo Pannonius intellegeret, verba dixisse." Emperor Aurelian (A.D. 270–275) did not understand Greek and so Apollonius was compelled to speak Latin.

We know next to nothing about the schools in towns and larger settlements. There were schools for higher education at least in the two capitals, where we hear about *scolastici*. In Aquincum one *scolasticus* was the son of the *decurio* of Mogentiana, and the son of a *decurio* from Mursella who died young had studied at Carnuntum. In Sirmium the epitaph of a young equestrian speaks of him as *omnibus studiis preditiis*. Pens and inkwells found in children's graves, the jet statue of a teacher, inscriptions incised into bricks, alphabets, etc., bear witness to the existence of education on a larger scale. We find quotations from Vergil incised on bricks, proving that the poet must have been read in the schools. Mythological scenes are strikingly frequent on stone relics, proving again that the old stories were not only familiar to the local public but loved too.[5] The presence of a *pragmaticus* and an *advocatus* are proof of the activity of the Roman legal system.

Limitations of culture conclusively show that Pannonia had nothing to contribute to the intellectual life of the ancient world in literature thought, or the arts. As far as literature is concerned, there was some contribution in grave verses; but those which survived on stone monuments are rather poor, halting verses, and in content they hardly went further than reproducing commonplaces. The verse inscriptions of Emona and Carnuntum come from the first century; the others, from the third and fourth centuries.[6] In Aquincum and its vicinity a poet produced verses many of which have come down to us: "Hic iacet in tumulo Aurelia Sabina pientissima coniunx. Quem lapis iste tegit, rapta est de luce serena, quae magis debuerat fessos sepelire parentes. O dolor, o pietas, o funera tristia coniugis . . ." goes an epitaph at Gorsium.[7] "Hic iacet in lapide Aurelia Marcellina pientissima coniunx, quem lapis iste tegit, rapta est de luce serena . . .". This is what we can read on a second gravestone; a third one reads as follows: "Clausa iacet lapidi coniunx pia luce serena . . .". Another verse has preserved the name of the poet at Aquincum and Ulcisia Castra: he was called Lupus. These verses, without being works of art, preserve for us in a conventional manner the contemporary attitude to life and death. "Virito mortalis dum dumdant tibi tempora Parc[a]e", wrote the poet Lupus in an Epicurean verse: "Were you a farmer, a citizen, a soldier or a shipman, love the blossoms of Venus, pluck the blissful gifts of Ceres, enjoy the plentiful and fruitful endorsements of Minerva. Live a pure life, be just and cheerful as a child, a youth, also later as man and weary old man. So in thy grave, thou shall not be forgotten by those in heaven."[8]

Of the world to come, of the gods who personify supernatural forces, the conquering state and the first Italia settlers had formed new views and brought a new world to the province of Pannonia. As to the beliefs of the natives and their religion, we can form conclusions only from surviving relics from the Roman period; these, however, did not remain unaffected by a changing world. In a province of mixed population we can hardly speak of a uniform religion in the period prior to the Romans. Owing to the multiplicity of local cults, it was impossible later, when Romanization also began to affect religion, to put up any resistance against the new concepts.

The multiplicity of cults exerted itself even in the worship of Silvanus, who is believed by many to have been the Pannonians' chief deity. It is evident that he had been most intensely venerated in the province, and the number of his shrines exceeded even those of Jupiter. He was presented as an elderly man, fully dressed, with his dog, holding a pruning knife and fruit tree. On the inscriptions his byname was Domesticus or Silvestris. In this former aspect he protected the soil, the farm, the home; and on the family altars he was venerated in the company of Lares. As Silvestris, he was the god of the forest and appeared together with Diana as a god of the hunt — in this capacity he is often represented with buck's feet. In certain localities, such as Poetovio, he was also a god of fertility and as such he lacked any official features. His adherents were mostly simple persons — citizens and soldiers; yet there were comparatively few natives among them. Many of his altars were ramshackle and badly carved, with illegible inscriptions; they were mainly made of sandstone. His statues were equally primitive and roughly carved. His first relics go back to the first century, but most were made in the third. The local references to the god are made clearer by those coins on which, during the reign of Trajan, the symbol of the Danube provinces also appeared. This official recognition occurred more than once: during the third century he was the god of the Danube provinces and he also found his place among the gods of the state.

The feminine equivalent of Silvanus always occurs in inscriptions in the plural: Silvanae. Silvanus can be identified in the southwest with Vidasus, his female representative, and with Thana — these were obviously local names which were not the same in all parts of Pannonia. Gods of different characteristics are referred to by names which suggest these qualities, such as Bellator, Deus sanctus, Erbarius, Magla, Magnus, Mammula.[9]

Local differences of character are reflected in the cults of other local gods. The Quadriviae were celebrated in Carnuntum. Similar gods were the Nutrices, whose cult was best known in Poetovio. Despite the plural, all presentations show a single Nutrix. The same goddesses were known in Savaria under the name of Fatae Tenatiae, but they are also referred to under the name of Matres Pannoniorum. The cult of the goddesses was marked by *terra-cotta* and lead filigree *vota*.

Among the local deities represented in Emona were Aecorna, otherwise called Aequorna, and Laburus, as well as Minitra and Genius Ciniaemus. There is no Pannonian relic to prove the cult of Mars Latobius in Noricum; but the connection with the Latobici tribe is obvious, and it is possible that Pannonians venerated this god. Mars Marmogius's cult in Noricum also survived in Siscia. In both provinces the Sedatus cult was alive, the worship of this god being widespread among the Celtic population. Moreover, Sedatus's name was also known in the eastern part of Pannonia — for example, in Gorsium and in the *sigillata* of Pacatus in Aquincum. In the survival of the names and cults of local deities the same differences are observable between the western and eastern sections of the province as appear in other spheres of life. It is even possible that the question of why native gods and cults had disappeared from Pannonia can be approached through that angle. In the west Romanization, which to a certain extent was responsible for the worship of ancient gods under their own names, was more advanced than in the east.

We can recognize local deities in the gods of various rivers and other waters. A freed *socius* had raised an altar to Savus. The worshipers of the river gods were mostly merchants. Dravus had an altar in Poetovio; Danuvius had an altar in Vindobona together with Salacea, Neptune, and an unknown river god. We can see other river gods on the reliefs of the *nymphea* decorating the *forum* at Gorsium. The gods wear local garments, trousers, sandals; they are bearded and they are pouring water into the basin.[10]

The legions, the veteran settlers in the new towns, and Italia merchants and tradesmen disseminated the cults of the imperial gods in the province. In each town which obtained the rank of *colonia* or *municipium*, a *capitolium* was built with a temple devoted to the cult of the three deities, Jupiter, Juno, and Minerva, in the center — this being at the same time the symbol of the Roman state. Worship of the triad of gods, though occupying an official, central place, according to the evidence of inscriptions was not far-reaching. Their altars, in the period of the Severi, were mostly raised by soldiers.

Among the imperial gods, Jupiter Optimus Maximus was most widely worshiped. Governors, magistrates, soldiers, and officials paid tribute to this god who protected the empire. Most of his altars were erected at various official occasions. His by-names (Accio, Ar[rubianus?], Conservator, Custos, Culminalis, Depulsor, Fulgurator, Fulminator, Monitor, Nundinarius, Paternus, Prestitus, Salutaris, Tavianus, Teutanus) point to various features of the many-sided god; but local characteristics and associations with local deities can also be observed. The couple Jupiter–Juno, so generally revered in Aquincum and Brigetio, may also conceal native traditions.

The cult of Juno herself was more restricted, as was that of Minerva. The former was worshiped (under the name of Juno Regina) particularly in and around Aquincum. At Carnuntum Juno's cult was linked with that of Nemesis. Minerva received her altars mainly from soldiers. It seems that the official worship of Minerva was connected with special ranks; most frequently *cornicularii, tubicines,* and *immunes*. At the performance of the *votum*, presentations were more frequently made, so that we can assume a wider cult than was indicated by the evidence of altars.[11]

As regards the cults of other Graeco-Roman gods, considerable differences may be observed in character and dissemination. For these differences local roots may be just as much responsible as the so-called Pannonian spirit that had gradually emerged and had recognized in certain gods the expression of its own aspirations, while not wanting to establish direct contact with others.

The variable cult of Diana extended to the whole province and went back to local worship. In Aquae Iasae she emerged, together with the Nymphs, as the goddess of the spa; at other times she was goddess of the hunt and the forests — in this capacity she was teamed with Silvanus. Her worship was later complemented and colored with Thracian and other Eastern traits. In A.D. 202 the Syrians at Intercisa raised a temple to Deus Sol Elagabalus together with Diana Tifatina. This originally Capuan goddess appeared here as the female counterpart of Baal of Emesa. The figure of Diana also appeared on altars with four deities, the chief gods of the state religion, in the company of Silvanus and Liber, as the representative of the Danube provinces on the Trajan arch at Beneventum. As a goddess of the hunt, she was also associated with the amphitheater games, which brought her in contact with Nemesis and Fortuna. In Floriana she was worshiped in a cave shrine; earlier it was believed that she had a sanctuary at Gorsium.[12]

Mercury relics emerged in great numbers in western Pannonia, especially relics made by private persons. At Carnuntum, Brigetio, and Intercisa Mercury was respected as the god of trade. A large number of *terra-cotta* and votive objects throw light on the deep roots of his worship. The wide range of this worship in western Pannonia seems to suggest that this god was much respected by the Celts and had survived for a long time.[13]

In the figure of Hercules, too, local elements intermingled with the classical appearance. In western Pannonian cities several monuments from early times prove this fact. However, most of those who built altars to him were not soldiers. Relics that survive from Pannonia Inferior from the territories inhabited by Celts, indicate that the majority of his worshipers here were soldiers. It is not impossible that soldiers recruited from the local population had seen local elements in the figure of Hercules. Under the Severi the cult of this god was enriched by new features: worship of Hercules-Melkar adhered to the imperial house and the range of his worship produced new types. In the worsening military situation in the third century, the army in Illyricum emerged as the protector of the empire and culture. Hercules offered an adequate symbol. This produced on an altar in Aquincum the figure of Hercules Illyricus.[14]

The cult of Liber Pater, Liber, and Libera was different in West and in East Pannonia. Little survived in the west, and what did, bears the features of the Italia cult. In Carnuntum, the legionaries raised a Liber altar. In Arrabona Liber joined Neptune, Diana, and the gods of the Capitolian triad to fall in line with the deities of the state and the Danube provinces. However, in East Pannonia his cult assumed a specifically local character. The majority of his shrines come from smaller settlements, without any official character. A Liber altar has been referred to above in connection with viticulture. Later his cult mingled with Thracian elements which betray Dionysian-Bacchus traces — here Libera can be compared with Ariadne.[15]

Venus inscriptions are rarer, and even those that occur refer to a Syrian goddess under Venus' name, in the company of Jupiter Optimus Maximus Heliopolitanus. The altars and other stone monuments do not reveal the real character and range of the cult. Small *terra-cotta* objects and both fine and less excellent bronze statues prove a general esteem in all strata of society.

Vulcanus had a common shrine with Venus at Poetovio. His cult was connected in *vici, canabae* and spas with the risk of fire. Neptune, on the other hand, was venerated near water, together with the Nymphs. The *Collegium negotiantium* paid tribute to the god of waterways on the altars dedicated to him. As protector of the fleets sailing and fighting on the rivers he was included among the war gods. In Cibalae a large shrine was raised in his honor. Among the war gods Mars received the greatest tribute on the part of the legionaries. His attributes listed on the relics are Victor, Custos, and Victoria. He appeared in the company of Fortuna Redux and Minerva. The *Collegium negotiantium* tried to win his favor as the protector of the security of commerce.

Like the cults of the aforementioned gods, the worship of Ceres and Apollo was limited. The latter was appealed to because of his curing power; he was known as Apollo Granus and Apollo Conservator. He also appeared together with Hygieia. The latter and Aesculapius were primarily venerated by physicians, but they frequently appeared on reliefs too. The Nymphs were also granted their altars as curing deities in baths and near curing waters. Their epithets as healing deities were Salutares and Medicae, that as water nymphs, Perennes. We may also consider Dis Pater and Proserpina as healing deities.

Many relics survive from the sites of amphitheaters to prove the cult of Nemesis; in both Aquincum and Carnuntum there was a shrine next to the amphitheater. His cult was joined with those of other deities, such as Diana and Fortuna, while his appearance with Fors Fortuna reflects the revival of old Roman traditions.

Both the Lares Domestici and the Lares Augusti in the western province were of Italian origin and were often connected with imperial cults: "collegium magnum Larum et imaginum domini n. Caesaris". This inscription comes from Poetovio, where several relics survived to prove the local cult of Lar. There was a comparatively small number of inscriptions, but from the large number of bronze statues we can conjecture that the cult had reached considerable proportions. The home shrines and those in camps were similar in shape.[16]

Besides the cults of local and imperial deities, those of many other gods can be accounted for in Pannonia. This profusion was due to the soldiers ordered here, to merchants arriving in the province and settling here, and also to tradesmen. Altars, statues, and cultic objects occurring only once refer mostly to individual cults. To the Batavian goddess Dea Vagdavercustis an altar was raised by the tribune of the Cohors III Batavorum.[17] Altars consecrated to Sulevia in Aquincum and Sucellus contained offerings including the personal objects of soldiers and merchants who came from the west.[18] To smaller African veteran groups under the Severans can be attributed the cult of Tanit and Juno Caelestis. Epona was venerated in a wider circle, especially among mounted soldiers recruited from the eastern provinces, at the turn of the second and third centuries. Thracian mounted gods were venerated in Siscia, Intercisa and Gorsium and among Thracian soldiers and even Thracian settlers after the Marcomannic wars.[19]

These sporadically emerging cults, which did not become permanent parts of Pannonian religion, are of lesser significance than the Eastern religions, which from the second century onward spread very widely. The adherents were also partly new settlers who at the end of the Marcomannic wars became followers because of the arrival of the Orientals. However, there were other cults that had no ethnic basis. Among these can be mentioned the Egyptian cults which were adopted in early times by Aquileian merchants, customs officials to the wealthy citizens, and customs officials of western cities. Isis had her temple in Poetovio and Savaria but many relics of Serapis also survive, together with those of Jupiter Ammon. Interest in the Egyptian cults passed on, although to a lesser degree, to the eastern portions of the province. Along with relics from Aquincum and Mursa, the stone monuments with scenes of the Nile at Gorsium are worth mentioning.[20]

Syrian soldiers had brought to Intercisa the worship of Deus Sol Elagabalus. The auxiliary company looked upon him as the home god of the city of Emesa and also revered his female counterpart, who was generally associated with Diana — according to a temple inscription, Diana Tifatina. This cult was directly associated with the Intercisa division, worship elsewhere being only sporadic in towns inhabited by Orientals. No trace survives of the official cult offered to the god of the city of Emperor Elagabalus. A relatively smaller circle among Eastern soldiers and settlers worshiped Jupiter Optimus Maximus Heliopolitanus, Assus, Sabasius, and two Arab gods, Theandrius and Manaphus. It was due to the soldiers that the cult of Jupiter Optimus Maximus Dolichenus became popular and the greatest number of his relics emerged in Pannonia. This is corroborated by the earliest inscription under Hadrian from Carnuntum: "iuventus colens Iovem Dolichenum." The adherents of the cult were primarily the soldiers of Legion X and XIV Gemina; it was not important in the camps of Brigetio and Aquincum. However, Jupiter Dolichenus was much worshiped by the civilian population in Carnuntum, Mursella, and Brigetio, especially where the Syrian population preponderated. His shrine at Brigetio was established by a *decurio* from Zeugma; among his adherents at Carnuntum the name of M. Titius Heliodorus *augustalis* survived. Also Septimius Severus offered respect to this god, and during his visit in Pannonia the priest of the province had a shrine raised to him at Gorsium. Outside the *limes* and its vicinity, the cult of this god was obvious in Emona, Aquae Iasae (Topusko), and Savaria.[21]

The most important Eastern religion was nonetheless the cult of Mithra. The first traces of it emerged in the first century in the province, but no Mithra community had yet been formed. About the middle of the second century the first group was formed from customs officials at Poetovio, and they were responsible for the building of the first mithraeum. It was not much later that the slaves of the governor of Pannonia Inferior, T. Haterius Saturninus, began to form an organization: "Deo Invicto pro salute familiae T. Haterii Saturnii leg. Augg. pr.pr.

Arporcras pater posuit." But the cult proper did not develop on a large scale until the decades following the Marcommanic wars. From the end of the second century to the early fourth century the Mithras cult was the mightiest religious movement; in it participated soldiers and citizens, rich and poor, Orientals and locals. The cult enjoyed great favor among Pannonian soldiers in particular, and the shrine at Carnuntum was restored by them at the time of an imperial conference within the walls of the city. Diocletian, though hostile to the cult, permitted the practice of the religion of the Danube soldiers. Of three centers of this religion, Poetovio had three (and a conjectural fourth) shrines, Carnuntum had three, and Aquincum five, but there were also shrines at Brigetio, Campona, Intercisa, and several smaller places: Stixneusiedl, Fertőrákos, Modric, Zgornja Pohanca, Sárkeszi, etc.[22]

The municipal clergy were known under the names of *flamines* and *augures*, both positions filled by *decuriones*; and both formed part of the regular municipal career. It is possible that the *pannoniciani augures*, who are spoken of in connection with the battle of Lugdunum in A.D. 197, had kept up the traditions of the Celtic *vates* in Pannonia.[23] The *augur* mentioned on the *pro salute civitatis Eraviscorum* altar was eventually not a municipal but a local priest. There were only a few cults with nonmunicipal priests. The Nemesis priests in Savaria, Carnuntum, and Aquincum were called *antistes*; the Lares Augusti in Emona had *ministri*; the Dolichenus and Heliopolitanus cults had *sacerdotes* as their priests. The priests of Dis Pater and Aeracura were also known as *sacerdotes*. In the Mithras communities we know of *patres* and *lecnes*. Besides the capitolian temples, Nemesis shrines, *mithraea, dolichena,* and *isea*, only a few shrines have been systematically unearthed. At the Pfaffenberg in Carnuntum a larger holy district is known to have existed; in Aquincum a Gallic-type, round church has been excavated. Inscriptions speak of several temples and shrines, without specific data, most of them dating from the third century.[24]

The first Christian communities were formed about the middle of the third century under Valerian. It was about that time that Eusebius, bishop of Cibalae, died a martyr's death. During Diocletian's reign our sources mention Christian communities in Sirmium, Cibalae, Siscia, Poetovio, and Savaria. The earliest Christians had Greek names. Nevertheless, it is possible that the first Christians came from northern Italia, from Ravenna, where Orientals were in a majority in the Christian communities. During the persecution of the Christians under Diocletian, among the victims were Irenaeus, bishop of Sirmium, Quirinus, bishop of Siscia, and Victorinus, bishop of Poetovio. All of them were executed. Of the memorial places of the victims, Quirinus's is the most important; over the ground where he died in Savaria a basilica was built. After the Synod of Nicaea, bishoprics were established at Sirmium, Cibalae, Mursa, Siscia, and Iovia. On the basis of surviving relics it can be supposed that Savaria and Sopianae must also have been bishops. According to available sources, Arianism was the ruling trend in Pannonia until the Synod of Sirmium in 378 and that of Aquileia in 381, which did not annihilate it altogether.

After the triumph of Christianity, the heathens for a long time continued to exist. In the *mithraeum* of Kroisbach, Gratianus coins have been found; in the

Christian cemetery of Tricciana, a silver plate with a magic text has come to light. On the metal mountings of chests, which became fashionable during the fourth century, Christian and pagan designs and symbols occurred together. On the milestone at Mursa, Julianus had been referred to as "ob deleta vitia temporum prateritorum". Nevertheless, all this did not mean a peaceful coexistence. The Christians, if they were in a majority, smashed the cultic pictures and prevented the shrines from fulfilling their function. The cultic pictures on the *capitolium* of Scarbantia were broken up and walled in, and the shrine was converted to fulfil another function. The altars of the governors of Aquincum were also walled in, and the statue of the Hercules shrine at Ajka was smashed to pieces and buried in the ground. The heathens were compelled to hide their cultic objects. The full equipment of the *mithraeum* of Sárkeszi and the *lararium* of Nagydém had to be hidden underground.

The bishoprics mentioned above and the events known from the history of the church are all connected with the southern portion of Pannonia. The evidence of a great number of relics reveals that Christianity had also developed in the north. We know about Early Christian basilicas in Vindobona, two in Valcum, two in Kékkút, one each in Kisdióspuszta, Sümegcsehi, Alcsút, Aquincum, and two in Gorsium. Finds suggesting the existence of basilicas emerged in Donnerskirchen, Csopak (a marble altar slab), Leithaberg (columns belonging to an altar), Bonyhád (a large bronze Christogram, perhaps part of a lamp), and Carnuntum (a baptismal font). The majority of Early Christian inscriptions derive from Sirmium; others come from Mursa, Siscia, Savaria, and from the northern section of the province, Vindobona, Brigetio, and Aquincum. The incised initials of Christ occur on all sorts of objects: whorls, bone tokens, lead labels, garments, and *lucernae*. There are glass cups with inscriptions, a *fondo d'oro* from Lugio, another from Intercisa. More significant than these are the mountings of caskets which occurred in many late graves: on some, pagan and Christian elements appear together; however, those with only Christian motifs were more frequent. Among the relics of Christian art the wall paintings on vaults deserve notice. From the Old Testament appear the figures of Adam, Eve, Noah, Abraham and Isaac, Moses, David, and Joseph; from the New Testament, the three kings, the good shepherd, and Lazarus — all these subjects decorated the vaults in Sopianae. Both the architectural relics and other finds conclusively prove the triumph of Christianity throughout the whole territory of Pannonia, even in those areas where written records make no mention of it. Moreover, material relics can be found in greater numbers north of the Drava River.[25]

In every period funeral rites were closely associated with religious customs. In early times the graveyards generally were located outside the settlements, primarily alongside the exit roads. This was so in Emona, Poetovio, and Carnuntum. It long remained a rule to separate the cemeteries from the settlements, yet as early as the third century small family vaults had been built inside the walls. From the fourth century onward, there was no rule or custom for fixing the places of cemeteries. In the selection of a place, naturally, personal motives also intervened. Thus, for example, after the Eravisci abandoned their settlement on Gellért hill, they continued to use the graveyard on the western slope of the hill, in order to follow the old tradition. And the graveyard La Tène of Oggau was used as late as the imperial period. In addition to the graveyards the clusters of graves for particular groups can also be distinguished. In Aquincum's *canabae*, for example, in the cemetery there was a group of graves for inhabitants of Mursa; in the cemetery of Carnuntum, plot 309 was kept exclusively for Italians, Orientals, and western soldiers.

However, little is known of the funeral rites used in the graveyards. An endowment inscription of a collegium from A.D. 220 says: "ad Rosalia celebranda itemque ad sepulchrum." On another stone monument we read: "uti rosas Carnarias ducant." The idea of providing for the dead is demonstrated by the pipes made of imbrices, which slantingly led into the graves. Inside the graves, side by side with the usual food and drink, other useless objects also found their way. To lead-framed mirrors magic power seems to have been attributed.[26]

Even the natives adopted the Italian custom of placing gravestones above the graves. The monuments of the native population throw light in many respects on their beliefs concerning the life to come and on their funeral rites and symbols. Astral symbols, the crescent of the moon, and the disc of the sun occur on very primitive graves in North Pannonia and sporadically in the southwest as well.[27] These stones belonged exclusively to the native population. Their main area in the north was the eastern portion of *civitas Azaliorum* and the northwestern borderland of *civitas Eraviscorum*, a mixed population. This belief links with the Azali aspects or Pannonian character. In the neighboring Eravisci area, in the Lajta region, the characteristic picture on native graves comes from the late first century and depicts a coach scene: the dead person travels toward the world beyond in a two- or four-wheeled coach; there is a coachman, behind on the bench sits a servant, and the march is often conducted by a leader. It is proof of the strength and duration of the custom that coach funerals continued concurrently with and in the same areas as the advance of Romanization, and also among the native aristocracy in other parts of Pannonia.[28]

The relatively late emergence of coach funerals can hardly be separated from the fact that the tribal leaders were getting very rich. In their graves, in addition to the richly decorated coaches, there were other objects of funeral rites: the *tripus*, *patera*, bronze jars, *balsamarium*, *strigilis*, etc.[29] These display the custom of sacrifices which are depicted on gravestones with the coach scene and which are to be found independently of the latter on native graves from the late first century onward. In these scenes we can see in the middle a *tripus* filled with food and on both sides figures offering a sacrifice — in their hands either a *patera* or a *jar*.[30] These scenes, as well as the coach scenes, appeared side by side with the custom of erecting gravestones for natives, so that we can consider them as unbroken survivals of ancient fancies. The presentation of funeral feass which relate to the sacrifice scenes appeared somewhat later on Pannonian gravestones. The inscriptions corroborate the fact that these stone monuments were always made for strangers, who probably arrived from the eastern Balkan. The type itself came from the Balkan.

The custom of *tumulus* burials was due more to Illyrian than to Celtic or Roman traditions. It emerged in western Pannonia, particularly in areas inhabited by the

Celtic population, throughout the first and second centuries and briefly in eastern Pannonia, south of Aquincum as far as the Mecsek hills, at the turn of the first and second centuries. The *tumuli* in eastern Pannonia, although the finds are of native character and bear the marks of contact with the local population, cannot be connected with either the Eravisci or the peoples living south of them. The *tumuli* appeared along important points of the exit roads north of Sopianae, which leaves little doubt of an organized settlement policy. It seems probable that at the time of the building of the *limes*, provision was made to bring settlers from the western areas to protect the frontier against the Eravisci and their neighbors, who were not yet Romanized to any degree.[31]

In some native graveyards, as in those belonging to the Eravisci, the graves contain partly burnt and partly skeletal bodies from as late as the second century. Burning bodies was still the general form of funeral in the first and second centuries. However, funeral rites varied. In most cases the ashes were not placed in an urn or stone jar but were simply thrown into the graves. If the ashes were still smoldering, they burnt the sides of the grave; or it is also possible that the fire was made above the pit and the remains fell directly into the grave. Some of the objects were consumed by fire and some were put into the grave afterwards. In many graves burnt and intact objects occurred together. A common place for the burning of dead bodies, a *ustrina*, hardly existed in Pannonia. Graves were round, oval, or square. There were also graves with shoulders, where the ashes were down below and the objects on top. Iron nails found in the graves were perhaps used to make fast the pils; others may have been part of the wooden chests containing the ashes and the objects. There were many graves put together from roof tiles. There were also walled-in vaults, particularly in the cemeteries of Neviodunum.

Different customs may be observed at the urn burials too. The urns were generally made of clay, rarely of glass, primarily alongside the Amber Route. In small stone urns there was room for few objects, but larger ones could hold a complete grave outfit. The stone *ossuarium* was generally attributed to veterans, but it is also possible that it came from Illyrian funeral rites. In East Pannonian cemeteries urns were seldom used, and from the second century their number diminished also in the west.[32]

Skeletal burials can be found in practically all second-century graveyards. In these graves there were hardly any objects, sometimes none at all. Perhaps only the poorest class or the slaves were buried in this way. The rarely occurring rich skeletal graves, one may suppose, were due to family traditions. Only after the Marcomannian wars did skeletal burials become general. It is hard to decide today what part the newly settled Orientals had in establishing this custom. The sarcophagi all show an eastern type: their top simulates a house roof; it has an inscription on one side, and it is generally framed in by Genius and Attis figures placed in niches on both sides. In third-century graveyards there was still a considerable number of urn graves, but by the fourth century they seem to have disappeared completely. Their sporadic occurrence may be attributed to newly settled Barbarians.

Various customs can be summarized in the skeletal and the urn burials. As far as the positions of graves are concerned, disregarding a few exceptions, three different types can be found. In the first, the head rests in the west, as in Mursella, Valcum, Castra ad Herculem, Intercisa, Gorsium, and many smaller graveyards. In the second type, the head rests in the east, as in Sopianae and vicinity due to Barbarian settlers, in Intercisa and Vetus Salina, in third-century graves in Brigetio, and in Mursa graves. In the third type, the head rests in the south, as in fourth-century graves in Aquincum, Keszthely-Dobogó, and Intercisa. By noting the presence or absence of particular objects many different customs can be traced. In certain graveyards coins do not exist; in others a great number of coins were placed in the graves. Chest mountings between Aquincum and Gorsium are usual pieces of cemetery inventories, but from the graves of Sopianae they are absent. There are also differences in construction; there are brick graves, walled-in graves, grave chests knocked together from early stone monuments and gravestones, and wooden coffins — all found practically within the same cemetery. Most of the differences, if not all, go back to ethnic characteristics. Distinctions can today only be explained conditionally. Orientation may be due to the customary order of the cemetery. Objects are generally Roman, even in resettled Barbarian graves. More conclusive are the position of objects of clothing and the overall picture of funeral habits in each particular graveyard. So far, no more can be stated with certainty than that there is a basic difference between burials along the *limes* and those inside Pannonia. The former are characterized by a multiplicity of shapes which correspond to the late Roman array. In the interior of Pannonia the graves are more uniform, which can be explained by the settlement of large Barbarian masses and the staying behind of the old population. The Roman character survived primarily in larger settlements and in the cemeteries of the towns.[33]

In fourth-century cemeteries the graves with jars and glass objects inside them are held to be Christian. Pouring lime on the dead was a Christian custom, and so was mummy burial, which may derive from the mystery religions. Small grave chapels, underground grave structures, and walled-in graves are typical of Christian cemeteries. The cemetery of Sirmium developed around the graves of the two martyrs Synerotas and Demetrius. In the center of Tricciana's Christian cemetery was a chapel, and the graves were arranged radially around it. *Custor cymiterii* on a Savarian inscription makes us suppose that at least in the larger settlements the Christian cemetery was separated from the pagan one.[34]

The marking of graves with stones and other structures was naturally only customary with the well-to-do. Grave *stelae* with pictures of the dead and members of his family, or without them, were most frequent. There were inscriptions, with references to the life to come and pictures of the cult with scenes from mythology. Among many ancient tales, the inscriptions often referred to the story of Alcestis, who was brought back to her husband by Hercules from the underworld, or to the story of Orpheus and Eurydice, etc. Grave altars occur in large numbers only in the southwestern part of the province. A representative grave memorial was the *aedicula* with three relief ornaments and sometimes a pyramidal top. However, even large grave structures existed in Pannonia. In Carnuntum there was a big chapel; in Keszthely-Újmajor, a common structure for urns and graves; there were graves of different sizes and shapes, round and square, in all parts of the province; family graveyards; etc. Even in Christian cemeteries there were many underground vaults above which memorial structures were built, as in Sopianae, Valcum, Ulcisia Castra, Aquincum, and Tricciana. The *cella trichora* was used as a grave chapel. There was a cultic site in Sirmium for martyrs and also in Sopianae and Aquincum.[35]

The cultural life of the time included a variety of attractions other than those of religion. Inscriptions from Siscia and Aquincum refer to theaters, although none has been traced so far. In Siscia we know of a fourth-century *magister mimariorum*. In Aquincum a *monitor* (prompter) raised an altar to the Genius of the *collegium scaenicorum* during the third century. In *Passio Quirini*, a theater in Savaria is mentioned. Although these scanty data furnish no information about the plays which were acted, there is little doubt that they imitated the performances of Rome and other cities. An inscription from Aquincum mentions a married couple: the husband was an organ player (*hydraularius*) in Legio II Adiutrix; his wife, a singer (vox ei grata fuit), a cither player (pulsabat pollice cordas), an artist of the organ ("hydraula grata regebat"), she alone surpassed her husband (superabat sola maritum). From Aquincum emerged an organ which was given by C. Julius Victorinus to the *collegium centonariorum* in 228.[36]

The games in the amphitheaters attracted great crowds, as is corroborated by the excavations and the size of the amphitheaters. The amphitheater of Carnuntum's legionary camp, built in the second half of the first century, was a wooden structure; it was rebuilt in stone during the second century. Its arena was then 71.3 by 44.25 meters. The amphitheater of the *municipium* was built here during the middle of the second century, with an arena of 68 by 50 meters. The inscription of the legionary amphitheater of Aquincum dates from A.D. 145; its arena surpassed all other Pannonian ones, being 89.6 by 66.1 meters. The amphitheater of the *municipium* dates from the middle of the second century, with an arena of 53.36 by 45.54 meters and a relatively small auditorium. West of the amphitheater was the gladiators' garrison with a large yard. Field observations prove that Scarbantia also had an amphitheater (125 by 85 meters); Brigetio had one west of the camp; in Mursa, Savaria, and Siscia the Nemesis altars refer to gladiatorial games. The popularity of the games is attested by the many gladiator statuettes that have survived. In the first century a veteran of Legio XV Apollinaris became *magister ludorum* in Scarbantia.[37] In Mursa there is mention of a stadium. In Sirmium it is supposed that a *circus* existed on the basis of the *curulis* games which were held there during the visit of Emperor Julian.

Many relics survive to prove that provision was made for daily entertainment: a mill-game (merils) incised into brick, markers, and dice are well-known results of archaeological research. In Gorsium from the pavement of the *portici* surrounding two sides of the *decumanus* several playing dice have emerged.

[1] Vell. Pat. 2. 110.5.

[2] A. Mócsy, "Pannónia", *R. E. Pauly-Wissowa*, Suppl. 9, p. 771.

[3] Fronto, *Princ. hist.* 13.

[4] *Aur. Vict. Caes.* 39.26.

[5] *Hist. Aug. Aurel.* 24.3.

[6] Mócsy, p. 768.

[7] *CIL*, III, 3351.

[8] L. Nagy, "A szír és kisázsiai vonatkozású emlékek a Duna középső folyása mentében— Les monuments se rapportant à la Syrie et à l'Asie mineure, dans le cours moyen du Danube". *ArchÉrt* 52 (1939), p. 120.

[9] Mócsy, p. 741.

[10] J. Fitz, Gorsium: *A táci római kori ásatások (Gorsium: Excavations at Tác from the Roman Period)*, 3d ed. (Székesfehérvár, 1970), Figs. 28–29.

[11] Mócsy, p. 730.

[12] R. Egger, "Pannonica", *Omagiu lui C. Daicoviciu*, (Bucureşti, 1960), pp. 167–69; G. Alföldy, "P. Merlat, Jupiter Dolichenus". *AntTan* 8 (1961): 300–302. The inscription on the Sárpentele altar corroborates this supposition.

[13] J. Fitz, "Bronzestatuetten", in *Intercisa* (Budapest, 1954–1957), 2:165–71(*ArchHung* 36).

[14] J. Fitz, *Hercules-kultusz eraviszkusz területen*—Culte d'Hercule dans les régions eravisques (Székesfehérvár, 1957); J. Fitz, "Sanctuaires d'Hercule en Pannonie", in *Hommages à A. Grenier* (Bruxelles, 1962), 2: 623–38 (*Coll. Latomus 58*).

[15] I. Paulovics, "Dionysosi menet (thiasos) magyarországi római emlékeken—Der dionysische Aufzug (Thiasos) auf ungarländischen Denkmälern", *ArchÉrt* 49 (1936), pp. 5–7, 31–32.

[16] Mócsy, pp. 731–34.

[17] A. Alföldi, "Epigraphica I: Eine batavische Göttin in Pannonien". *Pannonia* 1 (1935): 184.

[18] J. Fitz, "Bronzestatuetten", in *Intercisa*, 2:168.

[19] D. Tudor, "I cavalieri danubiani". *Ephemeris Dacoromana* 7 (1937): 189–356; J. Fitz, "Bleigegenstände", *Intercisa*, 2: 383–97.

[20] V. Wessetzky, *Die ägyptischen Kulte zur Römerzeit in Ungarn* (Leiden, 1961); A. Dobrovits, "Az egyiptomi kultuszok emlékei Aquincumban—The Cult of the Egyptian Gods in Aquincum". *BudRég* 13 (1937): 47–75.

[21] Z. Kádár, *Die kleinasiatisch-syrischen Kulte zur Römerzeit in Ungarn* (Leiden, 1962); J. Fitz, *Les Syriens à Intercisa* (Bruxelles, 1972), pp. 177–97 (*Coll. Latomus*).

[22] M. J. Vermaseren, *Mithras: Geschichte eines Kultes* (Stuttgart, 1965); M. J. Vermaseren, *Corpus inscriptionum et monumentorum religionis Mithriacae* (Hague, 1960), vols. 1 and 2.

[23] G. Alföldy, "Pannoniciani augures". *ActaAntHung* 8 (1960): 145–46.

[24] Mócsy, pp. 745–46.

[25] J. Zeiller, *Les origines chrétiennes dans les provinces danubiennes* (Paris, 1918); T. Nagy, "A pannóniai kereszténység története a római védőrendszer összeomlásáig" (History of Pannonian Christianity) (Budapest, 1939) (*DissPann* 2:12); Mócsy, pp. 750–58.

[26] L. Nagy, "Temetők és temetkezések" (Graveyards and Burials) in *Budapest az ókorban* (Budapest, 1943), 1: 464–85; K. Sági, "Die Ausgrabungen im römischen Gräberfeld von Intercisa im Jahre 1949". *Intercisa*, 1: 61–120 (*ArchHung* 33).

[27] L. Nagy, "Les symboles astraux sur les monuments funéraires de la population indigène de la Pannonie". *Laureae Aquincenses* 2 (1941): 232–43 (*DissPann* 2:11).

[28] K. Sági, "Kocsiábrázolások Pannónia császárkori szepulchrális vonatkozású kőemlékein—Rappresentazioni del carro sui monumenti sepolcrali della Pannonia imperiale". *ArchÉrt* (1944–1945), pp. 214–48.

[29] K. Sági, "Adatok a pannóniai császárkori kocsitemetkezések ethnikumának kérdéséhez—Sur le caractère ethnique des enterrements à char à l'époque impériale". *ArchÉrt* 78 (1951), pp. 73–78.

[30] A. Burger, *Áldozati jelenet Pannónia kőemlékein (Sacrificial Scenes on Pannonian Stone Monuments)* (Budapest, 1959) (*Régészeti Füzetek* 2:5).

[31] K. Sági, "Császárkori tumulusok Pannóniában—Tumuli dell'età imperiale nella Pan-

nonia." *ArchÉrt* (1943), pp. 113–43; J. Fitz, *Zur Frage der kaiserzeitlichen Hügelgräber in Pannonia Inferior* (Székesfehérvár, 1958); S. Pahič, "Nov seznam noriško-panonskih gomil". *Razprave-Dissertationes* 7 (1972).

[32] Sági, *Intercisa*, 2: 61–123.

[33] V. Lányi, "Die spätantiken Gräberfelder von Pannonien". *ActaArchHung* 24 (1972): 53–213.

[34] Mócsy, pp. 722–23.

[35] G. Erdélyi, *A római kőfaragás és kőszobrászat Magyarországon* (Roman Stone Carving and Stone Sculpture in Hungary) (Budapest, 1974).

[36] Mócsy, pp. 771–72.

[37] Gy. Hajnóczy, "Pannóniai amphitheátrumok" (Pannonian Amphitheaters). *Épités–Épitészettudomány* 5 (1973): 127–50.

# RELIGION

### EDIT B. THOMAS

## NATIVE AND "CLASSICAL" GODS

My intention is to throw some light on the religious beliefs, religious world, and religious conceptions of the peoples living in the former province of Pannonia and its neighborhood and to acquaint our readers with them to the extent rendered possible by our archaeological finds. There are, however, areas and relationships which have not yet been clarified by scholars; nevertheless, I shall try to discuss certain features and molding factors of the religious life of the province.[1]

Before the Roman conquest the population of Pannonia consisted of ancient Illyrian and Celtic elements. Their beliefs became complemented after the conquest with the world of the Roman gods. The inner world and imagination of the inhabitants of the province became saturated with the threefold components (Illyrian, Celtic, and Roman) of a specific sphere of faith. However, this religious belief and the concept of the gods did not survive in a rigid form based on static laws. Preserving its main features, it changed and altered to satisfy the requirements of individual men and groups of people, conforming to their interests and occupations, and also adapting elements of other religions of more distant regions.[2]

To be exact, very few traces survive of the religious beliefs of the native population.[3] The names of certain gods which come down to us provide little information about their essential nature. Aecorna can be of Illyrian origin. Fata Tenatia was an ancient Celtic god, the representative of fate, with two other companions, Riodiaspe and Didone. Sedatus was the god of fire, always accompanied by Succellus carrying the hammer.[4]

The gods of the natives, however, only take shape within the religious conceptions of the Romans. Even specific cults emerge in a Roman form. From the Romans the Pannonians learned how to represent their gods in pictures, statues, reliefs, and on metal, and also how to raise shrines and votive monuments. This Romanized outward form renders it almost impossible to find the underlying native cults; nevertheless, it is only through these adaptations that we can form any ideas of the religious life of the Pannonian native population.

To start with, it is gravestones especially which help us in forming ideas about the religious life of the natives; mainly those gravestones which are incised with the symbols of the world to come, the disk of the sun, the sickle of the moon, the stars, as the "keys of heaven" (Pl. XL.1,2) — these are eternal symbols serving as objects of religious worship for practically all early agricultural nature worships.[5] At the same time, we find the names of the deceased in Latin characters on the stones, according to the accustomed formulas.[6] These gravestones (Pl. XCI)

12

belong to Pannonian grave monuments of the roughest execution; the material was easy to handle, mainly sandstone and limestone, both being soft. They were made in the first century or the early decades of the second, in places distant from the area occupied by the Romans, in Celtic Eravisci and Azali tribal settlements. They emerged east and south of Szőny[7] (Brigetio), almost as far as Adony.[8] This region also shows traces of an Illyrian population previous to the Celtic settlement, and we can find on gravestones with astral symbols ancient Illyrian names. However, the emergence of these symbols in the first century can be connected with the western Celts.

The cultic rites of the ancient Pannonian population often referred to the sun and moon as sources of fertility.[9] An outstanding group of relics of lunisolar worship was unearthed at Szalacska in County Tolna (Fig. 3 and Pl. CIV.1, 2). The finds conserved for us the cultic jewelry of a Sol-Luna priestess and other objects of the cult. Radial crowns adorned with *lunulae*, seven breast plates covered with the symbols of sun and moon, a larger plate representing the figure of Sol formed the silver ornaments of the priestess's habit. Fibulae to hold clothing together and ornamented plating on girdles were some of the jewels of native Pannonian women. Bronze sacrificial vessels with handles, bronze *lucernae*, and the vessel for smoking resin that simulated the shape of a foot were all used in cultic rites and sacrifices. The silver cultic symbols to adorn the priestess and the jewels formed a set and emerged from the same silversmith workshop. Considering the jewelry and metal finds which emerged at Szalacska, we can rightly presume that they were all made by the same local silversmith. The time of their making is determined by the Noricum Pannonian type of fibula[10] which represents the earliest example of the type from the third quarter of the first century. The lamp and the foot-shaped vessel come from the late first or early second century. The *patera* comes from the second century. The bronze kettle where the finds were hidden is a second- or third-century type.[11] The early imperial Sun-Moon cult of the ancient Pannonian population survived unchanged by external events up to the early third century, according to the evidence of the finds.

Pan (Pl. CIII) was the chief deity of the native Illyrian population, which may also account for the name of the province — Pannonia. The original meaning of the word is: plentiful, opulent. Indeed, Pannonia was immensely rich in woods. The province, especially the present Transdanubian region, was covered with extensive wild forests some 2,000 years ago. Shepherds grazing their animals often seemed to hear the fearful yells and frightening sounds of the lord and god of the forests and were seized by panic, fear, and terror. When they succeeded in pacifying the god with plentiful sacrifices, then he appeared with his gay retinue and blew his fine-sounding pipe; he kept away all evil spirits from the flocks and woodland homes.

Pan represents natural forces uncontaminated and uncorrupted by culture — the fertility and plentifulness of nature. The Roman settlers arriving in Pannonia seemed to recognize in Pan their native Silvanus, the defender of woods, lands, and meadows, who was represented on relics with a pruning knife in his right hand and a tree branch in his left. Silvanus (Pl. LXXXVIII), however, was less unruly than the Pan of the Illyrians and Greeks. In Pannonian portrayals he

178

Fig. 3. Sol-Luna priestess in full ornate, restored

figures as a peaceful, bearded god of the woods, dispensing opulence. The duality of his being appears in epithets like *silvestris* (of the woods) and *domesticus* (of the home): referring to the farmers' belongings entrusted to him — woods, land, pastures, herds, or the house and garden of the farmer he undertook to protect.[12] The humble plowman dedicated an altar to Silvanus in one corner of his small garden, whereas the rich landowner raised a shrine in his villa. Silvanus was perhaps the most popular godhead of Pannonia.

Inside the province the soldiers did not have such dictatorial powers as on the *limes* and in the border area, where those gods were venerated who fostered bravery and rewarded military accomplishments — Jupiter (Pl. CVI), Mars, Minerva (Pl. XXXIV), Victoria (Pl. CXI), Hercules (Pl. XXXVI), etc. The religious veneration of the peasant-farmer population of Pannonia was limited, but almost without exception the fertility deities were worshiped. They showed great versatility in the presentation of various fertility gods, calling to life numerous deities who hardly differed from each other. There is hardly any difference between Silvanus and Priapus (Pl. CXII) who is represented with a fruit-covered branch and basketfuls of fruit in his lap; his portraits made the phallus obvious as the symbol of primeval power and fertility. In the agricultural communities shrines were raised not only to Silvanus and Priapus, but also to Silvana, Diana (Pl. CXIII), the Nutrices, and Fortuna, the goddess of plenty whose cult often overlapped that of Isis (Pl. LXXXIV), the Egyptian goddess of fertility and the preserver of corn.[13]

Besides the cult of Silvanus and his female counterparts, it was Diana who enjoyed the greatest popularity inside the province. The native goddess who lurks behind the name of Diana was a forest deity, the protector of wild animals. In the shrines of the humble population of Pannonia — farmers, hunters, woodworkers, etc. — there are plenty of cultic objects referring to the Diana worship.[14] In reliefs she often figures dressed in a short tunic, with bows and arrows in her hand, and accompanied by dogs and deer. The shrines of this virgin goddess, who generally hunted at night, were numerous in the villages and farmhouses of Pannonia, and then have been most frequently unearthed by chance excavations of archaeologists. In the retinue of Diana can be found the Nymphs of trees, fields, and wells.[15] Moreover, Diana was also present in the amphitheaters where wild animals were hunted to death. From the votive stones placed in amphitheater shrines it appears that her figure was often conjoined with that of the bloodthirsty Nemesis, as epigraphs refer to Diana-Nemesis.[16] She too, like most of the natural deities of Pannonia, had both a savage and a mild counterpart.

For the population of Pannonia it was of primary importance to find wells which supplied cool, fresh, and plentiful drinking water. When establishing towns, villas, or farm settlements, it also was considered most essential to have a good water supply. The Romans, who at home enjoyed first-rate bathing facilities, established in Pannonia both private and public baths. A high cult of water can explain the Pannonian worship of the source goddesses and Nymphs. Near hot water sources shrines and statues were raised for the deities of health, Aesculapius and Hygieia.[17]

The shady hills of the Pannonian mountains provided medicinal wells, and on the sunny slopes vines were planted, even before the Roman settlement, in order to obtain "the divine juice" from grapes. Excavations carried on near Lakes Balaton and Fertő, in the Mecsek hills, and in Szekszárd and Óbuda unearthed many altars of the wine god so much venerated by the Pannonians. When the Greek gods arrived in Italia, Dionysos (Pl. CXIV), the god of wine and the passions, was identified with Bacchus (Pl. CXVIII) and Liber (Pl. CXV).[18] The name of the ancient Sabine god means "free" and is the symbol of the soul liberated in the passions. The female counterpart of Liber Pater is Libera (Pls. CXVI and CXVII). On altars raised in their honor they are generally represented together. The flowering of the Liber Pater and Libera cult in Pannonia is primarily due to an extensive viticulture. By erecting altars the owners of the vineyards entrusted their vines to the gods — making it the gods' duty to provide a good vintage.[19] It seems to have been all in vain. The gods did not help the Pannonian vineyard-owners when an imperial edict by Domitian prohibited planting new vines and forced owners to eradicate half of the old ones because the excellent Pannonian wines jeopardized the export of Italian wine. The miserable situation continued for 200 years until Emperor Probus, who was of Pannonian origin and sympathized with the plight of his countrymen and their financial disaster. He stopped the ban on viticulture and promoted the production of wine.

The Romanized natives and the Roman small farmers, as well as the soldiers of mixed blood, produced objects of worship, altars, statues of gods, and votive reliefs in executions which can hardly be called artistic. The landowners and the equestrian class, members of the administration, and the rich merchants, when settling down in Pannonia, brought with them their statues of gods conceived in a classical spirit and frequently done in highly artistic executions.[20]

We can reconstruct the process by which the classical gods first emerged in Pannonia in the home services of private individuals and also in state services. We are fortunate to have discovered several private shrines in Pannonia, with full equipment and objects of worship, so that we are no longer compelled to conjecture on the basis of a few excavated statues about the whole ensemble of gods that were placed in shrines.

Objects of family worship were arranged in several ways. It was customary after the Italian fashion to place the statuettes of gods and *lares* inside the house in small niches built in the walls. However, one or two structures also served as shrines which could accommodate stone altars and larger marble or limestone statues enabling worship at the time of sacrifices. In some houses altars of the favorite gods were placed in interior courts, side by side in the open air, sometimes with roofs raised above them. Outside the buildings, small semicircular niches were built into the fences of gardens for statues or altars to be placed in them. Further removed from the houses, Pannonian inhabitants, both settlers and natives, also erected stands for the pictures of their gods in the woods belonging to estates, and in groves, fields, and arbors. But the finest statues cast in bronze and best reflecting divine images stood inside houses, in the niches of shrines or in *lararia*, erected by newly arrived landowners. From these a few that figure among our illustrations preserve practically intact their 2000-year-old shapes.

In 1907 on the edge of the village of Nagydém, in the shrine of a villa in a Roman settlement, a concealed treasure trove was discovered (Pl. XXVI).[21] The find contained the finest bronze figures in Pannonia. Judging from their quality, they must have served as ornaments in the home of a well-to-do settler from Italia. In the bronze vessel where the objects were hidden there were also three *lucernae*, a jug with an ornamented handle, and statues of Lar and Apollo. All these formed the equipment of a *lararium*, a family shrine.

The Lar of Nagydém preserves a typical form of the family god of the Romans. Its beauty and artistic execution raises it far above any average piece. Its appearance was festive, peaceful, and sublime. In the middle of the floating tunic there was a gap in the casting of the drapery, and two perpendicular grooves indicate that enamel or bronze of a different color was wedged into the *clavi*, or purple strips, the insignia of knighthood. This very rare Lar Angusticlavus (Pls. XXVI and XXVII) is a definite proof that the statue had been brought along by a Roman settler who ranked as a knight, in the first century A.D.[22]

A bronze Apollo statue (Pl. XXVIII) in the find shows the god during a drink offering, a libation. It is a highly idealistic presentation. Its appearance and conception typify the Hellenistic artistic trend and concept of god. Its provenance is conjectured to have been a first-century workshop preserving Hellenistic traditions from the Balkans, perhaps Dalmatia.

The god statues of the *lararia* emerging from the village of Tamási in County Tolna (Pls. XXIX–XXXII), those of Nagydém, by far surpass in significance the cultic objects of shrines in the various provinces. The first group of finds from Tamási emerged in 1941 on the southeastern edge of the village at the bottom of Szőlőhegy, in the valley of the rivulet Koppány, alongside the old Roman road. The objects were carefully hidden in a pit; their owner, fearing some disaster, must have dug them in and never returned for them.

Among the objects of this household shrine was a Jupiter statue,[23] exquisitely shaped of bronze, 31.5 centimeters in length. The godlike image, well-marked muscles, fine face, and highly stylized hair seem to prove an Italian origin.

Another piece in the find is significant, not so much from an artistic viewpoint but concerning its divine essence. It is a 33-centimeter tall woman's figure (Pls. XXIX–XXXI), sitting majestically enthroned, wearing a diadem and veil. The cornucopia in her left hand confused earlier scholars, who wrongly considered her to be Abundantia, the goddess of plenty. The analogy of Roman coins, however, convinces us that the household goddess of Tamási is identical with Concordia, the representative of the ideal of agreement. In Roman religion Concordia was among the most venerated deities. Originally she represented agreement between the citizen parties; later she became more and more associated with the imperial house and agreement between the emperor and his wife or among emperors themselves, symbolizing sometimes even good spirit in the army. Empresses had often been represented by the figure of Concordia.

We believe that the Tamási statue is a Concordia Augusta type. The nearest analogies to this statue can be found on the coins of Emperors Vespasian and Domitian. This similarity also determines the time when the statue was made, which must have been the last third of the first century. The statue's execution, its portraitlike character, and its majesty has suggested the attempt to find the person it was meant to represent. Comparing the statue's features with those of

Empress Domitilla, wife of Vespasian, found on coins, reveals a very striking similarity. A pronounced nose, a protruding chin, and a small sunken mouth all seem to prove that the statue represented the empress personifying Concordia.

Summarizing all that we know about the household shrine at Tamási, we can state that it had formed part of a shrine belonging to a person of Italian origin who filled either a high-ranking military office or an administrative role. All the pieces are of first-century origin, dating from its last third.

The second Tamási find, a bronze statuette of Athene or the goddess Minerva(Pls. XXXIV and XXXV) emerged in 1952 at the northwestern border of the village during spring plowing. It was also an object of worship in a house whose walls were dislocated by the plow.

The gods represented in the Tamási finds are not projections of ancient Pannonian worship and religious feeling, but rather are entirely foreign. They stand nearer to the classical deities and represent the impact and molding force which transmuted the images of the gods of the conquering Romans. At a later stage the need for high style artistic works also developed in the local population. This is proved by the existence of many excellent, small, bronze statues of gods found in Pannonia which were mostly made in Italia, Gallia, and even Alexandria, but became objects of worship in Pannonia.

The gods adapted from foreign cults took root in Pannonia and became objects of worship, especially among the military and administrative officials or even the natives who worked in the administration. Telling evidence of the Romanization process is the altar raised to Jupiter by Flavius Pitianus Augustus for the benefit of the emperor and his family. The priest came from the Celtic Eravisci tribe which had lived on Gellért hill and raised the altar in the third century.[24] It was in such ways that natives assimilated the customs of the Romans.

The Italian settlers gave highest consideration to the unity of the empire and the deities which promoted its greatness and splendor: Jupiter, Juno, and Minerva, the Capitolean Trias.[25] Shrines of the Trias existed everywhere in administrative centers, cities, larger localities, camps and adjoining civil towns. The temple erected on the Capitolium became the religious center of the state. Romans believed that religion promoted the survival of the state. Rome's fate was administered by the gods and her doom would result if the gods turned away from her. Anybody wanting to remain faithful to Rome had to persist in religion. However, it was considered a person's private affair, so to speak, depending on his sphere of interests, as to which god he selected as a patron. Some groups celebrated the gods of fertility providing plenty, and others, the classical high deities.

Soldiers and administrative officials raised altars to Jove, lord of the universe and god of order and security in the state, and marked their ranks on the inscriptions. Jove's activity as chief god extended to nature as a whole and to all manifestations of human life (Pls. CVIII–CX). It was due to his will that day followed night; he was the lord of thunder and storms. As he had mastered the forces of unruly nature, he was responsible for the wars waged on earth. His weapon was lightning and he was generally represented on Pannonian relics with a bunch of thunderbolts in his hand. The eagle, the insignia of the legions, was dedicated to him.

In the shrines of the capitolia notably in the pompous ones of Savaria (Pl. CVII) and Scarbantia (Pl. XIII), which were established in Pannonia as early as the late first century and early second century, on the right of the chief god sat his wife Juno and his daughter Minerva.

The cult of Juno (Pl. CXXI) was general throughout the empire. She was the queen of the skies, the paragon of wifely virtues. Minerva was the goddess of sober wisdom, a virgin who sprang fully armored from the head of her father, Jove. In war she sided with the wise, sober warriors, and in peace she presented mankind with the gift of the arts and sciences. Minerva's attractive figure and character inspired many artists. Her statues in Pannonia display all the aspects of beauty, striving increasingly to render a more perfect semblance of the goddess's exquisite being.

Mars, Venus, Apollo, Mercury, Hercules, Aesculapius, Fortuna, and other Roman gods also had their roles in the religious worship of the inhabitants of Pannonia. On relics they appear in many shapes, displaying many aspects of character, depending on which feature would best impress worshipers.

Later, to safeguard the survival of the state, not only the gods but the emperors were worshiped all over the empire (Pls. CXXII and CXXIII). The emperor became the symbol in which was concentrated the majesty of the state. This worship naturally was extended to some extent to other members of the royal family, too. The emperor's portrait was identified with his personality and evoked not only honor but religious worship in his subjects; no military camp or official place could exist without it.

On the forum of each important town in Italia and the provinces, including Pannonia, the temple of the ruling emperor was erected near the *capitolium* — Templum Augusti or his altar, Ara Augusti. In the temple and before the altar were conducted the religious ceremonies of the emperor's cult. Here prayers were given for the welfare of the emperor, food and drink were offered, and processions were held. The welfare and prosperity of all the citizens in the empire depended on the emperor; therefore, in the first days of the year, on the high feasts of the emperor, communal prayer was offered for his health, welfare, and salvation. At these occasions games were organized in the circus and amphitheater and it was also customary to distribute honey cakes (Pl. CXXII) among the people. The cakes were poured into molds which represented the emperor, members of the imperial family, a triumphal procession, symbols of opulence and plenty, Fortuna, Mercury, and others. Such molds have emerged in great numbers from excavations in Pannonian cities that existed under Roman rule.

Such days were not only occasions for merrymaking among the people but also for the meeting of leaders of colonies and *municipia*, where native delegates and legates of the cities also concurred. In the first century the Pannonian center of the emperor's cult was the Roman city of Savaria, the present Szombathely. Here there existed a marble altar raised in honor of Emperor Claudius who gave the city colonial status, the same Divo Claudio who was declared god after his death. We also know the names of some high priests of the cult, such as Valvius Verus and Chorius Sabinianus. Several portraits of the emperor were unearthed at Savaria, the legends of which refer to the divine cult of the emperors.[26] The camps and civilian cities of Brigetio, Aquincum, and Intercisa settlements within the province have provided many examples of the Pannonian imperial cult, in both inscribed altars and statues.

From the territory of the camp at Brigetio was excavated a portrait made of silver plate, representing Trebonianus Gallus (A.D. 251–253) (Pl. CXXIII). This half-length portrait was probably preserved in the legion's shrine with other military insignia.[27] At Ajka emerged a statue of Hercules, the cultic statue of a shrine belonging to a villa (Pls. XXXVI and XXXVII).[28] The shrine was erected in a grove near the house, and it contained an altar dedicated to Hercules and several gravestones of Dalmatian owners. The statue is the typical work of a Pannonian stonecutter. When he carved the body he left the surface rough, whereas the head is an artistic production. The face is well worked out like a portrait, so that we can even name the person it represents — Emperor Commodus (177–192). Commodus was the first emperor recognized by the senate as Hercules Romanus who also adopted the epithet.

Changes in the religious life of Pannonia kept pace with the rapid development of the province. A primitive nature religion was soon replaced by the diverse Roman god cults, as the primitive method of farming was supplanted by the Roman plantation system. However, neither the former, nor the latter was slavishly imitated. The ancient religion of the natives mingled with that of the conquerors and simultaneously assimilated with developing social and productive conditions. The early phase of development culminated in the adaptation of the imperial cult that was the religious reflection of the province's full assimilation as an economic and political unit of the empire.

## ORIENTAL CULTS

In the religion of Pannonia two dominant factors can be observed in the first two centuries. Both are inseparable from the historical process which is summed up in the term of Romanization. On the one hand, the native population of the province became acquainted with the classical gods of the conquerors; on the other, the Romans got to know the native gods of the locals. The gods with similar characteristics but different names were increasingly assimilated, their features intermingling in new local deities corresponding best to the conditions, circumstances, and imagination of the local Pannonians, to the extent that it became immaterial under what names they might survive. What is essential is that several gods and godlike characteristics are united in one form under one and the same name. This process, which in outward form (as far as the god's name is concerned) simplifies but in content (the multiple character of the god concerned) enriches, is called syncretism.

Archaeological material also indicates that still other gods were imported from the distant eastern provinces of the empire to live in the Danube region and to enrich the religion of the province with assimilated new features. When speaking of Eastern cults in Pannonia certain religions are meant which arrived in Pannonia from Greece, on the one hand, and from the more distant Hellenistic East, on the other.

As early as the turn of the first century Eastern traders arrived in Pannonia, as did customs officials, slaves, and soldiers who had worshiped gods not known at all in the Pantheon of the Graeco-Roman world — gods indigenous in their distant homes which before their coming to Rome and the far-off provinces of the empire were worshiped only locally. Eastern religions reached Pannonia already in Romanized form, in a stage of development when their deities had more or less mingled with some Graeco-Roman godheads, without losing much of their special attributes.[29]

It is characteristic of the majority of the Pannonian relics that they present Eastern motifs in a transformed, transmuted shape, so much so that both archaeologists and historians have difficulty deciding which of the gods are dominant and which particular religious cult an object in question had served.

It is extremely difficult to understand the essence of Eastern cults and the nature of their ceremonies even from contemporary authors, because most of them had adhered to the classical Roman religion. Tacitus himself despised the foreign gods, and Ovid described any deity through his fancy, as far removed from reality as possible. The question becomes even more involved because many Eastern cults belonged to the so-called mystery religions, and only those who were initiated into the mysteries knew about the holy acts and the essence of the rituals of the sacrifices, without revealing their secrets to the uninitiated. The only means of learning about these mysteries and local cults are Roman stones with inscriptions, altars, votive tablets, and statues of deities that have been luckily preserved in the earth and excavated. Studying these and interpreting them brings us step by step nearer to a solution.

During the four centuries of Roman occupation in Pannonia, side by side with the well-known Graeco-Roman gods, deities of Egypt, Asia Minor, Syria, Iran, and sundry Eastern provinces were venerated.[30] Their reception and adoption in Pannonia can be explained by the fact that the oppressed, slaves, poverty-stricken freemen, and men on military service for many decades were yearning for peace, happiness, wealth, and prosperous citizenship. One of the basic features of the Eastern gods was to promise not only happiness in the world to come but happiness on earth of those who worshiped them. Since the concept of welfare and happiness has always been relative, the circle of their worshipers was very large. From the humble, unpretentious poor to the wealthy businessmen all expected a lucky venture in their dealings and offered sacrifices to these gods.

Strict rules provided for the services of the simple, almost rigid Roman religion. However, the Eastern cults affected the senses, and the believers were dazzled by the pomp of the ceremonies, the scented incense burners, the music, and riotous cultic dancing.

We can discuss some Eastern gods whose names were preserved in inscriptions, whose figures are conserved in museums in the forms of statues and reliefs. They are so numerous and so different that we can deal only with some outstanding ones. The names of some less significant gods, such as Baltis, Arvis, Dea Syria, Baal, Deus Dobrates, Ariman, Kendrisos, etc., have been preserved only by name

186

i n some inscriptions.[31] There are many data for a full acquaintance with Isis (Pl. LXXXIV), Osiris (Pl. CXXVIII.1), Serapis, Kybele, Jupiter Dolichenus (Pls. CXXIV and CXXV),[32] Sabazios (Pls. CXXVI and CXXVII), the mounted gods of the Danube region, Sol (Pl. CXXVIII.2), and the Pannonian cult of the god of the Jews (Pl. CXXXV.1, 2).[33]

The Egyptian Isis cult had its followers in Pannonia as early as the first century. A member of an Aquileian family of merchants in Savaria named Barbius, who enjoyed the highest ecclesiastical rank of provincial priest (Pontifex),[34] erected an altar to the "Gloriosus Isis". Altars in Pannonia dedicated to Isis were raised mostly by merchants, customs officials, and slaves who hoped to receive welfare and wealth.

At the turn of the second and third centuries, the Severus dynasty, which loved the East, fostered in the province the cult of Isis, which flourished particularly in the civilian cities among the civilian population. A shrine recently unearthed at Savaria[35] must have been the center of the Isis cult; even its ruined remains point to its former splendor. The building of the shrine was contained in a yard divided by columns. The inscription placed above the shrine's entrance tells us that the temple was dedicated to "Gloriosus Isis". On the façade a fine marble relief represented the goddess riding on her holy animal, the dog Sothis, and holding in her hand the bell called *sistrum*, used during the service (Pls. XIX, LXXXIV and LXXXV).

Memories of the Isis cult also survive in Scarbantia (Pl. LXXXVI),[36] and within the province as well. Humble votive relics made of lead evidence the respect of simple farmers. These were found at Gyulafirátót-Pogánytelek (Pl. CXXIX);[37] the goddess is represented, besides her own attributes, with the insignia of Fortuna. According to the teaching of the myth, Isis supplied man with the plow and taught him to till the land. She is often represented with the cornucopia. It is natural, therefore, that because of similarities in her functions and attributes, her figure often mingles with that of Fortuna.

A concise definition of the actual character of Isis comes from an inscription of a second-century shrine on the island of Chios. Here the goddess tells us briefly what her connection with mankind means. In the text there exist, side by side, ancient traces of her Egyptian roots and the basic concepts of human culture: "I am Isis, Goddess of the Universe. Since Hermes taught me to write, the same runes were not used for everything. I made laws for men and I forbade them to change them. I am the eldest daughter of Chronos. I am the wife and sister of King Osiris. During the summer my abode is the constellation Sothis. I am called their goddess by the women. It was for me that the city of Bubastis was built. I have separated the earth from heaven. I have marked out the course of the stars. I have joined the path of Helios and Selene. I am the founder of navigation. I have established the kingdom of mighty justice. I made men and women get to know each other. I confine the ten-month-old embryos of women in labour. I made children love their parents tenderly. I brought affliction on those parents who lived without love. Together with my brother Osiris we stopped cannibalism. I made ceremonies compulsory for humans. I taught man to revere the image of the gods. I have pointed out the boundaries of the gods. I have crushed the power of the tyrants. I made men love women. I made truth more powerful than gold and silver. I made truth similar to beauty . . ." From the simple wording of the text, but more particularly from the simplicity of the presentation, it is evident that during the second century there emerged a regular Isis Panthea and the rule of Isis became general, extending to all spheres of life.[38]

The cult of Osiris, who died and revived again and again as a result of the cult of Isis was celebrated yearly and formed part of the religious ceremonies. Heretofore we hardly believed, on the basis of existing Osiris relics in Pannonia, that this celebration was regularly renewed. Today, however, on the basis of the Savarian Iseum, we have every right to suppose that ceremonies and processions to celebrate the symbolic death and resurrection of Osiris were held here.

The figure of the revived Osiris together with his holy animal, the bull Apis (Ozarhapi-Sarapis) (Pl. CXXXI), was identified by the Egyptians with an enormous deity who arrived from the Ponthus region and whose name was Sarapis. The Greeks identified Sarapis with Zeus; the Romans, with Jove. In Pannonia, as a result of the Roman identification, Jove appeared as Sarapis with an eagle on his breast and a basket on his head, as a symbol of the crop. By nature Sarapis was a god of the netherworld, an oracular god who helped his worshipers by showing them in their dreams the right course to follow. On the faces of his statues the ruling features are attractive goodness and mysterious rigor.[39]

Among the Eastern gods in Pannonia, Cybele, identified with Magna Mater, enjoyed the greatest respect among the civilian population, just like the gods of the Egyptian cults. Cybele was a Phrygian mountain goddess who was also called the mother of the gods. Left alone on a hill as a child, Cybele was raised by lions and panthers. Her worship started among the shepherds who, for her benefactions, called her the good mountain mother. In the imperial period of Rome, Cybele was already a syncretized figure, combining Egyptian and Graeco-Roman features of various gods. She was an all-powerful goddess, the mother of all and sundry, the queen of heaven and earth and the underworld. The only written record of the Pannonian Cybele cult was recovered from a settlement inhabited by natives, the Eravisci, in the territory of a *civitas* on Gellért hill in the second century. In the cult of the goddess the hill and its rocks played a great part, so that the columns erected on the side of the hill may remind us of a Pannonian mountain shrine.

On Pannonian gravestones and on the reliefs of sarcophagi, there are the recurring figures of young men, wearing short shepherd's garments, with cloaks on their shoulders and pointed caps on their heads — in them we may recognize Attis (Pl. CXXX). His story closely follows the Cybele myth. Respect for Attis belongs to the circle of mystical cults. Cybele fell desperately in love with the handsome young god who perished in a very cruel way. At Cybele's intervention the gods arranged for him to come to life in three days. His death was bitterly bewailed and mourned by his followers. Early in the spring celebrations were held in honor of Cybele and Attis that lasted for several days; they intermittently gave expression to deep mourning and unbridled joy.

The Attis worship was originally dependent on the Cybele-Magna Mater cult. While in Pannonia little survived from the cult of the great mother goddess, the

188

death cult of Attis found numerous expressions. His presence on gravestones is hardly the sign of the Attis worship. Rather he became the symbol of death and resurrection, with a burning torch raised upward and another extinguished one downward.

The conquest of Syria brought Syrian merchants galore to all parts of the empire. The gods of the Syrian religion were imported by them, by slaves carried away and by the women accompanying soldiers and craftsmen. The Syrian gods assimilated into the Roman environment enlarged the Roman Pantheon, which was already teeming with an indefinite number of gods. In the second century temples of the goddesses Dea Syria and Baltis already existed in the civilian city of Aquincum. Actually only a simple and humble inscription was discovered, which indicates the existence of only a shrine.

When surveying the archaeological finds in Pannonia we can see that the goddesses venerated according to Egyptian, Syrian, and Asia Minor cults were mainly popular among women and the civilian population. Even slaves and liberated slaves had raised altars to the Eastern goddesses.

Among soldiers in the camps only those Eastern deities whose myth centered on fight, battle, and ensuing victory were venerated. In the military camps alongside the Danube where Syrian troops were stationed, Jupiter Dolichenus (Pls. CXXIV and CXXV), the Baal of the city of Doliche in Commagene enjoyed great popularity. Dolichenus, the armored god was not only popular among the Syrians but also among veteran Pannonian soldiers who had served long in the East and continued to worship the god with whom they had become acquainted there. We are well introduced to the Dolichenus cult, the manifestation of the god, and his iconographic appearance by the relief of a bronze emblem found at Kömlőd (Pl. CXXIV) in County Tolna. The inscription on the exquisite festive emblem testifies that it was dedicated by the centurion P. Aelius of the I. Alpine foot division to Jupiter Dolichenus.[40]

The triangular insignia, decorated on two sides, was fixed to a pole as a standard at the cultic ceremonies of Jupiter Dolichenus. Originally the bronze tablet was entirely covered with silver. On the cultic picture Dolichenus, in full military attire, stands on the back of a bull; in his right hand he holds a double-edged hammer and in his left a bundle of thunderbolts. Victoria offers to Dolichenus the wreath of victory. Next to the god there is an altar with a burning offering. Above his head are Sol and Luna; at the feet of the bull are half-length portraits of Hercules and Mars. Both the gods and the scenes have a symbolic meaning. On the reverse of the emblem, among other scenes, we can see the holy eagle of Jove. The connection is obvious between the main god of the empire, Jove, and the Syrian Dolichenus, whose origin may go back to some appearance of the Greek Zeus.

In the southwestern corner of the Roman camp in Brigetio a shrine of Dolichenus was unearthed at the end of the last century.[41] Among the ruined walls there were some cultic objects, statues commemorating the god, and reliefs which partly bring to life the ceremonies which were held there. On the Pannonian cultic relics of Jupiter Dolichenus the god is never presented alone. He is always surrounded with unmistakably classical Roman deities, particularly with those whose features he had assimilated. All the relics show godheads of a similar nature blurred, their qualities amalgamated either directly or indirectly in such a manner that this god is presented in a character of unification, halfway toward monotheism.

In the religious life of Pannonia this process toward syncretization is in fact the expression of a lack of confidence in individual godheads. The endeavor to respect all the gods no longer reflects a deep religious feeling, but rather a cautiousness typical of the late Roman religious attitude. At times it is only the figure of the two Jupiters, the Dolichean and the Heliopolean, which is amalgamated, as seen on an inscription at Aquincum. But later one altar was raised "for all the gods and goddesses". It is worth recording that the imperial procurator Publius Aelius Hannonius (in the time of Philippus Arabs) had an altar raised with the following inscription: "For the great and worthy Jupiter, Juno, Minerva, the twelve main gods, Salus, Fortuna, Apollo, the victorious Diana, Nemesis, Mercurius, Hercules, the invincible Sun-God, Aesculapius, and all the immortal gods and goddesses." This inscription no longer shows an unshakable confidence in the gods but rather a total confusion which gripped the unlucky official who depended on the whims of the countless gods, not knowing which one he should worship or which god would inflict anger on him if he omitted to do so. Such inscriptions reflect a disguised mistrust of the gods. A long list of gods suggests a striving toward respect for one which should embody the qualities of all.

The third century syncretism in Pannonia reached its culmination in the cult of the so-called mounted Danubian gods.[42] Perhaps this was the most obscure one among all the mystery religions of Pannonia. Research has so far not discovered either relics with inscriptions or shrines. This cult was so intricate and confused that we only know some of its symbols, without knowing the essence of the religion. All we know derives from simple lead tablets (Pl. CXXXI.2) and awkward marble slabs which were unearthed in comparatively large numbers in Pannonia. Their plentifulness suggests that the followers of this group of gods were also numerous, although the names of the gods remain unknown. The votive tablets made in honor of the gods were particularly numerous in the Danubian provinces of the empire, which inclined archaeologists and historians of religion to refer to these gods as "mounted Danubian gods". It was chiefly soldiers who worshiped the gods on horseback.

In the center of the illustrated lead votive tablet in honor of the so-called cavalry gods stands a goddess (Cybele?) and on both sides, turning towards her a mounted warrior crushing the enemy. On the lower edge was a scene from a feast, a youth identified with Dioscuri or Kabiri and a mystic scene from the killing of the ram. The serpent, a mystic vessel, and other symbols were familiar to those who knew the religion. The whole scene was dominated by Sol wearing a radial crown, going on a triumphal procession in a four-in-hand chariot. The scenes brought into unison by an arch resting on columns.

The Iranian god of light, Mithra (Pl. CXXXIII), was worshiped in Pannonia both by citizens and soldiers.[43] Both cultic pictures and altars bear witness to a wide worship, as happened in Pannonia in the case of most Eastern religions. Archaeologists have excavated many shrines of the Mithra cult in military camps alongside the *limes*, in civilian cities, and also in the interior of the province near

190

civilian settlements. These shrines were generally built near brooks and wells in order to have water needed in the ceremonies. The buildings were dug into small hills; or if hills were not available, they were sunk into the earth to make them resemble the caves which play an important part in the Mithra legend.

We know of four such shrines in the civilian city of Aquincum. Data and observations derived from excavations of these shrines give more or less a sketchy history of this religion in Pannonia. The earliest shrine in Aquincum was built in the industrial quarter of the civilian settlement which counted as a suburb; it was destroyed in the late second century during the wars Marcus Aurelius waged against the Barbarians on the opposite shore. According to the inscription of an altar stone found in the shrine, one leading citizen belonged to the adherents of this religion as early as the middle of the second century. But the shrine was never rebuilt. It seems that with the advent of more peaceful years in the early third century, a larger Mithra shrine was built in the center of the civilian city; in equipping it a city councillor, Marcus Antonius Victorianus, played a leading role. In the late third century, however, this Mithra shrine was also destroyed. Contemporaneous with the former, a congregation functioned in the northern edge of the city; and the worshipers, as the inscriptions prove, were liberated slaves of an Aquincum councillor, Caius Iulius Victorinus. The conditions at the excavation render it probable that the Mithreum had been formed from the premises of a luxurious private house. The finest Mithreum of the Danubian region was built under the Severi: the so-called Symphorus Mithreum, which received its name after a leading figure of the congregation whose magnanimity made it possible for the shrine to be well equipped and to have an altar piece erected.

At Brigetio, outside the camp, side by side with the above-mentioned Jupiter Dolichenus shrine was erected a shrine of Mithra. One of the finest Pannonian cultic Mithra pictures emerged from this shrine — a bronze plate, covered with silver, bearing a relief. In the center is the figure of Mithra, killing a bull; next to him are torch-bearing Phrygian youths, under the hind legs of the bull. Mithra is shown arising from a rock, and the relief contains animals and objects which had a part in the Mithra Mystery. In the upper corners and the bottom of the tablet, classical gods are placed, signifying that whereas the camp soldiers worshiped Mithra, due respect was granted to the gods of the state religion as well.

Mithra was always represented in the same way on the cultic pictures in his shrines. He was a young man, dressed in Phrygian clothes, the god of light who subjugated the huge bull which defied the light and the day and was the symbol of savage, ruthless barrenness. Despite the great number of relics, we do not know much of the Mithra religion. The mystery remained inaccessible for the uninitiated, and the initiated vowed secrecy. We do know that a person wanting to be initiated had to pass through seven stages, *corax, nymphius, miles, leo, Perses, heliodromus,* and *pater,* in order to attain the highest degree of *pater.* The Mithra religion did not indulge in externals only, but in deep insights and spiritual experiences as well. Through different stages of introspection the believers came to grasp the essence of the mythical drama of Mithra and to feel close to the lord of the universe, the victorious and redeeming god, who granted peace of mind and also relief concerning the world after death. Contrasted with polytheism, the Mithra religion was monotheistic and differed from the Roman state religion mainly in preparing the soul still on earth for the life to come. Originally, Mithra was a kind spirit in an Old Persian religion whose cult mingled with Egyptian and Chaldean features. According to the legend, he was born in a cave; he was delivered from a rock and had no human mother. His birthday was celebrated in Rome on December 25. Originally, he represented light and brightness; but he was also considered as the manifestation of purity and truth, the eternal enemy of darkness and falsehood. The Mithra believers saw in him a god dominating the universe and defeating all. The Mithra cult arrived in Pannonia in a highly developed form, when Mithra was already identified with the sun-god. This identification emerged in the early imperial period, in the first and second centuries.

In ancient religion the acceptance of sun-god worship was the first step toward the belief in one god. Sol, the sun, unites in himself the blessing of welfare, the devastating of evil, the warming and maturing but simultaneously destroying and scorching sun. He is invincible, and the later divine emperors identified themselves with his qualities.

Let us briefly survey the Pannonian worship of Sol (Pl. CXXVIII.2), the sun-god, who in the late imperial period obscured the glamor of all other gods on the empire's horizon. Even official religious policy had supported the eastern cults. The women in the Severus dynasty (193–235), Julia Donna, Julia Maesa, Julia Soaemias, Julia Mamaea, came from the Syrian family of the high priest from Emesa, and through them a great number of Eastern deities received official recognition throughout the empire. Elagabalus (218–222) made the sun the chief god in the empire under the name of Deus Sol Elagabal, who was worshiped in Pannonian camp shrines and also in Brigetio and probably in Aquincum. But the cult of Elagabal came too early. Roman faith striving toward monotheism had not matured before the late third century, when Aurelian (272–275) placed at the head of the Roman Pantheon the universal sun-god. From then on Sol became the center in the divine services of emperors and armies. Sol's emblem, the radial crown, and the epithet of "invincible", had also been adopted by the emperors.[44]

The followers of the historically best-known Eastern religion, the Jews (Pls. CXXXIV and CXXXV), after the defeat of Judea (A.D. 70), entered Italia in great numbers as slaves and prisoners of war. But they also appeared in the wake of the Roman army in Pannonia as merchants and traders. The Pannonian population with a bent toward monotheism got to know them in the early third century and to become familiar with their dogma, which received expression in the formula "There is but one God".[45]

In the Roman camp of Intercisa the armed forces were practically annihilated in the Germanic-Sarmatian wars and the camp was quasi destroyed. After the camp was rebuilt, about 176, a new troop arrived with Marcus Aurelius returning from Syria — Cohors I Miliaria Hemesenorum, recruited at Hemesa in Syria. The Syrian influence and reinforcement in Pannonia even increased under Septimius Severus, owing to the family ties of the emperor in that country. The imperial couple was very popular in Palestina and among the Jewry of the diaspora. This becomes evident from a Greek inscription from 197 in a synagogue in Palestina, erected in honor of the emperor and his family. The synagogue of Rome bore the name of Severus.

Among the Hemesians arriving in Intercisa there were also many Jews, so that they established a religious community. It seems certain, on the basis of stone monuments with inscriptions unearthed at Intercisa, that under Alexander Severus (225–235) there was a synagogue (Pl. XCVII) in the city. The inscription which establishes this remarkable fact is most interesting, since it reflects the sentiments of the Jews for the imperial family which protected them. The text of the stone tablet says, "To God the Immortal. For the salvation of our Lord, Alexander Severus, the gracious holy Emperor and his spouse, Julia Mamaea, the emperor's mother. Cosmius is happy to fulfil his vow, he the chief of the customs house and superior of the Jews' synagogue."[46]

Pannonian Jews are also commemorated on gravestones made after the fashion of Roman ones. Their inscriptions, which mixed Greek and Latin characters, prove the mingled nature of their culture. It is worth noting that not a single Hebrew character exists on the ancient Pannonian Jewish inscriptions.

From the Roman settlement of present Esztergom was excavated a large limestone gravestone with the inscription "To the memory of Father Juda and Cassia. Blessing." Jewish identity is proved not only by the names but also by a menorah represented with seven-pronged flames. The stone commemorates a father and mother, both members of the civilian population, and evidently full citizens.

There is in the Hungarian National Museum an interesting gravestone that is implicitly Jewish which probably came from Aquincum (Pls. CXXXIV and CXXXV.1). Though otherwise a regular Roman gravestone, it contains a half-length portrait of a man in the top right corner, a woman in the left, and a child in front. Later owners incised among the figures an inscription of secondary Jewish importance. The parents raised a grave memorial for themselves and their deceased young son, so that the effigies should correspond to the family situation. The inscription, with mixed Greek and Latin characters, tells us: "In memory of Anastasius, Decusanes and Benjamin, our son. There is one God. There is one God. There is one God." To point to the fact that they were Jews, for each person a seven-pronged menorah had been incised, and the three times repeated "There is one God" seems to indicate the Jewish faith — it is a formula which at the same time defies the polytheism of the Romans. Besides these objects, a Jewish gemma and a clay lamp with a menorah decoration are known from Savaria. (Pl. CXXXV. 2.)

On the part of the ancient Jews the most important social movement of the late imperial period also started from the East. Christianity emerged side by side with the decline of the Roman Empire and many archaeological relics speak of its Pannonian history.


## CHRISTIANITY

Christianity advanced in Pannonia from two directions. One of these was Italian. Rome's "orthodox" Christianity came to Pannonia through the mediation of the Aquileian church on the northern shore of the Adriatic, arriving first in the territories between the Drava and Sava rivers. According to contemporary sources, during the third century A.D. there was already a bishopric at Cibalae.[47]

Naturally social, cultural, and geographic conditions contributed to the spread of Christianity. On the other hand, the introduction of the Danube provinces into the Christian community also depended to a great extent on the intensity of city development. In the first and second centuries Christianity took root particularly in the cities, where its propagators fought hard for its introduction. The second half of the third century was the most important period for the introduction of Christianity in the Danube provinces. In Macedonia, Salona, and Aquileia Christianity took root during those fifty years, and church communities were organized during the same time. The beginning of this period was marked by the Gallienus Edict issued in 260. The succeeding comparatively peaceful years rendered possible the partial Christianization of Pannonia as well.

From the early years of his reign Diocletian endeavored to restore old Roman laws and ethnic cults and to renew the worship of the gods which were thought to be responsible for the welfare of the empire. In the spirit of this restoration policy, he tried to check the trend of ideas hostile to Roman religion, including the introduction of Christianity.

Despite the ideological differences between Christianity and the empire, some Christians at the end of the third century became reconciled with the imperial ideal and tried to compromise with polytheism too. This gave the government the impression that it was possible, without the use of force, to win back the believers in Christ and to return them to the worship of the old gods.

Instead of peaceful means, Diocletian tried to use persecutions to realize his aims in religious policy. His first hostile edict was issued in February 303. It debarred Christians from holding state offices in Pannonia, refused them the right of prosecution before the law, and deprived their churches of all their wealth. The second and third edicts turned exclusively against this clergy. In 304 appeared a fourth edict which made the worship of pagan gods and sacrifice to them compulsory for all subjects to the Roman Empire, including the Christians. Enforcement of the edicts had a disastrous effect on young Pannonian Christianity. Many martyr priests, according to contemporary sources which also preserved their names, were executed in 303. The first year of the persecution demolished the fragile organization of the Pannonian Christian church and deprived the laity of their leaders. From the passion narratives of the Pannonian Irenaeus and Quirinus we can form the conclusion that in 303 lay Christian believers had moved about freely and could attend the executions; and, as the sources bear witness, they appeared in great numbers.

A lucky coincidence led to the recent recovery of the historical documentation of a martyrdom[48] in the form of an inscription (Pl. CXXXVI.1) etched into a Roman brick that originated in Brigetio. The inscription of this important object reads as follows:

Hoc die Felice [f . . .]        Hoc die felices fratres 18
sunt persecut [i m orie]       Sunt persecuti morie 18
n tes quorum a nima            n tes quorum anima

194

[non est vict a et in deo]           non est victa et in deo
longius iu bilabit                    longius iubilabit

The completion of the text of the inscription makes possible two variations which both reveal its essence: "on this day Felice and . . . have been harassed, they are dying but their souls will not be conquered and will rejoice for long in God"; or "on this day happy brothers have been persecuted, they will die but their souls will not be overcome and will for long rejoice in God".

On the basis of typical signs, it is possible to state that the text on the brick was written either in the late third century or in the early fourth. For a nearer approximation of the date the stamp Leg I Ad has helped us, since it was stamped on the brick while it was made, while the inscription was being prepared. The company was stationed at Brigetio, where the brick was found. The evidence of the stamp allows us to conclude that the brick emerged from the kiln of Legio I Adiutrix and is a characteristic piece of late production from the late third or early fourth century. In order to determine exactly when the inscription was made and the time of the above-mentioned martyrdom, we have endeavored to shorten the gap between the dates and to rely on the historical data referring to the persecution of Christians in Pannonia. Historical events seem to suggest that the martyrs at Brigetio were executed in either 303 or 304. There is no indication in the text that it described the sufferings of members of the clergy, so that the year 304 is suggested as that of the martyrdom of lay believers.

If we consider the aim and purpose of the inscription we can see that it was not meant as an inscription for a gravestone, since the usual formulae are omitted. It seems to be the spontaneous expression of a tragic event which deeply shocked the onlookers, who commemorated it in lively terms. At the military brickmaking factory of Brigetio convicted Christians must have been employed, the work being hard without demanding expert skill. A few of them must have been picked out and forced to worship the pagan gods, and then persecuted. However, in the words of the text their souls were not conquered — they were killed and made to wear the crown of martyrdom, their souls rejoicing eternally in God. The experience of the seen event shocked one of the onlooking Christians so much that the text came to him spontaneously. He did not commemorate it in marble but preserved it for posterity in a relatively durable material, so that the dreadful predicament of the Brigetio martyrs should be made immortal. The relative intactness of the brick's inscribed surface seems to prove that the inscription etched into soft clay, in spite of several processes — picking up, placing on shelves for drying, putting into the kilns — involving several days, even a week, was carefully preserved up to the moment when it was removed from the kilns to be baked hard and red. All this indicates the fact that a fair number of Christians worked there.[49]

We can be certain of the presence of a Christian community at Brigetio in the last decades of the third century, a period favorable to Christians. In the case of the martyrs we can also see proof of an early Christian community settled along the Roman *limes* in Pannonia. The few Christians who were in Pannonia before A.D. 300 were followers of the orthodox trend. During the fourth century various heretic sects fought in the southern part of the province against orthodoxy, so that

195

in the Drava-Sava region there emerged a bipolar orthodox and heretic Christianity. The heretics, following orthodox methods, also performed missionary work. After 313 Christianity began to spread into areas north of the Drava in Pannonia. Missionary work began only in the territories unaffected or hardly affected by Christianity. From the twenties to the fifties of the fourth century Arianism captivated the whole province with overwhelming force.

Simultaneous with the earlier bishoprics in the Danube-Sava region at Singidunum, Mursa, Sirmium, and Siscia, the Christian communities of Sopianae, Savaria, Scarbantia, Gorsium, and Aquincum certainly had their own bishoprics. Besides the municipal bishoprics, there were numerous and smaller regional bishoprics in the province. It was customary among early Christians that each group, church, and basilica had a clergy with full numbers — i.e., all offices from the subdeacon to the bishop had been filled. One of the most significant relics from the early Pannonian Christian clergy is a shepherd's crook which emerged from a Christian grave at Brigetio and was used earlier as an augur's staff.

In village and village settlements inside the province basilicas and smaller shrines were built between 324 and 330. In this period the number of Christians had so increased and they so insistently claimed free practice of their religion, that outside the cities too it became necessary to build basilicas.[50] We imagine that at the head of each of the basilicas belonging to village and villa settlements, similar to the bishoprics of the Eastern provinces, there was a provincial bishop, the *chorepiscopus*. Early Christian remains of Csopak (Pl. LIX.1,2), Kékkút (Figs. 38 and 39; Pl. CXXXVI. 2), Sümeg, and Fenékpuszta (Figs. 40 and 41) render this belief more acceptable. In the life of late Roman villas and settlements not only economic factors but also Christianity greatly contributed to the continuity and the survival of Roman culture.

In Pannonia the earliest and most common Early Christian cultic places were those which were not built to fulfil cultic functions but were simply occasionally handed over for services and community meetings. It is impossible to state by archaeological means which premises served these purposes, since the meetings were held in rooms in private houses. It seems that the *coenacula* and *tablina* in the latter were most suitable.

Other humble structures in the Early Christian period were specially built for cultic purposes; they were generally placed in cemeteries, next to graves of martyrs or other holy men who were highly respected. The graves were surrounded only by weak railings, so that the believers could pay their homage inside them. We can find them at Brigetio, in Ulcisia Castra, Sopianae, and Matrica, and in practically all Early Christian cemeteries in Pannonia. At a later period the graves were protected by semicircular walls which, so to speak, secluded them from the surroundings. A small *martyrium* at the Intercisa cemetery serves as a good example.

With respect increasing, or owing to the fact that other persons were also buried near the original grave, the cultic place had proved insufficient, so that further apses were added. This is how the multifoiled *cella trichoras*[51] emerged. In Aquincum's Vihar road, in Pest under the Rókus hospital, and at Sopianae several of these were excavated. The most highly developed of this type, the *cella septichora* at Sopianae, was formed in an egg shape from seven adjoining foiled apses. The simple type of hall church without apses, similar to the *basilica rustica* as part of a villa settlement, was most common, (Basilica II at Kékkút, Figs. 38 and 39; Pl. CXXXVI.2, and the shrines of Szentkirályszabadja-Romkút).These basilicas were either undivided or divided by two rows of pillars into three naves; a small *nartex* was occasionally added. When they were rebuilt a small, square apse often emerged, such as that in the three-aisled basilica at Sümeg. Excavations have confirmed that even temperatures were maintained in the hall churches by means of heating ducts.[52]

Basilica I at Kékkút was a magnificent building. Adjacent to its sanctuary there were rooms for the congregation which could be heated and some smaller rooms as living quarters for the clergy. It is certain that there was a Christian cultic center here for the Balaton region and for the interior of the province. We suppose that a *chorepiscopate* existed at Kékkút.

Three-aisled, apse-ended buildings at Gorsium had the arrangement of regular Early Christian basilicas. One of these even had a christening font added outside. During excavations, part of an altar table also emerged.

In the large towns of Pannonia the number of believers increased so much in the fourth century that large basilicas had to be built for public services. Unlike earlier times, the Christian religion did not cater only to the poor and oppressed who after a miserable earthly existence expected from it the eternal bliss of the Heavenly Kingdom promised them for becoming followers of Christ. After 313 Christianity became the religion of the state — made such by Constantine the Great. Consequently high-ranking, well-to-do officials of the state were also members of the Christian church, as were landowners and merchants who gave large sums of money and thereby contributed to building basilicas and supporting internal and external pomp.

Among the Pannonian cities which had basilicas only that of Aquincum has been fully unearthed. Regarding the so-called Quirinus basilica in Savaria recent scholarship has established that it was originally built as a palace, its mosaics belonging to the palace period and not bearing the signs of a Christian basilica.[53] There is great probability that the Early Christian basilica of Savaria was concealed underneath the foundations of Saint Martin's church. Relics emerging from the earth bear witness to the fact that Savaria had another Early Christian basilica which is hidden under the contemporary Franciscan church; its excavation remains a future task.

The excavations of recent years in Sopianae have brought to light such a great quantity of significant Early Christian relics, grave buildings, cemetery chapels, and painted grave cells that we have every reason to expect the emergence of a great Early Christian basilica too.

On the basis of experiences gained in the Balaton region we can say about still extant and functioning churches in the Hungarian Transdanubian area that approximately 25 percent were built over Roman bases, walls, and ground works. This is particularly true of churches which were first consecrated in honor of Saint Michael or Saint Martin, or Saint George.

The interior decoration of Pannonian Early Christian churches was rather sparse. This is particularly striking if we compare them with their Italian and

Eastern counterparts, resplendent with the pomp of mosaics. The remains of a small mosaic floor have come down to us from Sirmium in South Pannonia. However, traces of wall paintings remain in larger quantities and on larger surfaces. The finest and most complete are the wall paintings in the grave cells in Sopianae, produced with the polychrome secco system. The apostles Peter and Paul standing next to a Christ monogram enclosed in the wreath of victory remind us of models from the city of Rome, just as the Mother of God holding her child is also inseparable from similar representations in catacomb paintings. Wall paintings imitating marble and trellis work are humbler pieces; fine examples can be seen in the Christian burial vaults at Sopianae which have been recently unearthed. The same decorative style can be found in other Christian buildings at Kékkút, Fenékpuszta, and Pomáz. Architectonic decorative elements simulating structural ornamentation on windows, capitals, bases, ledges, consoles, and carvings also display symbolical motifs with Christian content. The decorations of stone or marble altars make use of scenes taken from the Old and New Testaments. On the edge of the marble altar unearthed at Csopak we can see the hunter in pursuit of evil; on that of Intercisa, scenes from the story of Jonah are incised.

Besides the ground plans and decorative elements of Early Christian buildings, the earth of Pannonia has preserved some finds which help us to form a picture of objects contained in shrines and cultic places and vessels used in the liturgy. Among the fittings of churches can be mentioned bronze and terra-cotta members for holding lamps (Pl. CXXXVII) which were frequently fabricated from the Greek initials of Christ's name, XP. Their rounded shapes remind us of the crown of victory, *corona triumphalis*, the copies decorated with buttons of the steering wheel used by Christ to control the ship of life. A fine and large example of this type from Bonyhád is displayed in the Hungarian National Museum (Pl. CXXXVIII), but many similar ones emerged in other parts of Pannonia, too. Fine lampposts were made from the combination of Christ's initials, the shepherd's staff, and the rood. Simple lamps and *lucernae* have preserved for us scores of Christian motifs. We know of numerous clay *lucernae* decorated with Christ's initials from Aquincum, Brigetio, and Savaria. A bronze *lucerna* adorned with the XP sign symbolizing the Lamb of God in the shape of a lamb was taken from Kaszaháza to be preserved in the museum at Zalaegerszeg.

Among Early Christian liturgical vessels most significant are the silver paten (Pl. CXXXIX) and chalice which were unearthed at Kismákfa near Vasvár and are described here for the first time. The plate-shaped paten, with a diameter of 10 centimeters, was used for consecrating bread; it is richly decorated with Early Christian symbols. Its center was furnished by the seven-radial sun, Sol, the symbol of Christ represented in the shape of Salutis. The gaps between the seven fishes turning toward the symbol of the sun are filled in by symbols of bread represented in the abstract. The fish may symbolize Christ, whose identification with fish is well known in Early Christian symbols, but it may also refer to the Bible story of the miraculous increase of fish and bread. On the edge of the plate thirteen birds as symbols of the soul alternate with small pointed motifs, with circular finials joined to each of their peaks. These anthropomorphic motifs are symbols of the apostles, the thirteenth representing the figure of Christ, who is distinguished from the others by having two circles to symbolize his head — perhaps one of them a halo, to stress his personality. Unfortunately the chalice which was part of the paten was mislaid, but those who have seen it describe it as a simple, late Roman, silver cup, divided diagonally into zones with identical small figures incised into each.

The chalice and paten found at Kismákfa form one of the earliest eucharistic ensembles of the Christian world in the fifth century.

Among the finds in an Early Christian sarcophagus from Szekszárd (Pl. CXL. 1, 2) a glass cup with filigree work has eucharistic significance and is, at the same time, one of the finest examples of its kind (Pl. CXLI.1, 2). Its Greek inscription, "honor the shepherd, drink and you shall live", renders the cultic use of the vessel indisputable. The inscription is separated by lacelike filigree decoration from the fish and snails which decorate the bottom of the cup.

A silver spoon decorated with the cross and Christ's initials, which is preserved in the museum at Keszthely, also commemorates the Eucharist.[54]

Part of a fine glass vessel with Early Christian connections was found at Dunaszekcső. It contains a golden foil lining in which we can see the half-length portrait of a couple and the inscription "rejoice perpetually in the name of God" in Latin (Pl. CXLII). On a similar glass vessel with a golden bottom that was found in Intercisa we can also read a Christian inscription: "Innocentius, live in God with all thy kind".[55]

It seems that it was a special Pannonian branch of industry to produce caskets with embossed, illustrated bronze plates (Pls. CXLIV and CXLV). Among the pictures on Pannonian casket plates, we can find many which have a Christian theme.[56] There is telling evidence of the fact that Pannonian Christianity was still interwoven with pagan elements because the illustrations on the casket plates contain both heathen and Christian subjects. On the Szentendre casket, for instance, there are pictures representing pagan mythology, war scenes, and the emperor cult side by side with Christian scenes, such as Daniel among the lions, the resurrection of Lazarus, and the miraculous increase of bread. This kind of syncretism is also apparent on other Pannonian caskets. In the vicinity of the village of Császár a grave was unearthed containing a jewel chest with pictures that show side by side the figures of Sol, Luna, Mercurius, Jupiter, and Venus and include in medallions such Judeo-Christian scenes as the good shepherd, Daniel among the lions, the Sermon on the Mount, and the sacrifice of Isaac. In the middle of the scene, representing the Sermon on the Mount, underlining its Christian character, we find embossed the initials of Christ.

Early Christian graves in Pannonian graveyards are, however, relatively poorer in finds than pagan ones. Coins seem to be missing altogether. Glass finds, on the other hand, are very numerous, as seen also in the Christian graves of Ságvár[57] (Tricciana) cemetery (Pl. CXLIII). This can be explained by the presence in Christian graves of the characteristic jar and glass which may refer to the Eucharist.

If we consider the finds from graves, it becomes evident that during the fourth century Christianity had deeply penetrated into Roman and Pannonian society, so that the symbols of the faith were represented not only on sacred objects but also on household ones — with the cross, Christ's initials, and the saints all present — however, secular their use was. Fine examples of brooches and safety pins with gold and silver incrustations were preserved in the graves of Ságvár, Csákvár,[58] and Tihany's Early Christian cemeteries. There are rings from Brigetio's graves decorated with pictures of the apostles, like Peter and Paul and the christogram as well. On a belt buckle from Dombóvár we find a cross with silver and niello decorations.[59]

While these objects with Christian symbols were supposed to grant protection to the wearers, they also bore marks of their provenance. Such objects were considered very valuable, as exemplified by the disc brooches (Pl. CLXI.2) which were taken by Pannonian pilgrims on their journeys to the Holy Land. These emerged in great numbers, particularly at Fenékpuszta and Sopianae, both places where the late Roman population survived in a Christian environment.[60]

The Savaria Museum of Szombathely has preserved a rare, so-called Menas Ampulla which seldom existed in any European province of the Roman Empire. Saint Menas stands on this clay ampulla between two camels, his arms held out, making an *orans* movement (Pl. CXLVIII.1). According to the legend, after the martyr's death, he was taken from the place of his martyrdom in Syria to his grave in Alexandria in a miraculous way by two camels. Miraculous forces were attributed to the well which broke forth next to his grave, and pilgrims filled their small ampullae with its water and took it along to their faraway homes.[61]

The funeral rites of Early Christians did not differ greatly from those of the late Romans. They were buried in sarcophagi, in graves put together from stone slabs, or in graves made with large Roman bricks; but it also happened that they were put directly into the earth. Christian graves were marked by wood or stone tablets, in the same way as had been customary according to the funeral rites of the pagan Romans in Pannonia. Grave tablets with inscriptions, both with and without Christ's initials, were found in the graveyards of Christian communities in Sirmium, Djakovar, Aquincum, and Brigetio, though the largest number emerged from the Savarian Early Christian cemeteries (Pls. CXLVIII.2 and CXLVI). These telling inscribed stones offer insight into the organization of the church and clergy, the social status of the believers, and the fine forms of their faith and creed.

As far as Early Christian graveyards are concerned, we have to mention a specifically Pannonian group of relics — i.e., symbolical brick etchings with a Christian content[62] — which emerged from this province in large numbers unprecedented in any other Roman province. Their multiplicity offers an insight into faith and superstition as well. Side by side with the simple ChiRho initials (Pl. CXLIX) and those compiled from *alpha* and *omega* as references to Christ as the beginning and the end, there are also palm motifs in great numbers, representing the victory of the soul. Figural elements merging with symbolical ones inform us, similar to the *biblia pauperum*, about more complicated concepts: the fight of good and evil, light and darkness, executed martyrs, and Christ's Passion. From the late Roman cemetery of Oroszvár (Rusovce) comes down to us the simple but expressive figure of Christ, carrying the cross, incised into a brick (Fig. 4).

We must devote more words to one of the most interesting incised brick figures in Pannonia,[63] a representation of Arius, since anything connected with him is

200

Fig. 4. Incised brick from Oroszvár cemetery, figure carrying cross

highly characteristic of Pannonian Christianity, which developed in an Arius-inspired spirit. The Szekszárd Museum is preserving a brick which was unearthed in the nearby village of Kisdorog from a late Roman grave (Fig. 5; Pls. CL and CLI). Into this was incised a bareheaded figure of a man holding his arms apart in an *orans* gesture and wearing a *dalmatica* decorated with *clavi*; in his right hand there is a short crook. On the left, under his arm, there are the letters *P.* and *E.* abbreviations referring to victory (Palma, Emerita). Under the ligature of the two letters, there is the sign +. Beside the man, on the left, there is an anchor. An inscription over the figure's head, ARIO, gives the name of the figure (ARIVS) in votive dative. This object is the only surviving representation of Arius, father of Arianism, one of the christological heresies. We do not know of any other artistic rendering of the figure of Arius. We do not find any references to him in churches, where it would have been logical to portray him in the form of a statue, a wall painting, or a relief. The Arians obviously did not care for building and instead occupied orthodox churches, without developing a bent for expression in any kind of symbolism. Arianism did not care for externals but strove toward clarification of the basic principles of Christianity, the divine or

201

human character of Christ. The essence of the Arian teaching was to deny the divinity of Christ. In a vulgarized version, Christ was not a god but a man possessed by a higher spirit, the new founder of a religion, venerated by the Arians but not considered identical with God the Father. The teachings of Arius were constantly discussed during the fourth century by various synods, causing great anxiety.

Arius had continued his organizing activities and was not stymied by his condemnation in 319. He went on recruiting partisans, in particular among ambitious bishops seeking the favor of the court. In this way a theoretical debate developed into an ecclesiastic polemic, resulting in Arius's excommunication in 325 by the Synod of Nicaea. Arius was obliged to retreat into exile and he retired with his adherents to Illyricum, on the borders of Pannonia. From there he continued to pursue his activities. It was there that his personal history came into contact with Pannonia and it is evident that he launched his campaign from there to propagate his teachings in Pannonia.

Arian teachings continued to spread in Pannonia, notwithstanding the fact that between 317 and 322 Constantine the Great several times had visited Sirmium on the border of Illyricum and had made provisions to prevent the introduction of Arianism into Pannonia. He dismissed all suspect members of the clergy who professed Arian views. Nevertheless, Arius succeeded in acquainting two presbyters of the Sirmian church with his teachings; these two later became bishops, Valens of Mursa and Ursacius of Singidunum. In November 326 Arius appeared before the court at Sirmium, presenting to the emperor his modified creed and asking for his rehabilitation and the suspension of his exile. The orthodox bishops showed at times more lenience, at times more severity, during the synods of the succeeding years, until Arius was finally fully expelled. Arius lived on to the age of eighty, dying in 337, the same year as his great adversary, Emperor Constantine.

When Emperor Theodosius I succeeded to the throne in the East, he gave the Nicaean religion permission to function, returning to the orthodox followers of Christian homoousianism the churches in Constantinople previously taken from them. But, in an edict issued on 10 January 381, Theodosius prohibited all heresies, including Arianism. Arians were again persecuted, but they were supported by the Goths, whose first bishop, Ulfilas, was consecrated according to the ritual of the Arian creed. Nor was Gothic Arianism a short episode; it finally became a national church on Germanic soil and continued to exist for over three centuries.

Beyond a fragmentary survey of Arian ideology, contemporary authors recorded some information on the conduct and outward appearance of Arius himself. According to them, Arius was a tall man of ascetic disposition. His personality was attractive in personal contact; he was well versed in dialectic and strict in ethics. However, his virtues suffered because of his negative qualities: he was false, vain, conceited, and ambitious. According to these observers, as a means of propaganda he made use of singing. To the music of millers' and shipmen's songs he formed verses from his theses.

Added to these depictions, if we also consider his ideology, a rationalist but impulsive personality unfolds before us — one who propagated his ideas fanati-

Fig. 5. Arius figure, engraved in brick

cally and aggressively, did not tolerate contradictions, and yet had an extremely persuasive manner.

The provenance of the Kisdorog brick with the figure of Arius suggests that it came from a single grave or a group of a few graves. That the brick emerged from a late Roman brick grave can be confirmed by the evidence of other bricks which were found together with it. Comparing the circumstances determining the age of the grave, the cemetery, and the brick, we can see that the brick was not contemporaneous with the life of Arius but was made at a time when his followers were still living in Pannonia and his teachings were spreading. It is most probable that the brick was made in the second half of the fourth century, perhaps in the last third. But on the basis of the available data, it is impossible to state exactly when the brick etching was made, whether in the years and months of oppression and illegality or during the free cultic practice of the Arians.[64]

As seen from the sketchy survey of historical events, during the fight of Arianism and orthodoxy the position or official recognition of either group changed very rapidly. Not even after the imperial edict of 381 did Arianism completely end in Pannonia; but its followers were obliged to retire, more or less, into illegality. It was only in the center of state and religious politics that the daily developments in the disputes and the relation of forces evident to persons of importance and naturally to influential bishops who had contact with the former. Presumably the clergy in smaller churches, off the beaten track, had no idea what the officially recognized view should be concerning the essence of Christ. Nor could the majority of believers keep track of the changes in the partly unclarified doctrines; all they did was to practice their religion, no matter according to which creed. Ideological struggles were confined to the more highly educated clergy.

Missionary activity in Pannonia, the first encounter of the great masses with Christianity, coincided with the time when Arius and his collaborators, the Pannonian Valens and Ursacius, fought their battles — not without success — at the court of Sirmium. Later, owing to their geographical position, it was Singidunum, Sirmium, and Mursa whose clever Arian bishops strove to make the concept of Christianity in Pannonia beyond the Drava — i.e., in the present Transdanubian area — identical with Arianism. It is understandable, therefore, that side by side with the orthodox doctrines of the city of Rome, from the second third of the fourth century, Arianism, a Christian trend coming from the East and emerging more pronouncedly than orthodoxy, should have taken deep root for a comparatively long period.

## NOTES

[1] A. BRELICH, "Aquincum vallásos élete" (The Religious Life of Aquincum). *Laureae Aquincenses* 1 (1938): 20–142 (*DissPann* 2:10).

[2] G. ALFÖLDY, "Geschichte des religiösen Lebens in Aquincum". *ActaArchHung* 13 (1961): 103–24; G. ALFÖLDY, "Zur keltischen Religion in Pannonien". *Germania* 42 (1964): 54–59.

[3] É. B. BÓNIS, *Die spätkeltische Siedlung Gellérthegy–Tabán in Budapest* (Budapest, 1969) (*ArchHung* 47).

[4] A. ALFÖLDI, "Vallási élet Aquincumban" (Religious Life in Aquincum), in *Budapest az ókorban* (Budapest, 1942), 2:386–463.

[5] L. NAGY, "Asztrális szimbólumok a pannóniai bennszülött lakosság síremlékein—Symbola astralia inventa in cippis incolarum Pannoniae indigenarum". *Pannonia* (1935), pp. 139–51, 399.

[6] A. MÓCSY, *Die Bevölkerung von Pannonien bis zu den Markomannenkriegen* (Budapest, 1959).

[7] L. BARKÓCZI, "Brigetio I–II" (Budapest, 1944–1951) (*DissPann* 2:22).

[8] L. BARKÓCZI and É. BÓNIS, "Das frührömische Lager und die Wohnsiedlung von Adony (Vetus Salina)". *ActaArchHung* 4 (1954): 129–99.

[9] E. B. THOMAS, "Ornat und Kultgeräte einer Sol- und Luna-Priesterin aus Pannonien". *ActaAntHung* 11 (1963): 49–80.

[10] E. PATEK, "A pannóniai fibulatípusok elterjedése és eredete—Verbreitung und Herkunft der römischen Fibeltypen in Pannonien" (Budapest, 1942) (*DissPann* 2:19).

[11] A. Radnóti, "A pannóniai római bronzedények—Die römischen Bronzegefässe von Pannonien" (Budapest, 1938) (*DissPann* 2: 6).

[12] E. B. Thomas, "Ólom fogadalmi emlékek Pannóniában—Monuments vo tifs en plomb sur le territoire de la Pannonie". *ArchÉrt* (1952): 32–38.

[13] E. B. Thomas, *Römische Villen in Pannonien* (Budapest, 1964),pp. 42, 144, 147, 399.

[14] K. Sz. Póczy, "Diana aquincumi kultuszához—Alcuni dati sul culto di Diana di Aquincum". *BudRég* 19 (1959): 139–43; K. Kerényi, "Die Göttin Diana im nördlichen Pannonien". *Pannonia* (1938), pp. 203–21.

[15] Thomas, *Römische Villen*, pp. 21–22.

[16] A. Mócsy, "Pannónia", *R. E. Pauly-Wissowa*, Suppl. 9, p. 732.

[17] K. Sz. Póczy, "Aquincum első aquaeductusa — Le premier aqueduc d'Aquincum". *ArchÉrt* 99 (1972): 15–32.

[18] I. Paulovics, "Dionysosi menet (thiasos) magyarországi római emlékeken—Der dionysische Aufzug (Thiasos) auf ungarländischen Denkmälern". *ArchÉrt* (1935): 54–102, 248–53.

[19] D. Gáspár, "Római ládikák felhasználása—Die Verwendung römischen Kästchen". *FolArch* 22 (1971): 53–69.

[20] Mócsy, pp. 669–70.

[21] E. B. Thomas, *A nagydémi lararium—Das Lararium von Nagydém* (Veszprém, 1965).

[22] E. B. Thomas, "Lar angusti clavi". *FolArch* 15 (1963): 21–42.

[23] E. B. Thomas, *Rómaikori háziszentély leletek Tamásiból—Römerzeitliche Hausheiligtümer aus Tamási* (Szekszárd, 1963).

[24] Bónis, Die spätkeltische Siedlung Gellérthegy.

[25] A. Alföldi, "Kapitóliumok Pannóniában" (Capitols in Pannonia). *ArchÉrt* (1920–1922): 12–14; K. Sz. Póczy, "Scarbantia városfalának korhatározása — La datation de l'enceinte de Scarbantia". *ArchÉrt* 94 (1967): 137–54.

[26] L. Balla, T. Buócz and Z. Kádár, *Die römischen Steindenkmäler von Savaria* (Budapest, 1971).

[27] A. Radnóti, "Trebonianus Gallus ezüstlemez mellképe—Silver Bust of Trebonianus Gallus from Brigetio". *FolArch* 6 (1954): 49–61, 201–4.

[28] E. B. Thomas, "Hercules szentély Pannóniában—Ein Heiligtum des Hercules aus Pannonien". *ArchÉrt* (1952): 108–12.

[29] A. Dobrovits, "Az egyiptomi kultuszok emlékei Aquincumban—The Cult of the Egyptian Gods in Aquincum". *BudRég* 13 (1937): 45–75, 494–97; V. Wessetzky, *Die ägyptischen Kulte zur Römerzeit in Ungarn* (Leiden, 1961).

[30] J. Fitz, Hercules-kultusz eraviszkusz területen (The Cult of Hercules in the Eraviscan Area). *Székesfehérvár*, 1957, pp. 1–31.

[31] Z. Kádár, *Die kleinasiatisch-syrischen Kulte zur Römerzeit in Ungarn* (Leiden, 1962).

[32] F. Láng, "Das Dolichenum von Brigetio". *Laureae Aquincenses* 2 (1941): 165–81 (*DissPann* 2:11).

[33] S. Scheiber, *Corpus inscriptionum Hungariae judaicarum* (Budapest, 1960).

[34] Balla, Buócz and Kádár, *Steindenkmäler*.

[35] V. Wessetzky, "A felső-pannóniai Isis-kultusz problémái—Some Problems of the Cult of Isis in Upper Pannonia". *ArchÉrt* 86 (1959): 20–31; T. Szentléleky, *A szombathelyi Isis-szentély—Das Isis-Heiligtum von Szombathely* (Budapest, 1960).

[36] E. B. Thomas, "Bronzekanne von Egyed", in *Archäologische Funde in Ungarn* (Budapest, 1956), p. 198.

[37] Thomas, "Ólom".

[38] T. Szentléleky, "Das Iseum von Szombathely", in *Neue Beiträge zur Geschichte der Alten Welt* (Berlin, 1965), 2:381–88.

[39] Wessetzky, *Die ägyptischen*.

[40] E. B. Thomas, *Führer durch die Ausstellungen des Museums von Szekszárd* (Szekszárd, 1965).

[41] I. Paulovics, "Dolichenus-háromszögek tartója Brigetióból—Halter für dreieckige Dolichenus Reliefs aus Brigetio". *ArchÉrt* (1934): 40–48.

[42] J. Hampel, "Lovas istenségek dunavidéki antik emlékeken". *ArchÉrt* (1903): 305–65, (1905): 1–16 (1911): 409–25, (1912): 330–52; E. B. Thomas, "Die römerzeitliche Villa von Tác–Fövenypuszta". *ActaArchHung* 6 (1955): 79–152.

[43] T. NAGY, "A sárkeszi Mithraeum és az aquincumi Mithra-emlékek—Le Mithréium de Sárkeszi et les monuments mithriaques d'Aquincum". *BudRég* 15 (1950): 47–120; T. NAGY, "Das Mithras-Relief von Paks". *ActaAntHung* 6 (1958): 407–31; MÓCSY, pp. 737–38.

[44] J. FITZ, A Sárkeszi mithraeum (The Mithraeum at Sárkeszi). *Székesfehérvár*, 1957, pp. 1–18.

[45] F. FÜLEP, "New Remarks on the Question of the Jewish Synagogue at Intercisa". *Acta ArchHung* 18 (1966): 93–98.

[46] S. Scheiber, *Corpus Inscriptionum Hungariae Iudaicarum* (Budapest, 1960).

[47] T. NAGY, A pannoniai kereszténység története a római védőrendszer összeomlásáig — Die Geschichte des Christentums in Pannonien bis zu dem Zusammenbruch des römischen Grenzschutzes (Budapest, 1939) *(DissPann* 2: 12).

[48] E. B. THOMAS, "Martyres Pannoniae" *FolArch* 25 (1973): 131.

[49] THOMAS, "Martyres . . ." 134.

[50] THOMAS, *Römische Villen.*, 379.

[51] L. NAGY, Az óbudai ókeresztény cella trichora a Raktár utcában — Altchristliche Cella Trichora in Óbuda. (Budapest, 1931).

[52] THOMAS, *Römische Villen*, pp. 52-56.

[53] E. TÓTH, "Late Roman Imperial Palace in Savaria." *ActaArchHung* 25 (1973): 117.

[54] K.SÁGI, "Die zweite altchristliche Basilike von Fenékpuszta." *ActaArchHung* 9 (1961): 397-459.

[55] F. FÜLEP, "Early Christian Gold Glasses in the Hungarian National Museum" *ActaAntHung* 16 (1968): 401-442.

[56] L. NAGY, *"Pannonia Sacra"*, Szent István Emlékkönyv. (Budapest, 1938), pp. 67-76; H. Buschhausen, Die spätrömischen Metallscrinia und frühchristlichen Reliquiare, (Wien, 1971).

[57] A. SZ. BURGER, "The Late Roman Cemetery at Ságvár." *ActaArchHung* 18 (1966): 99-234.

[58] Á. SALAMON and L. BARKÓCZI, "Bestattungen von Csákvár aus dem Ende des 4. und dem Anfang des 5. Jahrhunderts." *AlbaRegia* 11 (1970): 35-80.

[59] L. NAGY, *"Pannonia Sacra".*, p. 47., Fig. 13.

[60] A. ALFÖLDI, "A kereszténység nyomai Pannóniában a népvándorlás korában (Traces of Christianity in the Migration Period in Pannonia). Szent István Emlékkönyv (Budapest, 1938), pp. 151-157.

[61] L. NAGY, *"Pannonia Sacra".,"* p. 96., Fig. 60.

[62] I. PAULOVICS, *Lapidarium Savariense* (Szombathely, 1943): 46.

[63] E. B. THOMAS, "Arius-Darstellung". *SzekszárdiMúzÉvk* 5 (197374): 77-116.

[64] THOMAS, *Römische Villen.*, pp. 395, 399.

# ROADS

## SÁNDOR SOPRONI

During Roman times several factors influenced the development of a network of roads in Pannonia (Fig. 6). Before then, certain commercial roads had been developed between the tribes who lived in this area and distant producers or consumers with whom they traded. These roads were also used by migrating peoples. The Roman conquerors probably also used these roads when they occupied Pannonia. These routes followed the most convenient course as the terrain dictated. For example, the Amber Route (Pl. XII), the most important road between the Baltic and the Mediterranean, was a commercial road in ancient times. It was probably originally developed by the great migrations of the prehistoric age. It became one of the most important instruments in the Roman occupation of Pannonia. Long after the pacification of Pannonia, this road remained the most important strategic thoroughfare.

Among the natural roads were those ancient causeways which crossed the basins of great rivers and played an important role in practically every period from a commercial and strategic point of view. The roads alongside the Danube, the Drava, and the Sava rivers, particularly the latter, were essential operational factors during the Roman occupation. The road in the Sava basin established an important overland connection between the western and eastern parts of the empire, i.e., between Italia and the Balkans.

Military strategy was an essential consideration in the development of the road system in Pannonia. In addition to roads that connected the army camps along the entire *limes* area, routes that directly connected the hinterland and Italia were imperative (Fig. 8).

The cities and settlements which gradually developed in Pannonia were also important in the development of the road system. Most of the connecting roads, however, followed pathways that had been previously developed. Roads in the interior differed in quality according to the nature of the settlements they connected.[1]

Available source material concerning the development of the road network in Pannonia does not give well-balanced information. This unevenness has resulted from a lack of coordination in research. The most authentic source materials have come from itineraries which described the most important roads, with the names of the stations and the distances between them. Four itineraries listed routes in Pannonia (Fig. 6). The most detailed of them, the *Itinerarium Antonini*, mentioned only the most important roads; the *Tabula Peutingeriana* listed five routes. The *Itinerarium Hierosolymitanum* mentioned only the Aquileia–Singidunum (Beograd), which was important in the late Roman period. However, the roads listed by the Geographer of Ravenna contain no new information whatsoever.[2]

The most common objects associated with the Roman road system were the milestones (Fig. 7). They were placed one Roman mile apart. (One Roman mile was one *mille passuum* which was equal to 1,480 meters.) The milestones, in addition to the data for distances, offered some information about road construction, bridge construction, and repairs.[3] More than two hundred milestones or fragments of milestones have been found in Pannonia, considerably irregular in distribution (Pl. VII. 1, 2). Most of the milestones came from the large militarn routes which connected the camps of the *limes*. Fewer milestones have beey discovered along the Amber Route and the routes along the Drava and Sava rivers. No milestones have been found along the interior routes. It is possible that these routes were not marked or that the markers were made of easily perishable wood.[4]

The time when the milestones were erected has not been evenly recorded. Although road construction was begun after the Roman occupation of Pannonia, the earliest milestones date from the reign of Nerva (A. D. 96–98). Although milestones were erected more regularly under Hadrian (117–138), very few from this era have been discovered. A great many milestones remain from the time of Septimius Severus (193–211) and from the first half of the third century. In the second half of the third century their number again decreased. Some stones survive from the Tetrarchy period. From the period after Constantinus just five stones are known and two of them date from the time of Valentinian (364–375). All five were discovered in the southern part of Pannonia. The milestones from the third century give information about building and repairs; in addition, by mentioning the name of the emperor, they served as imperial propaganda (Fig. 7). The name of the emperor indicated that the road (*via publica*) was the property of the Roman people. Some large milestones, mostly with red inscriptions, in addition to road information, heralded the glory of the emperor.[5]

Fragmentary sections of road track which have been discovered are the third source of information about the Roman road system. Both research and technical analysis in this direction have advanced our knowledge significantly. The road relics along the *limes* have been mapped, while the interior of Pannonia has been only sporadically explored. Several sections have been found, of varying lengths, but no coherent sections have been registered.

The technical construction of the roads in Pannonia was not uniform, although the planning of the road network points to extremely logical central control.[6] Roman road engineers planned the roads according to military, commercial, and geographical considerations. The bulk of the construction work was done by military units stationed in the neighborhood. In addition, the troops worked at removing earth masses, draining swamps, and similar projects. The routing of roads, especially through mountains, was quite remarkable; the engineers took great pains to keep the roads on the same level. They usually made so-called diagonal roads through mountains in order to avoid significant grade differences. An excellent example of this technique was the diagonal road between Aquincum and Brigetio.

208

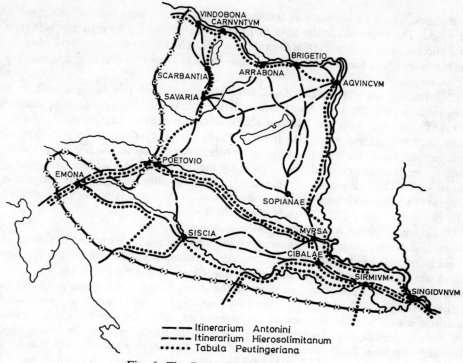

Fig. 6. The Pannonian data of Itinerarii

Examination of the structure and measurement of the roads in Pannonia has resulted in varying data. A stone foundation covered with sand and gravel, similar to our macadamized roads, was the most frequently used road-building method. The stone foundation usually had two layers. The bottom layer was gravel, with a limestone or basalt layer above it. Smaller stones were placed over larger stones. Finally, the top of the road was covered with sand and gravel. The classical Roman surface — large flat polygonal stones — was rare in Pannonia. It did occur between Carnuntum and Vindobona at the boundary of Mariaellend, and also in the larger cities. One of the most perfect examples which emerged intact is a section in Savaria (Fig. 19). The greater part of the local traffic was carried by these well-built highways and carriage and dirt roads. The smaller *vicus* and *villa romana* were linked by such roads with the main roads.

The width of the roads generally varied from 5 to 10.6 meters, but narrower roads were also built. *Limes* roads were wider, and, in addition, a boundary stone discovered in Emona referred to a *via publica* which was 50 feet (cca. 17 meters) wide. The average width was 5–6 meters. According to the twelve Tabular Laws, the narrower *via publica* was 2.4 meters in width. The road was flanked by several meters of shoulder which was also public property. These strips were necessary for road repairs and the milestones were ordinarily placed at their edge.

Bridges, which we consider as accessories to the roads, were built of either stone or wood. The bridge over the Sava at Emona was supported by a timber raft type of structure. The bridge itself was built of stone. The elements were interlocked by cramp irons. It is known that the Romans built bridges of varying spans entirely of stone.

Pile bridges were frequently used, too. The remains of pile bridges have been observed on the bank of the Danube River at Aquincum and near Aequinoctium (Fischamend) on the bank of the Fischa. Ammianus Marcellinus reported that pontoon bridges were also used in Pannonia. The supports for this type of bridge were boats placed parallel to each other. According to the historian Valentinian, the Romans crossed the Danube by means of such an *ad hoc* bridge near Aquincum. The milestones often recorded the building or repair of bridges. A fragment of an inscription which was discovered in Poetovio (Ptuj) commemorated the reconstruction of the bridge across the Drava.

We must consider watchtowers built to control the safety of roads, guard stations, and horse changing posts as accessories to the roads. Excellent examples of road guard stations have been discovered near Aquincum and Höfflein. Perhaps the road watchtowers between Intercisa (Dunaújváros) and Annamatia (Baracs) may also be referred to as forming part of the *limes* watchtower system. The road towers around Höfflein were built on a 6 by 6 meters ground plan. The *mutatio* and *mansio*, which were mentioned in the itineraries, have not yet been discovered. A *mutatio* served for changing horses and a *mansio* offered room and board accommodations. The building excavated at the boundary of Purbach, close to a 3-meter-wide road stretching from Scarbantia to Carnuntum presumably served such a purpose.

In Carnuntum (Deutsch-Altenburg), a group of buildings was discovered which may have been built as a station. Within an 18.7 by 17 meters garden was discovered a building built in the shape of a cross. Tracks of a horse stable were also discovered next to the building. The system of these stations and their distances from each other are not yet known.[7]

The plan of the Pannonian road system started from a triangle of Emona–Siscia–Poetovio.[8] The road which came from Italia through Aquileia and through the Alpes Iuliae branched out from Emona toward Carnuntum and Sirmium.

The Amber Route, which began at Aquileia, crossed the Sava River at Emona, going on to Atrans, then into the neighboring province of Noricum. Reaching Celeia, the road returned to Pannonia at Poetovio and the Drava valley. From this important traffic junction (Poetovio), the road continued along the western edge of the Transdanubian area (Dunántúl) through Salla (Zalalövő), Savaria, and, by way of Scarbantia on the western shore of Lake Fertő, up to Carnuntum. In addition to the roads in the valleys of the Drava and Sava rivers, a diagonal road along Lake Balaton to Aquincum also branched out from the Amber Route. A road led from Savaria to one of the earliest camps of the Danubian *limes*, Arrabona. Another branch of the Amber Route occurred between Scarbantia and Vindobona. With the exception of the diagonal road to Aquincum, these roads are all mentioned in the itineraries.

In the early period, but equally during late Roman times, one of the most

IMP·CAES·
M·IVL·PHILIPPVS·P·F·
AVG·PM·TRIB·POT·COS·
P·P·PRO·COS·ET·M·IVL·

PHILIPPVS·NOBIL·CES
PRINC·IVVENT·ET·MARCIA·
OTACLIA·SEVERA·SANC
TISSIMA·AVG·CONIVNX·
AVG·N

...P·XXIII·

Fig. 7. Inscription of milestone in Százhalombatta

important roads went along the Sava River. From Emona it reached the Sava at Neviodunum, continued on to Siscia, running close to the right bank of the Sava, crossed the river at Sirmium, and reached the *limes* route at Singidunum. Siscia itself was a very important junction. Three main roads branched off from this city. The first ran through Andautonia (Scitarjevo) to Poetovio. The second ran diagonally across the area bordered by the Sava River, passed through Cibalae, and joined the main road to Sirmium. The third main road followed the right bank of the Sava and connected with the side roads along the right banks of the tributaries of the Sava from the Dalmatian coast.

The second main highway from east to west also branched off from the Amber Route at Poetovio and, running through the valley of the Drava River, served as a main artery connecting the southern part of Pannonia with Moesia, Noricum, and

Raetia. The road came from Virunum in Noricum and reached the south bank of the Drava River east of Poetovio. It passed through flatland reaching Iovia and the Mursa on the west, where a direct connection was made with the *limes*. From Mursa, the most important crossing point and junction through Cibalae, an interior road led to Sirmium, the late Roman imperial seat.

The road along the Danube *limes* was in early times, long before the Romans, an important link between east and west. The road entered Pannonia and Cannabiaca (Klosterneuburg) and left it at Taurunum (Zemun). After Cannabiaca, the *limes* road reached on the right bank of the Danube the following important military camps: Vindobona, Brigetio, Aquincum, Intercisa, Lugio, Teutoburgium, and Malata-Bononia (Banoštor). Around the Danube bend, one branch of the road continued along the line of the river and connected Castra ad Herculem (Pilismarót) and Ulcisia Castra (Szentendre) with Aquincum. The other branch crossed the Danube bend area diagonally and connected Brigetio and Aquincum by a shorter route.

In the Transdanubian area, the road network between the Drava and Danube rivers was directly connected with the *limes*. The road which connected Poetovio with Aquincum was a very important part of this network. This road crossed the whole Transdanubian area diagonally through Valcum and Gorsium, passing along the south side of Lake Balaton. This road was not mentioned in the *Itinerarium Antonini*. North of the Drava, main highways radiating from Sopianae and Savaria reached the central part of Transdanubia. From Sopianae, one road bypassed Mecsek Mountain and followed the eastern edge of Lake Balaton to Arrabona. Another road which began at Sopianae crossed Valcum, connecting it with Savaria. The *Itinerarium Antonini* mentioned a road which connected Sopianae with Mursa to the south. Still another road led to Lugio, an important crossing point of the *limes*.

Another junction was Savaria, which the Amber Route crossed. In addition to the roads from Savaria to Arrabona and Sopianae, the *Itinerarium Antonini* mentioned a main road through Mogentiana (Somlóvásárhely?) to Aquincum. There was also a road toward Noricum.

Unfortunately, research on the road network of Pannonia is far from complete. Several routes are merely conjectured from a calculation of miles as mentioned in the *Itinerarium Antonini* or the *Tabula Peutingeriana*, not having been satisfactorily proved yet. Geographical identification of the data will be the task of future research.[9]

Although Barbaricum was never an organic part of Pannonia, we nevertheless ought to mention those roads which crossed the Danube to that region.[10] They were equally important to the Romans for commercial and military reasons. Each of the most important crossing points was defended by a legion camp. Carnuntum, which was built at the crossing point of the Amber Route, was connected with the Baltic Sea area through the valley of the Morava River across Silesia and Poland. A route started out from Brigetio which bypassed the guard station of Milanovce and followed the course of the Vág valley. From Aquincum a road through Barbaricum forked off in two directions. One branch through Hatvan, Miskolc, Kassa (Košice), and the Carpathian Mountains to the eastern steppe area. The other branch went through Szolnok to Dacia, cutting through the Great Hungarian Plain (Alföld). In the southern area of the *limes*, the crossing point was at Lugio; the road crossed the southern part of Barbaricum, entered Dacia through the guard station of Partiscum (Szeged), and then followed the course of the Maros River.

The Pannonian road system was naturally coordinated with that of the neighboring provinces. Along the Danube River, the *limes* road secured contact between the western and the Balkan provinces. The Amber Route supplied contact with Italia. The roads along the Drava and Sava rivers connected Pannonia with Noricum, Raetia, Germania, and Gallia. In addition, the itineraries referred to other important roads, such as those connecting Sirmium and Salona, Siscia and Senia. The western cities of Pannonia—Emona, Poetovio, and Carnuntum—were connected with the cities of Noricum by several roads.

The Roman roads, especially the main roads, survived after the end of Roman rule and were used during the migration period. They played an important role during the Avar rule, too. The early Avar cemeteries were placed close to the Roman roads in the Transdanubian area. In medieval documents, the *via magna* and *via antiqua* were possibly references to the still existing Roman roads.

# ITINERARIA

## Itinerarium Hierosolymitanum (Burdigalense)

560- 63:

mansio Longatico — mutatio Ad Nonum — civitas Emona — mutatio Ad Quartodecimo — mansio Hadrante — (fines Italiae et Norici) — [mutatio Ad Medias — civitas Celeia — mutatio Lotodos — mansio Ragindone — mutatio Pultovia —] civitas Petovione — (transis pontem, intras Pannoniam inferiorem) — mutatio Ramista — mansio Aqua viva — mutatio Popolis — civitas Iovia — mutatio Sunista — mutatio Peritur — mansio Lentolis — mutatio Cardono — mutatio Cocconis — mansio Serota — mutatio Bolenta — mansio Maurianis — (intras Pannoniam superiorem) — mutatio Serena — mansio Vereis — mutatio Iovalia — mutatio Mersella — civitas Mursa — mutatio Leutuoano — civitas Cibalis — mutatio Celena — mansio Ulmo — mutatio Spaneta — mutatio Vedulia — civitas Sirmium — (fit ab Aquileia mansiones 14, mutationes 39) — mutatio Fossis — civitas Bassianis — mutatio Noviciani — mutatio Altina — [civitas Singiduno — (finis Pannoniae et Misiae)].

## Itinerarium Antonini

129–31:

Longatico mansio — Hemona civitas — Adrante mansio [Celeia civitas — Ragundone] — Patavione civitas — Aqua viva — Iovia — Lentulis — Serota — Marinianis — Vereis — Mursa civitas — Cibalas civitas — Ulmos vicus — Sirmi civitas — Bassianis civitas — Tauruno classis — [Singiduno castra].

232–33:

Sirmi – Ulmos – Cibalis – Mursa – Antianis – Suppianis – Limusa – Silacenis – Valco – Mogentiana – Savaria – Scarabantia – Muteno – Vindobona.

241–48:

Item per ripam Pannoniae a Tauruno in Gallias ad leg. XXX usque: A Tauruno Lauriaco inde Augusta Vindelicum Argentorato Ad leg. XXX. A Tauruno Ritti in medio Aciminici Cusi – Bononia – Cucci – Cornaco – Teutiburgio – Mursa – Ad Novas et Aureo Monte – Antianis – Altino in medio – Lugione – Ad Statuas in medio – Alisca ad latus – Ripa alta – Lussunio – Annamatia in medio – Intercisa – Vetus Salina in medio – Matrica – Campona in medio – Aquinco leg. II adiut. – Ad Lacum felicis in medio – Crumero – Azao in medio – Bregetione leg. I adiut – ad Mures et ad Statuas in medio – Arrabona – Quadratis in medio – Flexo – Gerulata in medio – Carnunto leg. XIII gemina – Aequinoctio et Ala nova in medio – Vindobona leg. X gemina – (Comagenis).

259–61:

Item ab Hemona per Sisciam Sirmi usque – Praetorio Latovicorum – Nevioduno – Quadrata – Siscia – Varianis – Menneianis – Incero (sed mansio Augusti in praetorio est) – Picentino – Leucono – Cirtisa – Cibalis – Ulmos – Sirmi.

261–62:

Item a Vindobona Poetovione – Aquis – Scarabantia – Sabaria – Arrabone – Alicano – In medio Curta – Poetovione.

262:

Item a Poetovione Carnunto – Halicano – Salle – Sabaria – Scarabantia – Carnunto.

262–63:

A Sabaria Bregetione – Bassiana – Mursella – Arrabona. – Bregetione.

263:

A Sabaria – Acinquo – Mestrianis – Mogentianis – Caesariana – Osonibus – Floriana – Acinquo.

264:

A Sopianas Acinquo – Ponte Sociorum – Valle Cariniana – Gorsio sive Herculia – Iasulonibus – Acinquo.

264–65:

Item a Sopianas Bregetione – Iovia – Fortiana – Herculia – Floriana – Bregetione.

265:

Item a Siscia Mursa — Varianis — Aquis Balissis — Incero — Stravianis — Mursa.

265–66:

A Poetovione Siscia — Aqua viva — Pyrri — Dautonia — Siscia.

266:

A Sabaria Vindobona — Scarabantia — Muteno — Vindobona.

266:

Item ab Aquinco Crumero quae castra constituta sunt — Ulcisia castra — Cirpi mansio — Ad Herculem castra — Salva mansio.

266–67:

Item a Sirmio Carnunto — Ulmo — Cibalis — Mursa — Antianis — Sopianis — Ponte Mansuetina — Tricciana — Cimbrianis — Crispiana — Arrabona — Flexo — Carnunto.

267–68:

Item a Sirmio Salonas — Budalia — Spaneta — Ulmo — Cibalis — Cirtisia — Urbate — Servitti — [Ad Laddios].

273–74:

Ab Aquileia per Liburniam Sisciam — [Bibium] — Romula — Quadrata — Ad Fines — Siscia.

## Tabula Peutingeriana

[Citium] — Vindobona — Villa Gai — Aequinoctio — Carnunto — Gerulatis — Ad Flexum — Stailuco — Arrabo fl. — Brigantio — Lepavist — Gardellaca — Lusomana — Aquinco — Vetusallo — Annamatia — Lusione — Altaripa — Lugione — Antiana — Donatianis — Ad Labores — Tittoburgo — Cornaco — Cuccio — Malatis — Cusum — Acunum — Bittio — Burgenis — Taurono — Confluentibus.

*

Carnunto — Ulmo — Scarabantio — Sabarie — Arrabone — Ad Vicesimum — Petavione.

*

[In Alpe Iulia] — Longatico — Nauporto — Emona — Savo fl. — Ad Publicanos — [Adrante — Celeia — Ragandone — ] Petavione — Remista — Aqua viva — Populos — Botivo — Sonista — Piretis — Lentulis — Iovia — Sirotis — Bolentio — Marinianis — Seronis — Berebis — Iovallio — Mursa minor — Mursa maior — Ad Labores Pontis Ulcae — (Cibalae) — Causilena — Ulmo Spaneta — Sirmium — Bassianis — Idiminio — Tauruno.

*

Emona — Acervone — Ad Portorium — Crucio — Novioduni — Romula — Quadrata — Ad Fines — Siscia — Ad Pretorium — Servitio — Urbate — Marsonie — Certis — (Cibalae).

*

Marsonie — Ad Basante — Saldis — Drinum fl. — Sirmium.

## NOTES

[1] General: F. BERGER, *Über die Heerstrassen des römischen Reiches* (Berlin, 1882–1883), vols. 1 and 2. Wissenschaftl. Beilage z. Programm d. Luisenstädtischen Gewerbeschule); T. PEKÁRY, *Untersuchungen zu den römischen Reichstrassen* (Bonn, 1968).

[2] On the itineraries: K. MILLER, *Itineraria Romana*, (Stuttgart, 1916).

[3] On the milestones in general: O. HIRSCHFELD, "Die römische Meilensteine". *Sitzungsber. Akad. Berlin* (1907), pp. 165 ff.; K. SCHNEIDER, "Milliarium", *R. E. Pauly-Wissowa*, Suppl. 6, pp. 395–431.

[4] On the milestones in Pannonia: *CIL*, III. Recently: I. JÁRDÁNYI-PAULOVICS, "Palimpsestus feliratú és festett mérföldkövek Intercisából— Übermeisselte und bemalte Meilensteine aus Intercisa". *ArchÉrt* 76 (1949): 55–58; S. SOPRONI, "Kiadatlan pannóniai mérföldkövek— Unveröffentlichte pannonische Meilensteine". *ArchÉrt* 78 (1951): 44–48; E. B. VÁGÓ, "Új mérföldkövek az Intercisa és Mursa közötti útvonalon — Neue Meilensteine von der Wegstrecke zwischen Intercisa und Mursa". *ArchÉrt* 86 (1959): 73–75; P. PETRU, "K trem novim napisom s spodnjega Posavja — Zu drei neuen Inschriften aus dem unteren Savegebiet". *Arheološki Vestnik* 11–12 (1960–1961): 27–45; S. SOPRONI, "Römische Meilensteine aus Százhalombatta". *FolArch* 22 (1970): 91–112; W. WEBER, "Die römischen Meilensteine aus dem österreichischen Pannonien". *JÖAI*, 49 (1968–1971): 121 ff.

[5] On road techniques, repairs, and maintenance: J. FITZ, "Útjavítások Aquincum és Mursa között — Roman Road Repairs between Aquincum and Mursa". *ArchÉrt* 83 (1956): 196–206; A. MÓCSY, "Pannónia", *R. E. Pauly-Wissowa*, Suppl. 9, p. 655, with further references; S. PAHIČ, "K poteku rimskih cest med Ptujem in Središčem— Römische Strassen zwischen Ptuj und Središče". *Arheološki Vestnik* 15–16 (1965): 283–320.

[6] Recent literature on road techniques: M. KABA, "Római kori épületmaradványok a Király fürdőnél.— Römerzeitliche Gebäudereste beim Király-Bad". *BudRég* 20 (1963): 290; T. P. BUÓCZ, "Adatok Savaria topográfiájához— Angaben zur Topographie von Savaria". *ArchÉrt* 89 (1962): 181–187; T. NAGY, "Buda régészeti emlékei", in *Budapest műemlékei* (Budapest, 1962), 2:522.

[7] On guard stations along the roads and on bridges: J. SZILÁGYI, "Kutatások Aquincumból. I. Úti őrállomás a Csúcshegy tövében— Les fouilles d'Aquincum. I. Poste de sentinelles près d'une route au pied de Csúcshegy". *ArchÉrt* 76 (1949): 67–79; W. KUBITSCHEK, *Römerfunde von Eisenstadt* (Wien, 1926), pp. 39 ff.; A. BARB, "Angebliche und wirkliche Römertürme in Burgenland". *Burgenländische Heimatblätter* 11 (1949): 106–13. At Sirmium in the summer of 1968, A. Lengyel, together with G. Radan and V. Popović, formed a special task team to find the Roman stone bridge in the fast-flowing Sava River. They discovered a stone pillar of one of the Sava River bridges.

[8] On the Pannonian road network in general: A. GRAF, "Übersicht der antiken Geographie von Pannonien" (Budapest, 1936) (*DissPann* 1:5); A. MÓCSY, "Pannónia", pp. 658 ff.; *Tabula Imperii Romani*, vols. 33 and 34.

[9] Recent literature on the road network: PAHIČ; D. SZÉKELY, "Osones". *Antik Tanulmányok* 10 (1963): 50–55; P. PETRU, in *Dolenjski Zbornik* (1961), p. 202; A. NEUMANN, *Der Raum von Wien in ur- und frühgeschichtlicher Zeit* (Wien, 1961), pp. 19 ff.; I. PIRKOVIĆ, "Crucium". *Situla* 10 (1968): 8 ff.; D. PINTEROVIĆ, "Limesstudien in der Baranja und in Slawonien".

*ArchIug.* 9 (1968): 61 ff.; D. PINTEROVIĆ, "Problemi istrazivanja limesa na sektoru Batina–Skelo–Ilok". *Osječki Zbornik* 12 (1969): 53 ff.; *Veszprém megye régészeti topográfiája* (Budapest, 1966–1972), 4 vols. passim; G. PASCHER, "Römische Siedlungen und Strassen im Limesgebiet zwischen Enns und Leitha". *RLiÖ*, 19 (1949).

[10] On roads toward Barbaricum: A. MÓCSY, "Pannónia", p. 667; P. LAKATOS, "Funde der Römerzeit vom Gebiet der Szegediner Festung". *Móra Ferenc Múzeum Évkönyve, 1964–1965* (1966), I: 65–81.

# LIMES

SÁNDOR SOPRONI

## HISTORY

Pannonia was one of the frontier provinces of the Roman Empire. The borders of this province were defended by the *limes* (Fig. 8), a network of fortifications. The original meaning of the term *limes* was *trail* or *byway*; it then came to mean specifically the fortified line of the frontier.[1]

Research on Pannonia and research on the *limes* started at the same time. The first scientifically organized Pannonian excavations began in the army camps and adjoining settlements along the fortified line of the Pannonian frontier. It is no accident that the *limes* is the most fully documented area of Pannonia. The life, history, and topography of the fortified complex of the Danube area is also best known. However, information concerning the whole stretch of the *limes*, like Pannonian research in general, is unequal.[2] The Austrian and the Hungarian section of the *limes*, especially the Aquincum–Brigetio portion, has been investigated more thoroughly than the Yugoslav sector.[3] But in Yugoslavia intensive investigation of the frontier fortifications has been recently started.[4]

Pannonia was one of the most important frontier provinces of the Roman Empire for four centuries.[5] Especially after the abandonment of Dacia in A.D. 271, the pressure of migrations from the East affected primarily the Pannonian frontier. Life in Pannonia was closely related with the situation in the *limes*, which became the province's most important supply center. In the beginning, the armed forces stationed along the frontier in military camps were instrumental in Romanizing Pannonia. Later too they played an important role in the development of its economic, political, and cultural life.

In the history of the Pannonian *limes*, two periods can be distinguished. In the first phase of Roman rule, the fortifications were for offensive purposes — their primary function being to support and stabilize Roman expansion. For this reason, the *limes* was built in one line rather than being laid out in a series of parallel lines. At first, temporary camps constructed of earth and palisades controlled and defended the frontier. Later these camps were replaced by more solid structures built of stone. In the third and especially in the fourth century the need for reconstruction and the consequent decision to add to the *limes* reflected the precarious military situation and the shifting balance of political power. The role of the *limes* from this period on was primarily the defense of the frontier against Barbarian raids and invasions. This new defensive status of the *limes* demanded changes in the ground plans of the military camps: the fortified defensive lines were thus laid out in depth.

219

In the course of the gradual occupation of Pannonia, the pacification of the captured territory was secured by temporary earth and palisaded route camps. These route camps have not yet been discovered in Pannonia, although traces of them exist at Tác and Sárvár. Without question, they should be sought around the strategic points of the main roads which came from Italia and crossed Pannonia. During the second half of the first and especially at the beginning of the second century, when Pannonia received provincial administration, these camps were replaced by permanent stone constructions. Until the completion of these stone camps in the territory between the Drava and Sava rivers, and of the posts of the legions, mobile cavalry units carried out the territorial control and defense of the frontier. According to the latest research, a flotilla was also employed to patrol along the Danube River.

At the beginning of the Roman occupation, the Pannonian legions were stationed in the Emona–Poetovio–Siscia triangle. The Danube frontier was only provisionally occupied; only the crossing points were continuously guarded. The first more permanent type of camp on the Danube River was at Carnuntum, built as a permanent station for Legio XV Apollinaris where the Amber Route crossed the river. It is possible that during the reign of Claudius garrison troops were stationed at other important river crossing points, such as Arrabona, Brigetio, Aquincum, and so on. In addition to the camps along the Danube dating from this period, another camp in the interior of the province at Gorsium has been recently discovered.

Military diplomas (discharge certificates) dating from the middle of the first century also prove that a great number of auxiliary units were stationed in Pannonia. One military diploma issued in A.D. 60 listed seven *cohors* and several *alae* (equestrian units) in Pannonia. Shortly thereafter the military importance of the Danube frontier increased, causing more and more auxiliary units to be concentrated in this area. About A.D. 73, the first stone legion camp was built in Carnuntum, and the latest research has dated the building program of the first stone camp in Aquincum at about the same time.

A significant change in the history of the Pannonian *limes* occurred during the reign of Domitian. A military diploma from Beleg dated A.D. 85 listed five *alae* and fifteen *cohortes* in Pannonia. The following military earth and palisaded camps were built at that time: Cannabiaca (Klosterneuburg), Ad Flexum (Magyaróvár), Vetus Salina (Adony), Ad Statuas (Várdomb), Lugio (Dunaszekcső), and Ad Militare (Batina).

However, Trajan has been credited with the greatest development of the *limes*. The system was perfected but never entirely completed during his reign. The *limes* underwent transformations during the next centuries, and a few new camps or watchtowers were built, according to the changing strategic situation, but still the chain of fortifications was never completely joined.

The final system of legion camps and the following earth camps were built during the reign of Trajan: Gerulata (Oroszvár–Rusovce), Quadrata (Barátföldpuszta), Ad Statuas (Ács–Vaspuszta), Ad Mures (Ács–Bum–Bumkút), Azaum (Almásfüzítő), Solva (Esztergom), Cirpi (Dunabogdány), and Ulcisia Castra (Szentendre).

The system of fortifications which developed during the reign of Trajan, including the four legion camps (Vindobona, Fig. 9; Carnuntum, Fig. 10; Brigetio; Aquincum), consisted mainly of earth camps. The Romans transformed the earth camps of the auxiliary units, thirty-three in number, into stone structures during the reign of Hadrian. From this time until the end of the Roman rule, four legions were permanently stationed in Pannonia: Legio X Gemina in Vindobona, Legio XIV Gemina in Carnuntum, Legio I Adiutrix in Brigetio, and Legio II Adiutrix in Aquincum.

In addition to the legion camps, the following auxiliary camps were built along the Danube: Cannabiaca (Klosterneuburg), Ala Nova (Wien-Schwechat), Aequinoctium (Fischamend), Gerulata (Oroszvár-Rusovce), Ad Flexum (Magyaróvár), Quadrata (Barátföldpuszta), Arrabona (Győr), Ad Statuas (Ács-Vaspuszta), Ad Mures (Ács-Bum-Bumkút), Azaum (Almásfüzítő), Crumerum (Nyergesújfalu), Solva (Esztergom), Cirpi (Dunabogdány), Ulcisia Castra (Szentendre), the camp in Albertfalva (Roman name not known), Campona (Budapest-Nagytétény), Matrica (Százhalombatta), Vetus Salina (Adony), Intercisa (Dunaújváros) (Fig.11), Annamatia (Baracs), Lussonium (Dunakömlőd), Alta Ripa (Tolna), Alisca (Őcsény), Ad Statuas (Várdomb), Lugio (Dunaszekcső), Altinum (Kölked), Ad Militare (Batina), Teutoborgium (Dalj), Cornacum (Šotin), Bononia (Banoštor), Acumincum (Slankamen), Rittium (Surduk), Burgenae (Novi Banovci).[6]

The weakness of the Danubian fortification system, which was finished during the reign of Hadrian, became evident immediately. The enemy could easily break through a one-line system which was not reinforced by parallel defense lines laid out in depth. Most of the camps built of stone were heavily attacked during the second century A.D. Perhaps even during the reign of Hadrian, after the Sarmatian and Quadian raids in A.D. 136–138, it became necessary to rebuild certain military camps around Aquincum. It is remarkable that instead of stone masonry being used to repair the damage, the walls of the less important interior buildings were repaired with sun-dried bricks.

The devastation of the fortifications during the Hadrianic era, which certainly did not extend to the whole *limes* area, was followed by complete destruction of the greater part of the *limes* during the reign of Marcus Aurelius. The invasion of the Marcomannians and Sarmatians damaged every camp of the *limes* in some measure. Burned layers have indicated that the defense system was shaken to its foundations.

Reorganization of the *limes* was begun by Marcus Aurelius. He ordered the rebuilding of the fortifications and replacement and reinforcement of the destroyed military units. In order to match the Sarmatian force, he organized units of special equestrian archers. It is possible that the restoration program was carried out through several stages and was terminated only at the end of the century. In addition to the camps, stone watchtowers (*burgi*) and watch stations (*praesidia*) were built. The victory column of Trajan depicted quadrangular wooden watchtowers, but scanty archaeological evidence survives. From the period of Commodus thirteen plaques with inscriptions (Pl. XI. 2), partly in fragments, have documented the building of several watchtowers in 184–185 in Pannonia Inferior. In addition, it has been possible to document the origin of several small quadrangular watchtowers in Csillaghegy, Nyergesújfalu, Visegrád-Várkert, etc. During the building program of Marcus Aurelius, several camps seem to have been furnished with semicircular interior corner towers. This practice was continued during the third century. A great many bricks stamped with the Antoniniana sign, indicating the building program of Caracalla, have been found all along the *limes*.

During the second half of the third century the *limes* was again sorely tried. About 260 the Quadian–Sarmatian raids and in 270 the Vandal–Sveb–Sarmatian raids made severe inroads on the *limes*. The measure of the devastation is documented by the fact that the auxiliary camp at Albertfalva was completely abandoned after the raid of the Sarmatians in 259–260. Almost all the settlements along the *limes* and the military installations were heavily damaged.

The complete rebuilding of the *limes* and its military organization has been attributed to Diocletian. In addition to reconstructing the old military camps, he built new camps, small fortresses, and bridgeheads. He also reinforced the defense system of the province, which became extraordinarily important in the defense of the whole Roman Empire. The building program supposedly started in A.D. 292 with the modernization and reconstruction of the fortifications. Documents dated in 294 mention Contra Bononiam and Contra Acumincum as rebuilt counterfortresses. These reconstructions extended to the entire Pannonian *limes*; several new camps were even built on the right bank of the Danube, as, for example, Čortanovci, Rakovac (to defend Sirmium), and at the Danube bend Castra ad Herculem (Pilismarót) Diocletian and Constantine the Great reorganized the armed forces; centralized mobile units (*comitatenses*) were formed, overshadowing the importance of the frontier army (*limitanei*). Military service became hereditary, and the defense of the *limes* was turned over to settled mercenaries who received land as well as money for their services.

A further building program occurred during the reign of Constantine the Great. In order to meet the new defensive requirements, horseshoe-shaped towers were added to the walls of the camps. The corners of the fortifications were reinforced by fan-shaped donjons. The defenders were able to flank the invaders from these towers. This defensive tactic became extremely important after the abandonment of the province of Dacia.

Waves of migrations from the East became more and more menacing and required the constant reorganization and modernization of the defense system. Because of its geographical location, the Pannonian *limes* was most threatened. Only reconstruction which started during the reign of Diocletian in the Constantinian period was made into a complete defense system. Recent research has clarified the date and purpose of a network of several rows of mounds around the Great Plain of Hungary.[7] This fortification was built during the reign of Constantine the Great around the Sarmatian settlements. It was an important link in the chain of defense and was closely related to the network of the *limes*. The junctions of this wall system and the Danubian *limes* at the Lower Danube (Djerdap) in Moesia and the Danube bend were reinforced by new military camps. They connected with the wall system at Visegrád (Pone Navata) (Fig. 12), where a military camp which reinforced the Danube bend area was built. The mound system served as an outer *limes* to relieve direct pressure on the *limes*. The enemy was thus directed

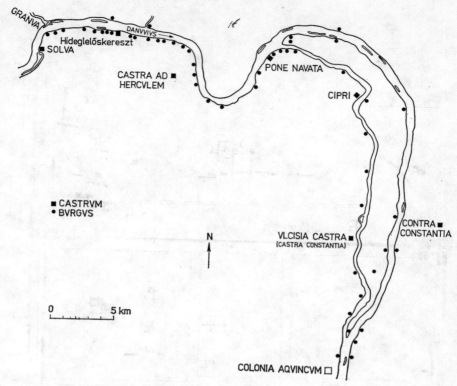

Fig. 8. Map of the *limes* section between Solva and Aquincum

toward the junctions of the wall system and the actual *limes*, which were well reinforced by the Romans.

During the fourth century, bridge-heads were built on the left bank of the Danube as a connection with the wall system or a supplement to it. Some counterfortresses had previously been built on the left bank of the Danube (for example, Celemantia-Leányvár), but a systematic network of bridgeheads, which was begun during the reign of Diocletian with Contra Bononiam and Contra Acumincum, was only completed during the fourth century. Along the Valerian section of the *limes*, the bridgeheads were connected with the reinforced *limes* built during the time of Constantius II (Fig. 18). The wall system of the Great Plain was completed at that time, as was Contra Constantiam at the juncture of the *limes* at Göd.

The following fortifications were built in the interior of Pannonia in the course of the fourth century: Kapospula-Alsóhetény (Pl. X. 1), Ságvár, Környe, Kisárpás, and Keszthely-Fenékpuszta. They served as an inner defensive line behind the *limes*. They are generally accepted as dating from the reign of Constantine.[8]

Finally, between 365 and 370 the Pannonian-born Valentinian attempted to reorganize and rebuild the damaged parts of the *limes*. First he ordered the rebuild-

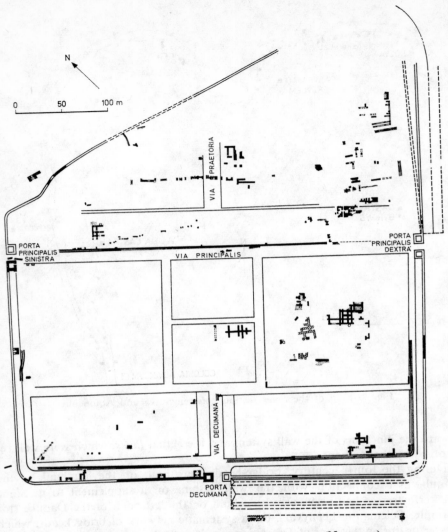

Fig. 9. Ground plan of the Vindobona camp (Neumann)

ing of damaged parts; this fact is documented by the use of stamped brick (Pl. IX. 2) and by an inscription from Carnuntum. New camps were also built. An inscription has been found in the Esztergom area which was probably associated with the building of the military camp at Hideglelőskereszt. Between 371 and 372, a chain of watchtowers was built between the camps (Figs. 13 and 14; Pl. VIII. 2). Although research on this network has not yet been completed, there is enough evidence to conclude that the line of towers was set only along sections most threatened by the invaders. There is also some evidence concerning small towers in the area of Brigetio and in the section of the Danube bend (Figs. 8 and 17;

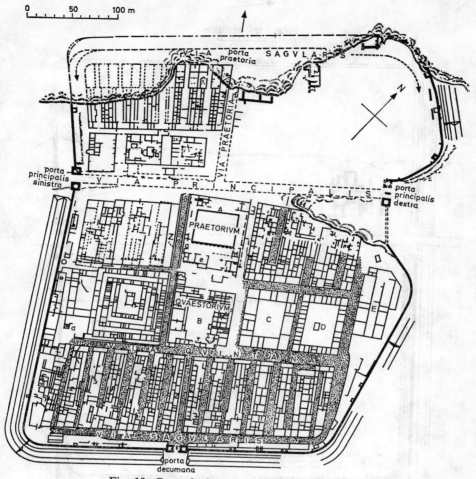

Fig. 10. Ground plan of the Carnuntum camp

Pl. XI. 1). In addition to the fortress at Milanovce, watchtowers were built around Brigetio, which were able to fend off any attacks from the Vág valley. The towers around the Danube bend were instrumental in eliminating the raids around the junctions of the *limes* and the wall system. Military stations dating from the Valentinian period have been discovered 60 kilometers east of the Danube *limes* at Hatvan.

Following the death of Valentinian, the Huns began a decade of expansion, again changing the balance of power. As they practically overpowered the *limes* and rendered it ineffective, gradually the function of the *limes* came to an end.[9]

Archaeological observations have documented the fact that the smaller watchtowers were abandoned before the end of the fourth century, and only the larger watchtowers, small fortresses, and the fortified military camps survived at the

225

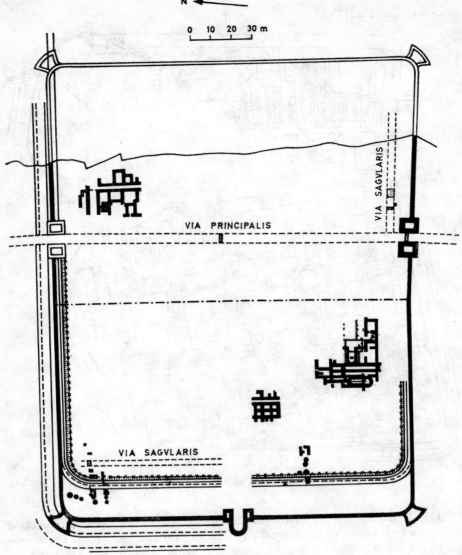

Fig. 11. Ground plan of the Intercisa (Dunaújváros) camp

beginning of the fifth century. However, from about 430 to 440 these military installations too lost their defensive significance. From the end of the fourth century *foederati* groups gradually settled down in Pannonia; these Barbarian settlements eroded the frontier line of the Roman Empire. In this way, the practical function of the *limes* ended.

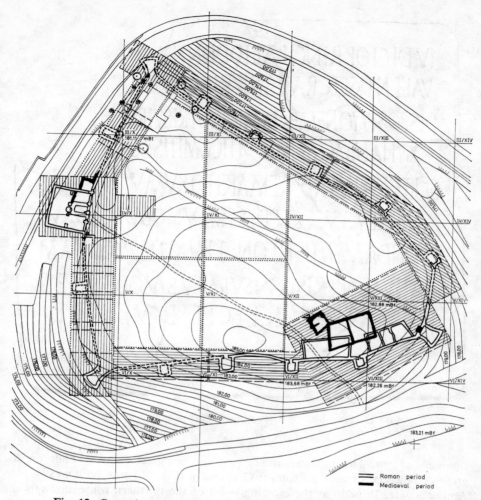

Fig. 12. Ground plan of Roman fortification of Visegrád-Sibrikdomb

## STRUCTURE

By examining the construction of the *limes* in Pannonia we may conclude that it was built as a loose network of military camps and watchtowers connected by a main route alongside the Danube River. Because the natural frontier line was the Danube, it was not necessary to build fortified ditches along the entire frontier as in Germania or heavy walls such as Hadrian's Wall in Britain and the walls on the frontier of Raetia.

About fifty military camps and nearly a hundred watchtowers, small fortresses, and bridgeheads have recently been discovered along the right bank of the Pannonian Danube section. The most important military installations were the

227

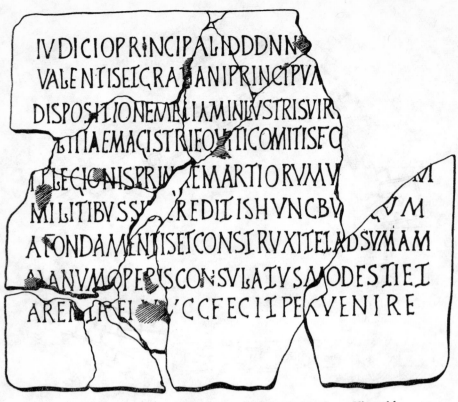

Fig. 13. Fragment of Roman *burgus* inscription from Visegrád

legionary camps. Three of Pannonia's four legionary camps were built on the great commercial or military routes: Carnuntum (Fig. 10) was built next to the Amber Route, Brigetio at the crossing point of the northern route through the Vág valley, and Aquincum defending the main east–west commercial route, which was simultaneously used as a main route for military maneuvers. Alongside the Danube, in the territory of Pannonia, the first camp was that of the Vienna (Wien) legion.

The site of the Vindobona (Fig. 9) legion camp coincides with the interior of present-day Vienna, and the city's network of streets still preserves the shape of the onetime camp.[10] It had the shape of a right angle trapezoid, the *decumana* and the adjoining principal sides forming a regular quadrangle. The *praetoria* front on the Danube side is not known exactly, but must have provided an irregular ground plan. The legion camp measured 450 by 300 to 500 meters.

The modern city has made excavations difficult. From the walls of the camp and the towers only small fragments have been unearthed. Of the camp gates, only the rectangular towers of the *porta decumana* and the *porta principalis sinistra* and some side towers between the two have come to light. It is notable that the rectangular

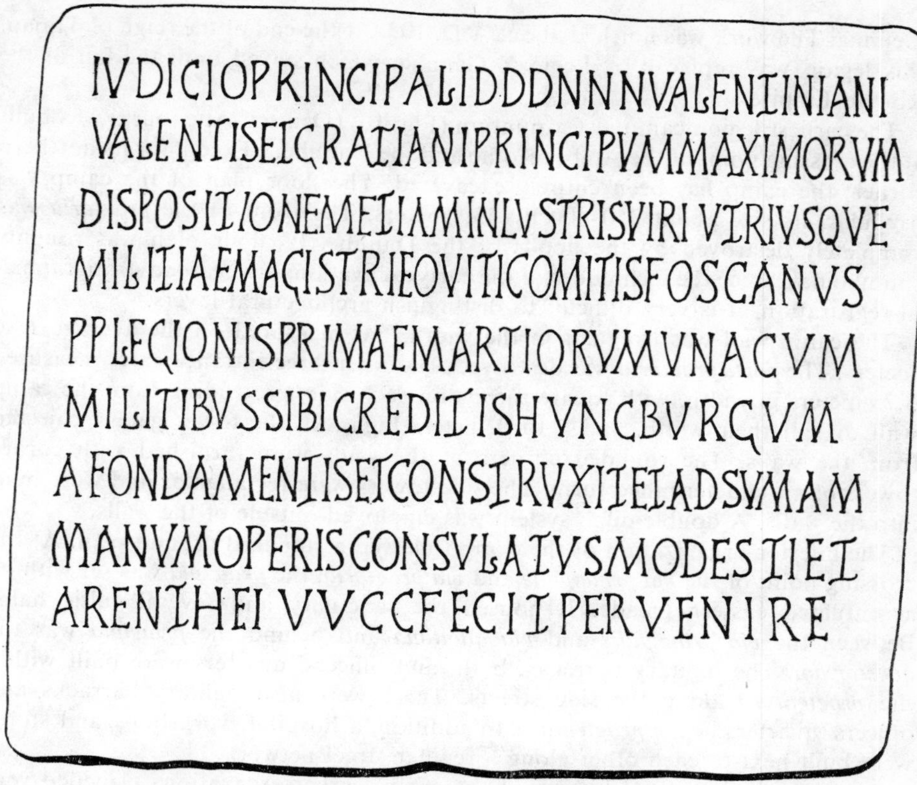

IVDICIOPRINCIPALIDDDDNNNVALENTINIANI
VALENTISEICRATIANIPRINCIPVMMAXIMORVM
DISPOSITIONEMETIAMINLVSTRISVIRIVIRIVSQVE
MILITIAEMACISTRIEOVITICOMITISFOSCANVS
PPLECIONISPRIMAEMARTIORVMVNACVM
MILITIBVSSIBICREDITISHVNCBVRCVM
AFONDAMENTISETCONSTRVXITEIADSVMAM
MANVMOPERISCONSVLAIVSMODESTIEI
ARENTHEI VVCCFECITPERVENIRE

Fig. 14. Remains of a *burgus* inscription and its complete text from Visegrád

tower of the *porta principalis* was built into the line of the wall, but the rectangular tower of the *porta decumana* projected from the wall (9.4 by 9.45 meters). The other towers were placed within the limits of the wall. Similar towers, which were 4.6 meters wide and about 8 meters long, were also located on the southwestern wall.

The fragments of the unearthed walls of the legion camp clearly indicate that after the devastation of the Marcomannic wars the wall of the reconstructed camp did not entirely follow the line of the old one. The thickness of the wall varied between 2.2 and 2.25 meters. The berm at the front of the wall was paved with large stone blocks. Traces of a double ditch system at the front of the walls have emerged only in the location of today's Graben.

Our knowledge of the interior system of the military camp and its buildings is fragmentary. The direction of the *via principalis* and *via decumana* are indicated only by the recently discovered gates. Few wall fragments have remained from the so-called legatine palace. In the *retentura*, fragments of military barracks remain; in the *praetentura* also, some building fragments were found.

The construction of the camp was begun before A.D. 98 by Legio XIII Gemina. The legion was transferred to Dacia in about 101 and replaced by Legio XIV

Gemina. The work was finished about A.D. 103. At the end of the reign of Trajan, this legion was replaced by Legio X Gemina, which stayed until the fall of the Roman Empire.

The earliest legion camp at Carnuntum (Fig. 10) (Deutsch-Altenburg) was built at the crossing point of the Amber Route and the Danube.[11] Except for its northern corner, the camp has been entirely excavated. The floor plan of the camp was irregular, its size about 350–400 by 450 meters. The front of the *praetoria* was completely destroyed by the floods of the Danube. Its floor plan was roughly rhomboidal. Since the techniques of the early excavations disregarded stratigraphical registration, it is very difficult to distinguish architectural layers.

The camp wall was irregular in the width, but measured on the average 1.80 meters. The *porta praetoria* was destroyed. The other three entrance gates measured 3.75 meters in width, with square turrets on each side, protruding from the camp wall to half their width, and 7 to 9 meters high. All the tower gates projected from the walls. The rounded corners of the walls were furnished with corner towers of an undetermined form. The side towers were rectangular and were built into the walls. A double ditch system was employed outside of the walls.

The interior organization of the camp followed a standard ground plan. At the crossing point of the *via principalis* and *via praetoria*, the *principia* was set with its accentuated eastern facade. It housed the sanctuary and several cultic halls. Between the *via principalis* and *via quintana* and behind the *principia* was the *questorium*. The military barracks, bath, and officers' quarters were built within the *praetentura* along the side streets. There were also military barracks and officers' quarters in the *praetentura*. In addition, a hospital, workshops, and stores were built next to each other along a regular street network.

As to the architectural history of the camp, earlier excavations provided very little documentation to satisfy contemporary standards. Recent excavations were able to differentiate eight cultural strata from the first century A. D. to the turn of the fifth century. However, there are no data concerning the earth camp which preceded the stone one. The building program here, according to the latest research, was possibly carried out during the reign of Vespasian. Significant reconstruction occurred after the Marcomannic wars and during the German invasion in the second half of the third century. An inscription from the Valentinian period has provided documentation of a building program which was carried out in the last third of the fourth century. Recent scholarship has discovered in the interior of the camp a comparatively late fifth-century shortened fortification.

An inscription from the Claudian era testifies that the first legion in Carnuntum was Legio XV Apollinaris. When this legion was transferred to the East in A.D. 62, it was replaced by Legio X Gemina. In A.D. 68 Legio X Gemina was in turn replaced by Legio XXII Primigenia, and then Legio XV Apollinaris returned from the East. This is the legion which built the stone camp and stayed until the second Dacian war, when it was replaced by Legio XIV Gemina. At the end of the Dacian war, Legio XV Apollinaris again returned to Carnuntum. Finally, at the end of the reign of Trajan, Legio XIV Gemina returned and remained until the end of Roman rule.

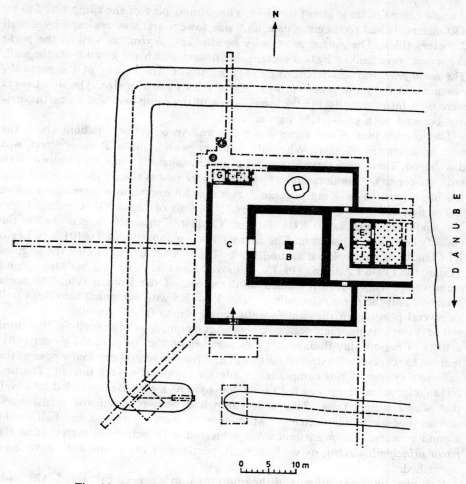

Fig. 15. Roman *burgus* at Pilismarót, in the fourth century

Similarly, the legion camp Brigetio at the site of modern Szőny defended a very important crossing point.[12] According to current research in Brigetio, three different military camps can be differentiated, built one on top of the other.

The earliest earth camp was situated on the bank of the Danube, but only its ditch system is known today, and measurements are still unknown. The second stone camp was built west of this location. Its eastern part was completely destroyed over a period of centuries by the floods of the Danube. Construction probably began in the fourth century. The orientation of the camp was northaest–southeast, but trial trenches made there have not uncovered an entirely coherent plan. Measurements, however, point to a legion camp.

The first stone camp was built around A.D. 100 south of the second stone camp. It is possible that constant floods of the Danube made it necessary to build the new camp farther south of the river. The ground plan of the camp was 540 by 430 meters. It had rectangular gate and side towers and was enclosed by a wall 2 meters thick. The *portae principales sinistra* and *dextra*, as well as the *porta decumana*, were flanked by two rectangular towers which projected from the wall. The *porta praetoria* was defended by only one tower. The towers of the *porta decumana* and the *porta praetoria* were 10.3 by 7.9 meters in size. The side towers were built into the walls. On the interior side of the camp wall was an earth corridor covered with grass, 1.80 meters wide.

The interior plan of the camp is not known. An oil refinery is built above the ancient *praetorium* today. When the refinery was built, a Roman street was discovered. There were three layers of pavement, suggesting that the roadbed often required repairs. A modern highway follows more or less closely the *via principalis*. During a modern building program, several small excavations were carried out.

The first stone camp was built by vexillatio units of Legio XIV Gemina, Legio XV Apollinaris, and Legio XIII Gemina. The last of these left Brigetio before the end of the building program. The first garrison troop after A.D. 101 was Legio XI Claudia; it was replaced around 105 by Legio XXX Ulpia. During the reign of Trajan, the Legio I Adiutrix (Pl. IX. 1) moved into this camp and built the second stone camp. This legion remained until the end of the Roman rule. The *porta praetoria* and the *porta principalis dextra*, defended with octagonal turrets, as well as several portions of the camp's walls have been so far excavated.

The fourth Pannonian legion camp, Aquincum, was stationed in the third district of present-day Budapest.[13] This area lies on the fringe of the modern city, making exact research extremely difficult, and for this reason our knowledge of the defensive system of this camp and its interior buildings is very limited. The first rectangular stone camp was built on a 460–by–430–meter site, encircled by a wall that was 2 meters thick. The stone camp had been repaired and transformed several times, especially after the Marcomannic wars, in the second half of the second century. The *porta praetoria*, protected with octagonal turrets, and the *porta principalis dextra*, as well as many portions of the camp wall, have been unearthed.

However, only a small part of the interior camp has been excavated. The most important discovery comes from the southwestern section, where several rooms from a large bath have come to light in the *retentura* (living quarters). Other discoveries include a *principia* (headquarters), fragments of the *valetudinarium* (hospital), a *horreum* (granary), parts of a water-supply system, and fragments of buildings.

The second stone camp, of the Constantinian period, was built east of the site of the first camp with horseshoe-shaped turrets. At the end of the fourth century this camp had been made smaller.

The building of the camp at Aquincum made headway during the reign of Domitian in the second half of the 80s when Legio II Adiutrix was on the scene, and was completed by Legio IV Flavia. In the early second century, Legio IV Flavia was replaced by Legio X Gemina. Toward the end of the reign of Trajan, Legio II Adiutrix returned again to Aquincum and built the second stone camp. This legion remained at Aquincum until the end of the Roman rule.

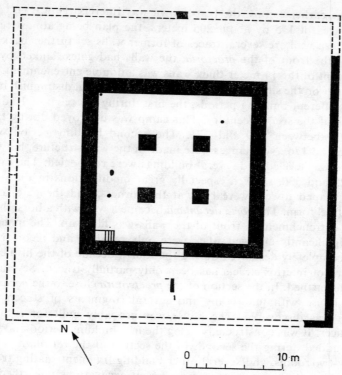

Fig. 16. Ground plan of Leányfalu *burgus*

During the period covered, near the four legionary camps there were auxiliary camps — the camps of cavalry and foot soldiers. Generally these auxiliary camps were situated between 15 and 20 kilometers apart. In strategically important sections which were more endangered by enemies, as in the Danube bend, the camps were built no more than 10 kilometers apart.

Many excavations have been made around the site of the auxiliary camps, but not a single one has been completely unearthed. The goal of the excavations in such camps as Ács-Vaspuszta, Dunaalmás, Szentendre, and Nagytétény was to clarify their architectural data and ground plans. In the camps of Intercisa (Duna-újváros) (Fig. 11) and Tokod, the excavations were more elaborate but not complete. Most of the auxiliary camps followed a regular plan. Only those built during the fourth century were irregular. The earlier camps were built on flat land close to the bank of the Danube. Later constructions were built to fit in with the topography of the heights or summits of hills and mountains. The irregular ground plan of these camps was also suited to the defensive situation. It is note-worthy that among the late auxiliary camps, interior buildings were constructed only for the defense of the walls. The center parts of the camp were left empty.

Among the earlier camps, the best-known was Intercisa (Fig. 11).[14] The camp covered an area of 175 by about 205 meters, the plan being almost rectangular. At the southern wall are several traces of former walls set further back. With the exception of the front of the *praetoria*, the walls had gates flanked by defensive towers. In front of the 14-meter-thick walls was a double entrenchment. The stone camp was built on the site of a former earth camp. One can distinguish in the stone camp three different building periods, the first during the reign of Hadrian in the first decades of the second century. This camp was destroyed during the Marcomannic wars between 169 and 171. The second building period was started between 175 and 176. No changes were made in the wall structure, but the area of the *retentura* was leveled and several buildings were remodeled. The third period, between 325 and 330, showed especially great modifications in the walls of the camp. Fan-shaped towers were built at the corners, and the gate towers were given their final form. The *porta decumana* became a wall with a horseshoe form of tower. The entrenchment in front of the walls was filled up. The inner buildings were not significantly changed; they were only repaired and reconstructed. The camp was completely destroyed during the fourth decade of the fifth century.

A network of interior streets has been only partially traced. Several buildings have been unearthed. In the section of the *praetentura* close to the *porta principalis sinistra*, a house with ten rooms and several fragments of stucco and fresco decorations was discovered. The house was probably built for a high-ranking military officer. It was constructed during the first building period and the decorations were added during the second. In the section of the *retentura*, a large bath, with twenty-two rooms, and several other building fragments dating from the end of the second century were discovered. Recent excavations unearthed the whole camp, and the *principia* was discovered at the crossing of *via praetoria* and the *via principalis*.

The earth or palisaded camp had been destroyed at the end of the reign of Trajan or at the beginning of the Hadrianic era. The camp had supposedly been built for Ala I Siliana and one of the units of Legio XIII Gemina. The first stone camp was supposedly built by Ala I Thracum Veterana Sagittariorum. This *ala* was replaced toward the end of the Hadrianic era by Cohors I Alpinorum Equitata, which remained in the camp until A. D. 176. From this date Cohors (Milliaria) Hemesenorum occupied the camp. During the time of Diocletian, the Equites Sagittarii were organized from this unit which, together with two other units, the Cuneus Equitum Dalmatorum and Cuneus Equitum Constantianorum, made up the garrison of Intercisa.

A significant example of the late Roman military camp was Pone Navata at Visegrád (Fig. 12).[15] It was placed on the summit of the mountain which rises at the bank of the Danube. The form of the camp followed the formation of the mountain and the plan was pentagonal. The greatest width of the camp was 114 meters; the length was 130 meters. The old walls were 1.15 to 1.30 meters thick and were reinforced at the corners by projecting fan-shaped towers (Pl. VIII. 1), and on the sides by horseshoe-shaped ones. The location of the camp's original gate is not yet known. In the second period, a two-arch gate was built on the Danube side, to replace a demolished horseshoe-shaped tower. When the camp was

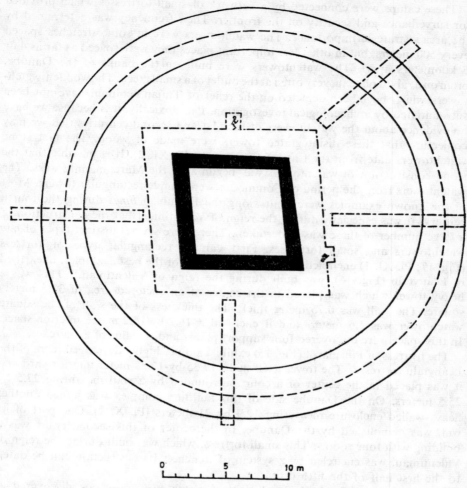

Fig. 17. Roman watchtower from the age of Valentinian,
Esztergom–Szentgyörgymező

abandoned, about 380, a watchtower 11.70 meters square was raised in front of the gate. The fan-shaped towers were 5.6 or 4 meters in width and 6 meters long. The horseshoe-shaped towers were 3.2 by 3.2 meters. The center of the camp was empty. The buildings were erected close to the defensive walls. The camp's name was preserved in the corrupt form Pone Navata in the *Notitia Dignitatum*. It was garrisoned by Auxilia Ursarensia. The date of the building construction was between A.D. 325 and 330. The camp may have been built in conjunction with the wall system of the Hungarian Plain. Between 378 and 380 the camp was evacuated, when a tower was built in front of the gate. This tower was probably destroyed in the first decades of the fifth century.

235

These camps were connected by a network of small fortresses which provided for surveillance and security on the frontier. Their frequency was determined by the area's strategic importance. The watchtowers were at some stretches spaced every 500 meters, but in other less important places they were spaced as far as 4 or 5 kilometers apart. The watchtowers were built on the banks of the Danube, often on a raised plateau, very often at the outlet of a small creek. The wooden watchtowers which have been depicted on the relief of Trajan's column have not been substantiated by archaeological investigation. Paradoxically, it is because we have no evidence about the watchtowers of the first and second centuries that we may conjecture that these disintegrated towers were made of wood. The first stone watchtowers date from the time of Commodus (Pl. XI. 2). It is possible that the stone construction of watchtowers was begun after the Marcomannic wars. The watchtowers from the period of Commodus were simple rectangular towers. Most of the known examples were built along the Danubian *limes* during the fourth century. It was especially during the reign of Valentinian that the construction of a large number of these was undertaken. There were several distinct types of late watchtowers and small fortresses. First came a rectangular type of fortress (Fig. 17; Pl. XI. 1) encircled by a heavy wall. One of the best examples, unearthed in Leányfalu (Fig. 16), was built during the reign of Valentinian.[16] This was a heavy tower which was placed in the center of an enclosed area of 32.5 meters square. The wall was 0.75 meter thick. The thickness of the wall of the square watchtower was 1.6 meters and it enclosed a 16.2-by-16.2-meter interior space. In this space were discovered four support pillars and remains of a staircase.

The fortress of Pilismarót (Fig. 15) exhibits a more highly developed form of the Leányfalu fortress.[17] The tower itself had a 12.35-by-12.25-meter interior yard and it was placed in the center of a court surrounded by a wall measuring 27.5 by 25.5 meters. On the Danube side of this building complex was joined another heavy-walled enclosure measuring 12.9 by 21 meters (Pl. X. 2). One part of its wall was demolished by the Danube. In the center of this second court was a building with four rooms. This small fortress, which was built during the reign of Valentinian, was encircled by a system of trenches. Its destruction can be dated to the first half of the fifth century.

Notable among the bridgehead type of watchtowers was one discovered at Szentendre-Derapatak (Fig. 18).[18] The watchtower itself was 20 by 20 meters, and two 30-meter-long walls connected it with a smaller tower of 7.5 by 7.5 meters on the Danube side. Many variations of this bridgehead type of fortress were built along the Danube River during the rule of Constantius II, in the mid-fourth century. Later, during the reign of Valentinian, they were rebuilt.

One of the late Roman rectangular simple watchtowers along the Pannonian *limes* was discovered at Visegrád-Kőbánya.[19] Its interior space was 8.9 by 8.9 meters, with walls 1.05 to 1.09 meters thick. In the center of the tower was discovered one pillar that supported the roof. This roof with the supporting pillar was a later addition. The watchtower was encircled with defensive trenches and at the back of the tower was a large door, 1.8 meters in width. Next to the door, an inscription on a plaque indicates the date of the tower as A.D. 372 (Figs. 13 and 14; Pl. VIII. 2).

236

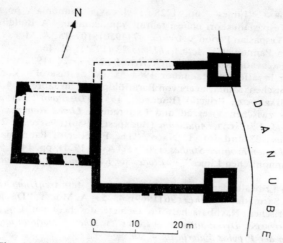

Fig. 18. Roman bridgehead from Szentendre, Derapatak

## NOTES

[1] E. FABRICIUS, "Limes", *R. E. Pauly-Wissowa* 13: 572–671; G. FORNI, "Limes". *Dizionario Epigrafico di Antichità Romana* 4:1074 f.

[2] FABRICIUS, "Limes"; A. MÓCSY, "Pannonia", *R. E. Pauly-Wissowa*, Suppl. 9, pp. 647 f; FORNI, "Limes", pp. 1226 f.

[3] Austrian portion of *limes:* "Der römische Limes in Österreich", referring to Pannonia; E. SWOBODA, *Carnuntum*, 4th ed. (Graz and Cologne, 1964), passim. E. SWOBODA, "Der pannonische Limes und sein Vorland". *Carnuntum Jahrb.*, 1959 (1961), pp. 17–30.
Hungarian section: A. GRAF, "Übersicht der antiken Geographie von Pannonien" (Budapest, 1936) (*DissPann* 1:5); I. PAULOVICS, "Il limes romano in Ungheria". *Quarderni dell'Impero* (1939); Czechoslovakian section V. ONDROUCH, *Limes Romanus na Slovensku* (Bratislava, 1938); T. KOLNIK, "Ausgrabungen auf der römischen Station in Milanovce". *Limes Romanus Konferenz Nitra* (1959), pp. 27–48; B. SVOBODA, "K dějinam římského kastelu na Leányváru u Iže, okres Komarno—Zur Geschichte des Römerkastells in Leányvár bei Iža, Bezirk Komarno". *Slovenská Archeologia* 10 (1962): 397–424; F. KRIŽEK, "Die römische Stationen im Vorland des norisch–pannonischen Limes", in *Studien zu den Militärgrenzen Roms* (Graz and Cologne, 1967), pp. 133 f.; T. KOLNIK, "Neskororimska vojenská stanica v Páci pri Trnave". *Arch. Rozhledy* 24 (1972): 59 f.

[4] N. VULIĆ, "Il limes romano in Jugoslavia". *Quaderni dell'Impero* (1937); J. KLEMENC, "Der pannonische Limes in Jugoslavien". *Arheološki Radovi i Rasprava* 3 (1963): 55–68; *Limes u Jugoslaviji*, vol. 1 (Belgrad, 1961); P. VELENRAJTER, "Castellum Onagrinum". *Rad Vojvod. Muzeja*, 1958 (1960), pp. 126–32; M. MANOJLOVIĆ, "Rimsko utvordjenje kod Čortanovaca". *Rad Vojvod. Muzeja*, 1962 (1964), pp. 123–25; D. PINTEROVIĆ, "Limesstudien in der Baranja und in Slavonien". *ArchIug* 9 (1968): 55 f.; D. DIMITRIJEVIĆ, "Istraživanja rimskog limesa u istočnom Sremu s posebnim osvrtom na pitanja komunikacija". *Osječki Zbornik* 12 (1969): 81 f.; M. BULAT, "Topografska istraživanja limesa u Slavoniji i Baranji". *Osječki Zbornik* 12 (1969): 39 f.

[5] A. RADNÓTI and L. BARKÓCZI, "The Distribution of Troops in Pannonia Inferior during the Second Century A.D.". *ActaArchHung* 1 (1951): 191–230; L. BARKÓCZI and É. BÓNIS, "Das frührömische Lager und Wohnsiedlung von Adony (Vetus Salina)". *Acta ArchHung* 4 (1954): 129–99; G. ALFÖLDI, "Die Truppenverteilung der Donaulegionen am Ende des 1. Jahrhunderts". *ActaArchHung* 11 (1959): 113–41; J. FITZ, "A Military History of Pannonia from the Marcomann Wars to the Death of Alexander Severus (180–235)". *ActaArchHung* 14

(1962): 25–112; FORNI, "Limes", pp. 1228 f.; MÓCSY, "Pannónia", pp. 647 f.; E. TÓTH and G. VÉKONY, "Vespasianuskori építési felirat Aquincumban—A Building Inscription in Aquincum from the Vespasian Period". *ArchÉrt* 97 (1970): 109–15; A. MÓCSY, "Zur frühesten Besatzungsperiode in Pannonien". *ActaArchHung* 23 (1971): 41–46.

⁶ J. FITZ, "The Excavations in Gorsium". *ActaArchHung* 24 (1972): 3–52; D. GABLER, "Előzetes jelentés a barátföldpusztai tábor 1964–65. évi ásatásáról — Vorbericht über die Freilegung des römischen Militärlagers von Barátföldpuszta im Jahre 1964–65". *Arrabona* 8 (1966): 67–98; L. BARKÓCZI, Brigetio (Budapest, 1951) (*DissPann* 2:22); S. SOPRONI, "Der spätrömische Limes zwischen Visegrád und Esztergom". *Limes Romanus Konferenz Nitra* (1959), pp. 131–43; J. SZILÁGYI, *Aquincum* (Budapest, 1956); *Intercisa*, 2 vols. (Budapest, 1954–1957) (*ArchHung* 33 and 36); T. NAGY, "The Frontier of Pannonia as Reflected by Recent Research". *Roman Frontier Studies*, 1967 (Tel Aviv, 1971), pp. 145 f.; T. NAGY, "Neue Forschungen am pannonischen Limes", in *Études sur les Frontières Romaines* (Bucharest and Cologne, 1974).

⁷ V. BALÁS, "Die Erdwälle der ungarischen Tiefebene". *ActaArchHung* 15 (1963): 309–36; S. SOPRONI, "Limes Sarmatiae". *ArchÉrt* 96 (1969): 43–53; A. MÓCSY, "Das letzte Jahrhundert der römisch–barbarischen Nachbarschaft im Gebiete des heutigen Ungarn". *Cumania* 1 (1972): 83–93; S. SOPRONI, *Der spätrömische Limes zwischen Esztergom und Szentendre. Das Verteidigungssystem der Provinz Valeria im 4. Jh.*

⁸ A. RADNÓTI, "Római tábor és feliratos kövek Környéről—Le camp romain et les monuments épigraphiques de Környe". *Laureae Aquincenses* 2 (1941): 77–105 (*DissPann* 2:11); A. RADNÓTI, "Pannóniai városok élete a korai feudalizmusban" (Life in Pannonien Cities during the Early Years of the Feudal System), *MTA II*. OK. 5 (1954): 489–534; K. Sz. PÓCZY, "Scarbantia városfalának korhatározása—La datation de l'enceinte de Scarbantia". *ArchÉrt* 94 (1967): 137–54; K. SÁGI, "Das Problem der pannonischen Romanization im Spiegel der völkerwanderungszeitlichen Geschichte von Fenékpuszta". *ActaAntHung* 18 (1970): 147–96; S. SOPRONI, "Die spätrömische Festung von Iovia". *Études sur les Frontières Romaines* (Bucharest and Cologne, 1974), pp. 181 f.; S. SOPRONI, *Der spätrömische Limes.*

⁹ A. MÓCSY, "Pannonia"; SOPRONI, *Der spätrömische Limes.*

¹⁰ A. NEUMANN, "Vindobona", *R. E. Pauly-Wissowa*, 2nd ser., 17th half-vol., pp. 53–80; A. NEUMANN, *Vindobona–Die römische Vergangenheit Wiens. Geschichte, Erforschung, Funde* (Vienna, Graz, and Cologne, 1972).

¹¹ SWOBODA, *Carnuntum;* H. VETTERS, "Zur Problem der Kontinuität im niederösterreichischen Limesgebiet". *Jahrb. f. Landeskunde von Niederösterreich* n. s. 38 (1968–1970): 48 f.

¹² I. PAULOVICS, "Funde und Forschungen in Brigetio". *Laureae Aquincenses* 2 (1941): 118–64 (*DissPann* 2:11); L. BARKÓCZI, *Brigetio* (Budapest, 1951) (*DissPann* 2:22); L. BARKÓCZI, "New Data on the History of Late Roman Brigetio". *ActaAntHung* 13 (1965): 215–57.

¹³ A. ALFÖLDI and L. NAGY, "Budapest az ókorban" (Budapest in Antiquity), in *Budapest története* I (Budapest, 1942); J. SZILÁGYI, "Aquincum", *R. E. Pauly-Wissowa*, Suppl. 11, pp. 83 f.; T. NAGY, *Budapest története* (History of Budapest), vol. 1 (Budapest, 1973).

¹⁴ *Intercisa*, vols. 1 and 2; J. FITZ, "Intercisa", *R. E. Pauly-Wissowa*, Suppl. 9, pp. 84–103; E. B. VÁGÓ, "Ausgrabungen in Intercisa 1957–69". *AlbaRegia* 11 (1971): 109.

¹⁵ S. SOPRONI, "A visegrádi római tábor és középkori vár—A Roman Camp and Medieval Castle in Visegrád". *ArchÉrt* 81 (1954): 49–54.

¹⁶ A. ALFÖLDI, *Leletek a hunkorszakból és ethnikai szétválasztásuk—Funde aus der Hunnenzeit und ihre ethnische Sonderung* (Budapest, 1932) (*ArchHung* 9); SOPRONI, *Der spätrömische Limes.*

¹⁷ S. SOPRONI, "Spätrömische Töpferöfen am pannonischen Limes". *ActaReiCret* 10 (1968): 28–35.

¹⁸ A. MÓCSY, "Die spätrömische Schiffslände in Contra Florentiam". *FolArch* 10 (1958): 89–104; A. MÓCSY, "Eine spätrömische Uferfestung in der Batschka" *Osječki Zbornik* 12 (1969): 71 f.

¹⁹ S. SOPRONI, "Burgus-Bauinschrift vom Jahre 372 am pannonischen Limes". *Studien zu den Militärgrenzen Roms* (Cologne and Graz, 1967), pp. 138–43; SOPRONI, *Der spätrömische Limes.*

# PANNONIAN CITIES

## KLÁRA PÓCZY

In Italia and in a certain sense in various provinces both during republican and imperial times, as far as the understanding of city ranks is concerned, there existed a basic difference. "The ordinary secondary town was not necessarily an important residential center in itself. It was a civilised nucleus for those who were able to live away from the soil, which also offered its facilities to the inhabitants living outside the town."[1]

In Pannonia, where the native population had not already created city units, the colonies produced by the Italia veterans and merchants represented the most advanced social and cultural communities in the province. The *municipium* had reached the same level only after four or five generations.[2] Provincial *municipia* tended to transform into colonies as soon as possible, since this type of formation represented a more advanced Roman way of life.

After the military occupation, the Romans surveyed the entire province and chose the locations of some future cities.[3] Cities emerged first in the territory between the Drava and Sava rivers and in the western border region. It was impossible to recognize the exact street networks from the early period of the cities of Siscia, Sirmium and Andautonia. During the course of excavations only the sequential changes in the city orientations were discovered. However, in Savaria and its territory the traces of the early network of streets (*centurisatio*) were clearly discovered. Recently scholars have established, for instance, that Savaria rose in the corner of four contiguous *centuria*. The units comprise *acta* of 16 by 25 and the area of Savaria was 570 by 820 meters.[4] When a city establishment had been planned, the extension of its *territorium* (area) was determined; in the first century the direction of the roads in Savaria diverged by 16 degrees from the north, in Scarbantia by 18–20, in Gorsium by 12, and in Aquincum by some 9 degrees. In most instances, the area was identical in scope with the land possessed by the tribal organization upon which the city was erected. The Romans created these artificial conglomerations, or *civitates*, by splitting up powerful tribal communities. In Pannonia the Boiian and Eraviscus tribal settlement systems are known most extensively. In the area of Aquincum the Gellért hill *oppidum* has been unearthed and it can be dated to the last decades of the first century B.C. and the first decades A.D. Houses from the living quarters of the *vicus* which emerged from the territory of Gorsium may be dated somewhat later, in the early second century.[5]

According to our present information, twenty-five or twenty-seven cities had been established in Pannonia, each with the diameter of about 50–60 kilometers (30–40 miles) and with each *colonia* and *municipium* administering its own territory (Fig.

19). It has been possible to trace with close accuracy the extension of the municipal territories of Emona, Savaria, Scarbantia, Carnuntum, Vindobona, Gorsium and Aquincum.[6] It was then that the land was divided and distributed with the best plots going to the veterans and city leaders and the less fertile parcels to the local inhabitants. Mention should also be made of new types of settlements where no trace exists of city foundations in an organized form, but which nevertheless gave rise to the settlements of Roman citizens. At the beginning of the first century Scarbantia may serve as an adequate example of what has been called in the written sources an *oppidum*.[7] This type of settlement possessed self-government and could rise occasionally to the rank of a city.

## URBANIZATION

Urbanization developed gradually in Pannonia, most particularly under the Flavians, Hadrian, and the Severi.[8] Under the Julio-Claudian dynasty urbanization went on simultaneously with the process of Romanization, mainly alongside the Sava River and along the Amber Route (Fig. 20). Along these two important routes three legions and several small auxiliary units were stationed to provide security for the peaceful development of the provinces. In this well-protected southern and western strip of Pannonia were concentrated army supplies and the commercial goods of the merchants; it was mainly from here that the newly occupied areas of Pannonia received their supplies. As a result, merchant cities developed, Emona and Savaria became colonies, Sirmium and Siscia became *municipia* (later colonies), and Scarbantia, an *oppidum* (later a *municipium*).

Emona, alongside the Sava River, was established in A.D. 14 on the abandoned site of the camp of Legio XV Apollinaris vacated a year before and inhibited by military veterans of the departed legions and other Italics. At the same time another city, Savaria, which was also alongside the Amber Route, was established by Emperor Claudius and here the veterans of Legio XV Apollinaris had settled. Under Tiberius, Italia merchants and retired auxiliaries established an urban center in Scarbantia (Sopron). The settlement had a certain amount of autonomy from Tiberian times. Designated to become *oppidum Scarbantia Iulia*, it was modeled after the organization of the *conventus civium Romanorum* with *decuriones*. In Flavian times it was raised to municipal status.

Outside these cities some 90 per cent of the population between the Danube and Sava rivers continued to live during the first century just as they had before the Roman occupation.

In the southern part of the province where cities developed earlier than in the north, the two *civitates* of Varciani and Latobici were organized with an indigenous population into *municipia*. Varciani was called Municipium Andautonia, and Latobici received the name of Municipium Latobicorum Neviodunum. The most important original *civitas* centers became autonomous *municipia* during the Flavian period, developing even further under Emperor Hadrian.

The last military veteran settlement to emerge as a city was the colony Ulpia Trajana Poetovio. Initially it was established in a vacated legionary campsite by

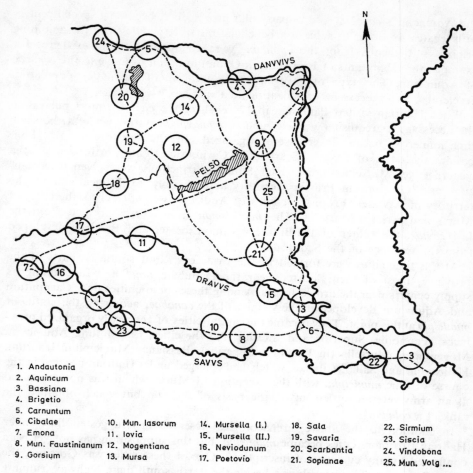

1. Andautonia
2. Aquincum
3. Bassiana
4. Brigetio
5. Carnuntum
6. Cibalae
7. Emona
8. Mun. Faustinianum
9. Gorsium
10. Mun. Iasorum
11. Iovia
12. Mogentiana
13. Mursa
14. Mursella (I.)
15. Mursella (II.)
16. Neviodunum
17. Poetovio
18. Sala
19. Savaria
20. Scarbantia
21. Sopianae
22. Sirmium
23. Siscia
24. Vindobona
25. Mun. Volg ...

Fig. 19. Map of Pannonian cities

the retired members of Legio I Adiutrix, Legio II Adiutrix, and Legio IV Flavia with a double *missio agraria* and a single *missio nummularia*.

By the time of Emperor Hadrian the monumental urbanization program had fulfilled its expectations and sundry *civitates*, maturing to a comparatively sophisticated level of Romanization, were ready to progress to more complex dimensions of community development. Most of them evolved into urban or urbanlike status during the process of their hundred-year existence. Their city authorities included Roman military veterans and various Italia, South-Gaulish, and Rhenish immigrant merchants and artisans who were the prime disseminators of Roman culture.

During the rule of Trajan, settlements called *canabae*, inhabited mainly by immigrant settlers, began to flourish around the legionary camps. Carnuntum's and Aquincum's *canabae* rose to particular prominence, becoming special centers with legates. In order to achieve a beneficial mixture, people from the adjoining original settlements, from the *oppidum*, were encouraged to move there; for example, the inhabitants of Eraviscorum Oppidum began to move to the *canabae* of Aquincum. This type of governmental relocation of the native population accelerated the process of urbanization in Pannonia.

When Hadrian visited Pannonia in 124 each of the *civitas* centers possessing the necessary prerequisites was elevated to *municipium* status, while at the same time numerous urban centers were organized. In the territory of the Boii, besides the cities of Savaria, Scarbantia, Carnuntum to its north, and on the eastern edge, Mursella were formed. Similar distinctions were given to Aquincum and Gorsium in Eravisci territory and possibly to Mogentiana in the territory of Arviates. From the *civitas* of Andizetes were detached the cities of Mursa and perhaps Mursella. The *municipium* Iasorum was formed from the Iasi *civitas*, the center of the Breuci became Cibalae, and Bassiana developed from the territory of the Scordisci.

At this time cities were formed in the west, east, and southeast parts of the province, most of them located near the Danubian *limes* and functioning as supply centers near the armed forces. Among these communities only Carnuntum and Aquincum developed in the vicinity of the *canabae*, acquiring the status of *municipia* at this point. They became the capital cities of the newly organized provinces, Pannonia Superior and Inferior. All the cities, Carnuntum, Aquincum, Mogentiana, Mursella (in Pannonia Superior), Bassiana, Municipium Iasorum, Halicanum and Salla,[9] Cibalae, which had emerged under Hadrian became in the course of time *municipia*, with the exception of Mursa which was not organized as an army veteran settlement on the basis of *deductio* but which obtained the rank of a *colonia*.

In addition to the above, archaeological research has suggested that under Hadrian two more cities became *municipia*, but this supposition has not yet been confirmed by any evidence of inscriptions. One of these cities was Gorsium, the other Sopianae. They developed beside the north–south main highway, running parallel with the *limes* road; by the second century Sopianae already possessed a number of monumental buildings. Later, the city eventually became the capital of Valeria.[10] From Gorsium the name of a *decurio*, that of Aelius, has come down to us; however, no date survives for the initial designation as such. Formerly, the city had been a settlement situated on the site of a military camp, inhabited by Ulpii, with some Papiria Tribus. Excavations have discovered a richly ornamented forum, an immense capitolium, and several buildings from the Trajan period. Inscriptions and the arrangement of the sanctuaries, the relationship between assembly rooms and the temple, all seem to prove conclusively that after the division of the province into two parts, the provincial assemblies of Pannonia Inferior were held here in Gorsium.[11]

In the development of Pannonian cities the crossroad was marked by the so-called Marcomannic–Quadian–Sarmatian wars waged under Marcus Aurelius. It is remarkable that the losses were heavier in the cities along the Amber Route than near the *limes*. In Savaria and Scarbantia, it was observed that the number of

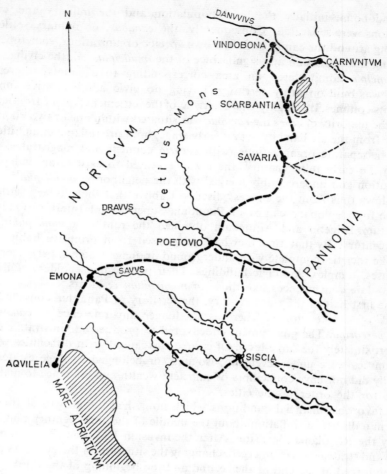

Fig. 20. Map of Amber Route

the local population decreased noticeably. It is possible that many prisoners were taken, and the plague also took its toll. It is a fact, however, that the cities did not recover for many generations. In the settlements near the border, on the other hand, building activities were resumed in a comparatively short time, because the wars made necessary the strengthening of the *limes*. Civilian cities and the *canabae* near the military camps increased in size. This was due to benefits, raised soldier's pay and the presence of a new Oriental merchant population which, in the same decades, had caused speedy deterioration in the city life in Italia and the interior provinces of the empire. The standard of living in the Pannonian border towns showed, for the time being, a rapid development.

Alongside the *limes,* as a result of the presence of the legionary camps, a double, sometimes even triple form of settlement emerged. Thus, for example, in Aquincum

243

and Vindobona, initially the local population and the military and civilian populations were separated. Consequently, the *canabae*, or military settlement, expanding around the camps had their own specific organizations, sometimes even surpassing the dimension and significance of the *municipia*, i.e., the civilian cities. The *municipia* administered an area corresponding to a county, whereas the *canabae* was built on army territory, having no civic administrative functions beyond its confines. It is highly characteristic of the settlement type of frontier cities that in the majority of cases the *canabae* was situated within one or two kilometers' distance from the civilian city. At an early stage, both urbanlike communities had their own separate organizations, with *decurio* councils and magistrates. Moreover, in the early third century, the *canabae* ceased to exist as an independent organization and imperceptibly merged with the neighboring *municipia*.[12]

It follows thus from the process discussed above that the military settlements near the four legionary camps of Pannonia Superior and Inferior, Vindobona, Carnuntum, Brigetio and Aquincum obtained the rank of a *municipium* in the second century. By that time they had progressed from primitive habitations to urbanlike societies complete with underground drainage systems, permanent storage houses, temples, and civic buildings. Their private householders could boast of up-to-date conveniences like those of *municipium* residents.[13]

In the first half of the third century, the territory of Pannonia consisted of the network of city territories. There is no longer any reference to *canabae* and *civitas peregrina*. The government endeavored to produce administrative districts on approximately the same level. Of course, this caused great difficulties when the communities were underdeveloped, isolated, or belonged to large estates; these naturally did not have councillors of sufficient wealth and influence to make them eligible for the order of the cities.

The favorable financial conditions which nourished the growth of these cities ended in a disastrous inflation about the middle of the third century and then, in 260, by the Roxolanian invasion. After the invasion had been repulsed, new economic and strategic plans emerged, changing the situation in the cities. As the *limes* was reinforced at the end of the second and the beginning of the third centuries, the government attempted to create satellite states along the Danubian defense *limes* among friendly Barbaric tribes. The Barbarian invasion of 260 brought an end to this effort, and the abandonment of Dacia left the way open for constant and direct attack from the Germanic and Sarmatian tribes.

During the administrative reform of the Tetrarchy, the imperial government tried to reinforce and make new investments in the border provinces of the empire in Asia, Africa, Britannia, and in the Rhine–Danubian provinces. After 260 a mint was established in Siscia, weapons were manufactured and coins were minted at Sirmium. Bassiana was protected by a city wall to fend off the recurring Barbarian attacks from the southeast. The same type of bastion was raised around Scarbantia in the northwestern section of the province which was also constantly harassed.[14]

During the fourth century in several strategically important towns military and civilian settlements were revitalized; deserted cities were repopulated and new building programs were launched. The new provincial capitals were rebuilt and encircled by irregular, monumental defensive walls. In Savaria, Siscia, Sopianae, and especially Sirmium, massive building programs were carried out. Behind the walls of those cities, bishoprics were established; in Savaria and Sirmium there were imperial palaces and the administrative offices of the province. During some fifty-old years Sirmium developed into an imperial city on the area of some 120 hectares.[15] At the same time the intersections of the Vindobona–Sirmium highway were reinforced. This highway, which crossed Pannonia diagonally, carried most of its traffic and secured the connection among the four provinces of Pannonia. The building program of the new capital of the unified Roman Empire, Constantinople, gave added importance to this highway as part of the extremely important Treveri–Vindobona–Sirmium–Constantinople route, the main imperial highway. Two of the cities along the Pannonian section of this highway, Scarbantia and Bassiana, were rebuilt with grandiose fortifications which also date from this period.[16] However, only the centers of both towns had been reinforced, on the area of some 20–25 hectares.

The excavations seem to prove that after the disaster of 260, Gorsium was completely rebuilt and received a new name, Herculia. This new city had an important part to play in the administrative life of Pannonia when it was divided into four parts. In the center of the settlement emerged a *palatium* to house the emperors visiting the city, along with their retinues and high-ranking officials. For the adherents of the new religion, Early Christian basilicas were raised. Huge granaries and storage depots were built for various commodities and food supplies.[17]

During the middle of the fourth century a belt of settlements was built parallel with the Danubian *limes*, as the defense of Pannonia moved from a one-line system to an in-depth three-step one. These fortified agricultural settlements never obtained city autonomy, but, with a newly settled population, developed slowly into industrial centers. To date five such fortified settlements are known in the area north of the Sava River: Mursella (Morichida-Kisárpás), Környe, Valcum (Keszthely-Fenékpuszta), Iovia (Kapospula-Alsóhetény), Felsőheténypuszta (near Dombóvár), and Tricciana (Ságvár). These settlements were fortified by three-meter thick walls in a rectangular arrangement, circular towers at the corners, gate donjons, with the fortifications extending some 300 meters in length.[18] Above the level of the simple residential buildings in each settlement rose an Early Christian basilica and a large *horreum* (granary). These fortified settlements and the fortified *villa romana* or other villa centers cannot be considered as cities.

At the end of the fourth century, while the fortification system of the interior part of Pannonia was accelerated, large civilian and military settlements along the *limes* were in ruins as a result of constant enemy attacks. With the civilian cities virtually deserted, the remaining population concentrated around and in the well-functioning fortifications.[19] The situation was practically the same in the four legionary camps (Vindobona, Carnuntum, Brigetio, Aquincum); the same could be observed in the cohort camps also, as, for example, in Intercisa. During several Barbarian invasions, certain fortified settlements and capital cities acted as emergency fortifications. The remaining ones later became the centers of the neighboring population, which radiated from there as from satellite suburbs. Except for the administrative centers of the divided Pannonia which were supported by the economic structure of the four new provinces, the standard of living in the cities at the end of the fourth century and the beginning of the fifth diminished significantly. With the original population nearly gone, the vacuum was filled with new settlers from Barbaricum. These declining cities came under the newly developing power and influence of the Church, the only force that still maintained order. While new buildings were rare, most of the larger buildings constructed were, of course, churches, which were erected from community donations of the cities.[20] The defensive requirements of the cities also made imperative the building of granaries and warehouses. Private buildings degenerated to a primitive level, with log and dirt-wall shelters and the ruined public buildings serving to house the population.

At the beginning of the fifth century the civilian population in the districts along the *limes* occupied itself with border defense. The civilian cities about one to two miles from the legionary camps were abandoned and the *canabae* near the camps were partially in ruins. However, near the fortresses smaller settlements cropped up. In fact, the fortifications along the *limes* became the refuge of the diminished military district, as did the fortified settlements in the interior of the province or the district centers. While new settlements were no longer built, the viable ones still existing at important locations were restored — among the old cities were Vindobona, Bassiana, Scarbantia, and Savaria at the junction of important roads, and among the new fortified agricultural settlements was Valcum (Fenékpuszta) at the swampy area of Lake Balaton near the confluence of the Sala River. These population centers in the new wave of Barbarian invasions defined the styles and standards of living for those arriving during the centuries of the migration period.[21]

During the great migration period from the sixth to the ninth centuries, Germanic and Avar invaders once more made Pannonia a stronghold for a while, providing a certain continuity by occupying the former Roman settlements and using the same cities as administrative centers. The Roman roads too were in continuous use and archaeological research has established that the Barbarians rebuilt the former Roman fortifications. The list of finds confirms contact — with short nterruptions — between southwest Pannonia and Italia, which remained unbroken up to the late sixth century. Imported goods had arrived from Italia through the trade routes, and the city population was able to buy these expensive commodities, — like glass-ware — for example, in Vindobona, Sopianae, Valcum, Savaria, and Scarbantia.[22] Sirmium is not included, owing to its special privileged status.[23] The Roman urban system seems to have survived even during medieval times, when a few of the former Roman cities were repopulated. The transformation of these onetime Roman cities to comply with the new medieval standard had occurred from the eleventh to the thirteenth centuries (Wien, Sopron, Szombathely, Óbuda, Pécs).[24]

In Pannonia — similar to other provinces throughout the empire — essential differences were apparent in the settlement structure of cities established at various times. In several instances specific features are observable in the formation of city centers, as well as in the orientation of buildings, the constructional changes of the city walls, the network of streets, the location of the forum, and the system of the burial sites.

Emona and Savaria army veteran settlements were created as products of efficient and systematic planning and were designed as quadrangular urban networks whenever the geographical circumstances permitted. Savaria, according to some theories, the capital of Pannonia in the second part of the first century, illustrated this form of arrangement, as opposed to the *insulae* system, by providing its forum with a capitolium and other spacious public buildings conforming to special requirements, such as the curia and basilica. The urban plan of the two colonies Emona and Savaria, followed the well-known pattern of other colonies throughout the empire during the Augustan and Claudian periods. However, there were certain noticeable differences between the two, as far as the proportions of architectural structure are concerned. The protective wall of Emona, mentioned in Tiberian inscriptions, had been part of the first-century urban plan of the city. In Savaria, as far as we know, the city wall built in Claudian times encircled a larger area. Also the breadth of the street networks was different in the two cities; the *insulae* of Emona were smaller than those of the more spacious Savaria, although the ratio of the side of an *insula* in both cities was 6 : 5. The rectangular street network of Emona occupied an area of 522 by 435 meters. In Savaria the city quadrangle in the first century — measured from the distance of surrounding roads — was 570 by 820 meters.[25]

Since the *municipia* were built upon early native settlements, they did not follow the same way of urban planning as the newly created colonies.[26] In Scarbantia, among the Flavian-established *municipia*, the street network, the ratio of the *insulae*, and the location of the forum were planned according to the pattern used in Claudian colonies. The plan, however, was never fully realized. The main street along the north-to-south axis of the city, in the vicinity of the forum, was cut by roads at right angles. A similar tendency was observable in the urban plans of the *municipia* Flavia Solva, Virunum, Iuvavum, in the province of Noricum.[27]

In the *municipia* established by Hadrian in the interior of Pannonia, diverse city formations emerged as a result of well-conceived city planning, though Gorsium represented a special case. About the middle of the first century, for example, in the same place where Gorsium was planned, an *ala* camp had been built to house 500 mounted soldiers, whose task was to watch over the security of the Poetovio–Aquincum road. When, toward the end of the century, the situation became more stable, the military camp was liquidated and a civilian settlement was established in its place. Recent excavations unearthed in the center of this new civilian settlement six *insulae* with the same measurements, 55 by 63 meters.[28]

In the border areas of the Danube, next to the legionary camps at Aquincum (Fig. 21) and Carnuntum, civilian settlements also arose as a result of city reorganization. Except for the forum area, long houses similar to barracks were built along the main street.[29] The external districts and the suburbs were developed on a smaller scale which still retained a provincial character. Until the end of the Marcomannic wars, most domiciles in these quarters were built of sun-dried brick rather than stone or brick manufactured by fire heating. The rest adhered to the traditional mode of habitation, the dwelling pit.

Until the middle of the second century at least, each of the autonomous cities had a forum, with a capitolium located on it, and curia and basilica, official buildings for city administrations. Naturally, this city center could not exist without a public bath, to which was added in the second century – when the interior of the city had not yet been too crowded – a *palaestra*. In addition, none of the city centers lacked public utilities, and on the border of the settlement there was an amphitheater. Large-scale building plans had been financed from the offerings of the city leaders and from taxes.[30]

By the middle of the second century the central city areas had been more highly developed than the surrounding residential areas; public utilities were not yet applied to private buildings. However, in the interior of Siscia the remains of water pipes were recovered; these had been introduced into a public well standing on a public square. In Savaria water supply could be traced along a stretch of 36 kilometers, leading from the hills into the city. In Carnuntum and Scarbantia the main track was discovered, starting from the wells, and in Aquincum a five-kilometer water supply of the aqueducts was found. In this city, several waterworks had been built in the course of time.[31] Side by side with a regular water supply, drainage systems had also been introduced into the cities.

Among excavated capitolia, the remains of those in Savaria and Scarbantia are the most remarkable. In both cities huge marble fragments were recovered of statues representing the capitolium deities, as well as a capitolium temple in Scarbantia. Archaeologists have unearthed a capitolium of magnificent proportions at Gorsium with temples on a higher plateau. Evidently the temple had collapsed in the late imperial period and the rubble covered up the entire coin collection of the treasury. Imperial palaces were found to have been erected at Sirmium and Savaria, while the ruins of Carnuntum and Aquincum have revealed governors' residences of extraordinary proportions.[32] After the decades of reconstruction following the disastrous Marcomannic wars, no more than the most needed renovations, accessory investments, and small building programs were carried out. However, private and public buildings were erected around the *limes* zone. Reliable evidence confirms that more artistry was present in the decorations of private buildings than in the public ones. The patrons of public buildings and the donators of public festivals and entertainments often came from the middle class, the military, and Oriental merchant families. From the border zone of Pannonia the largest number of building inscriptions have survived from the period of the Severi. Most of those who had raised monuments came from Oriental military families. They tried to express their gratitude in some sort of way, and they also had sufficient financial means to express it in a sumptuous manner to show their adherence to the Syrian dynasty.[33] In almost every Pannonian city, shrines for various Oriental cults were built, and Mithras, Dolichenus, and Magna Mater

248

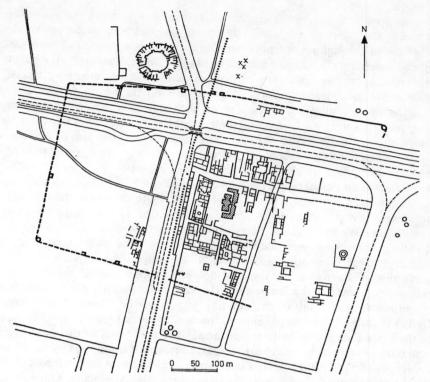

Fig. 21. Ground plan of Aquincum Civilian City

had found their places beside the classical deities, separated from the forum. Most significant among the buildings unearthed so far is the mithrea of Poetovio and the Isis sanctuary at Savaria.[34]

With well-designed streets, complete city utilities, stone buildings, and several-storied public buildings, these *coloniae* and *municipia* in Pannonia reached their peak of prosperity in the first half of the third century. Just to mention one characteristic fact about the well-designed interior arrangement of the buildings: in the public buildings, the wooden doors and window frames had been exchanged for carved stone and every room was painted with several patterns.[35] In the last third of the third century and for a long time to come, the impulse for building projects and the need for town developments had ceased. The political and economic troubles referred to above had taken a heavy toll on the civilian population of the Pannonian cities. Many families were ruined; leadership had changed hands causing effects for several generations.

The transformation and reorganization of the fortifications along the Danubian *limes* affected to a great extent the life and security of the cities for several decades, from the reign of Diocletian and Constantine I. The building programs of the first half of the fourth century give us a clear indication of the standard of living in these cities. During this time there was not a single town where the city structure did not undergo changes, indicated primarily by the new system of city orientation. This means that new centralization, new partition, and new systems of surveying were prepared throughout the province. Ruins and destruction had been cleared away and a definite rise in the standard of living became evident in all settlements in the late imperial period. New houses and constructions are characterized by great dimensions and spaciousness. At that time on the forum of several cities, among the public buildings, a new type of huge luxurious palace was given prominence, as in Sirmium, Gorsium, Savaria, and Carnuntum.[36] The structure of cities surrounded by walls did not recall any characteristic features of first-century colonies and precisely measured standard cities. Most of the streets followed the line of the polygonal or arched section of the city walls. Even more striking is that often two city centers emerged, one around the Early Christian church, the other on the large city square at the other end of the town, where there were large open spaces. It is possible that in periods of siege, the animals were driven there for protection. We know that the city walls were reinforced with additional towers and in large cities sizable public buildings were built. The utility system was also rebuilt and modernized for the last time. New triumphal arches were raised in the larger cities of Pannonia by imperial decree to commemorate the monumental building programs of the emperors, for example, the so-called Heidentor in Carnuntum was the result of the building project of the Constantine dynasty.[37]

By the middle of the fourth century, the border cities began to decline rapidly, with their population gradually disappearing. The best proof of this fact is offered by the burial sites which, from year to year, encroached upon the city walls; in the last decades of the century burials continued in abandoned and ruined buildings. With the exception of Vindobona, Savaria, Scarbantia, Aquincum, Gorsium, even perhaps Sirmium, where the total decay took place somewhat later, the situation was the same in each city. The inhabited area dwindled to a minimum and the decreased number of the population occupied but a small section of the city.[38]

The decay of urban life continued in various places with different intensity and manner. Valeria declined most rapidly, whereas Pannonia Prima, the Drava–Sava region continued longest to survive.

In the *limes* sector of Valeria, for example, the civilian population took part in the defense activities as early as the late fourth and fifth centuries. The civilian cities situated in the vicinity of earlier legionary camps were gradually abandoned, the *canabae* around the camps were relegated to destruction and only close to the fortification some small-scale settlements may have struggled on. The camps alongside the *limes* offered the safest shelter to a larger district. Recent excavations show that as far as Vindobona, Carnuntum, and Aquincum are concerned, the population survived longest behind the walls during the latest period, in late Antiquity, even during the migration period.[39] The fortified centers inside the province had fulfilled the same function as settlements surrounded by walls. Even for centuries to come, the walls represented some sort of protection, often keeping away uncertain and risky enemy attacks.

In Pannonia the most densely populated area developed to the west of the Danube, Drava and Sava rivers; the eastern extent of this area coincides with the border between Pannonia Inferior and Pannonia Superior. Least populated was the Transdanubian section of Pannonia Inferior, an area north of the Drava River. During the second century, three-fifths of the Pannonian cities were located in the area between the Drava and Sava rivers, the other two-fifths in the Transdanubian area: one-third of those cities were *coloniae*, two-thirds were *municipia*. At the same time that Pannonia had about eighteen cities, Noricum had only nine, Dacia twelve, and Moesia Superior thirteen. Taking into consideration dissimilar geographical conditions and different economic and political factors, the above-mentioned data unequivocally prove that the process of urbanization throughout Pannonia showed a healthy development in the second century. The number of cities indicates the intensive Romanization of Pannonia. According to statistical research, the average population of the cities at this time was 500 inhabitants per hectare (10,000 square meters or 2.47 acres), and the average size of the cities was 30 hectares (74.1 acres).[40] Savaria extended to 30 hectares, the civilian city of Aquincum to 50 (123.5 acres), the civilian city of Carnuntum to 60 (148.2 acres), Scarbantia (Fig. 22) to 23 (56.81 acres), and Vindobona and Brigetio each to about 20 (49.4 acres).[41] Calculating on the basis of an average of 30 hectares per city, we find that no more than half a million people lived in all the *municipia* and colonies. The distances between the cities built along the most important roads were generally about two days' journey by carriage on the plain, about 40–50 kilometers, or 25–30 miles. Distances were different around the Lake Balaton area, where the system of large estates with villas, rather than cities, existed. In this case the large estates had absorbed the working force, thus preventing a healthy urban development.[42] Such was the case in Caesariana, a conglomeration of estates near the north end of Lake Balaton, and Valcum (Fenék-puszta) near Keszthely, at the south edge of the lake.

In the third century the number of cities in Pannonia reached twenty-six; twenty-two have been located by excavation or trial trenches, but the location and general situation of the others are still in doubt. Among those located the topographical remains of eleven which have been entirely excavated provide excellent analogies for the archaeological identification of future discoveries. The sporadically emerging archaeological remains and epigraphic findings of the others give some clues to administrative questions and to dating. In Pannonia Superior, among the fifteen cities south of the Drava, twelve cities can be located. Among the nine cities of Pannonia Inferior south of the Drava, the location of seven may be cited. In South Pannonia the cities were situated some 30 kilometers from each other, while in North Pannonia the distance between cities was about 60 kilometers.

The area occupied by the cities was ordinarily 50–60 kilometers, about 40 miles, in diameter. According to the customary divisions, each territory was apportioned into *pagi* which in turn were divided into *vici*, such as *Pede Sirmiense pago Martis vico Budalia*. One of the *vici* bordering Aquincum was Vindonianus. The Vicus

Fortunae was a section of the city of Poetovio, and the Vicus Voleucionis was in the neighborhood of Savaria. Emperor Dacius was born in Vicus Budalia, which was in the territory of Sirmium.[43]

The boundaries of the urban territories are very difficult to draw. In aiding the reconstruction of the former boundaries the discovered stone boundary markers of the *magistrati decuriones* and the inscriptions on the milestones provide the most reliable information. Using these guides, one can safely establish the boundary lines separating Emona from Neviodunum, Savaria from Scarbantia, Scarbantia from Carnuntum, Carnuntum from Vindobona, Gorsium from Aquincum. The territory of Aquincum is marked in the east and in the north by the natural borders of Pannonia. The locations of the findings of inscriptions with names belonging to the *ordo* of Aquincum do not cross southward over the Aquincum–Poetovio roads.[44]

In the third century, the focal point of urban development shifted to the area of the Danubian *limes* and the eastern part of the territory between the Drava and Sava rivers which was to become Pannonia Secunda. Its capital, Sirmium, soon became one of the four imperial cities of the Roman Empire.[45] Pannonian urbanization emerged in a relatively late period when slave labor was no longer profitable. It is noticeable that in the western border of the province, in the land owned by the cities, slave labor was still used in the first century, with four to eight men being employed for tilling the land of middle estates. On the other hand, in the third century, it was particularly in the boundary of border legionary camps that more slaves were employed in city households and the home industries.[46]

Diverse colonies and *municipia* developed at different rates and intensities with settlements and cities varying greatly. From the first to the fifth centuries, during their transformation, a gradual change had taken place even in their function and as a result the significance of individual cities had diminished or increased. These changes became particularly conspicuous during the preliminaries of important occasions, such as the visits of the Roman emperors in Pannonia. The memorable stay of Hadrian in Pannonia in 124 occasioned a building activity in Sirmium, Savaria, Vindobona, Aquincum, and Gorsium. Temples on the capitolia had been enlarged and adorned with artistic decorations.[47] During the visitation of Septimus Severus in 202 and particularly of Caracalla in 214 the camps were restored and Hercules and Victoria temples were raised. Even public buildings serving military needs had been rebuilt, such as the governor's palace and villas for the emperor and for high military personnel. Nevertheless, it was primarily in the military towns that building activities had been carried out feverishly; statues and sanctuaries were consecrated, as confirmed by many instances in Brigetio, Mursa, and Aquincum.[48] Business houses with porticos and ornate gardens were built to decorate the city centers for these festive occasions. At that time the citizens could still afford to pay for these luxuries.

During the stay of Constantine I in Pannonia in 307, massive building plans were carried out. The mere fact that new imperial palaces had to be raised, which has been confirmed by archaeological evidences in Sirmium, Gorsium, Carnuntum, and Savaria, seems to prove that buildings serving this purpose had been badly neglected. It was in the same decades that many warehouses and granaries were

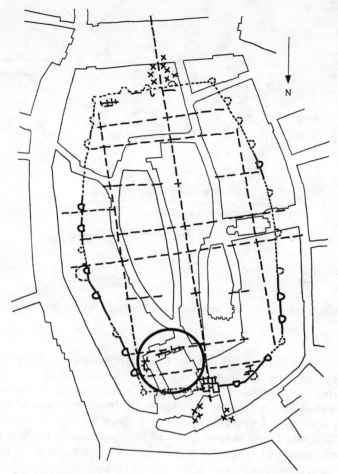

Fig. 22. Ground plan of Scarbantia with location of capitol

built in these cities — a fact confirmed by inscriptions and other archaeological evidences in Savaria, Sirmium, Scarbantia, Siscia, Aquincum, and Gorsium. These new establishments were so immense that they occupied the space of several blocks of houses.[49] It is very significant also that the expenses for these investments were not raised by the city authorities or private individuals but by the state, and the building program was conducted by the army.

By the time of the visit by Valentinian I in 374–375, these expensive palaces and offices in Pannonia raised some seventy-five years earlier were in a sorry state of neglect or even wholly destroyed. It is well known that this emperor did not find any suitable building for his winter quarters in Aquincum and the palace in Savaria was also in a desolate state. In Sirmium money was collected to build a theater but it was finally used to restore the crumbling defenses. The settlements were reorganized for self-sufficiency and defense.[50]

253

Fig. 23. Reconstruction of Fortuna-Augusta shrine, Aquincum Civilian City

By the late fourth century city life had been fully divested of its original character, and everything was in a state of neglect and desolation.

Cities during the first three centuries had owned areas corresponding to the extent of counties. Each city had its own individual function and by means of their autonomies they formed part of the Pannonian city network. However, by the fourth century their autonomies had been very much restricted: each city was uniformly governed from the provincial center, leading to a kind of uniformity. Nevertheless, the situation underwent fundamental changes in the fifth century, when central government ceased to exist, either gradually or immediately and the cities were left to their own devices. For a long time to come, even to the end of the sixth century, the bishoprics in Sirmium, Siscia, Savaria and Scarbantia had continued to sustain order, controlling the life of smaller communities, and for many generations the fortified settlement had fulfilled the function of shelter and protection.

## RESULTS OF ARCHAEOLOGICAL INVESTIGATION

The picture that emerges as a result of excavations in Pannonian cities is far from complete. We were unable to rely on all the data, however, since only those referred to in various publications were available. Moreover, from the details published in literature only those data have been used which proved necessary in the context. Thus the picture presented of Pannonian cities is still fragmentary:

Fig. 24. Reconstruction of *macellum*, Aquincum Civilian City

our results are not final, research is still going on which will necessitate further corrections.[51]

### Andautonia

While the city plan of Andautonia has not yet been entirely reconstructed, building inscriptions tell us that during the second century A.D. the city was densely populated. In that century the waterworks and some significant public buildings were built. Similarly, inscriptions prove that at the border of the city an amphitheater was erected.[52]

### Aquincum

In Aquincum the area of the civic settlement of craftsmen from the Flavian period was unearthed (Figs. 19 and 21; Pls. XIV.2, XV, XVI.1,2; XVII and XVIII). At the center of the dwellings in the midst of sunken adobe huts, there were solidly built public buildings. The living quarters stretched in a north–south direction, containing about four rows of streets in an east–west direction. This settlement arose to the rank of a city at the beginning of the second century.

The *municipium* occupied a 50-hectare area built up during two decades in Trajanic–Hadrianic times. The buildings were situated on a north–south axis, about 12 to 10 degrees toward the east. At the center of the remains of the Hadrianic settlement, some 72-by-90 meter (6 : 5) *insulae* are recognizable. At the intersection of the *cardo* and *decumanus* was a forum complete with a shrine of the capitolium, a curia, a bath, and a marketplace (Figs. 23–24). Shop followed shop under the covered arcades along the main streets; near the forum, public buildings such as a basilica, bath, and marketplace formed the *insula*. During the same period the main channels of the aqueduct and a complete network of drainage system were also built: the warm water was carried to the city from hot springs 1.5 kilometers north of the city over arch-supported aqueducts which traversed the city in a north–south direction.

In the second century an amphitheater with a 2,000–3,000 seating capacity was built. At this time the exit roads outside the city boundary were flanked by cemeteries and by industries which were banned from the interior of the city. On the western and eastern sides of the cemeteries, factories were conveniently placed. A great pottery (4 hectares large) has been discovered at the bank of the Danube, close to the harbor, having twenty-two kilns, dryers, wells, and warehouses. The number of shops and *tabernae* is unusually high in this city (about one-third of the buildings). Aquincum was considered an industrial and commercial center during the second century. After the Sarmatian wars in 194 Aquincum was elevated in rank to a *colonia*. At first, it referred only to the territory of the former *municipium*, but it soon included the *canabae* as well. This change in status strongly influenced the city's architecture. A document from the first decades of the third century reveals a contract between the city government and one of the rich tenants who paid for arched *portico* with double gates. At the beginning of the third century, the inner area of the *colonia* of Aquincum was entirely covered with buildings. The *diversorium* (hostelry), a blacksmith's workshop, potteries and many other factories stood outside the walls.

The buildings on the forum also were affected. The sanctuary was enlarged into a temple with a front portico, and the basilica itself was changed to correspond to the new floor arrangement of the forum. Several small baths were also built, their size indicating the possibility that they were health spas, especially since Aquincum was considered at this time to be a hot springs resort area. Headquarters were built for the Collegium Centonariorum and for the Collegium Iuventutis, and temples were erected to Mithras, Venus, Diana, and Nemesis. Overpopulation and shortage of space transformed the city of Aquincum. The former oblong courtyards behind the narrow houses were entirely built over with stripe-like long dwellings which were separated from each other only by narrow paths. Houses throughout the entire city were built of stone and furnished with floor heating systems, while most of the public buildings were also decorated with sculptural decorations, wall paintings, and floor mosaics. The waterworks and drainage system extended throughout the entire city; side branches and new sections of the underground channel system which emerged during the excavations give strong indication of the extension of the city.

At the same time a general period of development and monumental building programs were carried out at the *canabae*, which, during the third century, was considered the most important section of Aquincum. In 214 the governor's headquarters doubled its staff, necessitating a new building to house the working force. As traffic in the city increased, old sections were torn down to make way for more roads. This systematic reorganization of the city road network was followed by the organization of the roads of the *canabae*. Craft industries, wooden

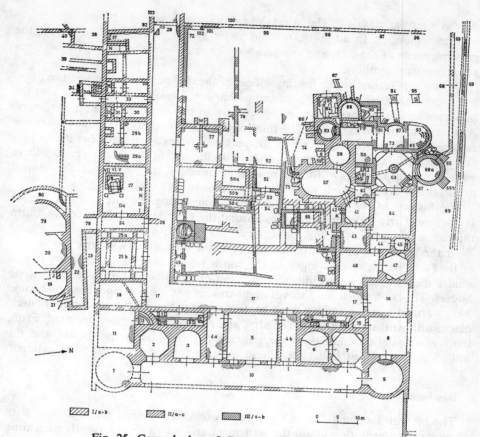

Fig. 25. Ground plan of Governor's Palace, Aquincum

bodegas, and *tabernae* disappeared from the area around the gates of the military camp, while the border road along the walls was flanked by new stone buildings.

The first rebuilding after the Marcomannian wars included the governor's palace (Fig. 25). In its interior court a temple for the imperial cult was erected, complete with statues of the emperors. During the second century new public buildings were erected at the north *canabae*, toward the civilian city. During the middle of the third century, the formerly hazardous industrial plants, such as potteries or brick workshops, the cemeteries of the *canabae* between the south gate of the military camp, and the amphitheater were leveled. At the foot of the western hills new cemeteries were opened as the rapidly expanding city pushed the cemeteries farther out.

At the southeastern side of the *canabae* the port gave room to new storage houses. The place behind the western gate of the legionary camp was enlarged and a sanctuary section with a row of temples decorated with Jupiter–Juno, Diana, and Pluto–Prosperina reliefs as well as with inscriptions and statues of Eastern gods was erected here. After the Marcomannic wars several buildings, such as those

257

named after the mosaic of Hercules, were rebuilt; this is evident from wall paintings and mosaics. In addition, the old water system was unable to supply the growing needs of the city, and the necessary quantity of water came from springs at the western foothills.

During the Tetrarchy the leading offices of the civil administration of the province were transferred from Aquincum, perhaps to Sopianae or Gorsium, and the public offices in Aquincum were converted into apartments for the newcomers. After a great fire during the middle third of the fourth century, the houses were not repaired, but intact wings of partly dilapidated houses were used. Several houses served es Early Christian assembly places. By the end of the fourth century there were burials at the sites of former residential quarters, the graves in a fan-shaped arrangement around small Early Christian cemetery chapels.

The history of the civilian city of Aquincum during the late imperial period is obscure. Early fifth-century graves around the Danubian watertowers do affirm that some life was left in the ruined city.

At the beginning of the fifth century even the military city shrank in the area of the Danubian ferry-crossings. At the south side of the former legionary camps mud-walled houses were built on the ruins of third- and fourth-century stone houses. They contained mud stoves and furnaces with funnel openings, and some were warmed with a primitive flood heating system. Among the residential houses, blacksmith workshops, a metal foundry, and a bread furnace also operated. From the former public buildings of the military city only the amphitheater and the walled port remained intact, but these were to form part of the fortification.[53]

### Bassiana

The city of Bassiana has not yet been excavated. Aerial photography gives accurate documentation about the settlement around its forum, clearly indicating its walls and streets, its horseshoe type of donjons, and the position of the city gates. The city wall was probably built during the Constantinian period.[54]

### Brigetio

Built along the western exit route outside the fortified legionary camps of Brigetio rose a civilian settlement, the extent of which is indicated by the location of early burial sites and hazardous industries. In the second half of the second century stone houses were built, the center part of which had water and drainage systems. As systematic excavations have not been carried out, conclusions can be drawn only on the basis of archaeological finds.

From the beginning of the second century to the Marcommannic wars, most construction took place in the civilian part of the city, but during the third century the military city developed in a far larger scale. About 200 meters from the south gate of the legionary camp a long underground water supply channel with two connecting branches was unearthed. One of these branches may have supplied the mills and other industrial workshops with water. At the confluence, the Temple of Apollo Grannus was built and in 217 was enlarged with a portico and some *tabernae.* Near this temple was probably the shrine of Fons Salutis. According to the inscription at the south-western quarters of the *canabae*, there were Mithraeum and Dolichenum shrines, and possibly a Magna Mater temple.

During the second century, shops were built along the exit roads of the military city leading toward Savaria and Aquincum, and at the end of the second and the third centuries, some luxurious villas were constructed here. At the same time, new industrial plants including a pottery factory, brick kilns, and even glassworks were established on the eastern, western and southern sides of the *canabae*. Although the affluence of the *canabae*'s population in the first third of the century is demonstrated by new burial customs, other funeral rites show also that they were impoverished by the second half of the century. As the city was depopulated during the fourth century, graves from this time were placed not only within the city confines but within the ruined residential quarters.[55]

## Carnuntum

Because excavations have been made only in the center of the city of Carnuntum, the full extent of the civilian city and its city walls are not yet known precisely. As the first century turned into the second, the mudhouses were destroyed and the terrain leveled. Later perhaps, before the city received *municipium* rank, stone buildings were erected. A forum with an *insula* of approximately 100 by 150 meters in a strictly east–west direction is detected. The block was enclosed with a portico, and the thickness of the main walls of the important buildings indicate that they were several stories high. At the south side of this rectangular block was placed a bath with a garden in front; later gardens were built inside the houses. The marketplace, which was connected to the bath by a narrow covered gallery, had a fountain at the center and polygonal sanctuaries. Rows of *tabernae* with portico encircled the central court. At the eastern part of the *macellum* was a large room containing the seat of the Collegium Fabrum Scholas: an inscription on an altar stone serves as evidence. The water supply, drainage system, and central heating system were built when the stone buildings were erected.

To date, three *insulae*, oriented 10 degrees off from the north–south direction, have been unearthed south of the forum. Above the leveled mudhouses a bath of stone and brick and so-called centrally corridored apartment houses were built, although the floor heating for these houses was completed at a later date. The drainage system and the main water-supply line were built during the second century, and only the baths of the public building were supplied with running water. At the southern edge of the city stood an amphitheater with a shrine of Nemesis.

During the Marcomannic wars the civic city was partially destroyed by fire. After the burned area was leveled, the whole network of streets was reorganized, and new roads and streets were built on the ruins of the former stone houses. The streets were paved with large flat stones and bordered with channels. In the residential area the apartment houses were enlarged, the porticos along the street side were fitted with small shops, and workshops, and storage areas were attached behind the living quarters. The apartment houses were divided from one another by narrow paths.

259

We know less about the settlement and city plan of the *canabae* of Carnuntum. One can observe streets with porticos and the existence of some 500 meters of exit streets coming from the city walk; they were flanked by houses, several of them built in the second and third centuries. At the east side of the military city was erected an amphitheater.

Until the first third of the fourth century both the *canabae* and the civic city of Carnuntum were rebuilt. One monumental building is left, a storied triumphal arch with four entrances, next to the amphitheater; it may have been constructed during the reign of the Constantine dynasty. Little documentation survives from the last imperial period, but there is evidence that the city was visited several times by Roman emperors, who stayed at the *palatium*.

We have evidence that in 378 a group of Marcomannians who had settled on the left bank of the Danube was baptized in Carnuntum. At the end of the fourth century a baptismal fountain was placed at the south gate of the amphitheater of the civilian city, and a niche in the gate itself was transformed into a chapel. The amphitheater may have been used as a meeting place for Christians or generally for big crowds.

A few loghouses and mud huts dating from the beginning of the fifth century have been found on the dilapidated ruins of stone houses on the main street of the military city. These houses may have been built by certain Germanic Alani groups living on top of the ruined city.[56]

### Cibalae

The city of Cibalae has not yet been excavated; only a few trial trenches exist. Significant among the findings are the pottery kilns. During the late imperial period the city had a large population. We have some documentary evidence about the life of the city and the activities of the Early Christian bishopric.[57] There must have been some significant churches and assembly places for Christians in the bishoprics.

### Emona

The Tiberian settlement of Emona is a 522.3-by-435.5-meter rectangle. The 10- to 16-meter wide streets ran from east–west and north–south directions and so divided the entire settlement into forty-eight *insulae*. Each *insula* was about 65 by 73 meters. Among the forty-eight *insulae*, thirty-one have been reconstructed. At the beginning of the second century the stone buildings within the *insulae* were built, where formerly mud huts had stood. The city had no waterworks, but every house had its own well until about the middle of the second century, when public utilities were built. At the center of the city were some luxurious houses, highly ornamented with wall paintings and mosaics, dating from Flavian and Hadrianic times. The building program also included baths and sanctuaries. As the city expanded early in the second century cemeteries were opened along the eastern and western exit roads. Inscriptions indicate a diverse network of landownerships in Emona. At the end of the second century, Emona was attached to Italia, but as an important city on the Amber Route, it was still tied economically to cities in the western part of Pannonia.

We know little of city life in the third century. Recent excavations prove that in the second half of the century the defensive system was overhauled; the new 3-meter-wide walls were protected by squared donjons, gates, and towers.[58]

### Municipium Faustinianum

Only the name of Municipium Faustinianum has survived, discovered on an inscription in Siscia. It is possible that this city was located at the eastern side of the junction of the Drava and Iszna rivers.[59]

### Gorsium

About the middle of the first century A.D. a small military camp stood on the sloping area of the riverbank of Sárvíz. The form of the camp determined the form of the center of the later city, Gorsium. The connecting roads of the four entrances of the camps became the main roads of the city, the *cardo* and the *decumanus*; the forum was placed at their intersection. After 106, when Pannonia was divided into two parts, Gorsium became the religious center of Pannonia Inferior and the yearly *concilium* for the province was held there (Pl. XIV).

In the center of the settlement a forum was built during the reign of Trajan and possibly completed during the Hadrianic period. As the result of excavations, the existence of six *insulae* about 55 by 63 meters was confirmed. On the eastern side of the assembly place carved fountains were placed, one with nymphs and the other with the figure of a river god. Three steps led from the forum toward the buildings of the capitolium which consisted of three colonaded large walls, one assembly room, and several sanctuaries. Near the intersection of the *cardo* and *decumanus*, shops and a sanctuary, possibly dedicated to Minerva, were discovered. Several other houses decorated with wall paintings have come to light, but their function has not yet been determined. Even in the richest houses, the wall paintings were applied to plastered reed-covered walls. It is interesting to note that while lavishly decorated stone houses and public buildings were built in the center of the city, the rest of the city still contained log- and mudhouses half sunk into the earth.

About 260 Gorsium was completely destroyed, but rebuilt soon after. A few decades later, a new city was built on a very thick layer of the wreckage of the old. During the fourth century the city was entirely reorganized, with the axis of the main streets remaining as it was, but the houses differently oriented. According to various itineraries, the name of Gorsium was changed to Herculia in the late imperial period.

In the middle third of the fourth century, two villas were built in the center of the city: one above the shrine of Minerva serving as a *palatium* (Pl. XXI) and the other near the eastern border of the city. Each was furnished with a floor heating system and a bath in one of the wings of the villa. Where the forum had stood was a monumental Christian basilica in front of which was built a stone square. At the eastern part of the main street, behind an entrance with a wooden colonaded portico, a pottery shop and a blacksmith workshop were in operation. At the south side of the same street was a longer house with a 55-meter-long row of *tabernae*. During the same period, the waterworks were restored; it is evident that the cistern was continuously in use. The second villa area dates from the late imperial period. In the late fourth century, the dead were buried in demolished and abandoned houses (Pls. XXII and XXIII). In most of the *tegula* graves Early Christian cultic objects were found next to the bodies.[60]

### Municipium Iasorum and Iovia (formerly Botivo Civitas)

Since these cities have not been excavated, their urban plans are still unknown.[61]

### Mogentiana

The urban system of Mogentiana, believed identical with the present-day city of Sümegcsehi, has not yet been discovered. Its territory extended to the bank of Lake Balaton and included the Villa Romana at Fenékpuszta.[62]

### Mursa

Some seventeenth-century prints and measurements of the ruins made at the beginning of the second century provide some idea about the settlement system of Mursa, a city located on the right bank of the Drava River. The sketches indicate the extension of the city, and its long distance roads which are near the city walls are flanked by cemeteries. An inscription has been discovered which reveals that one of the magistrates of the city decorated the bridge of the Drava River and the main street of Mursa with a double *porticus*, within which he placed fifty *tabernae*. It is well known that during the second century artistic *terra-cotta* and bronze statues were produced in several workshops of Mursa and sold in the shops near the bridge. In the fourth century there was an Early Christian bishopric here.[63]

### Mursella I in Pannonia Superior and Mursella II in Pannonia Inferior

Since they have not been excavated, these two cities remain unknown; not even the outlines of the settlements are known.[64]

### Neviodunum

At the beginning of the Roman occupation, Neviodunum was a very important settlement at the foot of the Alps. Since very early times control of the roads to the Alps had given a determinative role to this key settlement, it was secured by a Roman garrison right from the start of the occupation. Soon Italia merchants came, and from the end of the first century a *beneficiarius* station was built. The latest excavations have revealed the street network from the Flavian period and the remains of a warehouse which must have been at the port. The aqueduct, partially

262

unearthed, suggests the significance of the city. It was built in the Early Imperial period and supplied water for at least 10,000 inhabitants.

During the Constantinian period the strategic importance of Neviodunum became again evident; near the city was built a large 300-by-300-meter citadel. According to the present state of the excavation, the only new stone building from this time was an Early Christian church.[65]

## Poetovio

It is possible that when Poetovio received the rank of *colonia*, the early aboriginal settlement, the camp and the *canabae*, had grown together into one large city. Poetovio was probably built on the southern border of the Drava River (on the territory of present-day Igornja and Hajdina) over the abandoned legionary camp. A bridge was built over the river and shortly thereafter a new quarter of the town emerged on the northern bank. The center of this *colonia* was built on the bank of the Drava River, the forum possibly near the north side of the stone bridge. The distributing center of the aqueduct, which took the water from the nearest hills to the city, was also located here. The discoveries, statues of gods, richly decorated fragments of building decorations, inscriptions, and wall-remains suggest an affluent population on the right bank of the city during the first half of the second century.

Although the legion left the city in Trajanic times, the city remained an important merchant center and junction. From Poetovio the economic organization of the two Pannonias was directed by the procurator Augusti, who headed a large central bureau. The merchants of the Amber Route and the employees of the custom office built their houses largely in the territory of the former *canabae*, which, under the name Vicus Fortunae became a satellite town. From the second century a customhouse and warehouses were built. At the same time, a Venus and Vulcanus temple and a Mithraeum were established in the sacred quarter, the temples being built by Oriental employees of the customs bureau about the middle of the second century.

While excavation has not entirely revealed the urban plan, it is possible that the forum and its immediate surroundings were designed and built in a chessboard pattern. Some long houses divided by narrow paths were built along the other streets. At the outskirts of the suburbs were the cemeteries and the kilns of the pottery and brick industries. The outer territory of Poetovio was surrounded by the landed property of the city magistrates during the second and third centuries.

The sanctuary of Jupiter was discovered on the hilltop; from there the streets and rows of houses ran to the Drava River in a raylike arrangement. Perpendicular to the Drava River, villas were built in large gardens; other villas were built on the hilltops and had panoramic views. Still others were built at the western edge of the city which arched with the merchant quarter of Vicus Fortunae. The third century also saw the building of the second Mithraeum, and supposedly shrines of Isis, and of Serapis.

During Diocletian's reign the reorganized Poetovio was detached from Pannonia and became part of Noricum. During the fourth century, when the left bank of the river was gradually depopulated, a citadel was built on the right bank of the hilltop to house the diminished population during enemy attacks. After the crucial battle of Adrianopolis in 380, several new building programs took place. When at the end of the fourth century a Christian basilica was built on the left bank of the river, a new residential area grew up around it and the whole area was surrounded by fortified walls and defensive channels. At the same time, the settlement on top of the hill was also fortified. Other building programs were carried out in the city during the fifth and sixth centuries by various groups of people.[66]

### Sala

The name of this city, Sala, was found in recently discovered inscriptions in Aquae Jasae.[67] Recently excavations have been going on in the center of the settlement, on the site of public buildings. The Amber Route, flanked by stone houses from the second century, passed through the axis of the city.

### Savaria

Savaria was built during the middle of the first century in a regular *insula* system on an area of 580 by 720 meters. The rectangular network of streets divided the center of this *colonia* into 75-by-90-meter blocks. Several attempts have recently been made to mark the street-network of the city, relying on the results of new excavations. The *insulae*, according to the Roman scale, were 30-by-25 *acta* square. The relation to the side of the house blocks was 6 : 5, but backstreets shortened the distances. For a long time the register network made in Claudian times remained unchanged. Except for a few mudhouses, there are no remains of the very early houses of Savaria, for at this time durable materials were used only for the city center.

Savaria soon emerged as one of the most important junctions of the numerous highway systems, thus becoming an important commercial center. Most of the merchants came from Italia, mainly from Aquileia. The *Ara Provinciae* was placed in the city, first for the whole of Pannonia, then after 106 only for Pannonia Superior. Monumental public buildings, including a palatium and a big bath, and the headquarters of pontifical and other religious organizations were constructed here. In fact, the remains of a customs office has recently been discovered on a basalt-paved street. At the end of the first century, a pottery and bronze casting workshop were in operation at the same location, but they were demolished during the later development of the city in order to make room for new buildings, the customs bureau and the warehouses. The aqueduct of the city — 36 kilometers long — was built in this period. It was fed by the springs of the neighboring mountains. Its most important sections have been already discovered during the excavations.

During the Marcomannic wars, Savaria, like other cities on the Amber Route, suffered more than the other Pannonian cities. The rebuilding program of the city, starting at the forum, can be dated to the reign of Septimius Severus by inscriptions and coins, while the outer quarters were renovated only at the beginning of the third century. Greatly changed by this new building program, the city settlement of Savaria then extended in a north–south direction. The southern sacred quarter was leveled and rebuilt, several shrines of Eastern gods being placed next to the Temple of Isis. The remains of statues and inscriptions of Jupiter Dolichenus and Luna Lucifera have been discovered on the pilgrimage site.

In the center of the city, temple gods at the forum have been reconstructed. There were paved streets of about 6 to 10 meters wide. On the northern border of the east–west main road there was a row of *tabernae*; at the eastern extension of this road was discovered an inscription, in which the two Decurios of the city hailed the Emperor Septimius Severus for reconstructing the *cryptae*, the underground warehouse which had been destroyed by fire. This square, paved and encircled with colonnades and a fountain house at the center, may have been the site of the *macellum* (marketplace). From the western side of this square a staircase led to the *pronaos* of a podium temple. A note dated from 1897 states that a "gray" mosaic was discovered here in a 420-mile area which extended toward the east. Nineteenth-century descriptions mentioned great granite columns which correlate in size and materials with the columns of the portico of the Isis temple from the Severian period. In modern Szombathely (ancient Savaria) new monumental Roman ruins have come to light in the past twenty years during periodic renovations of the water supply system; these include several relics from the forum area: the Jupiter and Juno relief, the Jupiter altar stone, inscriptions of the Genii of Pannonia Superior.

After 294 Savaria became the headquarters of the *praeses* of Pannonia Prima; from this time documents mentioned the city walls, but whether this indicated a completely new construction or the partial incorporation of the existing wall into a new fortification system we do not know. The excavated 3-meter-wide wall fragments indicate a polygonal fortification of the streets. Descriptions dating from the fourth century mention the west and south city gates, possibly next to the Iseum (Pl. XIX). Between important palaces and the baths, present excavations at the forum may yield more. Recently, remains of a cross-vaulted building from the first half of the fourth century have been discovered next to a 6-meter-broad paved road near the center of the forum. Only inscriptions tell about the triumphal arch of Constantius II and the office of the *Praepositus Thesaurorum Sabariensium Pannoniae Primae*. The location of the *horreum* is not known; however, an inscription from the fourth century commemorates its repair. From what we know of the type of main buildings in Savaria, it seems that after the administrative division of the Pannonias, the supply center of the new province was placed securely in the hinterland.

Discoveries from the fifth century, during which time the city survived several enemy attacks, reveal some primitive huts on the ruins of the former stone houses. Such primitive houses, which prove the use of ruined material mixed with wood and mud, have been discovered behind the "basilica of Quirinus," on the mosaic floor of a former bath, and on the former pavement of the street near the Iseum. A cemetery from the late imperial period has been discovered at the perimeter of the city. During the fifth century the dead were buried in wooden coffins above fourth-century graves. Some graves from the fourth and fifth centuries have also been discovered in the Early Christian cemetery at the eastern gate of the city.[68]

### Scarbantia

What survives from Scarbantia as an *oppidum* are some sunken mud huts; buildings at the lower level of the settlement have not yet been identified. During the Flavian period, about 71–75 the city was elevated to the rank of *municipium* (Fig. 22). The street-network was oriented towards the east an 18-to-21 degree angle, and the 5-meter-wide paved streets intersected at right angles. From the first to the third century the *insula* area encircled by two creeks and their washland, consisted of about twenty building blocks (Pl. XX). On the basis of the excavations now in course, it is conjectured that a regular *insula* division in the center of the city existed only in the quarter of the forum. At the highest point of the city, at the intersection of the *cardo* and *decumanus*, stood the forum. In front of the sanctuary of the capitolium was a square covered with large stone slabs and bordered by a row of *tabernae*. The sanctuary itself was enlarged with a colonnade which was decorated with sculptures. Several cultic buildings may have been in the center of Scarbantia, but to date only the floor plan of the capitolium sanctuary has been recovered. A stone building dating back to the second century was unearthed; its altar stone west of the capitolium in the *insula*, dedicated to Isis and Bubastis, was removed from the area.

We know of two public baths with a central heating system at the northern and southern ends of the city-center and of a large stone house dating from the end of the first century, but showing evidence of several building periods. This house, containing some large halls which were probably used as assembly places, was built on the ruins of the demolished mudhouses. South of this, an altar stone dedicated to Silvanus Augustus and a building inscription have been discovered.

On a hillside area at the north end of Scarbantia was the amphitheater, with the auditorium sunk into the ground of the sloping side. Next to the building was placed the shrine of Nemesis and Diana. Flanking the exit roads, going north toward Vindobona and Carnuntum and south toward Savaria, were the cemeteries. At the intersection of the main highways are the remains of kilns and of pottery, blacksmith, and metal casting workshops. Some 12 kilometers (8 miles) south of the city, at the springs, the ruins of a second-century water channel system were discovered.

During the fourth century, Scarbantia was encircled by a system of walls larger than 3 meters. The new fortification system, which defended only the inner core of the residential area, crossed houseblocks from the second and third centuries. The elliptical walls was defended by thirty-two horseshoe-type donjons, with two monumental gates opening toward the city on the north and south sides. Then, at the beginning of the fifth century, parts of the public buildings were repaired and transformed. The ruins were partly leveled or, if necessary, filled

266

in to bring them level with the top of other ruins. In fact, there was little uniformity in height among the streets. New loghouses and mudhouses, characteristic dwellings of the late Roman population, were built amid the ruins of former monumental stone buildings. It is evident that people were buried near the gates and outside and inside the city walls for centuries as the latest excavations prove.[69]

### Sirmium

Sirmium, which is located approximately fifty miles southwest of ancient Singidunum (Belgrade), is mentioned in the *Itinerarium Antonini* as a city equal in importance to Milan, Aquileia, Nicomedia, Antiochia, and Alexandria (*Itinerarium Antonini Augusti*, ed. Cuntz, *Itineraria Romana* 123,8 and 124,4. Vol. 1. Leipzig, 1929). It was considered an important commercial city from the second century on.

Some important data concerning the history of the settlement of Sirmium came to light in the last few years of the excavations (1966–1971). Under the foundations of well-built stone houses were discovered the ruins of primitive dirt walls. In one of the *insulae* a large city government building was erected about the end of the first century and was later transformed into an imperial palace. Like other cities located between the Drava and Sava rivers, Sirmium entered its golden period after the abandonment of the Roman Provincia Dacia. At the end of the third century, there were a weapons factory and a coin mint in Sirmium.

During the Tetrarchy, Sirmium became the capital of Pannonia Secunda and one of the four capitals of the Roman Empire. About the end of the third century and during the fourth century, the city was enclosed by walls. The road from the city center crossed some suburban satellite cities which, although they became part of metropolitan Sirmium, developed separately rather than following the development of the main city of Sirmium. The names of these suburban cities have come to light from inscriptions.

In the center of Sirmium was discovered the imperial palace, which had been transformed several times during the fourth century. Near this palace was a *forum*, and alongside that a monumental *horreum* and a row of *tabernae* with decorative porticos. A *therma* discovered in the neighborhood was perhaps the *therma Liciniae*.

Ancient sources tell us that a stadium and a theater were built, and it is proved that the money allocated for the theater was used by Valentinian I for the reinforcement of the city walls. However, archaeologists could not locate those buildings. Around the port area on the Drava River, primitive mudhouses and other living accommodations were detected. The Romans had built a permanent bridge on the Drava River.

Licinius, who was defeated by Constantine the Great, lived in Sirmium between 306 and 314, during which time he built an imperial bath on the Island of Brac. According to Meneandrus, there were two islands in Sirmium, Cassia and Carbonaria, but today we have no knowledge of their exact location.

In Sirmium, a large Early Christian community flourished. There is a basilica which was dedicated to Saint Sinerus who may have been martyred in Sirmium in 306. On the south bank of the Drava River a fourth-century martyrium was discovered; it was probably built for Saint Ireneus, who was martyred in Sirmium in 304. Above the fourth-century level the ruins of three more churches were discovered. The first, not yet dated, may possibly be early medieval, and the second a Byzantine structure from the tenth or eleventh century, and the third a typical Benedictine church dated 1229–30.[70]

## Siscia

In 34 B.C. a Roman legion was placed in this *oppidum*. Later the *canabae* and the civilian settlement developed individually, the inhabitants of the latter being merchants who made contact with the military settlement and the local population for business reasons. Since there are remains of the settlement on both banks of the Sava River, the bridge and the customs house at the bridge must have played an important role in the development of the city.

Only recently has the systematic excavation of the city begun. Ornamented building fragments and monumental statues which have been uncovered show that large public buildings were already erected under the Flavians in the first century A.D. First-century cemeteries were discovered along the right bank of the river, the second-century graves clearly indicate the expansion of the city. Inscriptions from the second century inform us about the existence of a curia, basilica, a bath, a marketplace, a porticus; these buildings were possibly placed around the forum. Some important institutions of city life are mentioned in inscriptions, including the pontifical body and the organization of the Collegium Centonariorum, all of which must have had buildings at the civilian center headquarters. According to epigraphical evidence, the waterworks of this *municipium* were completed during the second century. On a find from the sacred quarter the names of Caracalla and Gordianus are visible, and some inscriptions tell us that the building programs here were underwritten by the state functionaries.

The city became more important during the time of the military emperors because of the proximity of iron mines. In fact, an armaments factory was built in the city; the necessary coins for the pay of the Pannonian legions also emerged from here, minted from 260 in Siscia. At the beginning of the fourth century, Siscia became a provincial center and developed into a bishopric (episcopal city). Public offices, a treasury built for the provincial administration, and a new residential section, different from the previous one, extended in the city toward the Sirmium road. The late Roman cemeteries flanked the road outside the city boundary.[71]

## Sopianae

Systematic excavations in the area of Sopianae have only recently begun. It is evident that from the second century monumental buildings were erected at the intersection of the main roads, but of earlier buildings we know little. The bath and the sophisticatedly decorated second-century buildings were totally covered in this latest building program. All the buildings in the city had collapsed about

268

260 and a new densely populated city was built on the debris of the ancient one. The new walls were weaker and the new floors were covered with a poorer quality of terazzo.

After the reorganization of Pannonia, when Sopianae became the seat of the civilian administration of Valeria, new public buildings were constructed. Several public buildings of the fourth century have been discovered in the main streets and their vicinity but no private buildings are yet known. Although we have no documentary evidence to indicate the existence of a bishopric in Sopianae, several articles unearthed in excavations and the presence of significant Early Christian buildings, painted grave chambers, and lavishly decorated grave chapels suggest that an Early Christian chancery may have existed there. A Christian cemetery was placed at the hillside of Sopianae.[72]

### Vindobona

A settlement of the native population near the legionary camps, entirely separated from the *canabae*, developed in the civilian city of Vindobona, which, until the reign of Caracalla, had no autonomous city rights. Since no extensive excavation has been carried on, a complete picture of the city is not yet available. It is very likely that the forum and the public utilities were erected about the middle of the second century in the center of the city. The settlement is bordered by pottery shops and brick-making workshops, and the exit roads are flanked by decorated graveyards.

The civilian settlement and the *canabae* of Vindobona were burned down during the Quadian–Marcomannian wars, to be rebuilt gradually during the Caracalla period. Soon the two cities expanded and merged. Near the city well-decorated villas with wall paintings and mosaics were built.

At the end of the fourth century and the first years of the fifth century, some houses in *canabae* in the military city of Vindobona were discovered; possibly they had been built by the Barbarian settlers. They were poorly constructed, but made of good quality stone and some mud; utensils discovered around those houses give evidence of the presence of the Barbarian invaders. After the collapse of the Roman administration these peoples lived among the ruins of the city and were buried near the walls.[73]

### Municipium Volg . . .

A fragment of an inscription was discovered near Intercisa; experts suppose that this city was between the Danube and the eastern edge of Lake Balaton.[74]

### NOTES

[1] A. ALFÖLDI and L. NAGY, *Budapest története I. Budapest az ókorban* (History of Budapest, I: Budapest in Antiquity) (Budapest, 1942), pp. 258–74; E. KORNEMANN, "Coloniae", *R. E. Pauly-Wissowa*, 4: 511–88.

[2] A. MÓCSY, *Die Bevölkerung von Pannonien bis zu den Markomannenkriegen* (Budapest, 1959), p. 114; E. KORNEMANN, "Municipium", *R. E. Pauly-Wissowa*, 6:570–638.

[3] E. FABRICIUS, "Limitatio", *R. E. Pauly-Wissowa*, 13: 685 ff.; A. PIGAGNIOL, *Les documents*

*cadastraux de la colonie romaine d'Orange* (Paris, 1962). (Suppl. à *Gallia*), pp. 44 ff.; A. GRENIER, *Manuel d'archéologie gallo-romaine, VI, L'archéologie du sol* (Paris, 1934); A. MÓCSY, "Savaria utcarendszerének rekonstrukciójához—Zur Rekonstruktion des Strassensystems von Savaria". *ArchÉrt* 92 (1965):35.

⁴ W. KUBITSCHEK, *R. E. Pauly-Wissowa*, 3:1960 ff.; A. SCHULTEN, *R. E. Pauly-Wissowa*, 3:1587 ff.; A. SCHULTEN, "Decumanus", *R. E. Pauly-Wissowa*, 4:1314; A. MÓCSY, "Pannonienforschung 1964–68". *ActaArchHung* 21 (1969):348; E. TÓTH, "A savariai insularendszer rekonstrukciója—Rekonstruktion des Insula-Systems in Savaria". *ArchÉrt* 98 (1971): 148, 152; K. PÓCZY, "Die Anfänge der Urbanisation in Scarbantia". *ActaArchHung* 23 (1971): 95; J. FITZ, "The Excavations in Gorsium". *ActaArchHung* 24 (1972): 14.

⁵ MÓCSY, *Die Bevölkerung*, pp. 105 ff.; MÓCSY, "Pannónia", pp. 604 ff.; A. MÓCSY, "Zur Geschichte der peregrinen Gemeinden in Pannonien". *Historia* 6 (1957): 488 ff.; A. MÓCSY, "Decurio Eraviscus". *FolArch* 21 (1970): 59 ff.; É. B. BÓNIS, *Die spätkeltische Siedlung Gellérthegy–Tabán in Budapest* (Budapest, 1969); FITZ, "The Excavations", p. 36; É. KOCZTUR, "Újabb adatok Gorsium őslakosságának háztípusaihoz—Neuere Daten zu den Haustypen der Urbevölkerung von Gorsium". *FolArch* 23 (1972): 43; K. SÁGI, "Adatok a pannóniai császárkori kocsitemetkezések ethnikumának kérdéseihez—Sur le caractère ethnique des enterrements à char à l'époque impériale". *ArchÉrt* (1951), pp. 73–78.

⁶ MÓCSY, *Die Bevölkerung*, pp. 30, 32, 45; MÓCSY, "Pannónia", pp. 600–601; MÓCSY, "Pannonienforschung", p. 340; FITZ, "The Excavations", p. 12; A. NEUMANN, *Vindobona: Die römische Vergangenheit Wiens: Geschichte, Erforschung, Funde* (Wien, Graz, Köln, 1972), pp. 24 ff.; J. ŠAŠEL, in *Kronika* 3 (1955): 110 ff.; J. ŠAŠEL, "Emona", *R. E. Pauly-Wissowa*, Suppl. 9, pp. 576 ff.

⁷ Plin. NH. 3. 147; MÓCSY, *Die Bevölkerung*, pp. 44, 76.

⁸ MÓCSY, "Pannónia", pp. 598–99; MÓCSY, *Die Bevölkerung*, pp. 16–96.

⁹ A. MÓCSY, A Borostyánkőútnál nyíló középületek—Public buildings at the Amber Route. *ActaArchHung* 26 (1974): 95 f.; the location of Halicanum was identified by S. Soproni. (No publication appeared.)

¹⁰ F. FÜLEP, *Pécs római kori emlékei—Die Römerzeit in Pécs* (Budapest, 1964), pp. 6 ff.

¹¹ FITZ, "The Excavations", p. 30.

¹² L. BARKÓCZI, "Beiträge zum Rang der Lagerstadt am Ende des II. und Anfang des III. Jahrhunderts". *ActaArchHung* 3 (1953): 201 ff.; A. MÓCSY, "Das Territorium Legionis und die Canabae in Pannonien". *ActaArchHung* 3 (1953): 191 ff.; L. ECKHARDT, "Die Zivilsiedlungen der Legionslager Lauriacum und Carnuntum". *Mitt. d. Vereins der Freunde Carnuntums* 7 (1954): 12–28.

¹³ NEUMANN, *Vindobona*, pp. 13 ff.; L. BARKÓCZI, "Brigetio" (Budapest, 1951), 2:9 ff. (*DissPann*, 2:22).

¹⁴ MÓCSY, "Pannónia", pp. 693 ff.; K. SZ. PÓCZY, "Scarbantia városfalának korhatározása—La datation de l'enceinte de Scarbantia". *ArchÉrt* 94 (1967): 137–53.

¹⁵ T. P. BUÓCZ, *Savaria topográfiája* (The Topography of Savaria) (Szombathely, 1967) pp. 9 ff.; E. TÓTH, "Late Antique Imperial Palace in Savaria". *ActaArchHung* 25 (1973): 117 ff.; DJ. MANOZISI, in *Zbornik Radova Narodnog Muzeja* (Beograd) 4 (1963): 93 ff.; V. POPOVIĆ, "A Survey of the Topography". *Sirmium* 1 (1971): 119 ff.; F. FÜLEP, "Neuere Ausgrabungen in der Cella Trichora von Pécs". *ActaArchHung* 11 (1959): 399 ff.

¹⁶ M. GRBIĆ, "Bassianae". *Antiquity* 10 (1936): 475; K. SZ. PÓCZY, "Városfalmaradványok a soproni Fabricius-ház alatt—Stadtmauerreste unter dem Fabriziushaus in Sopron". *Arch-Ért* 89 (1962): 47–67.

¹⁷ FITZ, "The Excavations", p. 34.

¹⁸ A. RADNÓTI, "Pannóniai városok a korai feudalizmusban" (Pannonian Towns in the Early Feudalism), *Magyar Tudományos Akadémia II. Oszt. Közl.* 5 (1954): 498; K. SÁGI, "Die spätrömische Bevölkerung der Umgebung von Keszthely". *ActaArchHung* 12 (1960): 254; K. SÁGI and K. BAKAY, "A népvándorláskor építészete Magyarországon" (The Architecture in the Period of Migrations in Hungary), *Építés–Építészettudomány* 2 (1971): 405 ff.

¹⁹ E. SWOBODA, *Carnuntum: Seine Geschichte und seine Denkmäler*, 3rd ed. (Graz, 1958), p. 176; K. SZ. PÓCZY, "Aquincum a IV. században—Aquincum im 4. Jahrhundert". *BudRég* 21 (1964): 55–77; L. BARKÓCZI and Á. SALAMON, "Remarks on the Sixth Century History of

270

Pannonia". *ActaArchHung* 23 (1971): 139, 153; V. LÁNYI, "Die spätantiken Gräberfelder in Pannonien". *ActaArchHung* 24 (1972): 53 ff.

[20] R. NOLL, *Frühes Christentum in Österreich* (Wien, 1954); H. VETTERS, "Zum Problem der Kontinuität im Niederöst. Limesgebiet" *Jb. d. Vereins f. Landeskunde v. Niederöst.* 38 (1968–70) : 48; L. VÁRADY, *Das letzte Jahrhundert Pannoniens 376–476* (Budapest, 1969); E. TÓTH, "Vigilius episcopus Scarbanciensis". *ActaArchHung* 26 (1974): 270 ff.

[21] A. KÁROLYI and T. SZENTLÉLEKY, *Szombathely* (Budapest, 1967), p. 20; BUÓCZ, *Savaria topográfiája*, p. 60; K. SZ. PÓCZY, "Scarbantia városfalának korhatározása", p. 152; NEUMANN, *Vindobona*, p. 54; L. BARKÓCZI, "A Sixth Century Cemetery from Keszthely-Fenékpuszta". *ActaArchHung* 20 (1968): 275 ff.; A. MILOŠEVIĆ and O. MILUTINOVIĆ, "Zaštitna arheološka iskopavanja u Sremskoj Mitrovici—Archeological Excavations in Sremska Mitrovica". *Gradja za Proučovanje Spomenika Kulture Vojvodine* 2 (1958): 5–45.

[22] BARKÓCZI and SALAMON, "Remarks", p. 139.

[23] POPOVIĆ, "A Survey", p. 119; M. PAROVIĆ-PEŠIKAN, "Excavations of a Late Roman Villa in Sirmium". *Sirmium* 2 (1971): 15.

[24] H. PETRIKOVITS, "Das Fortleben römischer Städte am Rhein und Donau im frühen Mittelalter". *Trierer Zeitschrift* 19 (1950): 73; K. BÖHNER, "Zur Frage der Kontinuität zwischen Altertum und Mittelalter". *Trierer Zeitschrift* 19 (1950): 92; ŠAŠEL, "Emona", p. 562; TÓTH, "Vigilius".

[25] W. SCHMID in *BRGK* 15 (1925): 202 ff.; M. DETONI and T. KURENT, "Modularna rekonstrukcija Emone—Modular Reconstruction of Emona". *Situla* 1 (1963): 1–71; MÓCSY, "Savaria utcarendszerének rekonstrukciójához—Zur Rekonstruktion des Straßensystems von Savaria". *ArchÉrt* 92 (1965): 27–36. "A savariai insularendszer", pp. 143, 152.

[26] MÓCSY, "Pannónia", p. 695.

[27] W. SCHMID, in *JÖAIBeibl.* 19–20 (1919): 140; A. SCHOBER, *Die Römerzeit in Österreich* (Wien, 1953), p. 68; W. MODRIJAN, *Schild von Steiermark* 9 (1959–1961): 13; H. VETTERS, in *R. E. Pauly-Wissowa* 17: 250; E. WEBER, *Die römerzeitlichen Inschriften der Steiermark* (Graz, 1969), p. 177; PÓCZY, "Die Anfänge", p. 93.

[28] FITZ, "The Excavations", p. 30.

[29] J. SZILÁGYI, *Aquincum* (Budapest, 1956), p. 25; E. SWOBODA, *Carnuntum*, 3d ed. (Köln–Graz, 1964), p. 288.

[30] D. GABLER, "Munera pannonica". *ArchÉrt* 93 (1966): 20 ff.

[31] K. SZ. PÓCZY, "Aquincum első aquaeductusa—Le premier aqueduc d'Aquincum". *ArchÉrt* 99 (1972): 15 ff.

[32] A. ALFÖLDI, "Kapitóliumok Pannoniában" (Capitols in Pannonia), *ArchÉrt* (1920–1922), 12–14; J. SZILÁGYI, "Az aquincumi helytartói palota—Der Stadthalterpalast von Aquincum". *BudRég* 18 (1958): 53 ff.; T. NAGY, in *Budapest Története* (Budapest, 1973): 120; FITZ, "The Excavations", p. 16; POPOVIĆ, A Survey, p. 119; BUÓCZ, *Savaria*; K. SZ. PÓCZY, *Sopron római kori emlékei* (Die römischen Altertümer in Sopron) (Budapest, 1965).

[33] J. FITZ, *Il soggiorno di Caracalla in Pannonia nel 214* (Roma, 1961); GABLER, "Munera", pp. 24 ff.

[34] W. SCHMID, in *BRGK* 15 (1925): 212; M. ABRAMIĆ, *Führer durch Poetovio* (Wien, 1925); T. SZENTLÉLEKY, *A szombathelyi Isis szentély—Das Isis-Heiligtum von Szombathely* (Budapest, 1960).

[35] K. PÓCZY, "Wandmalereien d. Statthalterpalast in Aquincum". *Bud. Rég.* 18 (1958): 101–138; E. B. THOMAS, *Römische Villen in Pannonien* (Budapest, 1964); I. WELLNER, "A magyarországi római kori épületek belső díszítő művészete—The Inner Decorative Art of Roman-Age Buildings in Hungary". *Építés-Építészettudomány* 2 (1971): pp. 327 ff.; Á. KISS, *Roman Mosaics in Hungary* (Budapest, 1973).

[36] A. ALFÖLDI, in *Századok* 70 (1936): 44; M. POPOVIĆ, "Sirmium, ville impériale". *Studi di Antichità Cristiana* 27 (1969): 665 ff.; E. TÓTH, "Late Antique", pp. 117 ff.; M. MIRKOVIĆ, "Sirmium, Its History from the First Century A.D. to 582 A.D.". *Sirmium* 1 (1975): 5 ff.

[37] E. SWOBODA, *Carnuntum*, pp. 152 ff.; N. FETTICH, "Colonia Claudia Savaria". *Vasi Szemle* 6 (1939): 128.

[38] MÓCSY, "Pannónia", p. 697.

[39] NEUMANN, *Vindobona*, p. 82; M. KANDLER, *Die Ausgrabungen im Legionslager Carnun-*

*tum*, 1968–1973 (Wien, 1974), p. 37; K. Sz. Póczy, "Aquincum a IV. században", p. 55.

⁴⁰ P. Lavedan, *Les villes françaises* (Paris, 1960), pp. 11 ff.

⁴¹ Tóth, "A savariai", p. 152.

⁴² A. Graf, "Übersicht der antiken Geographie von Pannonien" (Budapest, 1936)(*Diss-Pann* 1:5); Mócsy, *Die Bevölkerung*, p. 53; A. Mócsy, "Scribák a pannoniai kisvárosokban— Scribae in den Kleinstädten Pannoniens". *ArchÉrt* 91 (1964): 16 ff.

⁴³ Mócsy, "Pannonia", p. 600.

⁴⁴ Mócsy, *Die Bevölkerung*, pp. 44 ff.; G. Alföldy, "Municipális középbirtokok Aquincum környékén" (Municipal Medium Estates around Aquincum), *AntTan* 6 (1959): 19 ff.; Fitz, "The Excavations", pp. 35 f.

⁴⁵ A. Lengyel, "Excavation in Roman Sirmium", in *Etruscans*, (1970–1972), 2:6.

⁴⁶ A. Mócsy, "Die Entwicklung der Sklawenwirtschaft in Pannonien zur Zeit des Prinzipates". *ActaAntHung* 4 (1956): 221 ff.; G. Alföldy, "Augustalen und Sevirkörperschaften in Pannonien", *ActaAntHung* 6 (1958): 433 ff.

⁴⁷ Mócsy, *Die Bevölkerung*, p. 134; Mócsy, "Pannónia", p. 554; Szilágyi, "Az aquincumi", pp. 53 f.; Póczy, "Die Anfänge", p. 101; Fitz, "The Excavations", pp. 16 ff.; Neumann, *Vindobona*, pp. 23 ff.; Buócz, *Savaria*, p. 20.

⁴⁸ J. Fitz, "Der Besuch des Septimius Severus in Pannonien im Jahre 202 u. Z.". *ActaArchHung* 11 (1959): 237–63; Fitz, *Il soggiorno*.

⁴⁹ Swoboda, *Carnuntum*, pp. 63 ff.; Póczy, "Aquincum", pp. 55 ff.; Popović, "Sirmium, ville", pp. 665 ff.; Fitz, "The Excavations", pp. 5 ff.; Tóth, "Late Antique", pp. 117 ff.

⁵⁰ Amm. Marc. 30.5.14; Neumann, *Vindobona*; Buócz, *Savaria*; Károlyi and Szentléleky, *Szombathely*; Fülep, *Pécs római kori—* ; Popović, "A Survey", pp. 119 ff.

⁵¹ A. Alföldi, "Epigraphica III", *ArchÉrt* (1940), p. 202; Alföldi and Nagy, *Budapest története*, p. 301; Barkóczi, "Beiträge", p. 201; Mócsy, *Das Territorium*, pp. 191 ff.; Mócsy, "Die Entwicklung", pp. 231 ff.; Mócsy, *Die Bevölkerung*, pp. 16–76; Mócsy, "Pannónia", pp. 693–98; Mócsy, "Pannonienforschung", pp. 360–63.

⁵² Saria, B., *R. E. Pauly-Wissowa*, 22:1638 f.; Mócsy, *Die Bevölkerung*, p.21.

⁵³ *BudRég* 18 (1958): 1–21, passim; B. Kuzsinszky, *Aquincum: Ausgrabungen und Funde* (Budapest, 1934); A. Alföldi, L. Nagy and J. Szilágyi, *Budapest története I–II* (Budapest, 1942); L. Nagy, *Az Eskütéri erőd, Pest város őse* (Budapest, 1946); J. Szilágyi, *Aquincum* (Budapest, 1956); T. Nagy, "Buda régészeti emlékei", in *Budapest műemlékei* (Budapest, 1962), pp. 34–63; J. Szilágyi, "Aquincum", *R. E. Pauly-Wissowa*, Suppl. 11, pp. 61 f; K. Sz. Póczy, "Aquincum a IV. században— Aquincum im 4. Jahrhundert." p. 56; "Anwendung neuerer Ausgrabungsergebnisse in der Bürgerstadt Aquincum." *Acta Techn. A. S. H.* 67 (1970): 177–194.

⁵⁴ Mócsy, *ArchÉrt* 92 (1965): 39; Mócsy, "Pannonienforschung", p. 349.

⁵⁵ L. Barkóczi, "Brigetio" (Budapest, 1944–1951) (*DissPann* 2:22); L. Barkóczi, "Die datierten Glasfunde aus dem II. Jahrhundert von Brigetio". *FolArch* 18 (1966–1967): 67–89; L. Barkóczi, "Die datierten Glasfunde aus dem 3–4. Jahrhundert von Brigetio". *FolArch* 19 (1968): 59–86; L. Barkóczi, "Merkurflaschen mit Bodenstempel im Ungarischen Nationalmuseum". *FolArch* 20 (1969): 47–52; L. Barkóczi, "New Data on the History of Late Roman Brigetio". *ActaAntHung* 13 (1965): 215–57.

⁵⁶ Swoboda, *Carnuntum;* R. M. Swoboda-Milenović, *Carnuntum-Jahrbuch* (1958–1964); H. Stieglitz, "Beriche über die Grabungen in Carnuntum". *ÖJh* 48 (1966–67): 26 ff; 49 (1968–69): 34 ff. "Carnuntum". *RE* Suppl. XII. 1970. 1575 ff. "Palastruine" *PAR* 20 (1970): 2 ff; M. Kandler, Die Grabung 1969 der "Limeskommission" im Legionslager Carnuntum, *PAR* 20 (1980): 9 ff.

⁵⁷ J. Brunšmid, in *Vjesnik Hrvatskog Arheološkog Društva* 6 (1902): 117 ff.; Mócsy, *Die Bevölkerung* p. 76; B. Raunig-Galić, in *Arheološki Pregled* 7 (1965): 148.

⁵⁸ W. Schmid, "Poetovio", *BRGK* 15 (1925): 202 ff.; J. Šašel, in *Arheološki Vestnik* 4 (1953): 304 f.; *Vodnik po Emoni* (Ljubljana, 1955), pp. 42 ff.; L. Lesničar-Gec, *Antična Emona v scru moderne Ljubljane* (The Antique Emona in the Hearth of the Modern Ljubljana, Ljubljana, 1961), pp. 12 ff.; "La nécropole romaine à Emona". *Inventaria Arch* 10 (1967); *Arheološki Pregled* 5 (1963), 7 (1965): passim; M. Detoni and T. Kurent, "Modular Reconstruction of Emona". *Situla* 6 (1963): 1–71.

[59] Mócsy, "Pannonia", p. 600; Mócsy, "Pannonienforschung", p. 350.

[60] C. Patsch, "Herculia", *R. E. Pauly-Wissowa*, 8:612; J. Fitz, "Gorsium". *Das Altertum* 8 (1962): 155–73; J. Fitz, "Gorsium", *R. E. Pauly-Wissowa*, Suppl. 9, pp. 73–75; J. Fitz, *Gorsium: A táci rómaikori ásatások* (Székesfehérvár, 1964); J. Fitz, "Gorsium". *Der kleine Pauly* 2 (1967): 854 ff.; J. Fitz, in *ActaArchHung* 23 (1971).

[61] V. Hoffiller and B. Saria, *Antike Inschriften aus Jugoslavien. I. Noricum und Pannonia Superior* (Zagreb, 1938), pp. 269 ff.; G. Alföldy, "Epigraphica". *Situla* 8 (1965): 95 ff.; Mócsy, "Pannonienforschung", p. 349.

[62] Mócsy, *Die Bevölkerung*, p. 53; A. Kiss, in *Veszprém megyei Múzeumok Közl.* 6 (1967): 37 ff.

[63] D. Pinterović, "Prilog topografiji Murse—Topographical Studies about Mursa". *Osječki Zbornik*, pp. 55–94; D. Pinterović, "Prilog topografiji Murse", in *Limes u Jugoslaviji* (Beograd, 1961), pp. 35–42; Mócsy, *Die Bevölkerung*, p. 74.

[64] Mócsy, *Die Bevölkerung*, p. 53 (Mursella I); E. Biró, "Kisárpási későrómai temető—Das spätrömische Gräberfeld in Kisárpás". *ArchÉrt* 86 (1959): 173; Mócsy, *Die Bevölkerung*, p. 74 (Mursella II).

[65] Hoffiller and Saria, *Antike Inschriften* p. 109; B. Saria, "Die römische Wasserleitung von Neviodonum". *Serta Hoffilleriana* 1940, pp. 249–56; P. Petru, in *Arheološki Pregled* 6 (1964): 73 ff., 7 (1965): 123 ff.; P. Petru, in *Arheološki Vestnik* 17 (1966): 491.

[66] M. Abramić, *Poetovio: Führer durch die Denkmäler der römischen Stadt* (Wien, 1926); W. Schmid, "Poetovio". *BRGK* 15 (1925): 212–25; I. Mikl-Curk, "K topografiji rimske mestne četrti na današnjem Zgornjem Bregu v Ptuju—Supplément à la topographie du quartier urbain à l'emplacement de l'actuel Zg. Breg à Ptuj". *Arheološki Vestnik* 15–16 (1964–1965): 259–82.

[67] A. Mócsy, excavation report. *ActaArchHung* 26 (1974): 95–98.

[68] I. Paulovics, Savaria–Szombathely topográfiája (Szombathely, 1943); A. Alföldi, Jr., "Adatok Szombathely településtörténetéhez—Zur Entstehung der Colonia Claudia Savaria". *ArchÉrt* (1943) pp. 71–86; A. Mócsy, "Korarómai sírok Szombathelyről—Frührömische Gräber in Savaria (Szombathely)". *ArchÉrt* 81 (1954): 167–91; E. Türr, "Savaria aquaeductusa—Der Aquädukt von Savaria". *ArchÉrt* 80 (1953): 129–33; A. Mócsy, "Savaria utcarendszerének rekonstrukciójához—Zur Rekonstruktion des Strassensystems von Savaria". *ArchÉrt* 92 (1965): 27–36; T. Szentléleky, *A szombathelyi Isis szentély—Das Isis-Heiligtum von Szombathely* (Budapest, 1960); Z. Kádár and L. Balla, *Savaria* (Budapest, 1958); T. Buócz, *Savaria topográfiája* (Szombathely, 1968).

[69] K. Praschniker, "A soproni kapitóliumi istenségek—Die Kapitolinische Trias von Sopron". *ArchÉrt* 51 (1938): 29–44; F. Kenner, in *Mitt. d. K. K. Centr. Comm.* (1967) pp. 381 ff.; Gy. Nováki, in *Soproni Szemle* 9 (1955): 143 ff., 10 (1956): 344 ff.; A. Radnóti, "*Sopron és környéke régészeti emlékei*" (Archeological Monuments of Sopron), in E. Csatkai, *Sopron és környéke műemlékei* (Budapest, 1953), pp. 14–35; M. Storno, "A római amfiteátrum és nemeseum Sopronban—Das römische Amphitheatrum und Nemeseum zu Sopron (Scarbantia)". *Soproni Szemle* (1941), pp. 201–16; K. Sz. Póczy, "Scarbantia városfalának kormeghatározása—Kronologie der Stadtmauer von Scarbantia." *ArchÉrt* 94 (1907): 137–147; "Die Anfänge der Urbanisation in Scarbantia." *ActaArchHung* 23 (1971): 93–99.

[70] V. Hoffiller, "Prolegomena zu Ausgrabungen in Sirmium". *Bericht über den 6. Internat. Kongr. für Archäologie* (Berlin, 1939), pp. 517–26; V. Popović, "Sirmium, ville impériale", *Studii di Antichità Cristiana* 27 (1969): 665 ff.; S. Atanacković, "Arheološko proučavanje Sirmijuma". *Kalendar Matice Srpske* (Novi Sad, 1961), pp. 154–56; O. Brukner, "Arheološka iskopavanja 1959 godina u Sremskoj Mitrovici". *Arheološki Pregled* (1959): 118–25; O. Brukner, "Sirmium–Sremska Mitrovica–Naselje". *Arheološki Pregled* 2 (1960): 100–106; O. Brukner, "Iskopavanja u Sirmium 1957–1960 godine", in *Limes u Jugoslaviji* (Beograd, 1961), pp. 77–81; A. Lengyel, "Excavations in Roman Sirmium", in *Etruscans* (1970–1972), 2: 5–16.

[71] Hoffiller and Saria, Antike Inschriften, pp. 237 ff.; Mócsy, *Die Bevölkerung*, pp. 24 ff.

[72] Gy. Gosztonyi, "A pécsi hatkarélyos ókeresztény temetői épület—Ein altchristliches Gebäude mit 7 Apsiden in Pécs". *ArchÉrt* (1940): 56–61; F. Fülep, *Pécs római kori emlékei—The Roman Monuments of Pécs* (Pécs, 1963).; "Neue Ausgrabungen in der Cella

Trichora von Pécs". *ActaArchHung* 11 (1959): 399–417" Neuere Ausgrabungen in der Römerstadt Sopianae–Pécs." *Régészeti füzetek*. 16 (1974): 1–95.

[73] C. PASCHER, in *RLiÖ* 19 (1919): 167 ff.; A. NEUMANN, Die römischen Ruinen unter dem Hohen Markt, 2nd ed. (Wien, 1957); A. NEUMANN, *Das Raum von Wien in ur-und frühgeschichtlicher Zeit* (Wien, 1961); A. NEUMANN, in *RLiÖ* 23 (1967), 24 (1968); H. LADENBAUER-OREL, "Archäologische Stadtkernforschung in Wien". *Jahrbuch d. Vereins für Gesch. d. Stadt Wien* 21/22 (1965–1966): 7–66.; A. NEUMANN, *Vindobona. Die röm. Vergangenheit Wiens* (1972).

[74] MÓCSY, "Pannonia", p. 600; MÓCSY, "Pannonienforschung", p. 360.

# VILLA SETTLEMENTS

### EDIT B. THOMAS

Research to date has not been concentrated on the buildings of a private character in Pannonia. The *limes*, camps, roads, and cemeteries—the primary targets of research—have captivated the imagination of archaeologists. Yet, for a familiarity with Pannonian culture, a good knowledge of the circumstances in which people lived, how they shaped their surroundings, and what their habitations looked like is also essential. The different kinds of buildings which served various purposes resulted from differences in the professions and social circumstances of the people. Thus an organic connection arose between habitations and forms of culture. Differences also existed for other reasons. The most ancient of all purposes of building is to defend the property of its inhabitants against the weather, fire, robbery, and other potential dangers. Influenced by differences in soil and climatic conditions, available building materials, and ethnic factors, houses developed in different forms and shapes in various regions.

Flóris Rómer, the famous Hungarian scholar mentioned in the chapter on the history of Hungarian archaeology, noted as early as 1863: "as everywhere else, little attention has been focused on buildings and roads." He then called attention to the need for such research. In the 1890s Bálint Kuzsinszky urged that research at Papföld (the area known today as Aquincum) should include private houses "the more so from a scientific point of view, since few such buildings were ever unearthed on this side of the Alps". In spite of Rómer's and Kuzsinszky's endeavors, very little organized research was subsequently directed towards the excavation of villas, because only modest financial support was available. As a result, only a few villas are known and only the most sumptuous of these have been described and are thus available to scholars.

Such systematic excavations of villas within the province have revealed important facts about life in Pannonia.[1] First of all, the largest groups of Roman villas were situated around Lake Balaton and its immediate neighborhood.[2] The relatively densely populated Balaton area was crossed by three roads which ran parallel with each other and with the lake as well. The oldest one ran close to the shore and connected several greater and smaller settlements. The villas and groups of houses were built where possible very close to the roads. The Roman population preferred Lake Balaton and its surroundings because of its similarity climatically as well as geographically to Italia. Climate and geography have undoubtedly influenced the development of cognate cultures: it cannot be accidental that the native population of Pannonia quickly adapted to Romanism and to the forms of life dictated by Rome in the vicinity of Balaton and the Transdanubian area.

275

The second largest group of villas stretched west of Lake Fertő in Burgenland.[3] In addition to the favorable lake conditions, the Roman roads which ran along the borders of Pannonia and Noricum encouraged these settlements.

Behind the *limes* stretching along the Danube there were villas on the undulating hills, which granted a magnificent view and also had strategic advantages.[4] Naturally, a knowledge of the ground plans of these villas would be very rewarding from the point of view of the Roman defense system. At present, however, only about 40 percent of these villas have been excavated; the majority are known to archaeologists only from surface observation or surveys.

Villas situated between the Drava and Sava rivers are eloquent witness of the important role this area played in the life of the late Roman Empire.[5]

The fifth group of Roman villas was a scattering of villa settlements in the interior of the province.[6]

The adaptability of the native population to Roman ways manifested itself in every branch of art and in architecture as well. As early as the first century, the native population had already abandoned the custom of building in the manner of the Celtic Eravisci. Instead of primitive native one-room dwellings with walls of mud, they erected buildings in the Roman style made of bricks, stones, and mortared walls.

In most cases only the stronger foundations of the walls remain. In the best cases, walls to the height of 20 to 60 centimeters survive. Over the centuries the population often excavated even the foundations. So archaeology faces a more difficult task in Pannonia than it does in the southern provinces where walls remained almost intact, frequently to their original heights.

## CONCEPT OF THE *VILLA ROMANA*

The Roman villa was not an equivalent of the present-day summer resort.[7] Villas were detached buildings or groups which, remote from other larger settlements outside camps, cities, or villages, constituted an individual and independent economic unit. Since they were similar to modern farms, the word grange or farmstead more properly describes the *villa romana*. The tools which have been found in the buildings testify to the role villas played in the economic life of the provinces. Agricultural tools, equipment, and buildings which stored the products of the farm were integral parts of the *villa rustica*, but buildings of the *villa fructuaria* were found near the *villa urbana* only when the *urbana* was the living quarters of the landlord and had an agricultural residence attached to it.

In the period preceding the Roman occupation, we know of no agricultural production performed by the Illyrian or Celtic tribes resembling the system introduced by the Italian conquerors. Since literary sources do not discuss this subject, our conclusion about the methods of villas as productive units in Pannonia are related to what is known of the economics in general of the Roman Empire.

A pivotal ingredient of the Roman economy was the *latifundium*, the large estate which used slave labor. The conditions of the laborers and their relations

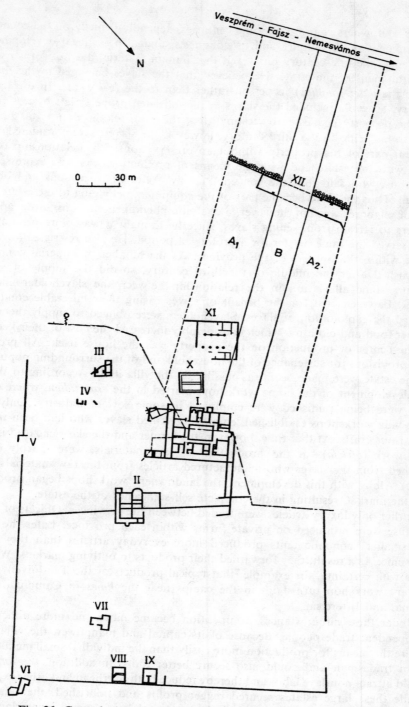

Fig. 26. Ground plan of Nemesvámos–Balácapuszta villa settlement

to the landowners changed over the centuries, depending chiefly on the economic development of the empire and its conquests, and also on legal developments.

Alfenus, the first-century jurist, felt that human material did not form part of the equipment of the estate. He believed that the slaves belonged to the *familia* (or to the *pater familias*) in person rather than to the real estate. In the second century, Scaevola included the slaves in the inherited property left by testimony. All the property bequested by testimony at the death of the donor was handed over to his heirs. In the third century, however, the slaves were considered as an integral part of the property without any reservations. The relationship of the slave with the estate and with the means of production was best expressed by Ulpianus, who said that the concept "family" referred to persons and objects as well. Thus the slave became part of the equipment belonging to the estate. The sale of a laborer was not, however, a common phenomenon. Constantine advised owners to refrain from selling slaves, since he thought it was more profitable if they served the same master for a prolonged period. He also restricted any sales made within the borders of the province. As the number of imperial conquests diminished after the middle of the third century, so did the supply of slaves, causing some alleviation in the relationship between the slaveholder and the slaves. Laws enacted for the benefit of slaves easing their miserable conditions limited the exploitation of slaves. Slaveholders were obliged to supply them with proper food and clothing, which considerably increased the cost of their keeping.

The center of production on the property was the house itself. All products and provisions for sustenance of the household and of the surrounding population of the estate were made, as far as possible, in the villa itself. According to Weber, the development of the home workshops started in the *ergastulum*, where slaves who were being punished were employed in some sort of industry. Only later were villa workshops established, employing skilled slaves who had been trained in various crafts. At that time, however, the farmer and the slave engaged in outside work (*officia*), or the production of raw materials, were sharply distinguished from the slaves who manufactured articles from the raw materials (*artificia*). Along with this development, the landowners' workshops began producing for the market, resulting in the complete self-sufficiency of the estate.

Earlier only luxury articles were traded between the provinces. Objects of everyday use were produced on private farms within each province. Later, however, the smaller economic units produced more everyday articles than they could consume. As a result, they forwarded their products to outlying markets. We may state with certainty, for example, that typical products of the Tác-Fövénypuszta pottery workshop turned up in the camps near the *limes* in Campona, Vetus Salina, and Intercisa.

Under these circumstances, competition became more and more difficult for independent tradespeople. Because of its capital and manpower, the villa could secure the means for production more easily than the individual small manufacturer or tradesman and could also secure better tools. In addition, the *latifundia* could attract nonslave labor and thereby reduce both overhead and other expenses. While these large estates secured bigger profits and flourished, the small and medium-sized estates, based exclusively on slave labor, showed increasing signs

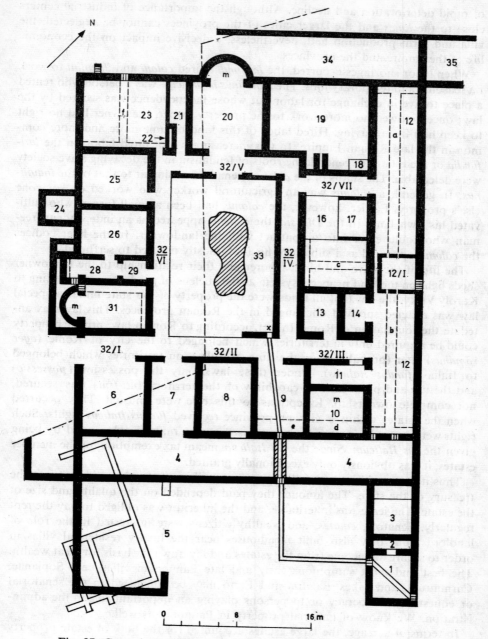

Fig. 27. Ground plan of central building, Nemesvámos–Balácapuszta

of rapid deterioration and decline. Although the importance of industrial centers close to the *limes* and the larger cities of the provinces cannot be neglected, the villa and farm production had, nevertheless, a decisive impact on the economic life of the empire and the provinces.

When labor shortages occurred, the *latifundia* hired *coloni* and *inquilini* to work on seasonal and temporary jobs. The *inquilinus in principe* was a person who rented a place to live in exchange for labor but whose independence was secured by the law; once he owed no more work to the property owner, the owner had no right to keep him from leaving. Hired labor of this kind became more and more common in the late Roman Empire. In this increasing use of hired labor on the *latifundia* or the estates in general, the roots of feudalism in the decaying slave society were detectable. The status of the *colonus* (tenant) was similar to that of the *inquilinus*. In general, a *colonus* was an agricultural worker who worked on someone else's property. Earlier, however, the *colonus* had been a small farmer who cultivated his own land. In the *Digesta*, the *colonus* appeared as an independent freeman who could enter into a contract with the landowner. In the later codices the *colonus* appeared as a subject to his lord, mostly reduced to serfhood.

The life of the slave and hired laborer and their relationship to the landowner sheds light on the land property system as the nucleus of production. According to Károly Visky, the law that all lands were the property of the state unless a special law was enacted survived unchanged in the Roman provinces. This law may antedate the foundation of Rome. In fact, according to Roman law, private property could be acquired only in territories which belonged to the city of Rome (*ager romanus*) later (probably from the time of Sulla), on territories which belonged to Italia (*fundus italicus*). Under these laws, only the possession (*possessio*) and the use and profit-making capability of the territory (*uti-frui*) was secured, not complete ownership. Exceptions to this rule were unusual. They occurred when the estates within a city or province received *fundus italicus* rights. Such rights were secured for cities by being raised to the rank of colony and by being given the *ius Italicum*. Since the *ius Italicum* meant tax exemption for the member estates, it was obviously only exceptionally granted.

Thus it appears that the landlords were actually tenants who paid rent into the treasury of the state. The amount they paid depended on the quality and size of the estate. The lease was inheritable, and the inheritor was obliged to pay the rent regularly. Senators, *equites*, and wealthy citizens were interested in the role of landlords. But they also built farmhouses near the luxury residential villas in order to secure an income from the estates and lay down the basis for great wealth. The best and most sumptuous early and late Pannonian villas near Sopianae, Carnuntum, and lakes Balaton and Fertő may be connected to the senatorial or equestrian aristocracy or to persons playing an important role in the administration. We know of imperial property in Pannonia as well.

In terms of acreage, the large estates accounted for the bulk of estate property in the provinces. Numerically speaking, however, smaller estates were more common. The earliest of these were probably lands granted to veterans. Besides the *colonus* and *inquilinus* settlements attached to the *latifundia*, leading citizens of the native population and well-to-do native aristocracy probably also established

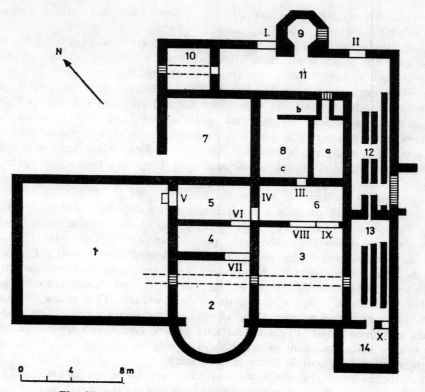

Fig. 28. Ground plan of bath, Nemesvámos–Balácapuszta

villas. In some cases, inscriptions have facilitated the task of determining who owned the villas.

Pannonia had many assets for the Roman Empire. In addition to supplying excellent soldiers, it had many natural resources. Its excellent soil and sunny hills favored grain and grape production; its forests and meadows insured a timber supply and promoted animal husbandry; its lakes and rivers rich in fish and its forests amply supplied with animals promoted fishing and hunting. Transportation within the province and to Italia was available via an excellent network of roads and riverways. The Danube provided a route from the middle Danube area to the western and southern provinces as well. Last but not least, Pannonian villas produced enough commodities to supply the province, including the camps along the *limes*, and surplus to send to Italia as well.

We may reconstruct the branches of production which were the basis of the economy from the finds in Pannonian villas. These finds do not, of course, include the large group of implements made of wood, which disappeared without trace. The great value placed on iron tools and implements was demonstrated by the fact that in many cases they were hidden as treasures. The variety of agricultural tools discovered showed that most farms did not follow a single aspect of agriculture, but were engaged in crop production, animal husbandry, fishing, hunting, and timber production.

Among the tools frequently found in the villas were iron plows, sickles, and forks, used for grain production. During Emperor Domitian's short reign, a decree forbade the planting of new vines and ordered the destruction of half of the existing ones, but evidence indicates that the decree was not always obeyed. The winepress found in the villa Winden am See and the many vine-cutting knives witness this evasion. Similarly, the wine cellars around Lake Balaton, where wine was kept in wooden barrels and ceramic containers, show that the decree probably had little effect and that bootlegging flourished in Pannonia. Grape production was again permitted — in fact, encouraged — by Emperor Probus for the provinces of Gallia, Hispania, and Pannonia, since the grapes of these provinces were best suited for first-rate wine production.

The abundance of pastures and of forests with acorns provided good conditions for animal husbandry. Unfortunately, since leather and horn quickly deteriorate, we know little about the products made from animals. The only remains found at excavations were the bones of domesticated animals. Cowbells, sheepbells, and thread-and-wool-processing equipment were other indications of this activity. The formation of the terrain close to various Pannonian villas indicates that artificial fishponds with dams once existed. This raises the possibility that some farms produced fish. In addition, fishhooks, anchors, and fishing spears have been found at villa excavations. Spears, lances, and arrowheads indicate both warlike and hunting activities. The wild animal content of the death pits is additional supportive evidence of hunting.

That timber production and woodworking were common is evidenced not only by the iron implements but also by the facts that wood and fuel was necessary everywhere and that log cabins and wooden buildings (cottages, sheds, and so forth) were attached to every villa. Since carpenters' instruments and stonemasons' tools were found mostly in larger villas, possibly only these larger villas could permanently employ carpenters and stonemasons for maintenance and repair.

Excavations in Pannonian villas have also uncovered brick-firing kilns, pottery workshops, metal foundries, blacksmiths' workshops, and weaving mills. The villas could also accomplish any work which was necessary around the farm but for which special workshops were not always provided. Thus, villa production included the work of carpenters, stonemasons, blacksmiths, wood-carvers, and even some traces of plate glass window-making have been found. In at least one instance, a wax plate with a form suitable for glass pouring was found.

As was pointed out above, the production of the villas exceeded the needs of their inhabitants, so that both agricultural and industrial products were marketed. We know that villas were engaged in a certain amount of trade; a more exact determination of the various functions of the villas in Pannonia is expected from more systematic excavations.

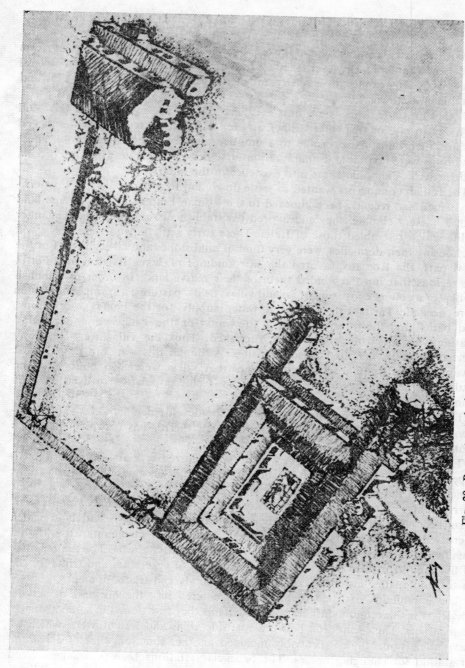

Fig. 29. Reconstructed view of Nemesvámos–Balácapuszta villa settlement

Hardly any real town existed in the territory enclosed within the roads and rivers bordering Pannonia. However, scattered villas, or colonies consisting of several villas, were often joined to a settlement of the regional inhabitants' houses and functioned as towns.[8]

Towns or settlements came into existence during ancient times as they still do today in one of two ways. Either an area suitable for settlement was picked out and then built up according to a uniform plan, or a settlement arose where conditions were favorable to immigration of settlers. The greatest number of villa settlements in Pannonia came into existence in the latter way.

The Italian conquerors wanted to settle down and to try their luck in this territory which had recently been annexed to the Roman Empire. In this foreign land with a harsher climate, they endeavored to establish surroundings and a manner of life which resembled those of Italia. These early settlers, who were very good at choosing their domiciles, were very fond of mild slopes and lake shores. For the most part, the Romans favored the surroundings of lakes Balaton and Fertő, the regions that most reminded them of their native land. In order to establish a villa (or a colony consisting of several villas) fields, pastures, a spring, and woods were needed. The spring offered a water supply for the homes, workshops, and baths, and the forests provided firewood for the extensive underground heating with which the rooms were furnished. Thus the villas were generally erected in places that had these requisites. Although this was the milieu most favored by the Roman settlers, it appears that villas were also built near the settlements and villages of the native Illyrian–Celtic population or even on their ruins.

The roads had also an important part in the development of villas. The directions in which they led determined more or less the regions to which the farms sent their products.

The villa settlements cannot be strictly systematized or classified according to types. In consequence of their particular location villas could perform various functions at the same time. Thus the villas established near the towns or in their surroundings were exclusively for relaxation, as were the generally sumptuously furnished *villae urbanae* of the urban population. Some also had land which increased the sources of income of town magistrates or merchants. In addition, smallholders' villas were located in the neighborhood of towns. The main source of livelihood for their owners was the production of vegetables and fruits for the townspeople; they also produced household goods and traded them.

The inhabitants of some roadside villas entered into the business of serving travellers with accommodations and food. Thus they functioned in some instances as *mansions*. The connection between roadside villas and watchtowers, which we have already mentioned, probably included another function, that of a *mutatio* or station for changing horses. The beneficiary stations (*beneficiarii* performed police and civil service functions) were also accommodated in villalike buildings. Several instances of villa settlements extending along both sides of a road were also found in Pannonia. These were *villae rusticae* of scattered settlement; they followed the road but were at a great distance from each other. This type was found in valleys and also near rivers.

The most characteristic location for a villa was a gently sloping hillside surrounded by valleys. These villas generally had excellent panoramic views. Building on a slope was favored not only for reasons of security, but also for ease in supplying water to the buildings. The location of the villa settlements was similar to that of detached farm settlements of today.

István Hahn has called attention to two settlement forms developed in the Eastern Roman Empire and also frequent in Pannonia. One was the village settlement built on the estate of a sole *"despot"* landlord, the other the "κομαι μεγαλαι πολλον εκαστη δεσποτον" — the village settlement including several smallholders. While literary sources bear witness to the existence of these types in the Eastern Empire, they have been ascertained also by archaeological methods in the West and thus also in Pannonia.

In our opinion, the originally native colonies in Pannonia belonged to the first type, which afforded the impoverished free population opportunities to work within the production organization of large estates. Later the population of these settlements became almost organic accessories of the large estates — living in cotters' villages — though they never came into slavery. The landowners who settled in the first and second centuries were presumably reduced to using the labor of the local poor free population because it was impossible to import a sufficient number of slaves to break the land of the large estates.

The so-called smallholders' village, the second type, was the most frequent form of villa settlement. We have mentioned this above as a grangelike form consisting of several scattered villas. These units may be groups of buildings close together or separated, depending on the geographic milieu, on arable forest land. This could probably be the type of the veterans' colonies.

In the course of the first century, settlements in the interior of Pannonia developed as *coloniae* of the original inhabitants. These soon came under the sphere of influence of the farms, the center of which was always a luxuriously constructed villa. Nevertheless, these early villas formed a rather scanty network; their stately buildings stood fairly isolated among the domiciles of the original inhabitants, the native dwellings which were built partly of stone and partly of packed soil, and were huts rather than houses. These early villas were the links which, along with the frontier fortresses, camps, and towns in the outskirts of the province, connected Pannonia to the empire and propagated the Italian spirit, way of life, and milieu.

In any overall account of settlements in Pannonia, veteran settlements must take an important role. It is mainly due to Mócsy's work that we know that the first- and second-century veteran settlements gave rise to towns in the southern and western parts of the province. The veteran settlements in the Balaton region were scattered and thus did not stimulate the foundation of towns, but they did determine the dominant type of settlement — the independent villa colony.

During the second and third centuries, the number of villas in the province and their significance increased. Fortified villas with projecting corner towers were built during the second half of the third century. Villa culture and villa society

285

Fig. 30. Mosaic from Nemesvámos–Balácapuszta

Fig. 31. Mosaic from Nemesvámos–Balácapuszta

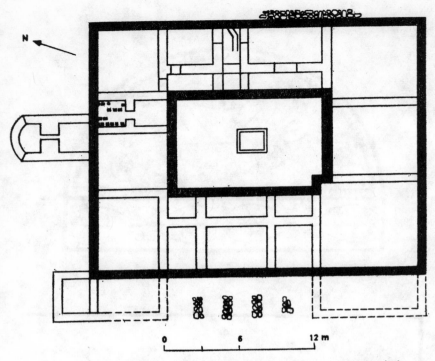

Fig. 32. Ground plan of Gyulafirátót–Pogánytelek villa, first period

flourished in Pannonia under Diocletian and Constantine, and many existing buildings were renovated in this period. Radnóti has listed the place names from the interior of the province that have been formed of Italian or Romanized native personal names: they end in *-ana, -iana,* or *-anis, -ianis.* Philologists suggest that the word *villa* or *mansio* was put before these place names; thus the villas or mansions would have been named after their owners. Unfortunately, the number of villas with their ancient names preserved is not very high.

## TYPES OF VILLA PLANS

The villa, as classically conceived by Roman authors, was not an architectural notion but an economic unit of production.[9] According to Cato, a major villa farm consisted of three parts: the *villa rustica,* the *villa urbana,* and the *villa fructuaria.* These categories reflect the purposes of the various buildings of the estate.

Cato designated the dwelling tract of the estate as the *villa urbana.* The origin of the *villa urbana,* i.e., the dwelling house of the farm, was, presumably, connected in some way with the urban villa. Vitruvius called the luxury villa "pseudo-urbana", an imitation of the city house, and considered it a city palace transplanted for country use. It was also related to the city palace architecturally, with

Fig. 33. Reconstructed view of Gyulafirátót–Pogánytelek villa, first period

its emphasis on privacy. It is well known that some villas had served economic purposes before having become luxury edifices. Varro mentioned this with some grief and ridiculed people who built villas to which neither land nor livestock belonged. In his opinion, such edifices should not have pretended to be villas. Villas of this kind were from the first luxurious detached buildings. In Italia, such luxurious edifices had been built by the members of the senatorial order or aristocracy.

In Pannonia, as in Germania and all the western provinces, a rustic part was added almost without exception, to the luxurious villas. Actually, there was little basic difference between urban and country dwelling houses, especially when they were buildings of a simpler layout. If any difference existed it was that several architectural units formed an *insula* in the towns and in the more densely populated localities in order to economize on space, while the dwelling houses of the villas stood quite detached. The outbuildings that belonged to the villas were built at a distance rather than being attached to the main building. May we remark that when discussing the villas of Pannonia we do not, and should not, use the classical nomenclature customary for villa types because, as it appears from our review of them, the designation *villa rustica* almost uniformly fits each of them.

The development of Pannonian villas was independent of influence from the architecture used by the original settlers of this province. From the excavations

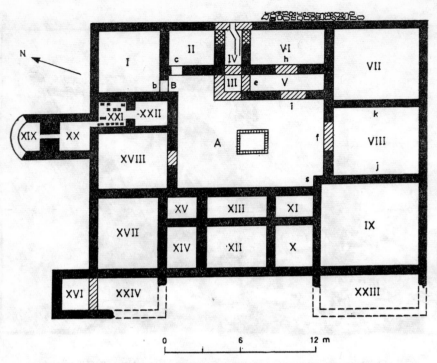

Fig. 34. Ground plan of Gyulafirátót–Pogánytelek villa, second period

of the last few decades, we are sufficiently acquainted with the relics of Illyric–Celtic–Eraviscan architectural activity to ascertain that the original inhabitants lived in small, *vicus*-like rural settlements, mostly in houses sunk half into the earth. Their architecture was most primitive — they hardly even knew the rectangle. Thus the Roman conquerors brought an entirely new culture and new tendencies to the architecture of the province. The dwellings of the native population could not have served as models for the villas erected as early as the first century, which had many buildings and a stately appearance and were the products of the early Italian settlers. The dwelling parts of these villas had been erected in compliance with uniform ideas and plans. Earlier buildings which had either decayed or had been deliberately pulled down in order to give room to build above them have often been found in the deeper layers of villa sites.

The influence of luxury villas in Italia on the larger and most luxurious buildings of Germania is evident. Similarly, one may observe that the civic buildings of urban character or other extensive settlements in Pannonia were often of a definitely villa-like character. The colonies along the *limes* which had developed mostly near the camps, often in places which had formerly been settlements of the original inhabitants (as, e.g., Albertfalva), showed an interesting mixture of primitive pre-Roman and the later Roman architecture.

Fig. 35. Reconstructed view of Gyulafirátót–Pogánytelek villa, second period

In the interior of the province where no big cities or fortresses had been built to the taste of the Roman conquerors, the poor, free population had kept to its traditional way of building in the ensuing centuries, while near the walls, on the estates of Roman settlers, most dwellings and outbuildings were of stone laid with a binding material. We are of the opinion that the simple *villae rusticae* of an unpretentious ground plan were built by the less wealthy of the new colonists and by soldiers who had served their terms, rather than by the indigenous population. Romanization — the advance of Roman civilization and culture, the acclimatization and adoption of the conquerors' modes of population — then eventually enabled the free-born original inhabitants to adopt and copy the way of life of the settlers.

The following presentation of villa types in Pannonia does not include buildings which had villa ground plans but were situated inside of towns or urban units of closed formation, because this would considerably extend an already vast material. We have also omitted the villa-like palaces, the luxury dwellings in the towns of Carnuntum, Sirmium, and Aquincum, and the legate's palace at Aquincum because, considering the purpose for which they were built, they cannot be grouped with the villas.

Systematizing Pannonian villas according to the known types is by no means an easy task. Archaeological reconstructions have in most cases not retained original character of the buildings, and only by means of a careful analysis of the ground plans can we determine the appearance of the original house.

The Italia atrium house was seldom erected; because this house faced inwards, it was only used in towns and in dense settlements. Yet it seems to us that this type is hidden beneath the ground plans of a few subsequently rebuilt villas.

291

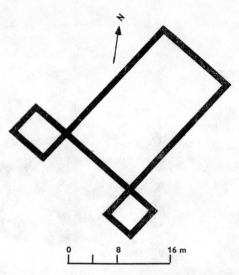

Fig. 36. Ground plan of building with corner projection from
Szentkirályszabadja–Romkút villa settlement

The type of building used most extensively in Pannonia was also the type used
most commonly in luxury villas in Italia. This type, the so-called peristyle villa
(Fig. 55), appeared in the province around the turn of the second century. It began
in the peristyled parts of refined Roman city residences which, in turn, were based
on Hellenistic models. Of the villa types, the atrium form was early abandoned;
the peristyle and portico forms (Fig. 56), which fit in better with the natural environment, were continued. Thus the type of villa with the dominating central
peristyle surrounded by a series of rooms was developed. Kropatschek believes
that the peristyle villas originated from the simple, quadrangular peasant cottages
with a courtyard in the middle. We, on the other hand, are inclined to believe
that these Pannonian peristyled villas were a more open form of the Italian private
houses. The relationship with the classic Italic house was especially conspicuous
in the arrangement of villas where the rooms opposite the vestibulum, their length
often increased by apses, were almost equivalent to the *tablinum*.

The peristyle villas in Pannonia were often combined with portico types. Yet
porticoes may also be found in the entrances or villas of other types. Several
variants of porticoed villas were found in the province. In addition to the peristyle–
portico kind, there were also side-portico buildings. In these, the rooms for dwelling were situated in a row along one longitudinal side; the entrance was built
either in the center or at one end of the line of rooms. In this type, the corridors
were bordered in most instances by colonnades or by a series of pillars and thus
transformed into porticoes. In general, the portico and peristyle villas bore the
richest inside decorations. These stately villas of Italic character were presumably
built by persons who intended to create an Italic milieu in the remoteness of
Pannonia and who wanted to feel at home far from their native land. These builders

292

Fig. 37. Reconstruction of building with corner projection from Szentkirályszabadja–Romkút villa settlement

might, in the majority of the cases, have been landowners of the equestrian class, military commanders, or rich merchants, the last mostly of Eastern origin.

The exact dates of these villa types in Pannonia cannot easily be determined. Yet on the strength of the known evidence, one may state that peristyle villas were the most popular as early as the end of the first century. They were numerous and in fashion during the second, third, and even fourth centuries. Around the turn of the fourth century, these villas were fortified by towers at the corners (Figs. 57 and 58).

In examining the influences that shaped the villa architecture of Pannonia, we must not disregard the effects of types developed in provinces conquered earlier. A new fashion spread as a consequence of third-century settlement in Germania. Projections at the corners started from the *Hallenhaus*, developed in the western provinces, and later spread all over the empire, causing variations in the ground plans of Pannonian villas. The earliest specimens of corner-projection villas in Pannonia were built during the last third of the third century. Nearly every one

293

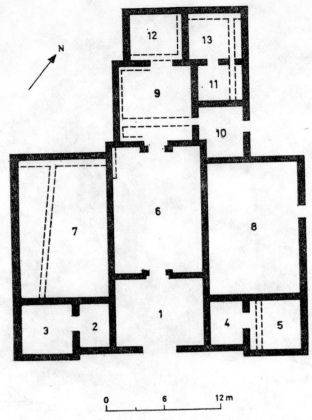

Fig. 38. Ground plan of Early Christian basilica, Kékkút

of these had its origin in an earlier, partly destroyed and reconstructed building. Reconstruction was probably effected in the interest of defense. The tower-like shape of the corner projections appeared for the first time in the Hellenistic *stoas* (or halls). Explorations in Italia thus far have revealed very few buildings with corner projections. Yet the style did exist there, even if in small numbers as compared with the frequently built atrium and peristyle villas. It was used by the Tuscans around the beginning of the second century according to a description by Pliny. We know from this description ("A capite porticus triclinium excurrit") that a dining room was located in the projection extending to the end of the portico. It had windows on two sides, both of which yielded a fine view. Consequently, this corner tower here did not serve fortification purposes.

The villas in Africa conformed to local geographic and economic conditions and thus differed from the ones in Italia. Their construction was similar to that of villas in other provinces. These were dwelling houses for the owner and his family rather than luxury buildings, and they often served as fortresses for the entire population of the estate against hostile invasions. In a mosaic found in

Fig. 39. Reconstructed view of Early Christian basilica, Kékkút

Carthage and described by Rostovtzeff, the combination of a dwelling house with a fortress was seen. The tower at the corner had a solid wall without openings on the ground floor while the next floor was encircled by a graceful loggia.

In his comments on the villas with corner projections built along the right shore of the Rhine, Grenier attached no special importance to the projections. The fact that these villas were found for the most part on the right shore of the Rhine clearly proves their primarily defensive character. These Rhineland villas were of a much earlier date than those in Pannonia. This can be explained by the fact that the Rhineland was conquered earlier and also needed to be defended sooner.

A type that occurred frequently in Pannonia was the portico villa with corner projections. In it, closed quadrangular corner rooms were joined to each of the two ends of the portico on the front side and projected from the building. The projection touched the building in some instances only at one of its corners; in others it was connected with the block of the house along one side. The role of the corner projections and towers was not merely aesthetic; in most cases they were erected for defensive purposes. Frequently, the buildings with corner

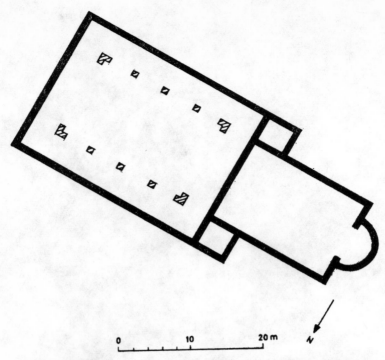

Fig. 40. Ground plan of basilica No. 1, Keszthely-Fenékpuszta

projections went through several reconstruction periods before they obtained their final forms. In general outline one may discern two types among them: the simpler smallholders' type and the more distinguished, profusely decorated building.

The origin of the corner-projection villa type is, because of the lack of a sufficient number of examples to study, not entirely clear. Contrary to earlier research which attempted to place the origin of this type in Italia, we now believe that these villas were characteristically provincial or, more correctly, were inspired by the necessities of the provinces. Italia was seldom subject to hostile invasion, while the provinces defending the frontiers of the empire naturally were. Thus when building their domiciles, provincials had to secure them to a high degree, especially from the third century on. One of the visible marks of this concern for security was the fortification of towns and settlements by walls.

In Pannonia, the villas in the estates allotted to former soldiers showed no uniformity in their ground plans. Thus we should not venture the opinion that, as in the Rhineland, a "standard house" for veterans is to be recognized.

Of the smaller dwelling houses in the villas with simple ground plans, even the most unpretentious had at least one room that could be heated, and in some instances the bath could also. The smaller villas consisted mainly of simple dwelling houses and some outbuildings of stone. Next to nothing has survived of the log outbuildings, since attention was seldom paid to them at excavations. This

296

Fig. 41. Reconstructed view of basilica No. 1, Keszthely-Fenékpuszta

was the category of villas that represented the smallholders' type. Such unpretentious buildings formed the veterans' roadside villas, one group of which was situated along the roads near the watchtowers in the Carnuntum section of the *limes*.

Another type of villa consisted of several buildings surrounded by walls. In addition to a spacious residence, this type of villa nearly always included a bath. The bath was often inside the main building itself but often it was planned apart as a detached bath building. The baths had various arrangements: in some instances, they were modest one- or two-room buildings, serving exclusively practical purposes; in others they were reduced copies of the public baths and had all the rooms of such baths as that at Baláca (Figs. 28 and 29). Even the simplest villa bath, Balatongyörök, contained the indispensable equipment of Roman baths — cold and warm basins and the heating chamber, along with comfort and elegance. Others had additions which depended on the pretensions and wealth of the owner.

Besides the manor house and the bath, several buildings suitable for living in were usually found inside the walls. We know of their existence primarily from the

297

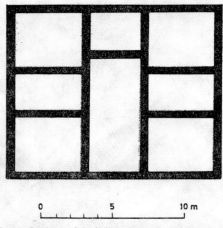

Fig. 42. Ground plan of building No. 8, Keszthely-Fenékpuszta

heating installations that have been found. Perhaps these buildings housed the domestic staff and the household slaves; sometimes they also accommodated the workshops belonging to the villas.

The crops grown in the villa estates were stored in stone granaries, drying towers, barns, or storehouses with air circulation through pipes within the walls. The most frequent type of dryer was the tower system found in the Balaton region (Figs. 32–35), Pogánytelek and Szentkirályszabadja (Figs. 36 and 37). But in villas of the Fertő area and the Leitha hills (Parndorf, Königshof), buildings aired by the western type of elevated floors were used. Also the stables and sheds were built of stone in villas that were surrounded by walls. There were threshing grounds at Pogánytelek and presses at Winden am See, but these were not often found in Pannonian villas. Although workshops have often been found at the excavations, they did not form a building type and have been identified only by their sets of tools.

Among the buildings found within the villas, the shrines deserve special mention. Some of these were simply household shrines, erected in a suitable place in the house. We know of their existence through the finds of gods' statuettes, shrine-lamps (Pl. XXV.1–3) and carafes at Baláca, Nagydém (Pls. XXVI–XXVIII), Tamási (Pls. XXIX–XXXV) and other places. There were also shrines which consisted of two or three rooms inside the main building where bases for gods' statues and altars have been found (Eisenstadt-Gölbesäckern, Tác). Finally, there were shrines situated outside the main building (Ajka, Balatonföldvár, Sopron-kőhida-Kecskehegy; Pls. XXXVII and XXXVIII), which in most cases were trellis-like structures under roofs supported by pillars, or in niches formed in the walls delimiting the estates. From the beginning of the fourth century, pagan places of worship were succeeded by Christian sanctuaries in many villa settlements. Their dominant architectural type was the so-called hall temple without apse (Kékkút: Basilica No. II, Figs. 38 and 39; Szentkirályszabadja, Figs. 36 and 37; Donnerskirchen; Sümeg, Figs. 53 and 54, etc.) which also had in some instances

Fig. 43. Reconstructed view of building No. 8, Keszthely-Fenékpuszta

a *pronaos* and a narthex. Broader spans were divided into three naves by two colonnades. The columns were generally made of wood and placed in front of pilasters.

From these various Early Christian basilicas developed the basilica consisting of several rooms for cultic and secular purposes. Building No. I of the villa settlement at Kékkút (Figs. 38 and 39) is one example. Basilica No. I of Keszthely-Fenékpuszta (Figs. 40 and 41) was converted from a villa with towers at the corners. The three-apsed room of the villa at Tác (Fig. 46) was, perhaps, an Early Christian cultic place reconstructed from the dwelling house.

The previous description of the character of the various settlements needs to be supplemented by an outline of the types of villa settlement in Pannonia, excluding villas within urban settlements.

Buildings erected near the towns, which were generally luxury villas, have been discovered for the most part in the vicinity of Aquincum, Carnuntum, Savaria, Poetovio, Emona, and Sopianae. They were often owned by merchants and persons performing important public functions. They were not exclusive nor in every instance based on land property. We also know of imperial villas and estates in the Sid and Sirmium regions of Pannonia.

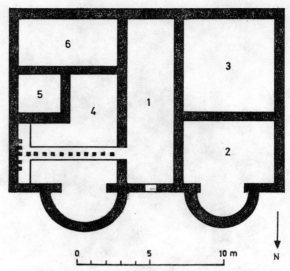

Fig. 44. Ground plan of building No. 7, Keszthely-Fenékpuszta

The villas that formed the centers of the vast *latifundia* were groups of buildings, fenced in or not, such as those at Parndorf, Baláca, Szentkirályszabadja, Pogánytelek, Hosszúhetény (Figs. 48 and 49), Šmarje, and Eisenstadt. The villas belonging to more moderate-sized estates had all the same characteristics but were on a smaller scale and had less luxury (e.g., Donnerskirchen and Deutschkreuz). A rather dense settlement is beginning to emerge from recent excavations at Tác. Its house groups consisted of numerous *insulae* built near each other. The villa settlements in the Balaton and Fertő regions consisted, on the other hand, of estates lying farther apart, yet not at too great distance from one another.

The province also had villa settlements extending along the roads — smallholders' villas and wayside veterans' villas highly resembling these.

The distribution of the villa and settlement types within the province was the result of necessity and practicality; no dominant type has been found in any of the regions. Although dominant architectural styles have been found in Gallia and Germania, no special form of construction was favored in Pannonia. It appears that the architectural styles of villas that developed in Italia and in the surrounding provinces met in this territory.

## TECHNOLOGY AND STRUCTURES OF BUILDINGS

African mosaics and Campanian architectural landscapes hint at the early forms of villas. Our only clues to the later (second- to fourth-century) villas, however, are the excavated foundations from which the ground plans may be reconstructed.[10]

It is characteristic of the Roman private house that exterior decoration was not particularly emphasized. Probably no opening besides the door appeared on the

300

Fig. 45. Reconstructed view of building No. 7, Keszthely-Fenékpuszta

facade. Although the Romans were acquainted with plate glass, their windows were primarily narrow slits integrated into the geometric shape of the building without encouraging the interplay of voids and solids. They served only for the necessary functional but rather subdued illumination of the rooms. On the other hand, the Romans put all their efforts into interior decoration, making the house as pleasant as possible for its inhabitants. Of the upper structures and facades of the Pannonian villas very little survived. In only a few cases have the excavations indicated that, besides the towers, the whole building or a section of it carried an upper story. Buildings used for agricultural purposes had towerlike projections or upper stories more often.

Most buildings were covered with pink mortar, the composition of which was more or less identical with that of terrazzo. Careful plastering or painting has been observed only on building "A" of the villa at Parndorf. It had a variegated, dotted marble pattern imitating a socle at the bottom of the building, above which were square blocks and quadratures, painted with red lines. This decoration gave the impression that the building was faced with stones.

Cornices, gateways, and other architectural decorations made of marble and stone have been found in excavations at Hosszúhetény (Figs. 48 and 49), Eisenstadt, Örvényes, and other places, but their number has been very modest compared to the number of buildings excavated. The stone carvers used porous limestone travertine in the Danube area, Margitbánya stone near Lake Fertő, and red sandstone near Lake Balaton. Corinthian marble was also used on the western borders of the province, particularly in the Mura area and around Poetovio and Emona.

A group of specific, typically provincial, architectural decorative elements was found near Lake Balaton. They were characteristically carved of red sandstone from the quarries of Balatonalmádi and Vörösberény. This reddish stone was easily workable on large uninterrupted surface planes, but was less useful when applied to smaller details in decoration or architectural carvings in the round, because of its soft and porous texture. Columns (Pl. XXXIX) made of this stone were crowned with a massive block capital composed of four or more leaves, upon which rested the architrave. Two types of columns were distinguishable: one type had a ring or collar form of astragal, while the other had a leaflike one. Their measurements were closely related to other characteristics. Columns 80 to 90 centimeters high were placed on a 1-meter-high plinth and used in peristyles or porticoes. Columns of 180 to 190 centimeters were erected without pedestals and carried the weight of the small half-roofs of the porticoes or peristyles. This taller type was, however, seldom used. In the red sandstone variety occurred smaller, delicate 50- to 60-centimeter columns, used for mullions dividing windows or supporting railings or copings or as balusters. These stone columns were sometimes employed for decorative purposes, as at Kékkút and Sümeg, where square pillars were used to support heavier beams. It seems certain that the columns were cut in the quarries near Balatonalmádi and Vörösberény. Since none of them were found in situ in the course of systematic excavations, the variety of forms has not facilitated the determination of their chronological sequence. Nor has an acquaintance with two or three types of columns been sufficient to determine the existence of workshops. Since the structure of the earlier villas did not include columns, one may conclude that square pillars were used to support the roof system. Red sandstone pillars must have been applied in the Balaton region for buildings of later dates only. Terra-cotta antefixes which adorned the villa at Ravazd must be counted among other exterior decorations.

On the ground of excavated villas, attempts at reconstruction took into consideration all the possibilities. In these reconstructions the span bridged by the tie beams may look somewhat exaggerated, but one must remember that the rich forests of Pannonia offered abundant timber for large beams. The material used for building the Pannonian villas did not differ from the materials used in other provinces or even in Italia. Local stones, roughly hewn as they were extracted from the quarries or shaped to similar sizes, were preferred for the walls. From the grooves of the doorsills (made of one or two pieces) frequently found in situ in the villas, the doorwings and lock system may still be reconstructed. Threshold stones were usually carved from stones available around the villas.

Among the known wall construction techniques, the *opus incertum* of irregular natural stones was most frequent. The *opus spicatum*, which used these same stones, was also found in many buildings, especially at places where a later filling of an earlier window opening occurred or where a wall of special strength was needed.

Fig. 46. Reconstructed view of *palatium*, Tác-Fövénypuszta

Bricks were used only mixed with roughly hewn stones and mostly for quoins, strengthening the corners of the buildings (as at Szentkirályszabadja and Romkút). The buildings at Budapest III, Csúcshegy, Dorog-Hosszúrétek consisted of alternating brick and stone courses. None of the villas were constructed entirely of bricks, but purely brick construction did appear where vaulting or canal coverings were needed or where walling material of regular dimensions was needed.

The roofing material used for the villas was exclusively tile, with *tegulae* and *imbrices* covering the rafters. Wood shingles have not been found anywhere, but we may be sure that they were used on auxiliary buildings. A 4- to 5-centimeter-thick lead sheeting and a gutter made of the same material were found at the Parndorf villa. At the Winden am See villa, the measurement of the projecting section of the eave could be determined.

Data pertaining to the ceilings also emerged at some excavations. Pieces of collapsed vaults prove the existence of vaulting (Pogánytelek II, Parndorf XIV). The most frequently used method of vaulting was by reeds which were tied on a lattice, which in turn was fixed to the rafters. The inside face of this ceiling was smeared and smoothened with mortar made of sand and lime. Such mortared ceiling remains with reed imprints have been observed at Tác, Hosszúhetény, Örvényes, and elsewhere.

Walls made of wooden panels divided the interior of some buildings, usually the sections of villas used for agricultural purposes.

The floors in provincial Roman buildings were often covered with terrazzo and there are traces of this in almost every villa. Even when the excavations unearthed

303

pounded clay floors, the floors were originally made of ship-type floor planks and had been destroyed by fire or other devastation; their imprint and texture remained in the clay. On the other hand, kitchens and auxiliary rooms of lesser importance had only pounded earth floors.

Floors covered with bricks also frequently occurred in villas. Brick floors show a truly great variety. Sometimes the bricks were identical with those used for walls and sometimes square bricks of smaller size were set in. In addition, sometimes special sizes and forms of brick covered the floors in a mosaiclike pattern. Most frequent were the hexagonal bricks which were manufactured in 3- or 4- to 15-centimeter sizes. Octagonal bricks also appeared in a variety of sizes. We also know of rhombic brick floors.

Mosaic flooring (Figs. 30 and 31; Pl. XLVII) occurred only in the most luxurious villas in Pannonia. We know of its existence from a few mosaic pebbles found in situ. According to the evidence of the pebbles found at the villa site of Nemesvámos–Balácapuszta, we surmise that the material used for mosaics was extracted from stones found nearby and cut on the spot; only those colors which did not occur locally were supplemented by *tesserae* and glass paste brought from abroad. We do not possess any data on the mosaic floor of the important Parndorf villa, although a few terrazzo-patched mosaics are known. Damaged and destroyed mosaic floors were frequently left unrepaired and covered with terrazzo.

Plaster, which served exclusively as wall covering, frequently remained in situ in various rooms. But traces of decorative wall paintings have also been found in many Pannonian villas. The artistic quality of these paintings ranges from primitive to sophisticated excellence. Even the simplest *villa rustica* usually had one or two rooms with painted walls.

Because Pannonia had a colder climate than Italia, its villas more frequently had heating systems than their Italian counterparts, where only the baths were heated. Although the heating systems in Pannonia varied greatly, they were all alike in one respect: the heating source was situated under the floor (Pl. XL). The hypocaust system which was supported with columns, the canalization system, or a combination of the two, provided heat through flues in the walls, as well as radiating heat.

Most hypocaust systems were built according to similar principles. Underneath the floors, the earth's surface was covered with terrazzo, and square bricks were placed equidistant from each other, constituting the base of the columns. The columns were made of different materials. Where they consisted of good fire-resistant stone, they were carved in one piece. At other places, smaller square bricks, placed one above the other like a pillar, served for the vertical support of the floor. Clay pipes with circular cross sections were used on rare occasions as supporting pillars. On the tops of the columns were placed square bricks or cut-stone squares, the edges of which were flush with each other to form a continuous platform. Above this platform was placed the mosaic floor covering in various layers.

The pipes (*tubi*) were made in various forms: oblong cross-sectioned, square with rounded edges, or flat. The openings cut into these pipes from which heat emanated were round, rhombic, or sometimes square. Instead of the *tubi*, the villa

Fig. 47. Reconstructed view from vestibule to peristyle, Tác-Fövénypuszta

at Ravazd employed large *tegulae* with four short legs, or projecting protuberances, which leaned against the wall. Several of these placed near each other gave the impression that the wall was covered with bricks. The *praefurnium*, or furnace area which fed heat through the system, was situated outside the building. The open *praefurnium*, which was in an area specially provided for this purpose, was covered with little temporary roofs, supported by beams, whose purpose was to keep the logs dry and also to serve as shelter for the servants. Heating was done exclusively with logs or with charcoal. At Csúcshegy and Eisenstadt-Gölbesäkern, the *praefurnium* was situated in a larger room, where utensils belonging to a kitchen were also found.

So far it has not been possible to ascertain the existence of kitchen ovens and cooking hearths in any of the villas. However, many patched fireplaces dating from the migration period were set up in the partially ruined villas. Pottery, brick, corn-drying, and other kilns have also been found in the villas.

Square, semicircular, and even circular bathtubs have been excavated from various villas, but toilets and other facilities serving hygienic purposes have not

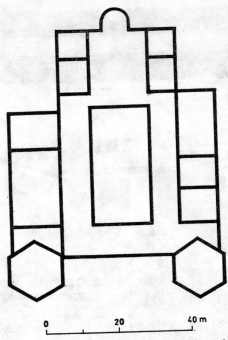

Fig. 48. Ground plan of villa at Hosszúhetény with multicornered towers

been found so far. In view of our knowledge of such installations in Italia and at some places in Pannonia, to spot them would be relatively simple. The existence of such facilities in temporary wooden shacks and sheds outside the villas, as well as the existence of latrines outside or inside the villas, is conjectural but very likely.

The water canals and sewers were mostly walled. Their bottoms were lined with *tegulae* and they were roofed over with large stones (as at Pomáz). Rainwater flowing from the roofs was led away in gutters consisting of *imbrices* (as at Szászhalombatta-Dunafüred, etc.). The type of vaulted canalization found in the cities did not occur in the villas. Because of the lesser amount of water handled, it was not needed.

The roads connecting the villas were paved with flat stone blocks. The pavement and walls outside the villas consisted of similar local flat stone blocks, or smaller stone rubble with bricks or brick fragments.

## INTERIOR DECORATION

Traces of at least one of the decorative arts may be found in most of the villas in Pannonia. Some villas were truly luxurious yet they must be considered as exceptions. Luxury was represented in most villas by one or two rooms with painted walls, or by one with tessellated pavement, or by some parget ledges. Of the

306

Fig. 49. Reconstructed view of villa at Hosszúhetény with multicornered towers

branches of the decorative arts, wall painting, stuccoes, and mosaics were represented in the villas. The covering of the plinths with marble slabs fell within the scope of technical science rather than of art. Examples of this ornamental method have been found mostly at the southern frontier of the province, in the Drava, Sava, and Mura regions.

## Wall Paintings in Fresco and Secco Techniques

Painted fragments have been discovered in nearly every villa explored in Pannonia.[11] Of moderate or rich, mediocre or first-class workmanship, they indicate that wall paintings were once present. Often such fragments were turned up by the plow or were found during viticultural work and marked the presence of a villa at the site. In the villas that have been explored, the design of wall paintings near the floor has been, as a result of careful work, preserved in situ on the walls. In certain cases, the stones of decayed or collapsed buildings buried their painted surfaces. When this occurred among favorable, relatively dry conditions, the fragments still showed traces of the former design. More frequently, however, the paintings, which were generally made by means of the secco technique, were dissolved by water and not even traces of the ornamentation were left. This, then, is the reason that, although we know the walls of several villas were painted, we do not have fuller particulars.

307

Our understanding of wall paintings in general, but especially with those in the villa at Nemesvámos–Baláca has partly changed, and there are only assumed datings of practically all Pannonian wall paintings. Early West Pannonian wall paintings were, at the turn of the first and second centuries, almost identical to those in Italia. During the reigns of Vespasian and Domitian, wall paintings in Emona, Poetovio, Savaria, and Scarbantia, and the earliest examples in the villa at Nemesvámos–Baláca, copied even the style and the shape of Italian ornaments.

The wall paintings at Baláca were made to satisfy the demands and tastes of settlers from Italia; they were most certainly executed by an Italian master. The date of this group was also determined by Lajos Nagy as belonging to the first century or to the first two decades of the second century. We do not know of any similar paintings in any other villas of the province, although the earliest paintings in the Parndorf villa, which are now known only by description, may be the same age as those at Baláca. The first group of Baláca wall paintings has been classed, in general, within the range of the fourth Pompeii style. Because these pictures have many of the characteristics of Pompeii styles two and three, we prefer to adopt the classification as "falscher zweiter" or "falscher dritter Stil" introduced by K. Schefold (Pls. XLI–XLIII).

Our opinion about the dating of the so-called second group of Baláca wall paintings differs from that of other scholars. Lajos Nagy has concluded, on the basis of unrestricted arrangement, the style of an old man's beard, etc., that the wall paintings with a white background were mostly subsequent to the age of Septimus Severus. Nagy laid special stress on the relationship of these pictures with the catacombs and, in general, with Early Christian wall paintings. He stated that those which date from the second century preserved the vivid phraseology of Early Christian art and were free of schematic representations. In the last analysis, he dated the group to the beginning of the third century. On the other hand, using the detailed descriptions of the pictures, he fixed the dates of the white-background, secco-technique wall paintings around the middle of the second century at the latest. They may even be late echoes of the white-background wall pictures of the Vespasian era (Pls. XLIV and XLV).

The few continuous fragments of wall paintings that we have from Szőny, Csák-berény, Hosszúhetény, and other villas, indicate that the unfinished, relatively rough way of painting surfaces was general during the third and fourth centuries all over the Empire. This method had an impressionistic effect, with plinth and panel paintings imitating marble or wallpaper.

### Stucco Decorations

In the interior decoration of villas in Pannonia, stuccoes often figured as orna-mental ledges which served to divide painted panels or frame them.[12] In only one Pannonian villa, in room No. VI of Budapest III (Csúcshegy) did stuccoes have an independent decorative function, filling the entire field of the wall. In some ins-tances, stucco alettes crowned by a console were applied as architectural dividing elements. Such finds have turned up at Baláca. Even there, the stucco alettes were secondary to the wall paintings and reflected the style of the paintings.

Fig. 50. Pattern of mosaic floor from room No. 8 of Nemesvámos–Balácapuszta villa

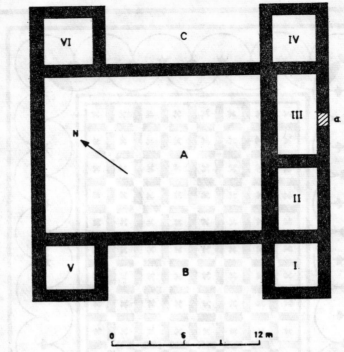

Fig. 51. Ground plan of four-towered villa from Gyulafirátót-Pogánytelek

Some of the stuccoed cornices found in the villas may be regarded as belonging to the artistic trend called West Pannonian by L. Nagy. One example is furnished by the stucco ledges of Nemesvámos–Baláca which belonged to the first or second century (Pl. XLVI). It had classic cornices in the yellow and violet rooms and fine-profiled ledges in the red and black chambers. L. Nagy argued that the inferior Baláca frescoes were the westernmost instances of the East Pannonian stucco trend. But from a minute examination of the interior decoration, we have concluded that the stuccoes at Baláca were in their artistic conception entirely independent of the styles of the Danube region and had been made, at least up to the Marcomannic wars, exclusively under the artistic influence and spirit radiating from Italia.

Among the stucco ornaments in a villa excavated in Brigetio on the *limes,* the ovolo motif was recurrent. Colored leaflets and petals were also found, although they have not appeared elsewhere in the province. The red, green, and white petals stood in twos and gave the impression of having been molded by hand. Their origin may have been due to a taste and demand for oriental elements that existed in Brigetio.

The broad frescoes of Intercisa reflected a truly oriental disposition for decoration. They were the characteristic representatives of stucco workmanship of the Danube region in eastern Pannonia from the third and fourth centuries. The fragments of a cornice in the Csákberény villa belonged to the rather conservative East Pannonian trend which preserved traditional decorating elements to a greater extent.

310

Fig. 52. View of four-towered villa from Gyulafirátót-Pogánytelek

Although stucco ledges had an organic connection with wall paintings, they outlived paintings of a later origin in several instances. Their styles were more constant and they were more lasting, as a result of their forms and nature, than the wall paintings, which varied in material and technique. This characteristic relates stuccoed cornices to mosaic where constancy was also secured by lasting material.

## Mosaics

Mosaic pieces and an occasional part of a mosaic pavement have been found in the remains of several Pannonian villas.[13] Since putting together mosaic floors was rather expensive, it is only natural that the most beautiful mosaics were found in the two most sumptuous villas, Parndorf and Baláca.

Tessellation was the most conservative of the interior decorative arts. The material used was durable and the square shape of the pieces unchanging. The pieces were generally 12 to 15 millimeters on a side, except for the tiny mosaic grains of the medallions in the centers. This constancy makes the dating of mosaic pavements a difficult task. It is not accidental that, while more and more refined dating methods are continually published in comparative works, summaries on tessellation in Pannonia include the designs and the pictures of a region only in quite certain and positive cases, which are most infrequent. As for dating mosaics according to style, analysis is absolutely impossible. More designs prevailed in this domain than in any other artistic genre. The patternbooks from which designs were chosen were collections of ornaments of the most diverse origins; out of them, the customer would choose the pavement decoration that suited his taste.

311

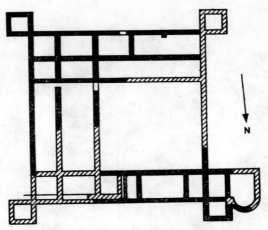

Fig. 53. Ground plan of four-towered villa from Sümeg

Unless the remainder of the building yields proof regarding its age, all statements concerning the mosaics are but baseless hypotheses.

For this reason, we cannot accept Ákos Kiss's dating of the mosaics of Nemes-vámos–Baláca (Figs. 30 and 31; Pl. XLVII). In his opinion, they date from the third century; yet the history of the building indicates that they could only have been made during the second century.

The Parndorf villa has disclosed a wealth of mosaic pavements unlike any other in Pannonia. Not having a full knowledge of these fine pavements for want of publications on them, we are not in a position to add further comments. The mosaics were probably set for the most part during the fourth century.

Relying exclusively on the material yielded by the villas, we cannot draw conclusions of general validity concerning the interior decoration of other buildings in the province. Yet we can state that the highest-quality, most brilliant representatives of all three decorating genres came from the villas.

The wall paintings and mosaics of the villa in Nemesvámos–Baláca, the stucco decoration of the Csúcshegy villa (Budapest III), and the tessellation of the Parndorf villa represent the pinnacle of these decorating methods within the province.

## FUNCTIONS OF LATE ROMAN VILLAS

### Fortification and Self-Sufficiency

The villas formed an integral part in the economic life of Pannonia during the late Roman period, i.e., at the end of the third and throughout the fourth century.[14] The Sarmatian and Quadian attacks of the third century had depopulated the province, sparing neither the villages of the native agricultural population nor the villas of the landowners. The raid in A.D. 260 was certainly crucial. Treasures

Fig. 54. View of four-towered villa from Sümeg

hidden at that time have been found in great numbers in the Balaton region at Kabhegy, Somogyfajsz, Kéthely, etc. These finds are unambiguous proofs of grave troubles during the last third of the third century. Furthermore, the land was not cultivated following this devastating attack.

This period saw several developments. Hoping for more security and protection, the peasants left their land and moved to the landowners' estates. Weakened economic conditions in the Empire also reinforced the drive towards self-sufficiency. A partial cessation of administration muted the differences which formerly had existed between municipal towns, frontier encampments, and the economic centers of inner Pannonia. Neither the settlements which had been preparing for self-sufficiency nor the villas were particularly shaken by the temporary stagnations in money circulation, for the lack of security had forced the villas to rely solely upon what they were able to produce within their walls.

In this period, people could hope to find security only behind thick walls. Yet the fortification process in Pannonia had begun long before. The statements of Rostovtzeff concerning the whole Roman Empire have been confirmed by research in Pannonia. Rostovtzeff believed that the scanty data discovered about Aurelian

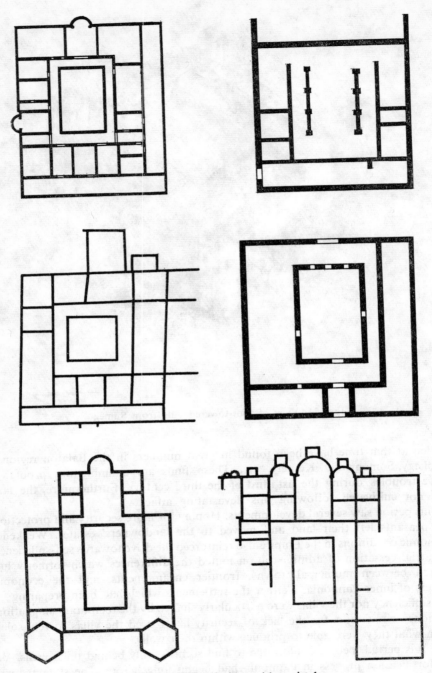

Fig. 55. Types of villas with peristyle

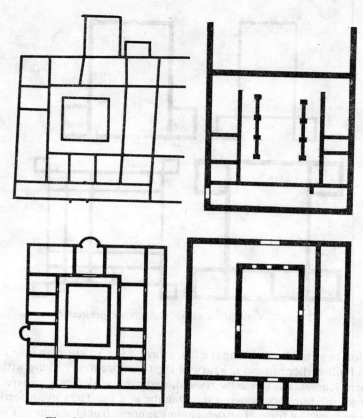

Fig. 56. Types of villas with peristyle and porticus

were sufficient proof that it was during his reign that the decisive transformation of the Empire into a military bastion took place. The entire Empire had a fortified appearance. The towns, settlements, villages, and villas of private owners within the fortified borderline all strengthened themselves for fear of enemy attacks. A decree of Constantine the Great ordered the landowners of the eastern provinces to enclose their grounds with walls. There was probably no need of issuing a decree of this kind for the western provinces, because the villas there had been walled in earlier times.

Even though our information is scanty, it reveals the outlines of a monumental defensive system. Vast edifices undoubtedly ensured shelter for great numbers of people. The fortification of the Pannonian farm centers; the development, building, and reconstruction of towered villas suitable for defense; the bulwarks of the major fortified settlements — all these helped to realize this defensive concept. The new settlers and the free peasants, as well as the slaves dispersed from the estates, hoped for protection behind the walls of the newly built fortresses. In addition to these fortified colonies, the villas were fortified separately. The corner towers which were built to fortify the villas in central Pannonia were not connected with the

315

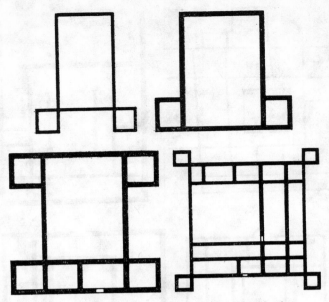

Fig. 57. Variants of villas with corner projection

construction of strongholds along the *limes* during the Valentinian era. They must have been built earlier around the turn of the fourth century. Along with fortified cities, where Romanized elements resided, and free settlements, where the Barbarians from the territory between the Danube and the Tisza lived, fortified villa colonies represented a form of settlement in Pannonia from the time of Constantius II onward. Károly Sági, in his contribution to A. Radnóti's lecture, regarded the smaller forts of approximately 30 to 40 meters in area and the watchtowers erected beside the roads as parts of the late Roman defensive system.

The purpose of these forts, watchtowers, and fortified villas was to secure peace in the interior of the late Roman Empire. An examination of these fortifications illuminates some particularly interesting facts. At the western shore of Lake Balaton and also on the extension of its longitudinal axis along a line diagonally crossing the Transdanubian area (Dunántúl) in a roughly northeast–southwest direction, fortified villas have been found. These buildings, as verified by numismatic finds, began almost uniformly under Aurelian and ended under Valentinian. It was in this period that, as a consequence of the increasing pressure of the Barbarians, the Danube *limes* did not seem to yield sufficient security. Thus the defense line in eastern Pannonia was forced back to Lake Balaton, which then became a second defense line for the western part of the province.

A. Radnóti believed that the walled towns were depositories of antique culture and that their population preserved great cultural values for the German peoples following the Roman conquest. Pannonian villas have often yielded relics of late Roman culture and life, as well as materials characteristic of Germanic peoples during the period of Hunnish invasions. These finds demonstrate that the fortified

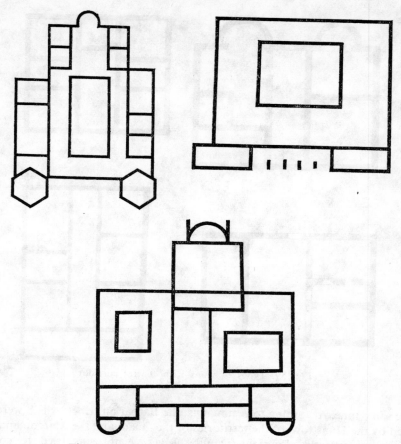

Fig. 58. Variants of villas with corner projection

villas and the urban settlements surrounded by walls were islands of cultural continuity in which the old ways of life continued to exist. The feudal estates which developed later were modeled after them.

## SURVIVAL OF THE ROMAN VILLA SYSTEM

We know from finds in Pannonian villas and from certain historical facts that life did not cease to exist in all buildings and settlements at the end of the fourth century when the Barbarians invaded and the Roman defence system and central administration collapsed.[15] As a result of the invasion many buildings and building complexes, especially behind the *limes*, were undoubtedly sacked and burned. But it appears that the fortified villas and settlements in the interior of the province escaped complete destruction.

The great Gothic invasion from Thracia at the beginning of 377 touched only the areas at the fringes of Pannonia. The Hunnish tribes made their appearance

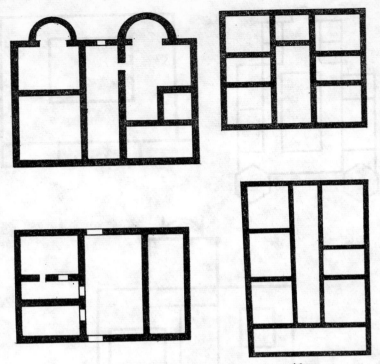

Fig. 59. Types of villas with central corridor

at the same time. According to Ammianus, the Barbarian tribes, which were being chased by the Huns toward the territories of the Quadi and the Marcomanni, kept moving on restlessly along the Danube River. Ambrosius tried to convince Gratian that the danger in the Danube area and in Valeria was God's way of punishing the Romans for heresy.

Although the Barbarians invaded Valeria, there was relative calm in West Pannonia. This may be attributed to the defense system which stretched along the eastern side of Lake Balaton and consisted of villas with fortified walls and towers, and of small fortifications. It was also true that the Goths did not appear in large enough numbers to annihilate all the Roman settlements and villas at once.

So the villa system not only survived but also continued to produce, thereby becoming a significant constituent in the economic system of the late Roman Empire. One may suppose that if the villas had come under the supervision of Barbarian masters, the landlords would have left their estates and escaped to safer regions. Even at the end of the fourth century, in the years from 387 to 394, the governor of Valerius Dalmatius Lugdunensis tertia retired to his villa at Beremend–Idamajor in the province of Valeria. Furthermore, Ambrosius mentioned that there was a rich harvest in 383 with enough surplus grain to be exported. We have learned from the epitaph of Amantius, bishop of Iova, that the migrant Gothic tribes settled down and that this bishop succeeded in converting them,

318

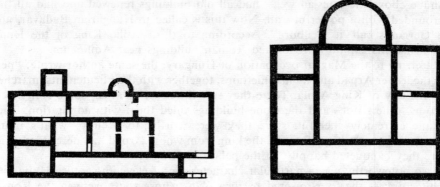

Fig. 60. Types of villas with side corridor

together with other Barbarian peoples dwelling in the same vicinity, to Christianity. Their leaders also listened to his advice.

Thus Roman life could still be seen in the last decades of the fourth century and even in the fifth century, especially in the western sectors of the area between the Drava and Sava rivers, east of the Sava, in the vicinity of Sopianae, and on the southern and western coasts of Lake Balaton. Finds from the fortified villas around Savaria and Lake Fertő also witness the presence of Romans in these late periods.

From the fifth century on, however, life must have assumed a different form in the villas of the province. A Romanized type of life continued so far as possible within the walls of the fortified villas and settlements, the inhabitants relying entirely on their own labor force. The smaller villas, whose residents had escaped to the larger, fortified villas, were now occupied by migratory people accustomed to a lower standard of life. The migrants sometimes fought and sometimes lived peacefully. They followed a Barbarian standard of life, however, never taking advantage of the surrounding Roman culture. Thus the waves of migrations endangered the existence of the Romanized inhabitants of Pannonia and of the Christians who had settled there in the second half of the fourth century. Archaeological finds point towards a strange mixture of Roman, Romanized, and Gothic elements in the villas and fortified settlements.

At some villa settlements, especially where the excavations of Christian basilicas have revealed a significant communal spirit and life (Csopak, Pl. LXXXIX. 1, 2), Kékkút (Pl. XXXVI. 2), Sümeg (Figs. 47 and 48), and Fenékpuszta (Figs. 40 and 41), Roman culture and ways of life, though gradually dwindling, continued actively until the ninth and even the eleventh century.

Buildings and villas of the Roman period were also mentioned by Hungarian medieval chroniclers. Referring to the Hunnish period, which followed the Roman, the historian Anonymus informed us in the first chapter of his work: "He (King Attila), leaving the land of the Szittyas (Scythians) descended on the territory of Pannonia leading an immense army in the 450s and, chasing the Romans away, occupied the land. And then the king himself established a capital near the

319

Danube above the artesian wells, had all old buildings renewed and had all this surrounded with a powerful wall. Now this is called in Hungarian Budavár, and the Germans call it Ecilburg". According to this, Attila, King of the Huns, established his capital in renovated Roman buildings near Aquincum.

Referring to the Magyar occupation of Hungary, the same author wrote: "next day the leader Árpád and all his chieftains, together with their lieutenants, marched into the city of King Attila. There they saw the royal palaces — some completely in ruins, others not — and the stone buildings filled them with admiration. They became extremely cheerful, since they were worthy to occupy — what's more, without a war — the city of Attila, the king from whom Árpád had descended. And there they banqueted happily, in the palace of King Attila".

Also noteworthy was a particular characteristic of eleventh-century Hungarian architecture — the four-towered basilica plan, whose roots were in the Roman architecture of Pannonia. The churches founded by Saint Stephen in the Dunántúl such as Székesfehérvár, Esztergom, and Pécs, and the basilica of Nagyvárad in the area beyond the Great Hungarian Plain were examples of this style. The four towers were not related to the liturgy and, in general, had no cultic significance. Their role as decoration was also minor; their purpose was to emphasize the fortress character of the church and its role in case of defense. It may not be too much to suppose that the four-towered villas and small fortifications, which still existed in the time of Saint Stephen, were the prototypes of these basilicas. The ground plans of the basilicas bear a closer resemblance to those of the villas and forts than to the prototypes of the Roman *castra*, as suggested by I. Henszlmann.

The villa form survived for centuries not only in these basilica plans and in palaces but also in scattered remains of villa buildings. Lands on which such buildings stood were donated to the dioceses which were established in Saint Stephen's time, and many of the diocesan farms undoubtedly used these buildings. The buildings were repaired but few changes were made in their outward appearance. In addition, the new buildings in the villages and settlements generally imitated the already existing ones. Thus the colonnaded and pilastered types of Hungarian porticoed peasant houses developed from porticoed villas.

The scattered capitals, cornices, and columns also had an important impact on the development of the architectural decoration of the early Romanesque in Hungary. It is conceivable that the provincial stone carvings of the Romanesque period, made from the same material as earlier Roman carvings, were inspired by the existing prototypes.

It is apparent that the standing walls of villas were often repaired and provided with roofs during the period of migration and during medieval times. Those situated near the roads where traffic was considerable were used as stations; those at other places were used as unpretentious farm houses. Where the walls were in good condition, they accommodated monastic communities or were used as the foundations for medieval churches.

[1] THOMAS, E. B., *Römische Villen in Pannonien. Beiträge zur pannonischen Siedlungsgeschichte* (Budapest, 1964).

[2] Ibid., pp. 13–127.

[3] Ibid., pp. 128–210.

[4] Ibid., pp. 211–69.

[5] Ibid., pp. 336–52.

[6] Ibid., pp. 270–335.

[7] Ibid., pp. 383–88.

[8] Ibid., pp. 379–81.

[9] Ibid., pp. 355–66.

[10] Ibid., pp. 367–71.

[11] Ibid., pp. 374–75.

[12] Ibid., pp. 376–77.

[13] Ibid., p. 378.

[14] Ibid., pp. 389–90.

[15] Ibid., pp. 395–98.

# ECONOMIC LIFE

JENŐ FITZ

Pannonia never played any considerable part in the economic life of the empire. Before the conquest, the population lived in humble circumstances, its production hardly exceeded its needs, and even trade relations had only a local significance. The conquest was undertaken for military reasons and even after the organization of the province, up to the end, military objectives were dominant. This circumstance determined the economic life of Pannonia. Its development and prosperity were dependent on the interests of the army. Rome, as we have seen, gave preference, even in the economic organization, to providing adequate supplies for the troops. In a province under military domination, the accumulation of wealth and the development of industry were rendered impossible by the uncertainty of the frontiers and the recurring raids; progress was achieved slowly and results were annihilated by the new and devastating wars.

Late literary sources speak of Pannonia as a fertile country well suited to agriculture. Side by side with the arable land, extensive woodland covered the middle of the province. There was also much marshland and many inundated areas. Many data confirm that in early times there were attempts at canal building, draining of waters — as aɪo undSirmium — and building of dams. From the latter more than one has survived. Galerian ordered woods to be cut down and marshes drained in the Balaton region.[1]

## AGRICULTURE

Agricultural production did not develop so much because of forest clearing and because of a change in land relations. It became the most important branch of economic life. Veterans settling in the western and southern parts of the province appropriated the best agricultural land from the native population. From early times these veteran-owned farms provided the basis of the new towns' economic life, and from them developed the typical municipal middle estate. During the first century, slaves cultivated the land. A villa constituted the center of each farm, its size, equipment, and luxury depending on the extent and success in management of the estate. Production in the province during the first century could not fully supply the army and this circumstance led to rapid development and prosperity in landed property. The increase and concentration of larger estates continued gradually throughout the Roman period, producing an even larger world economy. A villa was usually surrounded by a large area containing many farm buildings, and the free workers made up even larger settlements. The memory of this system

323

has been preserved in the inside of Pannonia by many place names: (villa) Bassiana, (villa) Varianae, etc. In those areas where no appropriation took place, the native population could retain their land for a longer time. In the beginning, land was tilled according to pre-conquest customs, in village communities. Later, the ancient order came to an end, probably as a result of Romanization, and the land was divided into small peasant farms. Even during the period of economic prosperity these farms were allowed to continue, although the municipal middle estate made great strides forward and this development occurred primarily at the expense of the peasant farms. It is impossible to trace the conditions in land relations during the second half of the third century, but we can do so in the fourth century when plentiful data refer to the development of large estates (latifundium), with enormous villas at their center.[2]

A special group of large estates constituted the imperial properties. In Pannonia, beyond the place name of Caesariana, we can form conclusions first of all on the basis of brick stamps. Such stamps are known from Mursa, Sopianae, Scarbantia, Vindobona, and Carnuntum. From a later period there is reference in Savaria to *silvae dominicae*. The centers of imperial estates, like those of the municipal aristocracy, were villas and palaces and many emperors were housed in these during visits to Pannonia. The earliest of these was built under Marcus Aurelius in Sirmium. During the fourth century in particular many references were made to these villas and palaces.[3]

The first agricultural products were millet and wheat. Excavations have recovered the seeds of wheat, rye, millet, pits of plums and other fruit, and almonds. Many grinding stones prove the survival of the pre-conquest types of agriculture. Viticulture also went back to pre-Roman traditions, but it became essential only in the period after the conquest. Probus employed military labor in viticulture, but an inscription also speaks of vine planting: ... *inseru*[*i*]*t* ... *vineae arp(ennes) CCCC ex his V Cupeni I Termini* ... *V Vallesibu*[*s*] ... etc. The altar dedicated to Liber Pater preserves an inscription which locates the vineyard on the south slope of Mons Aureus. Pruning knives and wine-presses are among archaeological finds which suggest that viticulture was on a high level.[4]

The wide expanse of forests made lumbering possible. Among the forest officials a single *servus saltuarius* is known on a Boian estate, as well as a *praepositus silvarum dominicarum* from later times. The woods and forests were rich in game; we know of bears, wildcats, and boars from the first century. In the Celtic *oppidum* on Gellért hill, the bones of deer, bears, stags, wolves, and birds of prey throw light on hunting before the conquest. However, it remained a favorite sport later too, since even hunting dogs are mentioned in the sources. We have proof that Emperor Hadrian, during his stay in Pannonia, hunted boar. The study of surviving animal bones in Gorsium suggests that in regions far remote from the forests, the population of the cities did not eat game.[5]

We can form ideas about animal husbandry from the bones uncovered by excavations. On the basis of the most complete list at Gorsium, ox, sheep, goats, and pigs can be regarded as the most important animals. In civilian settlements, comparatively few horses were used, although the number of asses domesticated in the Roman period was high. Among the animals kept in cities, primarily the bones of dogs have survived in great numbers, but these must have been small dogs, unsuited for hunting or for watchdogs. Tools in a grave at Keszthely indicate that leather processing was also known. In Aquincum wood-lined round pits were found which appear to be remnants of tanneries. One of the provincial centers of wool processing was at Siscia where the wool of the Karst region was collected and processed. From the basin of the Kuppa River, a large number of plumb labels were found, used to mark the wool bales.[6]

Larger estates also had fishponds. At Kikeritó there still survives in good condition a dam used to block off a valley. Elsewhere, field formations lead us to suppose that ponds must have existed. Such a pond bed is known north of Gorsium, although very few fish bones emerged from the area.[7]

## MINING

Augmenting a large organization of mines which extended as far as Noricum and Dalmatia were the mines of Pannonia, which are referred to on the coins of Hadrian: *Metalli Ulpiani Pann(onici)*, *Metal(lis) Pannonicis*. There was a significant amount of mining activity in the northwest part of the province. There we also have some new data on metal mines. In other parts of Pannonia mining must have meant stone quarrying. Near Lugio, there was a porphyry (red granite) quarry and from the Buda Hills lime was quarried which was used in the building schemes of neighboring towns as well as for stone carving. In the Balaton area red sandstone was used and is still favored. Larger stone quarries supplied the requirements of even more distant regions. In the western province there are references to iron metallurgy and in Scarbantia even furnaces are known.[8]

## WORKSHOPS

The survival of local traditions, the appearance of the more highly developed Roman practices and the mingling of the two can be observed in practically all branches of industrial production. In building, the use of stone, brick, and plaster was due to the Roman example. In earlier times, the native inhabitants of Pannonia lived in pits and in mud- and plaster houses. This form of building existed for a long time, even after the conquest, not only in villages but on a general scale, too. Until the middle of the second century it was common in the towns and *canabae*. In Gorsium, near the representative forum and *area sacra* the houses built in the early second century had foundation walls made of stone, whereas the rest of the walls were of mud. In the plan of the houses, especially those supplied with T- and L-shaped corridors, early local customs prevailed. Research still indicates this type of house is characteristically Pannonian. The emergence of local traditions is understandable since the names of master builders on inscriptions all came from the native population. From the inscriptions we know of a *mensor aedificium*, an *arci(tectus)*, of *ex subaedianis collegae lapidari* and *structores*. During the whole period, outside the mud brick which had been in use all along, stone was added, even when the quarries were a long way off. Walls were seldom made of brick, but it was used for the repair of stone walls and for rebuilding plans, especially in baths, canals, heating installations, and for the paving of floors. A large number of kilns were in operation near military camps and in imperial estates, but also in towns where they were run by private individuals. The owner of one kiln was Cassius Carinus, a *decurio* of Mogentiana. Yet even private merchants supplied brick for the great military building projects: we know of private stamps at building ventures in Barbaricum. The workers employed in the kilns, according to the legibly incised names, were probably slaves. The same inscriptions throw light on the workshops — their organization and daily output, where the norm was probably 200 to 300 bricks. Besides a find which lists carpenters' tools and the activity of carpenters employed at the buildings, there emerged from Aquincum a picture showing the tools used in making gravestones.[9]

Among the other branches connected with building, stone carving was the most significant, not so much because of its quality but because of the large number of relics. From the few stone relics which emerged from pre-conquest times, it would be impossible to form any conclusions about a systematic stone-building activity. The first stonemasons were attracted to Pannonia by the presence of the legions stationed there. In the beginning it was probable that North Italian workshops had sent out a master to comply with the orders. These men, naturally, visited several sites during their journeys. We can see the handicraft of such itinerant stonemasons during the Tiberius–Claudius period in the territory of Scarbantia.[10] In this way the requirements of the legions could be met as long as they were stationed close to Italia. When Legio XV Apollinaris left for Carnuntum during the reign of Tiberius, a local workshop soon emerged. Not much later another one started at Poetovio which began to work during the middle of the century. These local workshops also employed Italian masters who introduced the North Italian form of style to the new province.

Gravestones were not in general use by the native population. The earliest relics were raised by migrant Italian masters, probably for the local aristocracy which soon adopted the Italian way of life. Under the Flavii, however, for the ever-expanding local requirements, local masters also appeared. Their work is characterized by rough execution and with no tendency toward understanding the Roman forms and ornamentation. Nevertheless, the new custom of gravestones was soon adopted, so that the influence of Romanization is out of the question. It is most probable that the marking of the graves — perhaps by means of wood-carved memorials — was something that the local inhabitants had already practiced and only the use of stone was an innovation in the period.

In the decades when the *limes* had been constructed, the requirements of an increased army and municipal population gave enough work to numerous stonemasons, especially near the legionary camps, and in practically all the larger settlements. However, the activity of native masters did not come to a standstill in that period either. Their activities can be traced back until the middle of the second century, although there were considerable differences in standards. The differences, however, were no longer due to the backwardness of the natives but to differences between the local development and the general taste. The influence of a prominent workshop was felt also in distant parts: such was that of a stonemasonry in Sopianae and Mursa. The fashion of sarcophagi was mainly due to the work of a shop in Siscia. In the second century, in particular, the workshops supplied a wide area. To perform certain works, stonemasons were sent out, as were the North Italians at an earlier time, but as soon as transport facilities became adequate, as in the camps along the Danube, gravestones and even altars were delivered ready-made.

The Italian style which asserted itself in general was gradually affected by western and eastern influences. At the end of the first century, western stonemasons settled near the camp of Aquincum. The influence of eastern motifs was discernible from the last third of the second century which asserted itself under the impact of Orientals in all walks of life. The only stonemason whose name has survived in Pannonia was from the East; on an altar in Mursa the following inscription can be read: *Asclepiades fecit*. In Intercisa where the eastern influences were particularly strong, certain portraits show distinctly Syrian or Palmyrian influences.

The workshops of Noricum had exercised a direct and more lasting influence on Poetovio, Savaria, and, in general, on the stonemasons of the western cities. The Pannonian activity exerted even more influence on the workshops of Noricum: it is very likely that the enormous statues on the capitolium of Scarbantia and Savaria were made by Noricum masters.

Most of the local masters operated on the level of simple artisans. They generally pursued their work according to sample textbooks, and wherever possible, they even simplified these, leaving out the technically difficult parts. The gravestones in the cemetery, the statues ornamenting the squares and streets, reliefs, carvings on public buildings and temples in the eastern part of the province approached only under exceptional circumstances the standard of southern and western cities. In these the proximity of Italia and Noricum was felt; in the former the rustic spirit of the camp exerted itself to advantage.

The wall paintings adorning the interior surfaces of the buildings, or at least a considerable portion of them, can be attributed to itinerant painters. This is corroborated by similarities in style to other regions — as, for example, between Aquincum and Baláca. Somewhat later, during the third century, local masters were active in Aquincum. In painting, too, there was a striking difference between the western and eastern parts of the province. In keeping with the earlier development, the influence of Rome and Italia exerted itself directly on the western cities of Pannonia. In the eastern part of the province frescoes became fashionable during the building projects of the third century, an influence from the East. The surviving incrustation style arrived at Aquincum by means of Asia Minor and Antiochian mediation, and from there continued on its way westward.[11]

Stuccoes had been generally applied for closing down the painted upper walls. In Western Pannonia stuccoes had been applied generally to give an illusion of an architectonic support for the painted upper part of the wall. In Eastern Pannonia, however, stuccoes were also used independently. Stucco pillars, capitals, acanthus leaves, and clusters of grapes adorning the walls were all used to decorate ceilings in Aquincum and Gorsium. Frescoes were based on eastern models. The makers of stuccoes, as the provincial painters, also followed sample textbooks, even using stencils, rulers, and compasses.[12]

The floor mosaics of palaces and villas were also the work of itinerant masters. The mosaics of Baláca, the governor's palace of Aquincum, the Parndorf villa, and the Poetovio palace are all works of Italian masters. The mosaics of the imperial palace of Savaria were probably made by Aquileian masters. However, the claim for mosaics emerged only locally, especially under the Severi. Even then the local masters employed simple geometrical figures, and their work is characterized by a weakness of composition and an attempt for simplification.[13]

The activity of workshops dealing in bronze was affected by former native surviving traditions. They were busy in several towns — in Emona, Carnuntum, Brigetio, and Aquincum. The native population, however, went on producing bronze objects, particularly the characteristic pieces of clothing, such as winged fibulae, torques, and bracelets. In the city workshops fibulae, locks, key stems, and bronze mountings were made after Italian models. In later times bronze casting workshops active near the camps of Carnuntum and Aquincum developed into state *scutaria*. Certain workshops were engaged in casting statues, according to modest requirements. These statues are generally clumsy and rough. A workshop has been localized through its products in Southwest Pannonia and another one at Brigetio where the *artifex* Romulianus had been active.

From the third century we can observe an improvement in the Pannonian workshops and an extension of their activities. Festive mailcoats which had been produced here were much sought after outside the province, particularly in the Danube provinces. In the ornamentation, outside the survival of the Celtic style form the motifs of third-century military life and religion were also present, including Victoria, Mars, Minerva, and Hercules. Under the Severi, the workshops' activities extended to the production of various filigree and repoussé bronze mountings, cultic pictures, and even gold jewelry. During the fourth century this varied workmanship came to include chest mountings as well. In the decoration of chests a certain official character can be observed: it is possible therefore that they were produced in the *scutaria* as indicated by *Notitia Dignitatum*.[14]

The duality of the native population, on one hand, and the army and the new settlers, on the other, becomes clear even in the light of ceramic art. The native potters had continued to work even after the Roman conquest by following native traditions. These workshops could not participate in the supply of the army. In the beginning, the army met its requirements from Italian imports; later the potters' settlements which emerged near the *canabae* provided for its needs. In the early period the stamp of the legion or the auxiliary force marked on the vessels indicate that the army also had its own workshops when required. The earliest potters who settled in the *canabae* were Italics and it was through them that various Italian vessels of different shapes were introduced. Western influence appeared later, particularly in painting, ornamentation, and forms of reliefs.

In the activity of local potters in the late first century, the Italian and western influences brought considerable changes. Among these the most important was the appearance of a new ornamental vessel; the traditional gray Celtic vessels were decorated with stamped motifs, originally on the bottom of the vessels and later on the side. In East Pannonia one of the earliest masters was Resatus who marked his vessels with the stamp RESATVS FECIT and with large leaves. Research cannot yet identify his workshop but he worked somewhere in Eraviscan territory, in Aquincum or Gorsium, where his vessels emerged in the largest number. Later, this stamped ceramic imitated the Gaulish sigillata, both in shape and ornamentation, remaining true to Celtic taste only in color. Along with the gray vessels, yellow vessels were also made. This type of vessel was more widely used in a certain area up to the Marcomannian wars among the local population than the obviously more expensive western sigillata. Genuine sigillata were not produced in Pannonia outside Gorsium and Aquincum. In the second century production started under the influence of Lezoux and Rheinzabern; however, Italian and Hellenistic features can also be detected. The Aquincum workshop was connected with the name of Pacatus.

During the second century, the ceramic industry was able to supply local requirements, with the exception of ornamental vessels, and could develop independently under the impact of local, Italian, and western influences. Nevertheless, mass-produced articles dominated, and quality, particularly during the third century, deteriorated significantly. The use of metal glaze became general in the late third and early fourth centuries; however, in Pannonia, though to a lesser extent, it had been used already in the second century. From the centers of glazed vessels a workshop has been unearthed at Castra ad Herculem and a potters' settlement has been localized in Gorsium. New settlements in the fourth century produced Barbarian traits in the ceramic industry, and the earlier differences between the ceramic industry in the province and Barbaricum had decreased to a considerable degree.

The potters' workshops also produced oil lamps, illuminated turrets, and clay models — outstanding among which is the city gate found at Intercisa, with the inscription *Hilarus fecit porta(m) fel(iciter)* — as well as bakers' molds and terracottas. The traditional native kilns were round or oval, half sunk into the earth, made of clay or mud. Foreign potters had introduced a second type of oven which was square and made of bricks or stone.[15]

Among the branches of industry introduced during the Roman period was the glass industry. Simple vessels were produced locally. Real flourishing, however, did not come before the fourth century. On the basis of discovered glasses it is not possible to distinguish all glass-producing centers. It seems from numerous and specific types of glass objects that Gorsium was a prolific producer. On the basis of spoilt and unfinished fragments it is possible to state that a local production existed; large glasses with blue polka dots could not have emerged from elsewhere. Production must have continued until the end of the fourth century. Drawn glasses are known to have come from Arrabona and blown glass from Aquincum. There is also mention of a *specularius* in the ranks of Legio XIV Gemina.[16]

## TRADE

Local production and industry never reached the point of development which could have made it competitive in the markets of the empire. Very few commodities of Pannonian origin can be spotted in the other provinces. In pre-Roman times, the Danubian Illyrians had supplied the markets of Aquileia with slaves, oxen, and hides. Owing to the character of its local industry, Pannonia exported industrial products in a very small quantity. Some stamped vessels from Noricum and some marble Mithras tablets in the Rhine provinces are but rare traces of a westward-tending trade. Relations were rather established with Moesia and Dacia. It is possible to trace the exported goods of the workshop of Pacatus in Moesia.

During the flourishing period of the Pacatus workshop a branch was established in Mursa to get nearer the southern markets. Nor was there a considerable export of wild animals for the games in Rome; the export of agricultural products was also minimal. Contact with Barbaricum, however, was more lively, although most of the exported products were not Pannonian. Nevertheless, stamped ceramics were just as much favored among the Celts north of the Danube as among the Eraviscans. Transit trade supplied the Sarmatians with enamel fibulae and Germans north of the Danube with bronze vessels.[17]

A modest industrial and agricultural production along with a moderate export required a large import, especially as far as the better quality of commodities was concerned. During the first century, when local production was still on a small scale, the army also depended on imports, particularly wine and oil, which were brought in from Aquileia up to the early second century. During the early period most imports came from Italia, mostly through the mediation of Aquileia. Strabo stated that this great merchant city had supplied the Danubian Illyricum with wine, wood, oil, and maritime goods. During the first century there was also a considerable import of Italian vessels, late-Italian sigillata from the Po region, ornamental vessels, and lucernae, first for the army and then for the new civilian population. Fine Italian bronze vessels were in great demand in the southwest in the early first century and were favored by soldiers and settlers. During the middle of the century art penetrated even into the northwest and outside the Romanized population of the cities; it even reached the aristocracy of the Boii. Later, art was available in practically every part of the province. Up to the early second century Italian glassware arrived in great quantities, not only rich, relief-decorated examples and Millefiori glasses but also objects of everyday use. Glass urns and scent bottles were of Italian workmanship even during the second century. The products of the Aquileian amber industry were particularly popular in the western cities. There was a flourishing import of slaves through Aquileia, as can be seen from the names of the slaves arriving in the province: they were all of a North Italian type. Imports had reached their climax during the emergence of West Pannonian municipal estates; later their significance diminished. Works of art, marble and bronze statues, and reliefs which surpassed the facilities of local workshops all came from Italia or arrived in the province through Italian mediation. However, marble also came in as a raw material, even into the eastern parts of the province.

The large-scale import of Italian goods diminished partly as a result of local production, particularly in oil, simple ceramics, glass, and simple bronze objects. From the turn of the century when the Danubian waterways came into common usage, the otherwise diminishing Italian exports were counterbalanced by a rivaling western trade. It was not able to oust Italia altogether from the Pannonian markets, but it predominated in ceramic and bronze workmanship. Particularly during the second century western commodities arrived in tremendous numbers into Pannonia and they retained their supremacy until the great economic crisis of the mid-third century. The South Gaulish *terra sigillata* entered Pannonia in very small numbers, mainly from Aquileia. A larger import dates from the early second century with articles coming mainly from Lezoux, Rheinzabern, and Westerndorf workshops. Outside the *sigillata*, vessels from Raetia, terra-cotta from Cologne and vessels from Trier arrived in larger quantities. The Gaulish and Germanic bronze industry had provided mounting, vessels, coach ornaments, and enamel fibulae from the late first century onward. Glass import had not yet been significant; it reached its peak from Cologne, in the fourth century. Some of the bronze and stone production also came from western workshops.

Imports from the neighboring eastern and southern provinces were never significant on the Pannonian markets. Besides gold, some stamped vessels came from Dacia, first of all from the terminals of routes leading through the land of the Sarmatians. However, imports from the Balkans flourished in the third century which was mainly due to the settlement of Thracian soldiers who had served on the *limes*. Different types of marble came from the Greek islands to supply the materials for the interior decoration of the building projects.

It would be impossible to form an adequate picture of the dimensions of eastern trade on the basis of surviving finds. The main commodities were a Hellenistic sigillata with relief ornamentation from Intercisa, a sigillata chiara from Gorsium, lamps of Grecian style, bronze vessels, a *strigilis* with a Greek inscription, cake molds and gold jewelry. Yet there must have been a considerable import of glass from Asia Minor and Syria as well. In the glass material of Gorsium their number is strikingly large.

Commodities from Barbaricum consisted mainly of raw materials and livestock. An analysis of a barrel in Aquincum, however, proved that it had come from the territory of the Quadi. The slave trade also came from Barbaricum. Some of the slaves were transported to Italia and the other provinces. Barbarian trade also yielded amber: the rough amber pieces of the Nógrádverőce camp arrived in a similar manner in the fourth century.

A striking disproportion between exports and imports, that is, Pannonia's unending passive commercial balance, was more or less compensated by the money which arrived in the province. The soldiers' payments, state building projects, and the salaries of state officials under the Severi exceeded 100 million sestertia a year.

Booms in commercial life were dependent largely on the commercial imperial routes which cut across the province. The Amber Route starting north of Aquileia touched upon practically every important town in the western section of the province. The great commercial roads leading from east to west lent prosperity to all the towns between the Drava and Sava rivers, and the main road through Savaria, Sopianae, and Mursa provided contact between these cities and distant regions. Moreover, the *limes* road and the expansion of shipping on the Danube were responsible for the prosperity of cities near the military camps. In the interior of the province, however, only certain junctions played important parts, such as Gorsium which was the most outstanding one in North Pannonia, joining the imperial road from Poetovio and Sopianae.

At the outset the Italians played the leading role in trade and commerce. The commercial relations of pre-conquest times expanded in the western part of the province in the first century. Aquileian commercial houses which permeated Noricum in those days settled for the first time in Nauportus. Not much later, however, Emona took over the leading role. In the trade of the Sava basin the two cities held key positions. Later on, the Aquileians emerged as the stronger at each important section of the Amber Route and the Sava basin. For a long time, trade involved only the camps and the newly settled towns; however, with the progress of Romanization in the first half of the first century, it also affected the Boian aristocracy. The building of the *limes* contributed to an unprecedented expansion of the Pannonian market. In Aquincum, which developed into a town at that time, several commercial interests emerged, including Aquileians, merchants from the Rhine region, and South Gaulish ones. In the name of *collegium negotiantium*, a merchant from South Gaul fulfilled at that time the tribute vowed to Neptune, probably in gratitude for the successful transportation on the Danube. The merchants' center in both Carnuntum and Aquincum was the *canabae*. In Aquincum one of the *magistri canabarum* was Italic, the other a westerner.

Owing to the development of local industry there followed a decrease in the trade of mass products from distant parts; in certain commodities it stopped altogether. This was identical not only with a drop in the number of Italian and western merchants but also with the establishment of local contacts and a commercial network. In the finds from Vetus Salina and Intercisa the leading role was occupied by Aquincum workshops. Local trade reached its peak under the Severi; however, the crises of the coming decades and the wars inflicted on the province completely destroyed it. By the fourth century no trace survives of greater transactions.

Among external manifestations of trade were booths selling vessels and *sigillata* in Aquincum and Savaria. At Mursa a *decurio* established *tabernas L cum porticibus in quibus mercatus ageretur*. In Gorsium, on the *decumanus* near the forum, rose a row of *tabernae* with warehouses and in front of the *tabernae*, a *porticus*.

Trade with Barbaricum had been restricted for security reasons. Before the time of Marcus Aurelius market days were not fixed, but the sites of the markets were marked. These officially arranged markets, where Barbarians could also appear, had generally been near the legionary camps. Large-sized buildings — in Carnuntum 175 by 200 meters — allowed even for herds of animals to be led in. Altars dedicated to Genius Commercii also throw light on the fact that this type of foreign trade and the officially appointed markets were controlled by the *beneficarius*. During the fourth century, trade was more restricted. Valentinian built a watchtower near Solva, *cui nomen Commercium qua causa et factus est*.

## MONEY CIRCULATION

The use of money had been known to the Pannonian people some one hundred and fifty years before the conquest. Greek tetradrachmas, however, did not penetrate farther than the southern territories, circulating from the middle of the second century. During the first century, owing to the dominance of Roman trade, republican coins also appeared in this province. Boian and Eraviscan minting of coins began in the first century. The appearance of money is proof of a developed production of commodities. However, circulation did not extend far. In spite of extensive trade relations, its role was confined to local commerce.

In the early decades of the conquest the situation had hardly changed. Roman imperial coins did not reach Pannonia in any greater numbers than the republican coins had. Up to the time of Claudius the Roman life-style did not take root outside the camps and their vicinity, and money arriving as soldiers' pay did not circulate beyond those areas. The introduction of money corresponded with the growing military importance of the province, the increase of the number of formations, and the development of trade. However, changes in money circulation can be traced only by means of the surviving coins in a few cities such as Carnuntum, Aquincum, Gorsium, and Intercisa, where the excavations produced a plethora of coins which more or less proves the intensity of the money economy. In Carnuntum there was a systematic, though slow, circulation from the time of Augustus which reached its peak during the century under Vespasian and Trajan. In the east all this happened somewhat later: circulation began under Vespasian, the first rise occurring under Domitian and the second, under Hadrian. Subsequently a steady increase until the 180s denotes the economic strengthening of the province. Under Severus, after a recession during Commodus, the circulation reached its apex particularly in the years of Elagabalus. Recession was slower under Severus Alexander, later the economic crisis quickened. The increase in circulation and later a deterioration of the currency caused difficulties and the imperial mint could not keep pace with the requirements. During the second quarter of the third century, the colonial coins of Moesian and Thracian mints played an ever-increasing role in Pannonia. Viminacium in particular became important: during the time from Gordian III to Trebonianus Gallus, 51 percent of the total coinage of Carnuntum was colonial mintage from Viminacium; during the same time, 59 percent came from Aquincum. In order to supply the army with money, Gordian III established an imperial mint in Antiochia. It was probably Philippus who established the one in Viminacium. The latter, until its cessation in 258, provided the greatest portion of the Pannonian army's requirements. During a short time, in 258–260, a mint probably functioned in Carnuntum, producing *antoniniani* for Regalianus. To supplement the Viminacium mint, Gallien in 262 established a large mint at Siscia which, from then until the last quarter of the fourth century, supplied the largest portion of Pannonia's coin requirements.

A gradual increase in money circulation was largely dependent on the slow deterioration of the currency, but a temporary rise in the soldiers' pay also contributed to the increase. One of the major factors of the economic boom under the Severi was due to a considerable improvement in the financial situation of the soldiers, the rate of which surpassed the gradual deterioration of the money. Inflation under Gallien, in the course of which prices had skyrocketed within a few years practically a thousandfold, reached its apex under Claudius II. The money turnover of the two cities of Carnuntum and Aquincum showed considerable differences in the succeeding period. From Claudius Gothicus, despite an enormous deterioration, money circulation did not increase in Aquincum which reflects a catastrophic recession: it points to the great Barbarian inroads, destruction, and a stagnation in economic life. Carnuntum, on the other hand, had escaped a Roxolani raid. Inflation here as well was reflected by the accumulation of money. Nevertheless, the city remained far below the standard of prosperity of the early century.

In the fourth century inflation and recession altered periodically. The circulation of money became restricted, expressing to a lesser degree changes in economic life. Banks which had played such an important part in commercial life declined, industrial plants were in the hands of the state, the army, and the large estates and received in general only official orders. Workers were not given money as their wages and the *coloni* paid in kind. Up to the death of Valentinian the circulation of bronze coins of minor value in Carnuntum and Aquincum was very great. In Gorsium in its third flourishing period the use of valueless bronze coins was even greater. However, after the death of Valentinian, the supply of money to Pannonia was halted. From this time onward money decreased still further in circulation.

## NOTES

[1] *Hist. Aug. Probus* 21.2.

[2] A. Mócsy, *Die Bevölkerung von Pannonien bis zu den Markomannenkriegen* (Budapest, 1959), pp. 107 f.; A. Mócsy, "Die Entwicklung der Sklavenwirtschaft in Pannonien zur Zeit des Prinzipates". *ArchAntHung* 4 (1856): 227 ff.

[3] *CIL*, XIV, 176; VI, 790.

[4] Cass. Dio 49.36; B. Saria, in *Burgenländische Forschungen* 13 (1955): 27; *CIL*, III, 10276.

[5] A. Mócsy, "Pannónia", R. E. Pauly-Wissowa, Suppl. 9, pp. 669 ff.; É. B. Bónis, *Die spätkeltische Siedlung Gellérthegy–Tabán in Budapest* (Budapest, 1969), pp. 238 ff.

[6] A. Mócsy, "Ólom árucimkék Sisciából—Bolli romani da Siscia". *FolArch* 8 (1956): 5–87.

[7] J. Faller, "Római duzzasztó és kőbánya Veszprém vármegyében" (Roman Dikes and Stone Quarries in County Veszprém). *Vasi Szemle* 6 (1939): 294–300; Gy. Nováki, "Rómaikori halastavak" (Roman Fishponds), *Soproni Szemle* 12 (1958): 63–66.

[8] A. Mócsy, "Pannónia", pp. 673 ff.

[9] *CIL*, III, 11304, 13389; *CIL*, III, 3885; J. Szilágyi, "Inscriptiones tegularum pannonicarum" (Budapest, 1933), pp. 104 ff. (*DissPann* 2:1); A. Radnóti, "Római tábor és feliratos kövek Környéről—Le camp romain et les monuments épigraphiques de Környe". *Laureae Aquincenses*, 2 (1941): 89; *CIL*, III, 962 No. 3, 6487, 11378–11385, 11467, 11468, 14360[20]; W. Kubitschek, *Römerfunde von Eisenstadt* (Wien, 1926), p. 71, No. 7.

[10] M. Szabó, "A Badacsony–Lábdi kettősfej kérdéséhez—Zur Frage des Doppelkopfes von Badacsony–Lábdi". *ArchÉrt* 90 (1963): 69–74; H. Hofmann, "Römische Grabsteine aus Walbersdorf bei Ödenburg". *JÖAIBeibl* 12 (1909): 226; M. Abramić, "Archäologische Funde in Pettau", *JÖAIBeibl* 17 (1914): 136 ff.; I. Paulovics, *Lapidarium Savariense* (Szombathely, 1943), p. 22.; A. Sz. Burger, "Collegiumi kőfaragóműhelyek Aquincumban—Marbreries collégiales à Aquincum". *BudRég* 19 (1958): 9–26; J. Fitz, "Titulum memoriae posuit". *Alba Regia* 2–3 (1961–1962): 33–48; G. Erdélyi, "Steindenkmäler", in *Intercisa* (Budapest, 1954), 1: 169–231 (*ArchHung* 33); L. Barkóczi, "Brigetio I–II" (Budapest, 1944–1951), pp. 35 ff. (*DissPann* 2:22).

[11] L. Nagy, "Die römisch–pannonische dekorative Malerei". *RömMitt* 41 (1926): 79–131; J. Fitz, "Stuckverzierungen und Wandgemälde". *Intercisa*, 2:19–27 (*ArchHung* 36); E. B. Thomas and T. Szentléleky, *Führer durch die archäologischen Sammlungen des Bakonyer Museums in Veszprém* (Budapest, 1959), Parts 16, 26.

[12] L. Nagy, pp. 91 f., 101 ff.; J. Fitz, "Stuckverzierungen", pp. 19–27; E. B. Thomas, *Archäologische Funde in Ungarn* (Budapest, 1956), p. 216.

[13] M. KABA, "Az aquincumi helytartói palota mozaik-padozatai—Die Mosaikfussboden des Statthalterpalastes von Aquincum". *BudRég* 18 (1958): 79–101; Á. KISS, "The Mosaic Pavements of the Roman Villa at Baláca". *ActaArchHung* 11 (1959): 159–236; Á. KISS, "Mosaïques de Pannonie", in *La mosaïque gréco-romaine* (Paris, 1965), pp. 297–303; U. TRINKS, "Versuchsgrabung am Donauabbruch". *Carnuntum Jahrbuch* (1958), p. 70; I. WELLNER, *Az aquincumi mozaikok—Die Mosaike von Aquincum* (Budapest, 1962).

[14] E. PATEK, "A pannóniai fibulatípusok elterjedése és eredete—Verbreitung und Herkunft der römischen Fibeltypen in Pannonien" (Budapest, 1942) (*DissPann* 2:19); I. SELLYE, *Császárkori emailmunkák Pannóniából—Les bronzes émaillés de la Pannonie* (Budapest, 1939) (*DissPann* 2:8); T. NAGY, "Héraklész bronzszobrocskája Óbudáról—Une statuette de bronze d'Héraclès, découverte à Óbuda". *BudRég* 17 (1956): 9–44; I. PAULOVICS, "Római kisplasztikai műhely Pannóniában—De quadam officina in Pannonia olim florenti, qua artes effectivae minores exercebantur". *Pannonia* (1935), pp. 21–27.

[15] É. BÓNIS, *Die kaiserzeitliche Keramik von Pannonien* (Budapest, 1942) (*DissPann* 2:20); A. SCHÖRGENDORFER, *Die römerzeitliche Keramik der Ostalpenländer* (Wien, 1942); K. Sz. PÓCZY, "Keramik". *Intercisa* (1957), 2: 29–139 (*ArchHung* 36); K. Sz. PÓCZY, "Die Töpferwerkstätten von Aquincum". *ActaArchHung* 7 (1956): 73–138; L. BARKÓCZI and É. BÓNIS, "Das frührömische Lager und die Wohnsiedlung von Adony (Vetus Salina)". *ActaArchHung* 4 (1954): 129–99.

[16] A. BENKŐ, *Üvegcorpus* (Glass-Corpus) (Budapest, 1962) (*Régészeti Füzetek* 12:11); A. RADNÓTI, "Glasgefäße und Glasgegenstände". *Intercisa* (1957), 2:141–63 (*ArchHung* 36).

[17] Strab. 5.1.8.

# CURRENCY

## KATALIN BÍRÓ-SEY

Until now, information on Pannonia's currency has been extremely limited. We have no comprehensive works, such as a history of the currency used in large settlements, and no catalog of coins from sites. Even the few comprehensive articles on the currency of some towns and camps, such as Carnuntum,[1] Aquincum,[2] and Intercisa,[3] are inadequate; they deal with small quantities of materials, and they need to be updated. From time to time there have been news items about large hoards, most conspicuously from the fourth century, but the occasional stray coins present greater problems of collection and arrangement. We can hope that the work now in preparation on the currency of Brigetio, aiming at the description of nearly ten thousand coins, will produce several general statements which will be valid for the north Danubian *limes*, perhaps even for the whole province.[4]

If we examine Pannonia's currency before and during the Roman occupation, we find three groups of findings. First, we have hoards of coins from burial sites; these generally show the currency of the decades preceding their burial, sometimes going back to previous decades or even centuries, and occasionally containing the coins of earlier emperors. Naturally, the closer in time to the date of the hiding of the hoard, the more numerous the coins. In the second group are stray finds from sites. Most useful of all is the third group, the settlement material from excavations. Coins from cemeteries are not considered in these groups, since coins put in tombs were not always current.

In the future a complete statistical comparison should be made of the composition of all the coins from hoards and all those from museums. Naturally, collections of coins in museums generally come from hoards in the district; even if they are stray material, or if their exact sites are unknown, they must come from somewhere in the country. The collection of coins in Hungarian museums are rarely augmented by coins found abroad. The Greek collection in the Hungarian National Museum was augmented by purchases from abroad only because a find of Greek coins is very uncommon in Hungary. Comparing collections and hoards, including stray finds from sites and settlements, we can see that their distribution by emperors, or rather by periods, is similar. They can be grouped as follows:

1. Pannonia's currency before the Roman conquest.
2. Influx of Roman denarii and their Eraviscus imitations.
3. Currency of the province until the Marcomannian wars.
4. Currency of the province from Septimius Severus to Gallienus (260).

5. Period of great inflations and reforms (260–364).
6. The last great influx of money.
7. Appreciation of the two mints in Pannonia, Siscia, and Sirmium.

## BEFORE THE ROMAN CONQUEST

Several hoards found in Transdanubia and between the Drava and Sava rivers tell us much about the regular money economy of the peoples living in pre-Roman Pannonia. In the fourth century B.C. this area was invaded by different Celtic tribes who had close trade connections with the Macedonian Empire; these tribes knew and used the tetradrachmae of Philippus II (philippei) and the gold staters of Alexander the Great. The inflow of philippei into the area lasted until the fall of the Macedonian Empire in 168 B.C., when the shortage of money in trade was remedied by imitations of philippei. At first counterfeits were made only by the Scordiscus tribe; then others followed this practice. However, the Scordiscus tribe made the better imitations. The Scordiscans continued to make imitation philippei until 101 B.C. These early imitations, even if small in number, can be found between the Drava and Sava rivers and in Transdanubia.[5] (Pl. XLVIII. 1).

Tribal minting was not uniform; each tribe made its own coins, imitating the original ones in weight and quality. Coins were easily adulterated. In most cases we cannot localize tribes on the basis of the source material, nor can we ascertain exactly the dates of the beginning or end of tribal minting. These dates were probably different with different tribes.[6]

We can find, however, the tribes' own coins in gold, silver, and bronze. The coins of the Boii, who settled in the west of Pannonia and near the present Bratislava, are different from the philippei counterfeits, both in type and metrology. The Boii, who came from Bohemia around 60 B.C. at first had only gold coins (Pl. XLVIII. 2), although the other Celtic tribes had no gold minting. Here they began minting small silver coins (Pl. XLVIII. 3), which can be found in the hoards together with the gold ones. There are also large silver coins which weigh 17 grams; on one side they resemble a type of Roman denarius, and on the other they are reminiscent of Iberian mintages, as witnessed by one showing a dashing rider (Pl. XLVIII. 4). Roman influence is indicated not only by the choice of type, but also by the legend in Roman letters, BIATEC (Pl. XLVIII. 5), NONNOS (Pl. XLVIII. 6), etc.[7]

Connected in type with the Boii's small coins I mention the slightly heavier silver coins from the Tótfalu hoard (Pl. XLVIII. 7). In fact, they may be the prototypes of Boii change and were probably made about 70 B.C. Near the border of West Pannonia we can find coins of the Noricum type (Pl. XLVIII. 8), which are closely connected with coins of the so-called Apollo type (Pl. XLVIII. 9), silver coins adorned with the head of Apollo. Found together with a small amount of silver change, these silver coins may have been made about 90 B.C. From the middle of Pannonia, the area between Lake Balaton and the Danube, we have several buried hoards which contain only philippei imitations (Pl. XLVIII. 10). To the south of this area was found a group of hoards containing coins more roughly executed (Pl. XLVIII. 11).

338

The only type of coin that was made first of silver and then completely of bronze was the Regöly type, found in the counties of Somogy and Tolna and in the Kapos Valley (Pl. XLVIII. 12). A gradual deterioration in the style and composition of this coin can easily be noticed, proving that it had been used as a currency for a long time, perhaps even until the Roman conquest. These coins of the Regöly type are related to the coins emerging between the Drava and Sava rivers. This series, containing more and more bronze, may date back archaeologically to the end of the first century B.C.

## ROMAN DENARII AND ERAVISCAN IMITATIONS

Like the Boii coins, the Eravisci ones can be related to an ethnic group. Their minting was closely connected in metrology and style with that of the Roman denarius. The hoard from Lágymányos, Budapest, near the Eravisci Gellért hill settlement, contains Eraviscan imitations of denarii of the Roman Republic, minted between 60 and 40 B.C., and of Roman denarii from 80 to 70 (Pl. XLVIII. 13), as well as Augustan ones. Since its latest coins are from 10 B.C., this hoard comprises a long period, from the Roman prototypes of denarii to a range of coins leading to Eraviscan ones. Several hoards of this type, which still lack detailed description, close with denarii of Emperor Augustus, although a hoard from Bia[8] contains some coins of Tiberius and one of Caligula.

We know, therefore, that when the Roman denarii appeared, Greek coins, as well as imitations minted according to the Greek standard of coinage, were current in Transdanubia and the area between the Drava and the Sava. The influx of denarii dates later than the minting of the Eraviscan prototypes (80–70 B.C.) before the Roman conquest, i.e., the second half of the first century B.C. Eraviscan minting may have thrived in the twenties of that century, when a great shortage of money necessitated the substitution of Eravisci mint coins.[9] Hoards containing republican denarii, Eraviscan and Augustan mintage, may have been interred at the time of the Roman conquest because of a great Pannonian–Dalmatian revolt.

As we have seen before, Eraviscan minting was an intermediary between the tribal coinage following the Greek standard and the currency of the Roman denarii. In Pannonia there were not many denarii before the Roman conquest. We know only a few hoards, not many thousands of coins, most of them imitations and Eraviscan denarii. On the other hand, in Transylvania and in Barbarian areas, we know of hoards from the second and first centuries B.C. which contain nearly 20,000 denarii. We cannot speak of intensive trade connections between the native population in Pannonia and the Romans before the conquest. Hoards of denarii in Eraviscan areas do not prove real economic connections between the two peoples, as a Roman tradesman would never have accepted those imitations. Some connections certainly did exist between the two peoples; we come nearer to the truth when we suppose that these denarii show that even at an early time the Romans recognized the strategic importance of the Danube bend.

The influx of money into the newly conquered province of Pannonia occurred through several channels, with tradesmen being the first to bring goods and money in exchange for goods and raw materials. If industry in a province had been underdeveloped, or raw materials unavailable, the tradesmen would not have left their money there. Soldiers stationed here were paid regularly, though they were not paid much, particularly in early times — some 225 denarii before Domitian. Also the influx of money came in the form of regular sums received by government officials. Only if foreign tradesmen had the required industrial products did all this money go abroad; this is partly the reason for the scarcity of even stray coins from the period of the empire.[10] These coins are mostly from the area of the north Danubian *limes*.[11] In addition, minting had not kept pace with large-scale conquests: approximately the same number of coins were minted as during the days of the Republic, and now they had to be spread around a much larger area. Not only hoards from Pannonia, but also those from Britannia, western Europe (although to a smaller extent), Italia, the Balkans, and other Danubian regions give evidence of the shortage of money. Although most of these areas have no hoards interred past the end of the first century, the Pannonian hoards from the end of the second century present the same picture: Augustan, Tiberian, and Caligulan coins are generally missing.

Archaeological finds and excavations prove that in the early period of the empire the *limes* was not yet developed; there was no proper occupation, only the guarding and watching of strategically important places. So the new province of Pannonia was not completely consolidated militarily or economically. New settlements under Tiberius meant reinforcement of the province, and the Iazig uprisings under Claudius demanded the strengthening of the weak points of the defense lines. More soldiers had to be stationed in Pannonia and more money sent there, the effect of which can be seen even in the stray finds. The sestertii and dupondii of Claudius turn up more frequently in the province and in the western border areas, while denarii only rarely.

That no denarii before the middle of the first century are to be found in Pannonia or in the other middle Danubian provinces can be explained in several ways. Denarii of the early period of the empire were made of better quality silver than those of the days of the Republic. From the reign of Nero on, the quality of the silver used for denarii deteriorated noticeably. The intrinsic value of the denarii in the early empire was higher than that of the earlier or later coins, so they were taken out of circulation in a shorter time than the bad silver coins.[12] Therefore, the shortage of money in that period means a shortage of denarii as well.

During the Flavii the influx of money became even greater. This can be seen in the hoards interred during the wars in the second century; they contain coins of Nero and Vespasian, but also some denarii from the legions of Marcus Antonius the triumvir. This great number of triumvir's denarii is a common feature of the hoards found in other western and central European countries as well, because these denarii were coined in large quantities and dropped out of circulation more slowly. Silver and gold coins were used for 100–150 years or even longer.

340

We have indications that in the second century, until the outbreak of the Marco-mannic wars, there was a steady inflation, which explains the presence of a great amount of money from this period. Large numbers of gold coins were made in the period from Augustus to Marcus Aurelius, but in Pannonia we find only stray pieces on the side of the main roads. An aureus hoard like that from Brigetio[13] (118 pieces) is exceptional; in Carnuntum, for example, only 5 aurei from the first to the third centuries were found. Perhaps this great amount of gold in Brigetio was the hidden capital of a tradesman. The number of coins decreased during the Marcomannic wars. Frequently interred hoards indicate great devastation, and they also show the steadily rising influx of money in the previous decades.

Although archaeological and historical sources do not speak about important uprisings during the reign of Hadrian and Antoninus Pius, numerous hoards were interred, probably in connection with the invasion of the Carpathian basin, a consequence of the migration of the Goths.[14] These hoards, unlike those from the first half of the third century, contain a mixture of denarii and bronze coins of higher denomination. In the first two centuries of the empire coins were not worth hiding because money value, especially that of the bronze coins, did not change considerably, and later more coins were interred because of their content of precious metals. During the Marcomannic wars currency decreased. Frequently interred hoards indicate great devastation. They generally show the steadily rising influx of money in the previous decades.[15]

## FROM SEPTIMIUS SEVERUS TO GALLIENUS

Even in the second century there was a slow, nearly imperceptible decline in money value, evidenced in the diminishing quality of silver denarii. No great inflation was caused at that time, but by the end of the century the silver content of denarii had decreased by 70 percent. During the reign of Septimius Severus more and more denarii were minted partially or entirely of bronze (Pl. XLIX. 1), and increasingly more subaerati were made (Pl. XLIX. 2). During this time *usurpators* also did some minting, with resulting disorder, reminiscent of the two last confused decades of the Republic and of the days of the struggle for the throne after Nero's death. Minting conditions stabilized with the beginning of Augustus' rule, with issues by the emperor in Lugdunum and by the Senate in Rome; in A.D. 69 this stability vanished when coins were minted in Hispania and Africa as well.

The order that almost all coins be minted in Rome lasted from Vespasian to Septimius Severus. After the contests for the throne ceased, Antiochia had its own permanent mint, important above all from a military point of view. This occurred in the Danubian provinces as well. By the middle of the third century the military importance of the provinces alongside the Danube increased, since paying the soldiers in due time became a problem. Under Gordian III a mint was established in Viminacium and remained in existence until 260, when the Barbarian invasion necessitated its removal into the western part of the empire (Pl. XLIX. 3).

As a consequence of the instability of the denarius after the end of the second century, rough bronze imitations, originally silvered, appeared (Pl. XLIX. 4);

these can be found in large numbers all over the province. Other bronze coins, a few remarkable ones from the Greek colonies which also turned up in Pannonia, were only substitute money due to the shortage of Roman coins in the first half of the third century. There is another type of imitation, the so-called *limes*-imitation, which appears in such great numbers in Carnuntum that a counterfeiting workshop is thought to have operated there. These *limes*-imitations were copies of the bronze coins of several emperors, such as the sestertius and the dupondius. These counterfeit coins, in large numbers and dating until the middle of the fourth century, can also be found in Gallia under Tetricus, and in Pannonia in the Carpathian basin and on both sides of the Danube (Pl. XLIX. 5). Counterfeit types soon degenerated, since a minting die made of soft material was unsuitable for making a large number of copies, and a new die was only a copy of the old one. The counterfeits may have weighed more or less than the genuine ones, but it does not matter because the difference between the value of the metal and the denomination, in any way, was the profit of the counterfeiter. In the fourth century even gold coins were counterfeited, especially the types made in the minting period of the big solidus (Pl. XLIX. 6). Roman tradesmen most likely produced them for the Barbarians.[16]

Instead of, or in addition to, the deteriorating denarius, Caracalla initiated the antoninianus, of a weight of a denarius and a half. It came into general use only from the reign of Gordian III; almost all hoards from the first half of the third century show antoninianis nearly always substituting for the denarii. The hoards were interred around 204,[17] 230,[18] 252,[19] and 259[20] B.C., these years being connected with wars, the most tragic of them in 259–260. The province was temporarily weakened not only by the attacks of the Roxolans, but also by the anti-emperors, e.g., Ingenuus and Regalianus, the latter residing and minting coins in Carnuntum[21] (Pl. XLIX. 7). Archaeological research, the sources, and over twenty hoards of coins interred at that time are ample evidence of this. Because of the large number of hoards, we can analyze the currency of the third century. The mixed hoards of the previous century disappeared, and in their place we find collections of fine silver coins. Gone too is the poor quality denarius of the reign of Septimius Severus and the subaerati. Antoniniani, the new currency, are rare, with denarii being prevalent in the hoards interred around 230. However, during the 250s the majority of coins in the hoards are antoniniani.

The temporal distribution of hoards contains coins from the previous two hundred and fifty, one hundred, or fifty years. It is worth mentioning, however, that even in hoards like those from Ercsi and Börgönd, buried in the 230s but containing some denarii of Marcus Antonius the triumvir, the shortage of currency in the first century is obvious. Besides the denarii there are only coins from the end of the century or from the period of second-century emperors. However, the dating back of hoards to such early periods is rare; the majority of coins generally come from the previous thirty to forty or, at most, fifty years. We can therefore state that, in general, most of the coins interred up to the third decade of the third century come from Septimius Severus; these were hidden around 260, from the period of Gordian III.

342

In the province of Pannonia, 260 is a dividing line for both currency and hoards, mainly due to wars. The serious deterioration of money during the second half of Gallien's reign (antoniniani were first coined from billon, then from bronze) led to a decrease in the use of the currency and to a gradual increase in natural economy. In Pannonia relatively few hoards are known which extend over the last third of the century; the hoards came mostly from the Gallien period and closed with a few antoniniani of Claudius II, such as those from Szekszárd, Komin, and Kunfehértó, although these sometimes reached up to 270–280 (e.g., Vipava, Müllendorf).[22] Compared with the influx of coins in the first half of the century, this period shows a smaller number; in addition, these antoniniani were of different value since they were made of billon or bronze.[23] The gold hoard of Brigetio in the currency of the first and second centuries and the large silver collection from Siscia are exceptional.[24] It does not fit into the framework of the normal use of currency: the Siscian hoard consists of 1,415 silver coins. However, the total number of Pannonian silver coins which are in private and public collections from this period is only 533. Dealing with this hoard here would completely distort the picture of the silver currency of this period.

In order to stop devaluation culminating under Claudius II, Aurelian introduced silver-plated antoniniani of better quality. As a result of Diocletian's currency reform, the follis was introduced, and its reduced versions were current for some decades. This period, at the turn of the century, brought the ending of inflation and deflation and the increased importance of taxation. Currency reflects this; there was too little money, and coinage was almost worthless in value. We have no large hoards from this period; in the stray coins there is little gold or silver, but frequent follis.

Additional money reached the province during the Constantinian dynasty, and we even know about the hiding of some hoards. The hoard of Nagytétény,[25] for example, interred about 333, is interesting not only because of the more than ten thousand coins but also because it contains some contemporary counterfeits. Some smaller hoards and strays which come from the 350s[26] and hoards from the age of Valentinian show the essence of the late Roman mint system. Since each district got most of its money from the nearest mint, most of the coins in Pannonian hoards were minted in Siscia. The mint in Sirmium has little representation in these hoards, since it worked only periodically and generally minted coins of earlier metals, which occur only as stray finds. Accordingly, it can be stated that there are more coins in these hoards from mints near Pannonia and fewer from farther ones. The coins come mainly from Siscia, but also from Thessalonica, Aquileia, Heraclea, and Rome.

## ECONOMIC UPHEAVAL

The coins after 364 are dealt with separately, partly because the period was short but prosperous and partly because it is the vaguest period in the history of currency in Pannonia. There are several explanations for the disappearance or very occasional occurrence of coins minted after 375. This period between 364 and 375, however, is rich in large hoards,[27] and the large quantities of coins are the result of the rebuilding and reconstruction of fortifications during Valentinian's reign. From examination and analysis we know that there was a steady growth of currency in the period between 364 and 375. In 367 the number of coins suddenly became more numerous, declining somewhat until 370, then again decreasing significantly. The amount of incoming money rose to an ordinary level again in 374-375.

Coins from earlier years can always be found in these hoards. The validity of the coins was probably determined by the materials and the form, not by the name of the emperor inscribed. Since in the fourth century the largely Barbarian population of the province could hardly read, the legends and the portrayal of the emperors became completely schematic. The great amount of money coming in during the reign of Valentinian I was probably current after 375 as well. This can be proved by several large hoards, such as those from Veszprém and Árpás,[28] which consist of thousands of coins, only a few of which (fifteen pieces of the Veszprém hoard) were minted after 375. While the hiding of the hoards took place well after 375, there are hardly any coins in them from the intervening period. The same is true of the coins from settlements and cemeteries.

After 375 the regular influx of money stopped. Only a very few new coins were brought into the province, and after 385 none at all. In spite of this, archaeology seems to prove that life did not stop or change in the province; the province of Valeria was under Roman rule even at the beginning of the fifth century. It is more probable that the large amounts of money from before 375 continued to circulate in the province for several more decades, an assumption borne out by the fact that the coins in these hoards are often extremely worn. We believe that the province of Valeria had its regular currency until the province had fallen into the hands of the Barbaric tribes.[29]

## MINTS

From the middle of the third century inflation and war caused the proliferation of mints in the empire. This tendency can be seen clearly in the hoards. Coins that earlier had no mint mark and could be localized only by style or historical event were now marked distinctively. By the end of the third century nearly all minting places used their own initials or monograms as mint marks, together with the officina or the serial number given in Greek or Roman numerals.

Because of the inflation, mints began to work in the 260s in Lyon, Mediolanum, and Kyzicos. When the mint in Siscia was established during the reign of Emperor Gallienus in 262[30] (Pl. XLIX. 8), the minting of bronze coins had already ceased all over the empire, with the exception of bronze antoniniani made at different mints. The mint in Siscia worked uninterruptedly from its establishment to the end of the fourth century. From a military point of view, the importance of the mint increased in 268 as a consequence of the war against the Germans in the Balkans; at that time plenty of bronze antoniniani and even gold coins were minted. It is difficult to distinguish the first series from Siscia. Because the small mint with two officinas (later there were three, still later two more) could probably work only with staff and mint-dies from the large mint in Rome, its coins are similar in style and technique to the Roman ones. Lack of distinctive marks in the first series means that they can be separated only with difficulty. On the reverse the officina was indicated by such marks as P, S, I, and II. As the mint became larger and larger, it worked with four officinas for Claudius II,[31] with six for Aurelian, and seven for Probus.[32] For a long time, only Roman letters were used as mint marks; Greek letters appeared around the reign of Probus. Later on, and for a long period, the mark SIS was also used occasionally. When the seventh officina was organized under Probus, there may have been workmen in Siscia from the mint in Serdica, since that was the only mint in Europe using the Greek KA which in a short time appeared in Siscia. Three officinas worked for Carus. Not only the bronze antoniniani but also some gold coins were minted here for Gallienus, and later aurei for nearly every emperor.

At the turn of the third and fourth centuries, the mint of Siscia was of great importance in providing the Danubian regions with money. It minted great quantities of gold and silver coins as well as a steady number of bronze coins. In fact, 37 percent of the silver hoard from Siscia was minted there. After 305 the minting of silver coins in Siscia ceased for a long period, but in 326, 327, and 334 silver coins were minted at Siscia for the sons of Constantine, and some gold ones were produced for a short time in 345. Siscia was one of the first mints taken from Licinius by Constantine who began minting silver coins again, though not very intensively. The minting of bronze coins remained unbroken, five officinas having coined in the period between 324 and 346. Marks were changed at irregular intervals, one mark of a type being used for less than six months, another one for two years and again for another three years. For a short time in 345 the mint made gold and even silver coins for the sons of Constantine. As the coins became more and more complicated, a special symbol was added to the mark of the officina.

There were five officinas working in the days of Constans, Vetranio, and Constantius II, and in the days of the first issues of Constantius Gallus (Pl. XLIX. 9). From 365 to 395 there was an unbroken period in which were minted gold, silver, and bronze coins, the latter in three sizes. In 387 minting in Siscia was stopped for a short time, probably because of a Barbarian invasion not mentioned in the sources. However, the mint worked until 392, and then again, after a short interval, from 408 to 423 (e.g., the coins with the mark SM of Honorius and Theodosius II).

The other mint in Pannonia was established in Sirmium, which had been the capital since the campaign against Licinius, although the headquarters of Constantine was Thessalonica. Gold coins were minted only here and in Trier. The characteristic of the mint is periodicity, a few years' work followed by a long interval. Work began here in 320, and coins made of all three metals poured from the mint until 326. The beginning of the work in the mint is shown by a readily datable series of gold coins made in 320–321 for the quindecennalia of Constantine, and in 321 for the second consulship of Crispus and Constantine II. Variants of mint-marks make more exact chronological specification possible. Comparatively few silver coins were minted and only between 320 and 324. Bronze coins in the early period were made only from 324 on, with the minting of them going on in the following year, issued from two officinas.

From 326 on minting was done in Constantinople until 350, when minting began again in Sirmium. At this time bronze coins of different sizes were minted, as well as some silver coins for Constantius II, Constantius Gallus, and Julian. At first there was one officina, and before the death of Constantius Gallus there were two (A, B). The organization of the mint lasted until 367. The minting of gold coins began only in 364 under the joint rule of Valentinian and Valens, at the same time as the minting of bronze and silver coins (Pl. XLIX. 10). After some years' interval, work began again in 378, but only siliquae and solidi were minted until 383. Gold coins were made here for Arcadius and Honorius as well as for Theodosius I. Nothing certain can be said about the exact date when the mint stopped working, although it may have stopped before the death of Theodosius I. The mark of the last mintages was (S)(M)/COMOB.

## NOTES

[1] G. ELMER, "Der römische Geldverkehr in Carnuntum". *NZ* 26 (1953): 55–67.

[2] T. PEKÁRY, "Aquincum pénzforgalma— Money Circulation in Aquincum". *ArchÉrt* 80 (1953): 106–14.

[3] M. R. ALFÖLDI, "Der Geldverkehr von Intercisa", in *Intercisa*, 142–69 (*ArchHung* 33).

[4] B. KOCH, "Münzfunde dokumentieren den Geldumlauf im Burgenland". *NZ* 83 (1969): 82–97.

[5] A. KERÉNYI, "Gruppierung der Barbarenmünzen Transdanubiens". *FolArch* 11 (1959): 47–60.

[6] Ö. GOHL, *Gróf Dessewffy Miklós barbár pénzei* (Budapest, 1910–1915), 4 vols; K. PINK, "Die Münzprägung der Ostkelten und ihrer Nachbarn" (Budapest, 1939) (*DissPann* 2:15).

[7] Ö. GOHL, "A BIATEC-csoportbeli barbár pénzek" (Barbarian Coins of BIATEC-Group). *NumKözl* 8 (1909): 39, 99; Ö. GOHL, in *NumKözl* 10 (1911): 52; *NumKözl* 12 (1913): 41; *NumKözl* 20 (1921): 9; *NumKözl* 21–22 (1922–1923): 28; V. ONDROUCH, *Keltské mince typu Biatec z Bratislavy* (Bratislava, 1958); K. CASTELIN, "Biatec a Nonnos". *NumListy* 12 (1957): 157–68; K. CASTELIN, "Zur Chronologie des keltischen Münzwesen in Mitteleuropa". *J. f. N. Geldg.* 12 (1962): 199–207.

[8] J. HAMPEL, "Az eraviscus nép és emlékei" (Eraviscus Tribe and Relics), *BudRég* 4 (1892): 34 ff.; Ö. GOHL, "Eraviszkus pénzek a Lágymányoson" (Eraviscus Coins at Lágymányos). *BudRég* 8 (1904): 181; Ö. GOHL, "A budapesti eraviszkus éremlelet" (Eraviscus Coin Find in Budapest). *NumKözl* 1 (1902): 41.

[9] A. KERÉNYI, "A regölyi típusú barbár bronzpénzeink kronológiájához— Contribution à la chronologie des bronzes barbares du type Regöly". *NumKözl* 56–57 (1957–1958): 7–9, 71; J. KLEMENC, "Ostava u Ličkom Ribniku". *Vjesnik Hrvatskog Arheološkog Društva* 16 (1935): 83 ff.; A. MÓCSY, "A római pénz forgalmáról a római uralom előtti Pannóniában— Über den Umlauf des römischen Geldes in vorrömischen Pannonien". *NumKözl* 60–61 (1961–1962): 15–18, 101.

[10] A. ALFÖLDI, "Dunántúl felé terjeszkedő korarómai kereskedelem új nyomai— New Traces of Early Roman Trade Extending towards the Trans-Danubian Region". *Magyar Múzeum* 2 (1946): 52–57, 95–96.

[11] V. ONDROUCH, *Der römische Denarfund von Vyškovice* (Bratislava, 1934).

[12] E. JÓNÁS, "Két római denárlelet Aquincumból" (Two Roman Denarius Finds from Aquincum). *BudRég* 12 (1937): 278–88.

[13] L. BARKÓCZI and K. BÍRÓ-SEY, "Brigetioi aranylelet — Trouvaille d'or à Brigetio". *NumKözl* 62–63 (1963–1964): 3–8, 109.

[14] A. BARB, "Ein römischer Münzfund aus Wallern". *Mitt. d. Num. Ges. in Wien* 16 (1927): 10 ff.; A. RADNÓTI, "A zalahosszúfalusi ezüstlelet" (The Silver Find from Zalahosszúfalu).

*FolArch* 3–4 (1941): 102–15; E. PEGAN, "Najdbe novcev v Sloveniji". *Arheološki Vestnik* 18 (1967): 203–21.

¹⁵ K. BÍRÓ-SEY, "A szombathelyi koracsászárkori denárlelet—Ein Denarfund aus der frühen Kaiserzeit in Szombathely". *FolArch* 12 (1960): 75–89; R. GÖBL, *Zwei römische Münzhorte aus dem Burgenland: Illmitz und Apetlon II* (Eisenstadt, 1967) (*Wissensch. Arbeiten aus dem Burgenland* 37); M. KŐHEGYI, "Pótlás a szombathelyi koracsászárkori denárlelethez" (Contribution to Szombathely Denarius Find from Early Imperial Period). *NumKözl* 64–65 (1965–1966): 69–71.

¹⁶ A. ALFÖLDI, "Anyaggyűjtés a római pénzek Magyarországon készült utánzatainak osztályozásához" (Classification of Roman Coin Counterfeits Made in Hungary). *NumKözl* 25 (1926): 37–48; *NumKözl* 26–27 (1927–1928): 59–71.

¹⁷ I. SZ. CZEGLÉDY, "Császárkori denárlelet Szombathelyről—Trouvaille de deniers datant de l'époque des empereurs romains mise au jour près de Szombathely". *NumKözl* 60–61 (1961–1962): 19–22, 101–2; J. FITZ, "Septimius Severus-kori denárlelet Mór–Felsődobosról—Une trouvaille de deniers de Mór–Felsődobos datant l'époque de Septime Sévère". *NumKözl* 58–59 (1959–1960): 16–22, 84–85.

¹⁸ S. SOPRONI, "Az ercsi éremlelet—Deniers romains à Ercsi". *NumKözl* 62–63 (1963–1964): 9–17, 109–10; A. RADNÓTI, "A börgöndi éremlelet" (The Coin Find from Börgönd). *NumKözl* 34–35 (1935–1936): 24–27.

¹⁹ E. JÓNÁS, "A bajóti római ezüstpénzlelet és tanulságai (Roman Silver Coin Find from Bajót)". *Orsz. Magyar Régészeti Társulat Évkönyve* 2 (1923–1926): 137; K. BÍRÓ-SEY, "A pilisszántói éremlelet—La trouvaille de Pilisszántó". *NumKözl* 64–65 (1965–1966): 9–11, 109–10; A. RADNÓTI, "Néhány adat a consecratiós érmekhez" (Contribution to Consecration Coins). *NumKözl* 44–45 (1945–1946): 6; L. BARKÓCZI, "Újabb éremlelet Intercisából—Récente trouvaille de monnaies à Intercisa". *NumKözl* 54–55 (1955–1956): 3–6, 79; W. KUBITSCHEK, "Ein Fund römischer Antoninianae aus Serbien". *NZ* 23 (1900): 185.

²⁰ K. BÍRÓ-SEY, "A kistormási éremlelet—Le trésor de monnaies de Kistormás". *FolArch* 15 (1963): 55–68; M. R. ALFÖLDI, "A rábakovácsi római ékszerlelet—Roman jewelry of Rábakovácsi". *FolArch* 6 (1954): 62–73, 204–5; Ö. GOHL, "Éremleletek" (Coin Finds). *NumKözl* 7 (1908): 122; M. ALFÖLDI, "A IV. szalacskai éremlelet" (The IV Szalacska Coin Find). *NumKözl* 50–51 (1951–1952): 7 ff.; Ö. GOHL, "A korongi római éremlelet" (Roman Coins at Korong). *NumKözl* 1 (1902): 39 ff.; R. GÖBL, *Der römische Münzschatzfund von Apetlon* (Eisenstadt, 1954); M. ALFÖLDI, "A sztálinváros-(Dunapentele-)dunadülői éremlelet—La trouvaille de monnaies de Sztálinváros-(Dunapentele-)Dunadülő". *NumKözl* 52–53 (1953–1954): 5–8, 63; K. BÍRÓ-SEY, M. KÁROLYI and T. SZENTLÉLEKY, "A balozsameggyesi római ékszer- és éremlelet—Der römische Schmuck- und Münzfund aus Balozsameggyes". *ArchÉrt* 98 (1971): 190–204.

²¹ J. FITZ, "Ingenuus et Regalien". *Latomus* 81 (1966): 7 ff.

²² Magyar Nemzeti Múzeum Éremleletek 863-13-4 (1952); Z. BARCSAY-AMANT, "A komini éremleletek" (Budapest, 1937) (*DissPann* 2:5); T. PEKÁRY, "A kunfehértói éremlelet a III. századból" (Third-Century Coins from Fehértó). *NumKözl* 54–55 (1955–1956): 50–52; J. BRUNŠMID, "Nekoliko našašca novaca na skupu u Hrvatskoj i Slavoniji". *Vjesnik Hrvatskog Arheološkog Društva* 13 (1913–1914): 269; W. SCHMID, "Ein kleiner Weisskupfermünzfund aus Emona (Laibach in Krain)". *Berliner Münzblätter* 4 (1911), No. 110, pp. 2–4; J. BRUNŠMID, "Colonia Aurelia Cibalae". *Vjesnik Hrvatskog Arheološkog Društva* 6 (1901): 159–60; J. ŠAŠELJ, "O naojdbi rimskih denarjev v Mokronagu". *Slovenski Narod* (1880), pp. 9, 10; F. REDŐ, "Numismatical Sources of the Illyr Soldier Emperors' Religious Policy". (Budapest, 1973) (*Dissertationes Archaeologicae* 2:2).

²³ *Fundberichte aus Österreich* 5 (1959): 190 ff.

²⁴ A. JELOČNIK, *The Sisak Hoard of Argentei of the Early Tetrarchy* (Ljubljana, 1961) (*Situla* 3).

²⁵ A. ALFÖLDI, "Il tesoro di Nagytétény". *Rivista Italiana die Numismatica* 34 (1921): 113–90.

²⁶ A. ALFÖLDI, *NumKözl* 56–57 (1957–1958): 63; M. ALFÖLDI, "Későrómai kisbronzleletek Alsópáhokról" (Late Roman Bronze Coins from Alsópáhok). *NumKözl* 50–51 (1951–1952): 71; P. HARSÁNYI, "Éremleletek" (Coin Finds). *NumKözl* 20 (1921): 55; K. PINK, "Der Geld-

verkehr am österreichischen Donaulimes in der Römerzeit". *Jahrb. f. Landeskunde v. Niederösterr.* 25 (1932): 60–61; K. Bíró-Sey, "A perbáli éremlelet— A Hoard of Roman Coins from Perbál". *FolArch* 16 (1964): 63–74.

[27] *Fundberichte aus Österreich* 1 (1930–1934): 265; A. Barb, "Der Münzfund von Jabing (374 n. Chr.)". *NZ* 29 (1936): 61 ff.; A. Alföldi, "A keczeli lelet" (The Find from Keczel). *ArchÉrt* 39 (1920–1922): 99–102; M. Alföldi, "Fragen des Münzumlaufs im 4. Jahrhundert n. Chr.". *J. f. Num. Geldg.* 13 (1963): 75–104; M. Alföldi, "A szőkedencsi későrómai éremlelet" (Late Roman Coins at Szőkedencs). *Antiquitas Hungarica* 2 (1949): 1–7; A. Radnóti, "Későrómai éremlelet Öregcsémről" (Late Roman Coins at Öregcsém). *NumKözl* 41 (1942): 11 ff.; E. Jónás, "Az öcsödi éremlelet" (The Coin Find from Öcsöd). *NumKözl* 28–29 (1929–1930): 30 ff.; A. Sz. Burger, "Későrómai éremlelet Kazsokról" (Late Roman Coins at Kazsok). *NumKözl* 66–67 (1967–1968): 3 ff.; S. Soproni, "Valentinianus-kori éremlelet Hajdúnánás-Tedejről—The Hajdúnánás-Tedej Coin Find from the Period of Valentinianus". *A Debreceni Déri Múzeum Évkönyve*, 1966–67 (1968): 91–117; E. Polaschek, "Zwei römische Münzschätze aus Wien". *NZ* 18 (1925): 127 ff.; M. R. Alföldi, "Der Geldverkehr".

[28] W. Kubitschek and O. Voetter, "Ein Münzfund aus Veszprém". *NZ* 2 (1909): 117–36; K. Bíró-Sey, "A Zsófia-pusztai lelet—La trouvaille de Zsófia-puszta". *NumKözl* 60–61 (1961–1962): 29–48, 103; E. Jónás, "Az árpási római éremlelet" (Roman Coins at Árpás). *NumKözl* 32–33 (1933–1934): 3 ff.

[29] S. Soproni, "Über den Münzumlauf in Pannonien zu Ende des 4. Jahrhunderts". *FolArch* 20 (1969): 69–78.

[30] A. Alföldi, "Siscia I". *NumKözl* 26–27 (1927–1928): 50 ff.

[31] A. Alföldi, "Siscia II". *NumKözl* 34–35 (1935–1936): 3–8.

[32] A. Alföldi, "Siscia V". *NumKözl* 36–37 (1937–1938): 3–89.

# ARTS AND CRAFTS

### EDIT B. THOMAS

Roman life in Pannonia, as in the city of Rome and the entire Roman Empire, was saturated with love of arts. However, in the interior of the province and behind the *limes*, a different kind of Romanization existed from that in the *limes* and in the camps of the military forces and the adjoining civilian towns.

The Roman expansion which began during the reign of Emperor Augustus not only produced new social institutions but also fostered new religious beliefs which resulted in new conceptions of the gods. The new settlers also wanted to live in a wholly Romanized atmosphere.[1] Their buildings followed Roman models and were produced in their majority by Roman masters and their statues were copies of Italian counterparts, whereas the bronze statuettes of their household gods belonging to the families were brought along from the homeland.[2]

Certain products which arrived in Pannonia bore the imprint of Roman court art fashionable during the Augustan period, while others were deeply affected by more distant Hellenistic and Eastern influences.[3]

## EARLY IMPERIAL PERIOD

Early Roman finds in Pannonia from the period of Augustus, Tiberius, and Claudius—such as the Satricum sanctuary, the Mater Matuta terra-cotta group of statues—are the best examples of humble peasant votive objects.[4] Their traditional forms suggest the conclusion that they were produced in the third century B.C., or even in the late second, but accompanying finds and the circumstances of their emergence make it absolutely certain that they were objects of worship from the early first century in a Dunakömlőd shrine in Pannonia (Pls. LXXX and LXXI).

The *lar* of the *lararium* of Nagydém (Pls. XXVI—XXVIII) is a specimen representing the taste and trend of Augustan court art.[5] It must have been produced in the south of Italia in the early first century, as indicated by the evidence of a highly characteristic postament which forms part of it.

A second piece in the Nagydém lararium, Apollo,[6] (Pl. XXVIII), reflects Hellenistic Greek taste; nevertheless, we believe that it was also made during the Roman period, in the early first century.

Elements of Hellenistic taste in a Romanized conception also emerged later in first-century paintings made under Italian influences,[7] and in later works, as, for example, the floating figures in early wall paintings of the villa at Balácapuszta[8] or

349

the Dionysus child figure with a bunch of grapes and accompanying decorative masked bands (Pl. XLI). Hellenistic traditions seem to survive also in the studding of the Kisárpás casket (Pl. LXXXII. 1, 2)[9] and in the Dionysian *thyas* scenes on a bronze lamp from Intercisa. The relief with a woman opening a casket also from Intercisa seems to suggest a relationship with Pandora's box[10] (Pl. LXXXIII). The examples cited above suggest that during the four centuries of Roman rule in Pannonia the Hellenistic spirit and the influence of Hellenistic arts prevailed. This influence is evident not only in imported pieces from the said territories, but also in pieces produced in Pannonian workshops.

After Savaria,[11] Scarbantia,[12] and other cities in Pannonia Superior were granted the rank of *colonies* and *municipia*, it is evident that Roman institutions were introduced and art also developed accordingly. *Capitolia* were soon to be raised to celebrate the Trias (Pl. XIII). Statues of Jupiter, Juno, and Minerva were carved in marble twice the size of life. Carinthian marble statues emerged from an art center extant in Noricum during the reign of Domitian, at the end of the first century A.D. Differences in execution seem to indicate there were several artisans. Prototypes of the statues' forms can be found on the reverse of Domitian coins, so when dating them we can rely on solid evidence. This was the age which aspired to build *capitolia* everywhere throughout the empire as a result of an urbanistic trend and a desire to emulate Rome.

Recent excavations in Savaria seem to prove that in several Roman cities of Pannonia planning was systematically organized.[13] The study of sanctuary centers among the houses suggests a network of *insula* building clusters. Around the Isis sanctuary[14] (Pls. LXXXIV and LXXXV) arose a network of columned courtyards which, together with the sanctuaries of other Eastern gods, constituted an *insula*. The urban significance of Savaria in the province did not lose ground from the early Empire to late Roman times. In an early period the Amber Route,[15] (Pl. XII), and through it direct contact with Italia, had determined the trend of Savaria's development. From early urn cemeteries alongside the roads leading to Savaria fine glass vessels and early amber carvings have been unearthed.[16] Early gravestones carved in the Roman style were succeeded in the second and third centuries A.D. by grave sculpture which observed the finest traditions.[17] The best Savarian carvings survive in the ornamental marble section which adorn the facade of the Isis sanctuary[18] (Pl. LXXXV). Along with the architectonic elements the facade of the sanctuary excels in the presentation of the goddess Isis, sitting on the back of the dog Sotis (Pl. LXXXIV) and holding in her raised hand the sistrum. The figures of Mars and Hercules to her right and left reflect the characteristics of sculpture during the reign of Severus in the third century.[19]

North of Savaria lies Sopron, the former Scarbantia. Excavations in recent years revealed the full topography of the Roman city and the wall system surrounding it (Fig. 22).[20] Senetio's gravestone[21] commemorates the early occupants: it is a classic example of the narrative style in grave sculpture.

In Egyed,[22] not far from Sopron, was unearthed the finest treasure of the Isis cult (Pls. LXXXVI and LXXXVII). Into the bronze base of pitcher and *patera* were encrusted silver and gold inlay figures of Egyptian gods and goddesses, and there was a Nile scene as well. However, the laurel wreath encrusted on the edge of the *patera* places these treasures within the bounds of Roman art of the Augustan period.

In Gorsium, which was situated on the outskirts of the village of Tác,[23] excavations which have been going on for decades have uncovered the remains of a city with villa-type houses in an insular arrangement (Figs. 46 and 47). Recent excavations have unearthed the remains of a Roman nymphaeum[24] built in Hellenistic style, with cultic statues and rich architectonic elements from the third century (Pl. XIV. 1).

In the interior of the province, Roman villas and villa settlements played an important role.[25] Shrines either placed inside or outside the buildings were indispensable parts of the villas. Disregarding whether they were artistic or primitive in execution, the shrines reflected not only the beliefs but also the tastes of the inhabitants, their artistic sense and, last but not least; their financial capabilities.

The Hercules sanctuary at Ajka[26] preserved for posterity the full equipment of a shrine placed outside the building (Pls. XXXVI–XXXVIII). The statue representing the Hercules cult was made by a local artist; the altar stone and the tombstone refer to the Dalmatian or Italian origin of the family which erected it. Lead objects which emerged from a villa at Pogánytelek (Pl. CXXIX) conserved for us small votive pieces belonging to the Pannonian agrarian population, offered to gods and goddesses of fertility, Silvanus, Silvanae, Isis, and Isis Conservatrix.[27]

At this juncture we are not interested in the humble *villa rustica* which served exclusively economic ends, but rather we wish to devote our attention to the *villa urbana* with ornate peristyle and particularly the villa at Baláca and its paintings and mosaics (Pls. XLI and XLIII). The reconstructed walls of this villa display changes in style and taste from the late first to the late third century as evidenced in the wall paintings of a Roman house. It preserved all variations of style from the Pompeian pseudo second (Pl. XLI) to late frescoes displaying a faulty performance. The floor mosaics of the villa at Baláca accord well in their splendor with the wall paintings.[28]

The richly decorated stuccoes (Pl. XLI), frescoes, and mosaic floors of the early period clearly prove that the Italic owners of the *latifundia* in Pannonia attempted to produce a Roman setting even though living far from home.

While the Baláca villa secures a prominent place for itself with its interior decorations of Pannonian art, other villas — as, for example, those of Pogánytelek[29] and Sümeg[30] — are remarkable, since we consider them the forerunners of the four-towered Romanesque churches.[31] There seems to be no doubt that they influenced the emergence of the four-towered basilica in the time of King Stephen I. Nor can we leave out of account the *basilica rustica* which in villa settlements, especially in the Balaton region, is an important link in continuity. The Early Christian basilica[32] preserved for us other fine specimens of art, and it also helps us in reconstructing the altar table, lamp hangings, and cultic objects of late Roman provincial Early Christian basilicas.

The marble carvings of the altar table at Csopak[33] (Pl. LXXXIX) have preserved the symbols of Arian Early Christianity. The piece is also a westernmost representative of marble tables of this type prevalent in Arian circles. Its marble carvings suggest that the table is a noble example of Alexandrian art.

Cemeteries seem to have been much the same in the interior of the province as they were on the *limes*. The Pannonians adopted from the Romans the ritual of erecting tombstones, placing either simple or more ornate stones over the remains of their dead. Wealthy families had family graveyards with richly embossed buildings.[34]

On the *limes* the sparse unadorned astral tombstones[35] (Pl. XC) of the aborigines were soon followed by archaic stones of Roman style (Pls. XCI and XCII) preserving for us the picture of characteristic clothing and jewelry belonging to that style. The tombstones represented dead soldiers wearing the uniforms of their divisions, marking rank as well (Pl. XLIII), so that these stones provide important evidence on ancient weaponry.

However, we have other sources of information too on Pannonian armor[36] which we wish to mention now, since the last few years have produced very fortunate, unexpected finds from the Danube bed (Pls. LXXIV–LXXVI). Many weapons have been unearthed, among them a dagger from the first century, with gold, silver, and red enamel inlay; richly decorated shield bosses from the first and second centuries; and other objects as splendid examples of Roman military decorative art—the symbolic motifs on them serving imperial propaganda as well.

Among seven Pannonian helmets, the most sumptuous one emerged during the building of Elizabeth Bridge in Budapest (Pl. LXXIII). We have tried to date this exquisite piece in a study, fixing its date around 363 A.D., and its provenance from an imperial workshop in Sirmium, in South Pannonia.[37]

From an architect's point of view, the most significant building of the *limes* was the governor's palace at Aquincum, notwithstanding the fact that its dating in Hadrian's time has been recently regarded as too early (Pl. XVII).[38] The palace follows a centralized plan and is built around a central courtyard with an arched facade on the Danube side. Its mosaic floors and wall paintings and its architecture as a whole make this building among the best in the province. In the interior of the building has been found a statue representing Fortuna as Nemesis (Pl. XCIV) in accordance with the Eastern conception of the goddess.

Not far from the governor's palace, in Meggyfa Street (Pl. XVIII), one of the finest mosaic floors of Pannonia has been recovered.[39] It represents the first appearance in Pannonia in the third century of the smallstone technique, beloved in the East. The figure of Deianeira dominates the field, which is made up of small mosaic stones of a few millimeters.

Military buildings also emerged from the *limes*. A small bronze model of a camp gate (Pl. XCVI),[40] serving as decoration on a carriage, also came down to us. Solidly built objects have not changed with the passing of time. An outstanding record of military building survived in a terra-cotta (Pl. LXI) copy of a wide, arched gate with several doorways at the camp of Intercisa.[41] Its maker cut his name into the clay in the inscription: *Hilarus fecit portam feliciter*.

352

Numerous Eastern elements, remnants of the military units, survive on the Pannonian *limes*.[42] These Eastern motifs and religious features are welded into a harmonious whole by the transforming force of the all-embracing style of the Danube area. To give an example, in the Jupiter Dolichenus shrine at Brigetio (Pl. CXXVI) there is a cultic image of laminated bronze covered by white metal, an outstanding representative of the metal craftsmanship of the Danube region.[43]

We cannot fail to mention Romulianus,[44] the only name of a metalworker which has come down to us. He refers to himself as *artifex* on the base of the statue made in honor of Jupiter Dolichenus.[45]

In the numerous Mithras shrines in Pannonia,[46] the traditional rigid cultic picture is also marked by the artistic conception of the Pannonian mason.

A zither discovered at Intercisa preserved for posterity on the covering plate of the musical instrument the ivory figures of Isis (Pl. XCV), Mars, and Venus. Figures of a playful Eros are in a typical Alexandrian style.[47]

The Jews, the first Eastern monotheists arriving in Pannonia, left to us their creed carved in stone on a family tombstone. To express their homage they also erected a stone monument in Intercisa (Pl. CXXXIV).[48]

The Christians, last among the Eastern mystery believers, arrived in Pannonia at a time when it was essential to counteract the official religion of the Sirmium court and to carry on extensive missionary work, since against the orthodox trend, the Arian heretics enjoyed the full support of the court.[49]

In order to illustrate this later imperial style, let us look at a late Roman half-length portrait of an emperor (Pl. XCIX) wearing a laurel wreath which contains the seeds of the Byzantine portrait tradition.[50] Its insignia, the pearl diadem and the fibula, remained for centuries the distinguishing marks of emperors.

To summarize briefly the main artistic trends prevailing in Pannonia,[51] we can say that during the early first century a direct Italian influence determined the style of life and an attitude toward art products among the local inhabitants. Strong Italian influence during the Flavian and Domitian era expressed itself in monumentalism and in a Baroque mode of expression that typifies the period. Italian forms of architecture and Pompeian styles of decoration appeared everywhere in the province. Even Hadrianic styles reached the province, leaving some important monuments behind. Yet traces of the archaic style prevalent during the Antonines left their mark on Pannonian art.

In the Severan era the province underwent a period of flowering affecting the arts and leaving behind valuable relics. During the second and third centuries Eastern and North African influences were strengthening, while the Rhineland affected the applied arts particularly. The image of late Roman Pannonia and its artistic endeavors were strongly affected in the province by the proximity of the imperial court of Sirmium. Barbarian taste became more and more noticeable. The artistic expressions of the Christians, who were gradually gaining ground, also depended on these factors, which naturally sifted through their own ideologies.

When considering certain relics in sculpture, painting, and the applied arts, we come to the conclusion that the art in Pannonia which emerged under Roman influences, particularly as far as the outward forms of the artistic products are concerned, represents a wide range. From primitive art, through abstract trends, and to the emergence of truly classical forms, every variety of style and taste existed.

This great variety was due primarily to successive historical events, but the geographical position of Pannonia was also a factor. At the meeting point of north, south, east and west contacts and relations offered multiple opportunities which the people in the province never failed to exploit and weld into specifically Pannonian products.

## NOTES

[1] A. HEKLER, "Kunst und Kultur Pannoniens in ihren Hauptströmungen". *Strena Buliciana* (1923), pp. 107–18.

[2] E. B. THOMAS, *A nagydémi lararium: Das Lararium von Nagydém* (Veszprém, 1965); E. B. THOMAS, "A tamási Athéna-szobor" (The Athena-Statue of Tamási). *AntTan* 3 (1956): 167–74.

[3] HEKLER, Kunst und Kultur, p. 111.

[4] E. B. THOMAS, "Italische Mater Matuta-Votive aus Pannonien". *ReiCretActa* 10 (1968): 58–61; E. B THOMAS, "Itáliai Mater Matuta votivok Dunaföldvárról—Italische Mater Matuta Votive aus Dunaföldvár". *Szekszárdi Múz. Évk.* 1 (1970): 19–39.

[5] THOMAS, *A nagydémi.*

[6] E. B. THOMAS, *Römische Villen in Pannonien* (Budapest, 1964), pp. 282–87.

[7] E. B. THOMAS, "Italische Einflüsse auf das frühkaiserzeitliche Pannonien", in *Le Rayonnement des civilisations grecque et romaine sur les cultures périphériques* (Paris, 1965), pp. 364–75.

[8] E. B. THOMAS, *Baláca: Mosaiken, Fresken, Stukkos* (Budapest, 1964).

[9] A. RADNÓTI, "Möbel- und Kästchenbeschläge, Schlösser und Schlüssel". *Intercisa* (Budapest, 1957), 2:241–363 *(ArchHung* 36); D. GÁSPÁR, "Római ládikák felhasználása—Verwendung römischer Kästchen". *FolArch* 22 (1971): 53–69.

[10] G. ERDÉLYI, "Steindenkmäler". *Intercisa* (Budapest, 1954), pp. 169–231 *(ArchHung* 33).

[11] L. BALLA, T. P. BUÓCZ and Z. KÁDÁR, *Die römischen Steindenkmäler von Savaria* (Budapest, 1971).

[12] A. ALFÖLDI, "Kapitóliumok Pannóniában" (The Capitols in Pannonia). *ArchÉrt* (1920–1922) 12–14; K. Sz. PÓCZY, "Die Anfänge der Urbanisation in Scarbantia". *ActaArchHung* 23 (1971): 93–110.

[13] E. TÓTH, "A savariai insularendszer rekonstrukciója—Rekonstruktion des Insula-Systems in Savaria". *ArchÉrt* 98 (1971): 143–69.

[14] T. SZENTLÉLEKY, *A szombathelyi Isis szentély—Das Isis-Heiligtum von Szombathely* (Szombathely, 1965).

[15] T. P. BUÓCZ, *Savaria topográfiája—Die Topographie von Savaria* (Szombathely, 1968).

[16] A. MÓCSY, "Korarómai sírok Szombathelyről—Frührömische Gräber in Savaria". *ArchÉrt* (1954): 167–91.

[17] BALLA, BUÓCZ, and KÁDÁR.

[18] V. WESSETZKY, *Die ägyptischen Kulte zur Römerzeit in Ungarn* (Leiden, 1961).

[19] SZENTLÉLEKY.

[20] K. Sz. PÓCZY, "Scarbantia városfalának korhatározása—La datation de l'enceinte de Scarbantia". *ArchÉrt* 94 (1967): 137–54.

[21] E. B. THOMAS, "Antike: Die Römer", in *Archäologische Funde in Ungarn* (Budapest, 1956), pp. 181–261.

[22] THOMAS, "Antike"; WESSETZKY.

[23] E. B. THOMAS, "Die römerzeitliche Villa von Tác-Fövenypuszta".*ActaArchHung* 6 (1955): 79–152.

[24] J. FITZ, "Gorsium: Bericht über die Ausgrabungen in der römischen Siedlung bei Tác". *Alba Regia* (1960–1966).

[25] THOMAS, *Römische Villen.*

[26] E. B. Thomas, "Hercules szentély Pannóniában—Ein Heiligtum des Hercules aus Pannonien". *ArchÉrt* (1952): 108–12.

[27] E. B. Thomas, "Ólom fogadalmi emlékek Pannóniában—Monuments votifs en plomb sur le territoire de la Pannonie". *ArchÉrt* (1952): 32–38.

[28] Thomas, *Baláca*.

[29] Thomas, *Römische Villen*, pp. 34–49.

[30] Ibid., pp. 111–16.

[31] Ibid., pp. 395–98.

[32] Ibid., pp. 391–93.

[33] E. B. Thomas, "Bruchstück einer frühchristlichen Marmortischplatte mit Reliefverzierung aus Csopak". *ActaAntHung* 3 (1955): 261–82.

[34] Thomas, *Römische Villen*, pp. 201–10, 232–37.

[35] L. Nagy, "Les symboles astraux sur les monuments funéraires de la population indigène de la Pannonie", *Laureae Aquincenses* 2 (1941): 232–43 (*DissPann* 2:11).

[36] E. B. Thomas, *Helme, Schilde, Dolche: Studien über römisch-pannonische Waffenfunde* (Budapest, 1971).

[37] E. B. Thomas, "Der Helm von Budapest, Ungarn", in H. Klumbach (red.), *Spätrömische Gardehelme*, (München, 1973), pp. 39 ff.

[38] I. Wellner, "Az aquincumi helytartói palota építésének kora—Zur Frage der Bauzeit des Statthalterpalastes von Aquincum". *ArchÉrt* 97 (1970): 116–25.

[39] I. Wellner, *Az aquincumi mozaikok—Die Mosaike von Aquincum* (Budapest, 1962).

[40] E. B. Thomas, *Über Pannonien* (Veszprém, 1970).

[41] K. Sz. Póczy, "Keramik". *Intercisa* (Budapest 1957), 2:29–139 (*ArchHung* 36).

[42] J. Fitz, *Les Syriens à Intercisa* (Bruxelles, 1972) (Coll. Latomus); Z. Kádár, *Die kleinasiatisch–syrischen Kulte zur Römerzeit in Ungarn* (Leiden, 1962).

[43] N. Láng, "Das Dolichenum von Brigetio". *Laureae Aquincenses* 2 (1941): 165–81 (*DissPann* 2:11).

[44] I. Paulovics, "Dolichenus-háromszögek tartója Brigetióból—Halter für dreieckige Dolichenus Reliefs aus Brigetio". *ArchÉrt* (1934), 40–48.

[45] Láng.

[46] T. Nagy, "A sárkeszi Mithraeum és az aquincumi Mithra-emlékek—Le Mithréum de Sárkeszi et les monuments mithriaques d'Aquincum". *BudRég* 15 (1950): 47–120.

[47] M. R. Alföldi, "Knochengegenstände". *Intercisa* (Budapest, 1957), 2:477–95 (*ArchHung* 36).

[48] G. Erdélyi, "Steindenkmäler". *Intercisa* (Budapest, 1954), 1:169–231 (*ArchHung* 33); S. Scheiber, *Corpus Inscriptionum Judaicarum a temporibus saeculi III., quae exstant usque ad annum 1686* (Budapest, 1960).

[49] L. Nagy, "Pannonia Sacra", in *Szent István Emlékkönyv* (Budapest, 1938), pp. 31–148.

[50] E. B. Thomas, "Römerzeit", in *Die Geschichte der Völker Ungarns bis Ende des 9. Jahrhunderts: Führer durch die Ausstellung* (Budapest, 1963).

[51] E. B. Thomas, "Kunst in Pannonien", in *Évolution générale et développements régionaux en histoire de l'art*, vol. 1 (Budapest, 1972), pp. 61–69 (Actes du XIIe Congrès International d'Histoire de l'art (Budapest, 1969).

# POTTERY

ÉVA B. BÓNIS

## ORIGIN OF THE POTTER'S CRAFT AND EARLY POTTERY

At the time of the Roman conquest the autochthonous population, with a heritage partly Illyrian but mostly Celtic, had already developed a ceramic art in Pannonia, manufacturing their variously formed vessels on the potter's wheel. In the century preceding the Roman conquest, fortified and open settlements had come into being in mountainous regions and along the rivers. Best explored among them has been the late Celtic settlements on Gellért hill, situated in the center of the capital, Budapest, and in the Tabán district at the foot of this hill. Here a well-protected mountain settlement was established in the second half of the first century B.C., at the beginning of Emperor Augustus' rule.[1] The founders were the Eravisci, a tribe of the free Celts retreating eastward before the advance of the Roman armies and the onrush of the Germanic peoples from the north.

In the settlement and in the gently sloping Tabán region adjacent to it numerous traces of kilns have been uncovered among the remains of houses. The sherds and whole vessels yielded by these pits represent the highest technical standard of their age, superior to the pottery of contemporary settlements uncovered in Switzerland, Germany, Bohemia, and Poland. Made on iron-axled potters' wheels, the slate-colored vessels were well smoothed and had a fine, resounding tone. Their surfaces were covered with flattened ornaments, tastefully arranged. In fact, the noble forms and red and white stripes of the rather rare, egg-shaped, large vessels of the brownish material are acceptable even to modern taste. The most beautiful ware of this kind was manufactured by a master of the Tabán; his hallmark was formed by impressing into the soft clay of the bottom of his vessels a gem of Italian origin, depicting Victoria.[2] He also added a rich, brownish lilac ornament to his red and white striped vessels (Pl. L). The thousands of sherds yielded by the excavation testify to a very large production.

Besides the potters' settlement on Gellért hill, some kilns from the late Celtic period are known from Békásmegyer on the bank of the Danube. It is certain that potters were active in the major centers, at Velemszentvid, for example, where a large number of pots of graphitic materials with combed ornament have been found.

Pannonian potters, influenced by native ceramic art for a long time after the Roman conquest, perfected the ancient production of gray, thick-walled vessels, especially bowls (Fig. 65.16) and large pots. On the latter even the smoothed-wave ornament appeared now and then. The traditional thick-walled, gray bowls have been found among the vessels inhumed in the *tumuli* of the early imperial period, especially in those of Fejér County. The finest specimen was the so-called

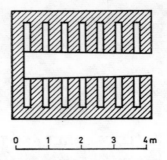

Fig. 61. Ground plan of potters' kiln from Aquincum

Pátka bowl (Fig. 65.12), whose shape and incised, rouletted ornament reveal provincial Roman influence. Following the native traditions, burial urns, thrown on wheels, were made of thick, brown material and were covered with combed decoration (Fig. 64. 28–29); these have been found throughout Pannonia, e.g. at Brigetio in graves dated as late as the end of the second century. Ancient Celtic traditions were also followed by the marking of the early Roman tripodal bowls (Fig. 64.27).[3] The so-called Pannonian gray ware showed the impact of the Roman provincial *terra sigillata* vessels. However, instead of the low-relief decoration of the *terra sigillata* bowls, the walls of the vessels bore stamped ornaments, easier to produce.[4] The origins of the stamped ornament are to be found in Asia Minor, in the orbit of late Hellenistic pottery, imitative of metal vessels.

The Celtic native population of Pannonia often applied impressed decoration, especially concentric ornament, to its vessels. The larger store of patterns and the general extension of this ornamental device may be explained only by a fresh wave of Romanization. The vessels with impressed ornaments in the southwest Pannonian cemeteries, large hemispherical bowls with inverted rims, did not follow the shapes of the Gaulish *sigillata* ware. Their system of decoration, simply alternating rosettes and foliage, differed from the practice of northeast Pannonia. The potters of this northeast region continued the manufacture of vessels with stamped ornament and brought it to its prime; this was prompted by western importation, becoming intense by the end of the first century, and by the importation of *terra sigillata*, first from southern Gaul.

From the second half of the second century on, vessels with a stamped ornament (Fig. 65.4) could hold the Pannonian market only as long as the mass products of the real *terra sigillata* factories (Rheinzabern, Westerndorf), which were gradually extending eastwards, did not push them out. In their red-painted variety, bowls with a stamped ornament were manufactured in several regions as late as the third century. The potters of the Aquincum region, mainly the master Resatus and the master Respectus and his companions, produced Drag 37-shaped bowls with a stamped ornament. The signature of Resatus was flanked by two small stylized trees as a trademark (Pl. LI). The latest excavations at Gorsium, which have uncovered additional settlements and numerous vessels with impressed ornaments, will greatly enrich our knowledge of this group of ceramic ware. However, the finds of western Pannonia have quite a different character. Here we find

358

Fig. 62. Potters' kilns from Aquincum (B. Kuzsinszky)

flat bowls with a trefoil or quatrefoil stamp on the inside of the bottom (possibly influenced by the black-gloss, so-called Campanian bowls of Italian origin) and with stamped ornaments. In the first century B.C. large numbers of them reached Noricum but stopped short of Pannonia, at that time unconnected with Italia.

A special group of stamped ware existed in later Pannonia. We know just a few pieces of the small cups with angular profile, a gray (Fig. 67.4) or a red sliping, and simple rosettes or S-shaped impressions.[5] Research has established their connection with Dacian materials bearing impressed ornaments. In view of the fact that this sort of late stamped ware was a general feature of the Mediterranean basin, belonging to the general category of *terra sigillata chiara*, we may suppose that it came to Pannonia via oriental elements immigrating in the wake of the troops from the Near East; I do not believe that it was a product of the Barbarian population.

The late Celtic vessels—red-white, striped, painted, and uniformly without handles—had a long existence in the provincial Roman material of Pannonia. Its earliest specimens were found in large numbers in Poetovio and its surroundings in the time of the Flavians. The types of vessels most frequently found in the material of the *tumuli* from the end of the first century and from the second century were vessels with horizontal rims, painted and striped, and shaped like pots or eggs (Fig. 65.14). The later varieties of the painted, striped vessels—earless and two-handled jugs, usually with undulating decoration—are known from Vindobona from the end of the second century and the beginning of the third century.[6] While relatively few specimens of striped ware occurred at Aquincum and along the eastern *limes*, striped pottery flourished mainly at Brigetio in the second century and at the beginning of the third (Fig. 65.13; Pl. LII. 1). This late flourishing of the pottery of local origin was a conspicuous phenomenon. Brigetio, in spite of its Celtic name, had no autochthonous precursors. However, the developing and expanding native population of its immediate neighborhood must have required more ceramic products than could have been produced by the number of craftsmen belonging to the legion. Therefore, the artisans of the surroundings, imbued

359

with the traditions of the Celtic potters, must have lent a hand in the production.[7]

Celtic traditions were preserved not only by the technical execution of the red-striped vessels, but also by artistic work on individual specimens. Graves of the middle Imperial Age yielded egg-shaped, painted vessels showing dark-brown-figure subjects. The scenes depicted jumping animals surrounded by various sun symbols and signs alluding to magical power (Fig. 65.15).[8] The application of related symbols on late Celtic and provincial vessels, even after the historical change marked by the Roman conquest, shows that the continuity of religious life remained unbroken. These portrayals were related to the cult of the Celtic sun god Jupiter Teutanus, of whom we have evidence from Aquincum as late as the third century A.D. The connection between the Gallic priestly families and the potters has been shown by French research. We may suppose that the *vates* and *augures* of the priestly families of Celtic origin in Pannonia may have influenced the potters in shaping the figured ornaments of vessels destined for cultic use or for graves. The southern Pannonian windowed urns served as grave goods, too (Fig. 64.30).[9] According to most recent Yugoslavian research, these vessels were the legacy of a rite, typical of the Latobici tribe, preserving ancient prehistoric traditions.[10]

The province of Pannonia, established in the first century A.D., received the strongest Roman cultural influence from northern Italia, the most important mediator of such influences being Aquileia, the great commercial center. As is common knowledge, the centers of Romanization were situated along the ancient Amber Route. The earliest finds, especially from the Claudian period, have come to surface along the Emona–Poetovio–Savaria–Scarbantia–Carnuntum route. The process of Romanization was faithfully mirrored by pottery finds here, too. The earliest Italian *terra sigillata* have been found in the cities mentioned, but *terra sigillata* specimens from Arezzo itself must have arrived at the earliest legionary camps, Emona and Poetovio.

Working in the period of the emperors Tiberius and Claudius, the very popular north Italian potter Gellivs also sent his products to Savaria and Carnuntum (Fig. 64.3). In addition to troops on the move, veterans and tradesmen settling in the new province were the solid pillars of Romanization. Where we do not possess epigraphical data, the ubiquitous pottery finds have testified to the early settlements. *Sigillata* of the Po region, found in groups at Keszthely-Dobogó, may have derived from the cemetery of a veteran settlement (Fig. 64.1–2). Wares of the masters of the Po region reached the inhabitants of Pannonia in the time span between Claudius and Domitian, and research in Pannonia is uncovering an increasing amount of Po-Valley *sigillata* which circumstances assign to the second half of the first century. Their most frequent stamps were LMV (Fig. 64.1), QSP, and CT SUC.[11]

Parallel with the *sigillata* of the Po region, small, thin-walled jars and mugs from the same region made their appearance. The small, finely wrought vessels were mainly gray, decorated with leaves and berries made of clay, *à barbotine*; they sometimes had a glossy black covering, and their central parts were emphasized by indented belts (Figs. 64.10, 11, 13–15). These small cups were frequently decorated with a row of embossed crescents (Pl. LII. 2). The small black beakers were ornamented by horizontal and vertical incisions (Fig. 64.12).

360

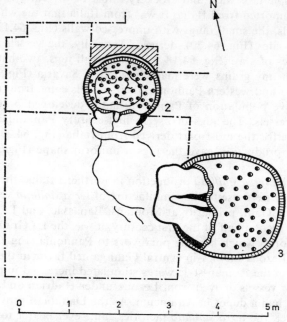

Fig. 63. Potters' kilns from Bicsérd (L. Barkóczi)

In the cemeteries of southern Pannonia, graves of the first century have yielded small, yellow-glazed mugs. The two-handled, small, yellow-glazed cups, decorated *à barbotine* belonged to the group of rare, early metal-glazed vessels (Fig. 64.5–6). The models of these vessels, which in the last analysis are of Hellenistic origin, are known from Asia Minor; they appeared in the Roman provinces in the period of Augustus. Representing a typical Italian imported ware in the Pannonian cemeteries of the Early Imperial Age, they reached the northern and eastern regions of Pannonia at the end of the first century and in the course of the second;[12] the finest specimens have come to light in Arrabona. The one-handled jar with an undulating profile and a crescent decoration (Fig. 64.8), the two-handled, brownish green-glazed jug imitating an amphora (Fig. 64.9), and the yellowish green-glazed skillet with a pecking bird and a ram on the handle (Fig. 64.7) were modeled on the best pieces of the glaziery and metallurgy of the Early Imperial period.[13] In the eastern half of Pannonia the production of metal-glazed vessels may be observed from the beginning of the second century. They included an ornamental vessel from Kiskőszeg with a yellowish green glaze and, formerly, a spout (Fig. 65.9). A row of animals ran among crescents on the shoulder of the three-handled can and among foliage on the lower part. The manufacturer of the ornamental vessel drew on the repertoire of forms of the Pannonian workshops, imitating the Gaulish *terra sigillata* factories which favored stags, lions, chamois, panthers, and boars running in a circle.[14]

As for the simple tableware and crockery, most of the early Imperial types in Pannonia were imported from Italia. It was from Italia that the white vessels with eggshell-thin walls, the small mugs with impressed walls (Fig. 64.17), and the jars with grooved profiles (Fig. 64.20), derived. Typically, the vessels were sprinkled with tiny granules of clay (Fig. 64.18–19). The small jugs, two-handled, painted, and covered with tiny grains, were manufactured in Savaria (Fig. 64.22). Known only in southwest and western Pannonia, their models came from northern Italia. Though the native population of Pannonia had a developed pottery, it did not know handled vessels. The jugs in use in first-century Pannonia, especially the rarer forms, were for the most part derived from Italia (Fig. 64.23–25). Later the local Pannonian production gave the jugs a uniform shape (Figs. 65.17 and Fig. 66.19).

Parallel with the *terra sigillata* production in northern Italia, the second decade of the first century gave rise to the manufacture of *terra sigillata* vessels in the factories of southern Gaul, primarily at Montans, Banassac, and La Graufesenque. Especially in the second half of the first century, it was the La Graufesenque (Condatomagus) factories which became purveyors to Pannonia (Fig. 64.4). Later, the products of the Lezoux factory in central Gaul gained favor in the province (Fig. 65.1) as interest in the Roman-style wares stimulated increased imports.[15] Most of the *terra sigillata* vessels arrived from Lezoux under Hadrian and the Antonines; the factory even had a depot in Aquincum. As the Danubian provinces proved to be a good market, the *terra sigillata* factories came ever nearer to the purchasing area. From the middle of the second century the Rheinzabern *terra sigillata* workshops also supplied imported ware to Pannonia. Large bowls and vase-shaped vessels, covered with fine-figured appliqué designs, were a feature of all the important settlements of the province (Fig. 65.3; Pl. LIII).

Raetian-style vessels were carried to Pannonia from Raetia and from the surrounding provinces. On these rather finely wrought vessels, which had a brownish-red paint and, often, a bronze-looking gloss, one sees typical appliqué ornaments (Fig. 65.7–8). Double sticks interrupted by relief dots and small horseshoe-formed ornaments were characteristic. The importing of these vessels lasted from the beginning of the second century to early decades of the third, the later types having even more belts with rouletted decoration and even darker, metal-like gloss (Pl. LIV. 1).

As we have seen, typically local Pannonian pottery was developed from the amalgamation of the practice of native potters and of forms and techniques of vessels imported from Italia and the western provinces.

## POTTERS' SETTLEMENTS AND KILNS

On the border of military and civil settlements, local potters' settlements arose everywhere. The rapidly developing workshops satisfied the demands of the population of the city and its neighborhood; some Pannonian workshops, however, transported their wares even farther. Best known to us are the potters' settlements of the provincial capital, Aquincum.[16] It was on the site of the gas tanks in the

Óbuda district of Budapest that the largest potters' settlement adjacent to the civilian city of Aquincum was uncovered: the so-called Gasworks potters' settlement (Figs. 61 and 62).[17] Although this is the best-known example of a settlement of this kind, the earlier potters' settlement of the Aquincum legionary camp has been unearthed in Kiscelli Street, Óbuda. Flourishing at the turn of the first and second centuries, it manufactured vessels imitating South Gaulish terra sigillata, those with impressed decoration, orange-coated household ware, and mortaria marked with the stamp of the masters. This settlement also produced bricks for the Legio X Gemina, Legio IV Flavia, and the final Aquincum garrison, the Legio II Adiutrix. Recent excavations here have uncovered large quadrangular kilns.

To the northeast of this Kiscelli Street workshop in Selmeci Street, two kilns used under the rule of Hadrian have been unearthed. Perhaps it was the owner of this workshop who did the handled jug containing 800 denarii, the last one of which was dated 137. These kilns were primarily used for the production of vessels with impressed ornaments. To the north of the northwestern corner of the wall of the Aquincum civilian settlement, railway construction has uncovered a potters' settlement which produced ornamental vessels and was active mainly in the period of Trajan and Hadrian. Its terra sigillata imitations and small-handled beakers with embossed ornament revealed Italian and southern Gallian influence on more recent Romanization.

The above-mentioned Gasworks potters' settlement, established in the first half of the second century, was much more extensive. In the beginning it manufactured household ware, but in the middle of the century an anonymous "first master" also began to make ornamental vessels. His initiative was followed by the master Pacatus,[18] whose molded bowls, showing the impact of Lezoux sigillata, have been unearthed in large quantities by the excavations. He shipped some of these to the South Pannonian Mursa, where numerous Pacatus bowls have been found (Fig. 65.5). Pacatus had no successor in Aquincum, but his tradition was probably continued by the potters of Siscia.[19]

Outside the southern wall of the civilian city, to the left of the limes road running southwards and near the "Schütz Inn", two potters' kilns have been found. At the beginning of the third century, probably in the course of some military operation, the potters' settlement was burned down, leaving the vessels prepared for baking still in the kilns. Vessels of various types were paralleled in assemblages from the early third century (Fig. 66.17–19); besides the various jugs, the kilns contained jars of metal gloss and impressed walls, copies of the vessels imported from Raetia and Germania. After the end of the so-called Gasworks potters' settlement, a few masters, among them Maximinus, worked further in a small workshop along the northern wall of the civilian city. Imitations of terra sigillata pieces, vessels with impressed ornaments, and household wares were produced here.

Several potters' settlements are known from the other legionary camp, Brigetio, of which the earliest was that of Kurucdomb, appearing at the beginning of the second century. It lay on the bank of the Danube, north of the camp; like the earliest potters' settlement of Aquincum, it manufactured fine ornamental vessels decorated with reliefs in the pattern of the ceramics of Italia and of the earlier conquered western provinces – ornamental earthen bowls, clay *paterae* (Pl. LV), and a mass of *terra-cotta*. The later potters' settlement, the so-called Gerhát settlement, which flourished in the time of the Antonines, produced quantities of household ware for the use of the army (Fig. 65.13–14). Second-century kilns which also manufactured simple household wares have been unearthed beside the wall of the military city. One was especially remarkable, since its supporting columns and grate were made of large adobe bricks, previously baked.

The production of the major camps, Carnuntum[20] and Vindobona,[21] were characteristically local wares of eastern Pannonian towns, a typical form being the "Carnuntum jug", a product of a local workshop.[22] The outer potters' quarter of Vindobona, known by a workshop found at III. Rennweg No. 81, manufactured gray pots and covers for everyday use. The settlement uncovered at III. Bahngasse Nos. 4–8 produced bowls, plates, incense burners, and *mortaria*. Kilns have been found in both the legionary camp and the civilian city.

Among the towns of western Pannonia kilns have been unearthed at Savaria, where the ware was sold in stalls standing at the crossings of the still-visible Roman roads. Beside the early, gritted, fine, small Savarian jugs found there, a deficiently burned, compressed, small vessel, derived from a small-handled jar with a turban-shaped body, has also been discovered (Fig. 66.12). According to recent Yugoslavian research, one of the most important pottery centers of southern Pannonian cities was Poetovio. Here all sorts of vessels were manufactured, from name-stamped burial urns to snake-ornamented vessels for use in the *mithraea* and *mortaria* with hall marks.[23]

We have already mentioned the production of ornamental vessels at Mursa. The potters' settlement of a *minor vicus* belonged to the region of Sopianae. From the turn of the first and second centuries, there were three kilns at Bicsérd (Baranya County) (Fig. 63) which were used for the manufacture of painted ware and *terra-cotta* specimens.[24] From the cities of the legionary camp of Intercisa we have no kilns; only the cast mold of a head of Hercules and molds of *lucernae* are evidence of local production. So far, research has been centered on the cemeteries, but excavations on the border of the legionary camp are being carried out at present.[25]

In inner Pannonia the excavations at Gorsium uncovered the local variety of pottery in a hitherto unknown quantity. Edifice No. 1 contained a third-century workshop for metal-glazed pottery where green-glazed, two-handled mugs, *mortaria*, and little bowls with horizontal, pre-ornamented rims were made (Fig. 66.24–25).[26] Two kilns of the Pilismarót camp came from the late Roman period, the end of the fourth century (ca. 376). Gray pottery with a smoothed ornament, typical of the Barbarization of late Pannonian ceramics, was manufactured there (Fig. 67.20).[27]

The enumerated potters' settlements have yielded different types of kilns. The LT D and early imperial potters' settlements of Gellért hill–Tabán, and the LT D workshop at Békásmegyer used the typical kiln of the LT D system. It was horseshoe-shaped, with a lengthened, double-furnace flue divided by a bearer bar. A thick, earthen grate, pierced by round holes, leaned on the bar, and the rim of the fire chamber was sunk into clay. After the vessels were put into the kiln, an earthen dome was raised above the grate. This type of kiln was so practical that it was used for a long time not only in Barbaricum, but also in the Roman provinces. The Bicsérd site, where large furnaces of an identical system were used, offers further evidence of the Pannonian practice (Fig. 63). In the so-called Gasworks potters' settlement of Aquincum most kilns were already square (Fig 62), but even if one or two of them were circular or oval, they differ from the late Celtic tradition in not having the double-furnace flue. In the kilns of the Gasworks settlement, brick walls were important. The two kilns near the "Schütz Inn" of Aquincum, dated to the beginning of the third century, were simpler than their precursors. The round kilns were dug into the clay soil, with only their domes reaching above the Roman level. The clay grate was supported by prefabricated adobe bricks, and the stokehole was carved into rough clay. These kilns needed restoration from time to time. Similarly, the kiln uncovered at Brigetio in 1959 was sunk into rough clay, its firing areas constructed of adobe bricks. The erection of late Roman kilns at Pilismarót was extremely interesting. Above the sunken firing chamber was a large-holed grating of clay, radially built and supported by a thick, cylindrical column in the middle. The dome was vaulted above the Roman level. Two small, round kilns recently uncovered at Balatonaliga, which also have grates supported by central cylindrical pillars, were used to manufacture smooth vessels during the last phase of Roman rule in Pannonia.

## POTTERY IN THE LATE IMPERIAL PERIOD

In the third century the provincial Roman pottery of Pannonia bore the mark of full uniformity, with native ceramic traditions barely discernible, and the usual forms constantly repeated. Gone was the rich array of early imported ware and the rich store of patterns of the *sigillata* factories. Yet even third-century pottery had its typologically characteristic forms, and new types were brought in with the increasing influx of wares from the Rhine region. At the beginning of the third century the Westerndorf factory was the largest supplier to Pannonia (Fig. 66.2–3);[28] the products of the potter Helenius were the most favored (Fig. 66.2). Now and then the Raetian-style jars of the early third century appear, especially in the graves at Brigetio. Equally fine were the small jars with impressed walls and dark metal coating imported from the neighborhood of Trier and Cologne in Germania (Fig. 66.6). Above the white clay ornaments, their glassy surfaces bore white-lettered good wishes and commands: *Ave, Bibite, Ave vita*, etc. (Pl. LIV. 2). Local workshops imitated these imported specimens of jars with impressed walls (Fig. 66.7–8). The large jars with impressed walls (Faltenbecher) followed the local traditions; this practical, characteristic Roman drinking vessel was made in red from the first century on (Fig. 66.11).

The third century types of vessels of local production are known mainly from the above-mentioned potters' settlement near the "Schütz Inn" at Aquincum and from the material of the Brigetio and Intercisa cemeteries. Vindobona also yielded a large number of third-century ware. The most favored forms were the red or red-striped jugs, jars with two or three handles, and pots and deep bowls with two handles (Fig. 66.13–15). A favorite style of ornamentation was the rouletting of the whole body, produced by cog wheels (Fig. 66.9, 17). Although it was known earlier, the adding of the so-called incense bowls as grave goods reached its widest extent in this period. The finest specimens of such bowls came from the cemeteries of Aquincum and Brigetio (Fig. 66.16).

Metal-glazed pottery also produced new varieties. In addition to the products of the above-mentioned Gorsium workshop, there appeared a group of high jars or jugs without handles and with embossed horseshoes and crescents illustrative of the impact of imported Raetian and Germanic ware (Fig. 66.20–21). A center for production of such vessels may have existed in western Pannonia, in Savaria, or in its surroundings. The vessels from the Savaria region were usually coated with an orange metal glaze.

Influences from Asia Minor, connected with the army and permeating Pannonia, affected jewels and glassware more than pottery. Vigorous local manufacture made the importation of pottery unnecessary. However, a late Roman grave yielded an *oinophoros* of Alexandrian origin (a two-handled wine jug with a cylindrical body) on which Dionysus and his retinue were depicted (Fig. 66.1).[29] This jug may have reached Arrabona through Italian sources.

Even more vigorous importation from the Rhine region was revealed by the material of the fourth century; this may be because the population had settled into Pannonia from the Rhine. In the graves of the Constantinian dynasty one finds white vessels made of pipe-stone from the Rhine region, especially Cologne (Fig. 67.1–3).[30] Besides these painted ornamental vessels, the grave goods of fourth-century pottery were gray plates with a scabrous surface, one-handled jars, and some glassware and metal-glazed types derivative of the Rhineland and imitative of metal vessels. In a form copying the third-century Rhineland glasses, but done with rough execution and course materials, handles arched from the neck of the jugs to the shoulder (Fig. 67.14). The one-handled, slender, collared jugs with funnel-shaped rims and green- or brown-glazed and red-painted surfaces also imitated glassware (Fig. 67.15–16; Pl. LVI). Similarly, glasses from Cologne and Trier were the models for numerous varieties of two-handled jugs with cylindrical bodies and metal glaze. These were primarily the features of military camps (Intercisa, Aquincum, Brigetio, Arrabona); they were entirely absent in Savaria, and the literature mentions no such vessels from Noricum. The noted wave ornament (Fig. 67.12), revealing the intermingling of Roman and Barbarian tastes, appeared on the latest variants.

Noteworthy numbers of one-handled jugs with green and brown metal glaze occurred in Pannonia (Figs. 67.11 and 14–15).[31] These were certainly manufactured in local workshops, as evidenced by a master from Aquincum. From a late grave in Arrabona (Figs. 67.13) we have a brownish green-glazed jug depicting a female figure, her head crowned by a rosette-ornamented headdress and surrounded by a string of pearls. Holding a glass in her right hand, a distaff in her left, she personified Clotho, one of the Fates. The notched inscription of the vessel, *Ienvarie Pie Zeses*, indicates that the vessel was intended as a gift, given with good wishes to a person called Ianuarius. The quality of the glaze and the shape of the jug characterize it as a late Roman product of Pannonia, and the formula "*pie zeses*" dates it to the second half of the fourth century.[32]

366

Some of the gray ware began to be decorated with a smoothed ornament, a typical feature of the Barbarians living in the areas east of the empire. These vessels were characteristic of the pottery of the Barbarian settlers of late Roman Pannonia and of the barbarized army (Fig. 67.20). The quality of the ceramics deteriorated, and the end of the fourth century saw the production of brownish black smoked vessels which flaked off in layers (Fig. 67.19). The cemetery belonging to the Pilismarót camp yielded a one-handled vessel which, were it not grave furniture, could not possibly be attributed to a Roman province (Fig. 67.22).[33]

## LUCERNAE

*Lucernae* (lamps) were the characteristic utensils of ancient life, having reached Pannonia through the Roman conquest. In the first century the population of Pannonia received ready-made the types of *lucernae* developed in Greece, Italia, and the earlier-conquered provinces.[34] In the course of the development of Pannonian pottery these forms were varied, and typical local shapes also came into being. The *lucernae* were made of clay, bronze, iron, and glass. In several cases the earthenware *lucernae* were imitations of the finely shaped bronze ones. Clay *lucernae* may be divided into several groups: those with reliefs, those with design whose outlines were filled in with paint, pieces without decoration, and the later variants, covered with the most diverse ornaments.

The noses of the embossed *lucernae* were executed in volutes. The decorations of their covers, figured or ornamental designs which embraced the whole gamut of Roman life, pictured gods and myths as well as scenes from everyday life — gladiators, theater masks, etc. The firm *lucernae* took their original forms from the bronze ones. Their bottoms carried the most frequent embossed hallmarks, such as Fortis. The hallmarks were later borrowed by other potters, but by that time they were depersonalized trademarks rather than the signature of the original potter.

*Lucernae* were manufactured in the potters' workshops of all significant settlements, but not all of these workshops have been uncovered. Workshops manufacturing *lucernae* are known from the areas of Aquincum, Brigetio, Savaria, Poetovio, Mursa, and Siscia. Evidence of local production is given by the discovery of molds of *lucernae*. Lamps manufactured in the earliest military potters' settlements imitated the early imperial forms faithfully; thus *lucernae* and molds in the form of a gladiatorial helmet were found in Aquincum and other Pannonian sites (Pl. LVII).[35] There are parallels known from the earliest camps on the Rhine. The workshop of the military city of Aquincum, where such *lucernae* were also fabricated, was active until the beginning of Hadrian's rule. From Aquincum and Brigetio come the ring-shaped *lucernae* with several lights (Pl. LVIII). In workshops of the civilian cities the names of masters were often found on *lucerna* molds; Victor, Petilius, Maximinus, Pacatus, Victorinus and Florentinus are the names most frequently found on second-century Aquincum *lucerna* molds. At the beginning of the third century a master called Fabius appeared, but the names of the fourth-century producers of *lucernae* are unknown. In the period of the Constantinian dynasty, Master Iustinianus worked at Poetovio manufacturing earthern vessels as well as *lucernae*. It was from this date that the Early Christian *lucernae* made their appearance in Pannonia; they had the Chi Rho symbol of Christ and one small specimen depicted the Good Shepherd. In the second half of the fourth century there were also simple, round *lucernae*, coated with green leaden glaze.

## EARTHEN LIGHTHOUSES, TOWER MODELS

Still undefined is the practical use of the tiny, baked clay turrets — fretwork ornaments which remind one of the windows of real towers in the camps. Their angular (Pl. LX) and round varieties are to be found sparsely all over the provinces on the Rhine and the Danube.[36] On the tower model from Intercisa (Pl. LXI), the potter tells us that he was happy to make it: *(Hilarus fecit porta(m) fel(iciter)*.[37] We suppose that he has portrayed the gate of a Pannonian city or camp, but without any practical purpose, although some suggest that these tiny turrets were illuminated from inside and placed on graves. However, such turrets have not yet been found originally situated on graves.

### TERRA-COTTAS AND CRUSTULUM MOLDS

In settlements producing vessels, baked clay figurines were also made. The Roman provincial art of small sculpture in clay was nourished on the traditions of Asia Minor, Alexandria, Greece, southern Italia, etc.; and local production was modeled on one or the other imported specimen. Mainly votive objects were manufactured of baked clay, with *terra-cotta* statuettes and reliefs usually portraying religious subjects.[38] Caricatures were also popular, and the stockpile of a workshop producing caricature statuettes was unearthed in Savaria long ago.[39] In Pannonia the Aquincum workshop produced mainly figurines of Venus; the Brigetio ones preferring representations of the Mother Goddess.[40] *Terra-cottas* were baked not only in the major cities, but also in the kilns of insignificant *vici*. In the potters' settlement of Bicsérd figures of Venus, Priapus, and pigeons have been found beside the kilns of the workshops.

On New Year's Day and other festivals especially during the jubilee celebrations of the emperors (*decennalia, vicennalia*) round flat cakes (*crustula*) with relief figures were distributed. The cakes were pressed in round molds of baked clay, the patterns of which served the imperial cult by eulogizing the victories and the glory of the emperor. Best known in Pannonia is the series of *crustulum* forms from Savaria; one of them depicted the city goddess, Tutella, who personified the Golden Age under the protection of Mars and Virtus (Pl. LIX). Framed by a *tabula ansata*, the inscription ran: *Salvo Aug(usto) aurea s(a)ecula videmus* — i.e., "If the emperor is safe, we are entering the Golden Age."[41]

It was the duty of the magistrates to distribute the presents among the people at the festive occasions. *Crustula* depicting the fighting gladiators were also distributed after the gladiator contests. As early as the second century, the cakes proclaimed the victories of the emperors; most of the known molds, however, date from the third century. The *crustulum* forms and the baked clay medallions applied to the vessels faithfully mirrored the leading official ideas of the imperial period.[42]

# CHRONOLOGICAL CATALOG

## Pannonian Pottery

(Fig. 64.1– 30)

1. *Terra sigillata* plate. Imported from the Po region. LMV stamp. Preserved in (hereafter cited as Pres.): Budapest, Hungarian National Museum. Inv. No. 1. 1942. 2. Site: Keszthely—Dobogó.
2. *Terra sigillata* cup. Imported from the Po region. Pres. Hungarian National Museum. Inv. No. 1. 1942. 4. Site: Keszthely—Dobogó.
3. *Terra sigillata* cup. Imported from the Po region. L. GEL stamp. Pres. Hungarian National Museum. Inv. No. 7. 1930. 1. Site: Szombathely (Savaria).
4. *Terra sigillata* bowl. Imported from Southern Gaul, from the factory at La Graufesenque. Pres. Hungarian National Museum. Inv. No. 1. 1942. 7. Site: Keszthely—Dobogó.
5. Yellow-glazed, pedestalled cup with ornament *à barbotine*. Pres. Ljubljana, Mestni Muzej. Inv. No. 3023. Lit.: É. B. Bónis, "Die kaiserzeitliche Keramik von Pannonien (Ausser den Sigillaten)" (Budapest, 1942), *(DissPann* 2: 20), Pl. XIX. 36.
6. Lemon-yellow-glazed, two-handled cup. Pres. Ljubljana, Mestni Muzej. Inv. No. 963. Site: Drnovo (Neviodunum). Lit.: Bónis, "Die kaiserzeitliche Keramik," Pl. XIX. 38.
7. Yellowish green-glazed, handled skillet. Pres. Győr, J. Xántus Museum. Site: Győr (Arrabona). Brewery hill. Excavated by E. Bíró. Lit.: E. B. Thomas, *Arrabona* 3 (1961), Figs. 5—7.
8. Lemon-yellow-glazed, one-handled jar. Pres. Győr, J. Xántus Museum, Inv. No. 53. 156. 4. Site: Győr (Arrabona), Calvary Cemetery. Lit.: E. B. Thomas, *Arrabona* 3 (1961), Fig. 1.
9. Brownish green-glazed, small amphora. Pres. Győr, J. Xántus Museum. Site: Győr (Arrabona), Brewery hill. Excavated by E. Bíró. Lit.: E. B. Thomas, *Arrabona* 3 (1961), Fig. 4.
10. Dark gray, small cup with a glossy coating. Imported from Northern Italia, Pres. Ljubljana, Mestni Muzej. Inv. No. 3153. Site: Ljubljana (Emona). Lit.: Bónis, "Die kaiserzeitliche Keramik," Pl. XX. 63/1.
11. Small cup with a dark gray coating. Imported from Northern Italia. Pres. Ljubljana, Mestni Muzej. Site: Ljubljana (Emona), Necropole septentrionale. Lit.: Lj. Plesničar-Gec, *Inventaria Arch.* Y 93 (2) 1. No. 6.
12. Small goblet with dark gray coating. Imported from Northern Italia. Pres. Graz, Landesmuseum Johanneum. Inv. No. 5200. Site: Ptuj (Poetovio). Lit.: Bónis, "Die kaiserzeitliche Keramik," Pl. XIX. 14/1.
13. Two-handled, small cup with glossy brown coating. Imported from Northern Italia. Pres. Bad-Deutschaltenburg, Museum Carnuntinum. Inv. No. 879. Site: Carnuntum. Lit.: Bónis, "Die kaiserzeitliche Keramik," Pl. XX. 36.
14. Small cup with grayish-yellow coating. Imported from Northern Italia. Pres. Ljubljana, Mestni Muzej. Site: Ljubljana (Emona), Necropole septentrionale. Lit.: Lj. Plesničar-Gec, *Inventaria Arch.* Y 93. 10.
15. Small cup with light brown paint and ornament *à* barbotine. Pres. Maribor, Gradski Muzej. Inv. No. 1025. Site: Ptuj (Poetovio). Lit.: Bónis, "Die kaiserzeitliche Keramik."
16. Flat plate, light yellow. Pres. Szombathely, Savaria Museum. Site: Szombathely (Savaria), Hámán K. Street, cemetery. Lit.: A. Mócsy, *ArchÉrt* 81 (1954): 171, Fig. 9, 49/3.
17. Small cup with impressed walls, white. Pres. Maribor, Gradski Muzej. Inv. No. G. B. 10. 1. Z. 798. Site: Ptuj (Poetovio). Lit.: Bónis, "Die kaiserzeitliche Keramik," Pl. XIX. 42/1.
18. Small cup with red paint, gritted with clay. Pres. Ptuj, Gradski Muzej, Inv. No. 57. Site: Ptuj (Poetovio). Lit.: Bónis, "Die kaiserzeitliche Keramik," Pl. XIX. 59/1.
19. Goblet with impressed walls (Faltenbecher) and reddish bronze coating, gritted with clay. Pres. Budapest Hungarian National Museum. Inv. No. 54. 6. 72. Site: Pannonia.
20. Small two-handled jar, light yellow. Pres. Ljubljana, Mestni Muzej. Inv. No. 5564. Site: Ljubljana (Emona). Lit.: I. Mikl-Curk, *Arheološki Vestnik* 20 (1969), Pl. 1. bottom left.
21. Rough-colored, one handled, small jug. Pres. Budapest, Hungarian National Museum. Inv. No. 54. 6. 26. Site: Pannonia.
22. Two-handled, small jug with glossy red coating, gritted with clay. Pres. Budapest, National Museum. Inv. No. 132. 1872. B. cs. 7. Site: Szombathely (Savaria).
23. One-handled jug, light yellow. Pres. Szombathely, Savaria Museum. Site: Szombathely (Savaria). Lit.: A. Mócsy, *Arch Ért* 81 (1954): 168, Fig. 4, 14/2.
24. One-handled jug with red paint. Pres. Graz, Landesmuseum Johanneum. Inv. No. 8351. Site: Ptuj (Poetovio). Lit.: Bónis, "Die kaiserzeitliche Keramik", Pl. XXVIII, 6/1.
25. One-handled jug, light yellow. Pres. Ljubljana, Mestni Muzej, Inv. No. 5359. Site: Ljubljana (Emona). Lit.: B. Vikić-Belančić, *Starinar* 13—14 (1962—1963): 104, Fig. 37, 20.
26. Thick-rimmed bowl, gray. Pres. Budapest, Hungarian National Museum. Inv. No. 35. 1923. 1. Site: Budapest, Gellérthegy, Mányoki Str. No. 16. Lit.: É. B. Bónis, *ArchHung* 47 (1969): 233, Pl. LIII.
27. Three-footed bowl, yellowish-gray. Pres. Zagreb, Narodni Muzej. Site: Stenjevac. Lit.: B. Vikić-Belančić, *Starinar* 13—14 (1962—1963): 106, Fig. 42, 3.
28. Dark gray urn with combed ornament. Pres. Graz, Landesmuseum Johanneum. Inv. No. 4733. Site: Ptuj (Poetovio). Lit.: Bónis, "Die kaiserzeitliche Keramik," Pl. I, 2/2.
29. Yellowish-gray urn with combed ornament. Pres. Ljubljana, Mestni Muzej. Site: Ljubljana (Emona), Necropole septentrionale. Lit.: Lj. Plesničar-Gec, *Inventaria Arch.* Y 97/11.
30. Dark gray, windowed urn. Pres. Wien, Naturhistorisches Museum. Inv. No. 35648. Site: Drnovo (Neviodunum). Lit.: B. Vikić-Belančić, *Starinar* 13—14 (1962—1963): 100, Fig. 28, 6.

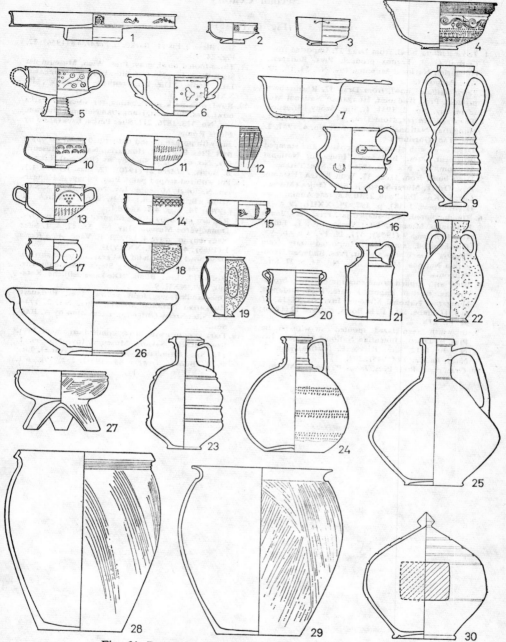

Fig. 64. Pannonian types of vessels from first century

# Second Century

## (Fig. 65.1—18)

1. *Terra sigillata* bowl, from Drag 37. Imported from Central Gaul, Lezoux product. Pres. Budapest, Hungarian National Museum. Inv. No. 54. 13. 30. Site: Pannonia.
2. *Terra sigillata* bowl, from Drag 37. Rheinzabern product. Pres. Budapest, Hungarian National Museum. Inv. No. 2. 1931. 1. Site: Szőny (Brigetio).
3. *Terra sigillata* jar, from Drag 54. Pres. Budapest, Hungarian National Museum. Inv. No. 4, 1937. 5. Site: Tác (Gorsium).
4. Deep bowl with glossy black coating and stamped ornament. Pres. Budapest, Hungarian National Museum. Inv. No. 310. 1876. 363. Site: Pátka.
5. Deep bowl, from Drag 37. With PACATI stamp. Pres. Osijek, Muzej Slavonije, Site: Osijek (Mursa). Lit.: K. Kiss, *DissPann*, 2: 10. Laureae Aquincenses II (Budapest, 1938), p. 197, Pl. XXIII, 29 a—b.
6. Rough-colored, head-shaped jar. Pres. Pécs, Janus Pannonius Museum. Site: Vasas. Lit.: F. Fülep, *ActaArchHung* 9 (1958): 377, Pl. IV. 4 a—b.
7. One-handled, small jug. Imported ware from "Raetia". Dark brown paint. Pres. Budapest, Hungarian National Museum. Inv. No. 54. 6. 27. Site: Pannonia.
8. Bowl with applied ornaments and brownish bronze coating. Raetian imported ware. Pres. Budapest, Hungarian National Museum. Inv. No. 55. 18. 17. Site: Halimba. Lit.: É. B. Bónis, *FolArch* 12 (1960): 94, Pl. XIX. 2.
9. Brownish green-glazed, spouted can with handle. Pres. Budapest, Hungarian National Museum. Inv. No. 30. 1912. Site: Kiskőszeg (Batina). Lit.: L. Nagy, *BudRég* 14 (1945): 293.
10. Gray plate. Pres. Pécs, Janus Pannonius Museum.
    Site: Bicsérd. Lit.: L. Barkóczi, *FolArch* 8 (1956): 72, Fig. 24. 7.
11. Three-footed bowl, gray. Pres. Wien, Museum der Stadt Wien. Inv. No. 905. Site: Wien III, Steingasse. Lit.: Bónis, "Die kaiserzeitliche Keramik", Pl. XXIV. 8/1.
12. Bowl with dark gray coating, the so-called Pátka bowl. Pres. Budapest, Hungarian National Museum. Inv. No. 318. 1876. 211. Site: Pátka. Excavation by Flóris Rómer.
13. Jug with no handle and red-striped paint on upper part. Pres. Budapest, Hungarian National Museum. Inv. No. 63. 15. 7. Site: Szőny (Brigetio). Lit.: É. B. Bónis, *FolArch* 21 (1970): 72, Figs. 1.1 and 4.3.
14. Pot with red-striped paint. Pres. Budapest, Hungarian National Museum. Inv. No. 4. 1946. 5. Site: Szőny (Brigetio). Lit.: É. B. Bónis, *FolArch* 21 (1970): 74, Figs. 4.8. and 8.1.
15. Red urn with brownish red-figured ornaments. Pres. Dunaújváros Museum. Inv. No. 58. 42. 4. 1. Site: Nagyvenyim, grave 4. Lit.: B. E. Vágó, *Alba Regia* 1 (1966): 46, 54, Pl. XXXVIII.
16. Deep bowl with high lid and gray, flattened surface. Pres. Székesfehérvár, King Stephen Museum. Inv. No. 6717. Lit.: Bónis, "Die kaiserzeitliche Keramik", Pl. XXII. 9.
17. One-handled jug, light yellow. Pres. Budapest, Hungarian National Museum. Inv. No. 2. 1946. 253. Site: Szőny (Brigetio). Excavation by A. Radnóti.
18. Dark gray urn with dense, combed ornament. Pres. Szombathely, Savaria Museum. Inv. No. 66. 1. 116. Site: Szombathely, Rumi Str. (Savaria). Lit.: P. T. Buócz, *ArchÉrt* 88 (1961): 232, Fig. 6/1.

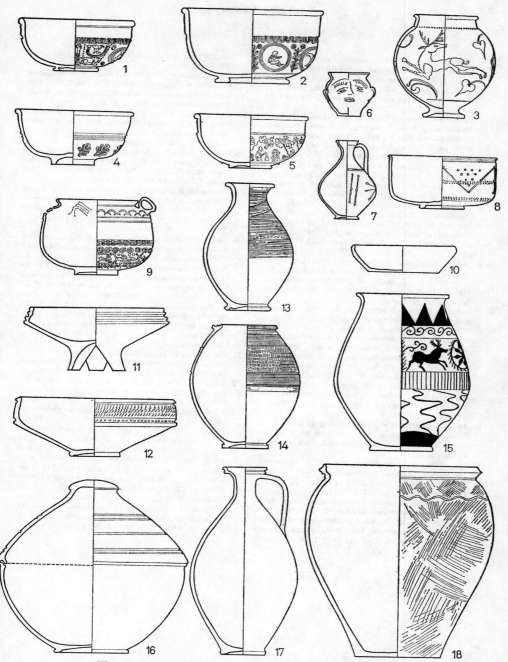

Fig. 65. Pannonian types of vessels from second century

373

# Third Century

## (Fig. 66.1—25)

1. *Oinophoros* with relief ornament, rough color. Imported from eastern basin of Mediterranean. Pres. Győr, J. Xántus Museum. Site: Győr (Arrabona), Városház Str. No. 423. Lit.: F. Rómer, *ArchÉrt* 3 (1870): 115, Figs. 1—2; É. B. Bónis, *ArchÉrt* 79 (1952): 24—25.
2. *Terra sigillata* bowl, from Drag 37. Imported from Westerndorf. Pres. Budapest, Hungarian National Museum. Inv. No. 10. 1951. 76. Site: Pannonia.
3. *Terra sigillata* cup, from Drag 33. Pres. Budapest, Hungarian National Museum. Inv. No. 66. 1966. 398. Site: Dunapentele (Intercisa). Lit.: K. Póczy, *Intercisa* 2: 100, Pl. XII. 6. Cat. 42.
4. Jar with bronze coating and applied ornament. Raetian imported ware. Pres. Budapest, Hungarian National Museum. Inv. No. 91. 1905. 1. Site: Dunapentele (Intercisa).
5. Jar with brownish bronze coating. Pres. Budapest, Hungarian National Museum. Inv. No. 2. 1946. 39. Site: Szőny (Brigetio). Excavation by A. Radnóti.
6. Motto jar (Spruchbecher) with blackish brown coating and white clay ornaments and letters. Imported from the Rhine region. Pres. Budapest, Hungarian National Museum. Inv. No. 29. 1927. Site: Szőny (Brigetio).
7. Jar with impressed walls, with five round imprints and black glossy coating. Pres. Budapest, Hungarian National Museum. Inv. No. 2. 1931. 173. Site: Szőny (Brigetio). Excavation by I. Paulovics.
8. Jar with impressed walls and glossy, dark gray coating. Pres. Székesfehérvár, King Stephen Museum. Inv. No. 4115. Site: Dunapentele (Intercisa). Lit.: A. Marosi, *ArchÉrt* 37 (1917): 5 .
9. Brown-painted jar with grooved surface. Pres. Székesfehérvár, King Stephen Museum. Inv. No. 4026. Site: Dunapentele (Intercisa). Lit.: A. Marosi, *ArchÉrt* 37 (1917): 5.
10. Jar with impressed walls and dark gray paint. Pres. Budapest, Hungarian National Museum. Inv. No. 6. 1940. 58. Site: Szőny (Brigetio). Excavation by L. Barkóczi.
11. Jar with impressed walls, gray. Pres. Budapest, Hungarian National Museum. Inv. No. 10. 1903. 141. Site: Dunapentele (Intercisa). Lit.: K. Póczy, *Intercisa* 2: 83, Pl. XIII, 8. Cat. 12.
12. One-handled jar with reddish brown paint. Pres. Budapest, Hungarian National Museum. Inv. No. 132. 1872. B. cs. 9. Site: Szombathely (Savaria).
13. Two-handled small pot. Pres. Budapest, Hungarian National Museum. Inv. No. 97. 1909. 25. Site: Dunapentele (Intercisa). Lit.: K. Sági, *Intercisa* 2: 589, Fig. 130.
14. Two-handled vessel with red paint. Pres. Budapest, Acquincum Museum. Inv. No. 50. 362. Site: Aquincum, so-called "Schütz" kiln. Lit.: K. Póczy, *ActaArchHung* 7 (1956): 114, Pl. X. 1.
15. Two-handled bowl, rough color. Pres. Budapest, Hungarian National Museum. Inv. No. 1909. 24. Site: Dunapentele (Intercisa). Lit.: K. Sági, *Intercisa* 2: 588, Fig. 129.
16. Incense bowl, rough color below, gray baked on the upper part. Pres. Budapest, Hungarian National Museum. Inv. No. 21. 1931. 33. Site: Szőny (Brigetio). Excavation by I. Paulovics.
17. One-handled jug, the rough surface decorated with indenting. Pres. Budapest, Aquincum Museum. Inv. No. 54. 313. Site: Aquincum, so-called "Schütz" kiln. Lit.: K. Póczy, *ActaArchHung* 7 (1956): 113, Pl. XI. 6.
18. One-handled jug, rough color with red paint. Pres. Budapest, Aquincum Museum. Inv. No. 41. 587. Site: Aquincum, so-called "Schütz" kiln. Lit.: K. Póczy, *ActaArchHung* 7 (1956): 113, Pl. XI. 10.
19. One-handled jug with spout, reddish yellow painted. Pres. Budapest, Aquincum Museum. Inv. No. 33. 455. Site: Aquincum, so-called "Schütz" kiln. Lit.: K. Póczy, *ActaArchHung* 7 (1956): 112, Pl. X. 15.
20. Jar with embossed, horseshoe ornaments and brownish green glaze. Pres. Budapest, Hungarian National Museum. Inv. No. 136. 1881.2. Site: Óbuda (Aquincum).
21. Jar with embossed, horseshoe ornaments and brown glaze. Pres. Budapest, Hungarian National Museum. Inv. No. 89. 1909. 3. Site: Dunapentele (Intercisa).
22. Small goblet with yellowish green glaze. Pres. Budapest, Hungarian National Museum. Inv. No. 54. 6. 156. Site: Pannonia.
23. Two-handled cup with yellowish brown glaze. Pres. Budapest, Hungarian National Museum. Inv. No. 75. 1911. 77. Site: Dunapentele (Intercisa). Lit.: K. Póczy, *Intercisa* 2: 120, Pl. XVIII, 4. Cat. 227.
24. Small bowl with brownish yellow glaze, a ring of pearls on the rim. Pres. Székesfehérvár, King Stephen Museum. Inv. No. 10295. Site: Tác (Gorsium). Lit.: E. B. Thomas, *ActaArchHung* 6 (1955): 121, No. 18, Pl. XLIX, 22.
25. Mortarium, lined with green glaze. Pres. Székesfehérvár, King Stephen Museum. Inv. No. 10298. Site: Tác (Gorsium). Lit.: E. B. Thomas, *ActaArchHung* 6 (1955): 122, No. 42, Pl. L. 12.

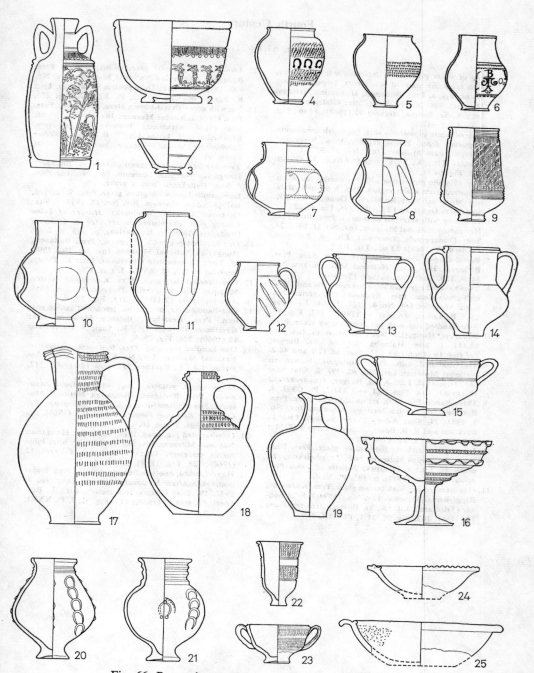

Fig. 66. Pannonian types of vessels from third century

# Fourth Century

## (Fig. 67.1—22)

1. Jug of white pipe-stone, depicting a female figure, with brownish painted ornaments. Imported from Rhine region. Pres. Budapest, Hungarian National Museum. Inv. No. 66. 3. 49. Site: Gödrekeresztúr. Lit.: A. Sz. Burger, *ArchÉrt* 95 (1968): 13 sq., Fig. 6 a—d, 11—17.
2. Cup of white pipe-stone with brownish ornaments. Imported from Rhine region. Pres. Szekszárd, Balogh Ádám Museum. Inv. No. 38. 933. 1. Site: Tamási. Lit.: K. Sági, *ActaArchHung* 12 (1960): 246. Fig. 64.11.
3. Cup of white pipe-stone with brownish ornaments. Imported from Rhine region. Pres. Keszthely, Balaton Museum. Site: Keszthely—Dobogó. Lit.: K. Sági, *ActaArchHung* 12 (1960): 246, Fig. 42.22.
4. Gray cup with stamped ornament. Pres. Budapest, Hungarian National Museum. Inv. No. 51. 1911. 24. Site: Dunapentele (Intercisa). Lit.: Á. Salamon, *FolArch* 20 (1969): 53 sq., Fig. 1.1.
5. Barrel-shaped vessel with dark green glaze. Pres. Budapest, Hungarian National Museum. Inv. No. 54. 6. 159. Site: Szombathely (Savaria).
6. Two-handled cup with appliqué crescents and brownish green glaze. Pres. Budapest, Hungarian National Museum. Inv. No. 4. 1927. 3. Site: Kisárpás. Lit.: E. Lovas, *ArchÉrt* 41 (1907): 197, Fig. 97.
7. Two-handled cup with yellowish brown glaze. Pres. Budapest, Hungarian National Museum. Inv. No. 62.411. 1. Site: Halimba. Lit.: A. Sz. Burger, *FolArch* 19 (1968): 83—93, Figs. 46 12/1 and 48.2.
8. Three-handled cup. Pres. Budapest, Hungarian National Museum. Inv. No. 62. 398. 2. Site: Ságvár (Tricciana). Lit.: A. Sz. Burger, *ActaArchHung* 18 (1966): 133. Figs. 121 and 331/2, Pl. CVII. 4.
9. Three-handled jar with yellowish green glaze. Pres. Budapest, Hungarian National Museum. Inv. No. 9. 1951. 24. Site: Adony (Vetus Salina). Lit.: L. Barkóczi and É. B. Bónis, *ActaArchHung* 4 (1954): 178, Fig. 24.16, Pl. LIX. 11.
10. Large plate with reddish brown glaze. Pres. Pécs, Janus Pannonius Museum. Site: Zengővárkony II, cemetery. Lit.: J. Dombay, *Bulletin of Janus Pannonius Museum* (1957), p. 193, Pl. V. 4.
11. One-handled jug with brown glaze. Pres. Kaposvár, Rippl-Rónai Museum. Inv. No. 11594/1. Site: Ságvár (Tricciana). Lit. :A. Sz. Burger, *ActaArchHung* 18 (1966): 103, Fig. 96, 31/1, Pl. CIX, 6.
12. Two-handled jug with greenish brown glaze. Pres. Budapest, Hungarian National Museum. Inv. No. 100. 1912. 304. Site: Dunapentele (Intercisa). Lit.: K. Póczy, *Intercisa* 2, 122, Pl. XIX, 2. Cat. 273.
13. Jug with greenish red-brown glaze, portraying Fate. Pres. Győr, J. Xántus Museum. Inv. No. 53. 159. 56. Site: Győr (Arrabona), Brewery hill cemetery. Excavation by E. Lovas. Lit.: E. B. Thomas, *Kölner Jahrbuch für Vor- und Frühgeschichte* 9 (1967—1969): 83 sq, Pl. 26. 1—3.
14. Two-handled jug with green glaze. Pres. Budapest, Hungarian National Museum. Inv. No. 66. 1904. 2. Site: Budakeszi, from a grave.
15. One-handled jug with green glaze. Pres. Szekszárd, Balogh Ádám Museum. Inv. No. 13. 1933. 1. Site: Gerényes. Lit.: M. Wosinszky, *History of Tolna County from Pre-history to the Age of Conquest* (Budapest, 1896, in Hungarian), p. 783, n. 2.
16. One-handled jug with red paint. Pres. Budapest, Hungarian National Museum. Inv. No. 101. 1908. 35. Site: Dunapentele (Intercisa). Lit.: K. Póczy, *Intercisa* 1: 117, Pl. XVIII. 1. Cat. 170.
17. One-handled jar, gray. Pres. Keszthely, Balaton Museum. Site: Keszthely—Dobogó. Lit.: K. Sági, *ActaArchHung* 12 (1960): 214, Fig. 58.10.
18. One-handled jar, gray, with smoothed lattice pattern. Pres. Keszthely, Balaton Museum. Site: Keszthely—Dobogó. Lit.: K. Sági, *ActaArchHung* 12 (1960): 208, Fig. 29. 9.
19. One-handled jug, gray, Pres. Budapest, Hungarian National Museum. Inv. No. 8. 1937. 3. Site: Pilismarót. Lit.: L. Barkóczi, *FolArch* 12 (1960): 112, Pl. XXII. 2.
20. Large vessel, yellowish gray with flattened ornaments. Pres. Budapest, Hungarian National Museum. Site: Pilismarót camp. Lit.: S. Soproni, *Rei Cretariae Romanae Fautorum ACTA X* (1968), pp. 28 sq., Figs. 7 and 10.
21. One-handled jug, gray. Pres. Budapest, Hungarian National Museum. Inv. No. 8. 1937. 80. Site: Pilismarót cemetery. Lit.: L. Barkóczi, *FolArch* 12 (1960): 128, Fig. 33, Pl. XXVI. 1.
22. Hand-formed, one-handled jug, brown. Pres. Budapest, Hungarian National Museum. Inv. No. 8. 1937. 19. Site: Pilismarót cemetery. Lit.: L. Barkóczi, *FolArch* 12 (1960): 114. Fig. 31. 3, Pl. XXVI. 2.

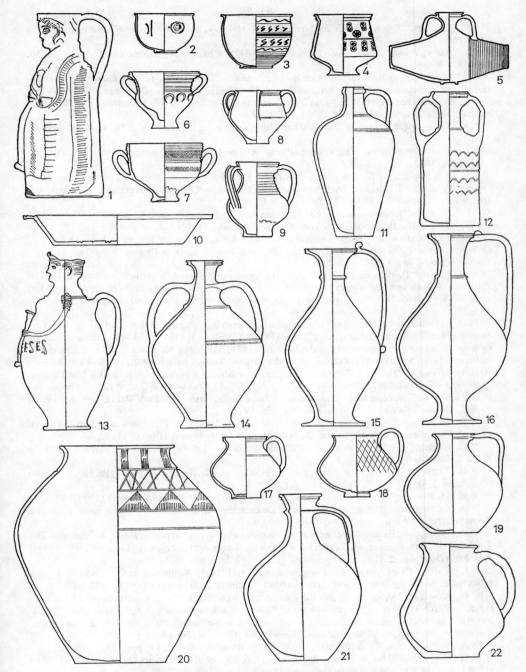

Fig. 67. Pannonian types of vessels from fourth century

# NOTES

[1] É. B. Bónis, *"Die spätkeltische Siedlung Gellérthegy–Tabán in Budapest"* (Budapest, 1969) (*ArchHung* 47).

[2] L. Nagy, *Budapest története I* (Budapest, 1942), Chapt. "Agyagművesség", pp. 251–58, 627–36.

[3] É. Bónis, *"A császárkori edényművesség termékei Pannoniában–Die kaiserzeitliche Keramik von Pannonien (Ausser den Sigillaten) I"* (Budapest, 1942), (*DissPann* 2:20): 184–88.

[4] L. Barkóczi and É. Bónis, "Das frührömische Lager und die Wohnsiedlung von Adony—Vetus Salina". *ActaArchHung* 4 (1954): 129–99.

[5] Á. Salamon, "Spätrömische gestempelte Gefässe aus Intercisa". *FolArch* 20 (1969): 54–62.

[6] A. Schörgendorfer, *Die römerzeitliche Keramik der Ostalpenländer*, 2 vols. (Brünn, München, 1942), pp. 38–41.

[7] E. Ettlinger and C. Simonett, *Römische Keramik aus dem Schutthügel von Vindonissa* (Basel, 1952); É. B. Bónis, "A brigetiói sávos kerámia–Die streifenverzierte Keramik aus Brigetio". *FolArch* 21 (1970): 71–90.

[8] E. B. Vágó, "Kelten- und Eraviskengräber von Nagyvenyim und Sárkeszi". *Alba Regia* 1 (1960): 43–62; É. B. Bónis, *Die spätkeltische Siedlung*, pp. 215–22.

[9] B. Vikić-Belančić, "Neka obiljezja ranocarske keramika u jugozapadnoj Panoniji—Quelques caractéristiques de la céramique du haut empire dans la Pannonie du Sud-Ouest". *Starinar 1962–1963* (1965): 89–112.

[10] P. Petru, "Poskus časovne razporeditve lončenine iz rimskih grobov na Dolenjskem in Posavju—Cronologia della ceramica delle tombe romane nella Carniola inferiore (Dolenjsko) e della valle della Sava". *Slovanska Akademija Znanosti in Umetnosti Razprave* 6 (1969): 197–212.

[11] D· Gabler, "Az importált terra sigilláták forgalma Pannoniában—Angaben zur Verbreitung der Sigillaten in Pannonien". *ArchÉrt* 91 (1964): 94–110; A. Mócsy, "Frührömische Gräber in Savaria–Szombathely". *ArchÉrt* 81 (1954): 167–91; K. Póczy, "Der Einfluss der spätitalischen Sigillata-Werkstätte auf die Fabrikation pannonischen Prunkgefässen". *ActaArchHung* 11 (1959): 143–58; D. Gabler, "Arrabona legkorábbi sigillatái—The Earliest Sigillatae of Arrabona". *Arrabona* 9 (1967): 21–53; Lj. Plesničar-Gec, "La nécropole romaine à Emona", in *Inventaria Archaeologica Jugoslavija*, Fasc. 10 Y 89-Y 102 (Ljubljana, 1967).

[12] É. Bónis, *Die kaiserzeitliche Keramik*, pp. 19–20.

[13] E. B. Thomas, "Rómaikori mázas agyag díszedények a győri múzeumban—Roman Glazed Ornamental Vessels in the Győr-Museum". *Arrabona* 3 (1961): 17–32.

[14] L. Nagy, "Zöldzománcos római dísztál Budáról—Le vase de luxe à glaçure verte de Buda". *BudRég* 14 (1945): 283–302.

[15] Gy. Juhász, *A brigetioi terra sigilláták—Die Sigillaten von Brigetio* (Budapest, 1936) (*DissPann* 2:3).

[16] K. Sz. Póczy, "Die Töpferwerkstätten von Aquincum". *ActaArchHung* 7 (1956): 73–138.

[17] B. Kuzsinszky, "A gázgyári római fazekastelep Aquincumban—Das grosse römische Töpferviertel in Aquincum". *BudRég* 11 (1932): 5–384.

[18] K. Kiss, "A Pacatus-féle aquincumi fazekasműhely gyártmányainak időrendje—Die Zeitfolge der Erzeugnisse des Töpfers Pacatus von Aquincum". *Laureae Aquincenses* 1 (1938): 188–228 (*DissPann* 2:10).

[19] L. Nagy, "Egy sisciai terra sigillata gyár termékei Aquincumban—Prodotti di una fabbrica di terra sigillata di Siscia ritrovati ad Aquinco". *BudRég* 14 (1945): 303–31.

[20] H. Kenner, "Ausgrabungen und Funde im Lagerfriedhof von Carnuntum: Die Kleinfunde". *RLiÖ* (1937): 81–98; E. Vorbeck, "Die römischen Funde bei der Petroneller Rundkapelle", in *Mitteilungen des Vereins der Freunde Carnuntums* (1950), pp. 7–8; E. Swoboda, *Carnuntum: Seine Geschichte und seine Denkmäler* (Graz, Köln, 1964), pp. 112–13.

[21] A. R. Neumann, "Vindobona", R. E. *Pauly-Wissowa*, Suppl. 11, pp. 1273–74.

[22] G. Reinfuss, "Die Keramik der Jahre 1958–59, 1952–54". *Carnuntum Jahrbuch 1959* (1961), pp. 74–99; *1960* (1962), pp. 54–95.

[23] I. Mikl-Curk, "K sliki naselbinske keramike rimskega Poetovija—Beitrag zur Siedlungskeramik des römischen Poetovio". *Časopis za Zgodovino in Narodopisje* (1965), pp. 1–12.

378

[24] L. Barkóczi, "Császárkori kelta edényégető telep Bicsérden—Celtic pottery kilns from the times of the Roman Empire at Bicsérd". *FolArch* 8 (1956): 63–87; F. Fülep, "Das frühkaiserzeitliche Gräberfeld von Vasas". *ActaArchHung* 9 (1958): 371–406.

[25] K. Sz. Póczy, in *Intercisa–Dunapentele: Geschichte der Stadt in der Römerzeit*, Vol. 2 (Budapest, 1957), Chapt. "Keramik", pp. 29–139 (*ArchHung* 36), p. 130.

[26] E. B. Thomas, "Die römerzeitliche Villa von Tác-Fövenypuszta". *ActaArchHung* 6 (1955): 79–152; J. Fitz, *Gorsium: A táci rómaikori ásatások—Excavations from the Time of the Romans at Tác* (Székesfehérvár, 1970).

[27] S. Soproni, "Spätrömische Töpferöfen am pannonischen Limes". *ReiCretActa* (Tongeren, München, 1968), 10: 28–32.

[28] K. Kiss, "A westerndorfi terra sigillata gyár mesterei és kronológiája— Masters and chronology of the terra sigillata workshop at Westerndorf". *ArchÉrt* (1946–1948): 216–74; H.-J. Kellner, "Zur Sigillata-Töpferei von Westerndorf". *BayVb* 26 (1961): 165–203; D. Gabler, "Westerndorfer Sigillata im Barbaricum ostwärts von Pannonien". *BayVb* 33 (1968): 100–110; H.-J. Kellner, "Die Sigillatatöpfereien von Westerndorf und Pfaffenhofen". *Kleine Schriften zur Kenntnis der römischen Besatzungsgeschichte Südwestdeutschlands* 9(1973): 1–24.

[29] É. B. Bónis, "Die Verbreitung einer Gruppe von späthellenistischen Reliefgefässen in den römischen Provinzen". *ArchÉrt* 79 (1952): 23–32.

[30] I. Čremošnik, "Über die römische Keramik in Jugoslawien". *ReiCretActa* (Zürich, 1958), 1:13–14; A. Sz. Burger, "Terrakotta ex voto Gödrekeresztúrról—Ein terracotta ex voto aus Gödrekeresztúr". *ArchÉrt* 95 (1968): 13–28.

[31] K. Sz. Póczy, *Intercisa*, vol. 2, "Keramik"; A. Sz. Burger, "The Late Roman Cemetery at Ságvár". *ActaArchHung* 18 (1966): 99–234.

[32] E. B. Thomas, "Der Parzenkrug von Győr: Glasierte Ware aus Pannonien". *Kölner Jb.* 6 (1967–1968): 83–85.

[33] L. Barkóczi, "Későrómai temető Pilismaróton—Ein spätrömisches Gräberfeld in Pilismarót". *FolArch* 12 (1960): 111–32.

[34] D. Iványi, "A pannoniai mécsesek—Die pannonischen Lampen" (Budapest, 1935) *DissPann* 2:2).

[35] T. Szentléleky, "Aquincumi mécseskészítő műhelyek—Lamp-producing workshops at Aquincum". *BudRég* 19 (1959): 167–203; A. Neumann, "Lampen und andere Beleuchtungsgeräte aus Vindobona". *RLiÖ* 22 (1967).

[36] L. Nagy, "Egy pincelelet az aquincumi polgárvárosban: A pannoniai agyag-világítótornyocskák kérdése—A Cellar Find in the Aquincum Civil Town: The Problem of the Pannonian Clay Lighthouses". *BudRég* 14 (1945): 157–202.

[37] Fr. Oelmann, "Die tönernen Porta des Ilarus von Intercisa", in *Festschrift für R. Egger I* (Klagenfurt, 1952), pp. 114–25.

[38] E. B. Thomas, "Italische Mater-Matuta-Votive aus Pannonien". *ReiCretActa* (Tongeren—München, 1968), 10:56–61.

[39] L. Castiglione, "L'influence orientale dans la plastique de terre-cuite de Pannonie", in *VIIIᵉ Congrès International d'Archéologie Classique, Paris 1963* (1965), pp. 361–64.

[40] K. Sz. Póczy, "A termékenység kultusz terrakottái Aquincumban—Terra-cottas of fertility cult at Aquincum". *BudRég* 20 (1963): 241–58.

[41] A. Alföldi, "Tonmodel und Reliefmedaillons aus den Donauländern". *Laureae Aquincenses* 1 (1938): 312–41 (*DissPann* 2:10); A. Alföldi, "Die alexandrinischen Götter und die vota publica am Jahresbeginn". *Jahrbuch für Antike und Christentum* 8/9 (1965–1966): 53–87.

[42] E. B. Thomas, "Medallion Applications on Pannonian Vessels". *Alba Regia* 1 (1960): 71–76.

# GLASSWARE

### EDIT B. THOMAS

A favorable geographical location in the middle Danube basin made Pannonia a very important trading center for the glassware produced in the southern, western, and eastern provinces of the Roman Empire between the first and fourth centuries A. D. To a certain degree, early research recognized and separated from the local Pannonian glassware those items which were presumably made in Italian workshops.[1] With the development of a precise classification of the glassware, modern research gives recognition to glasses made in Pannonia even if their styles and prototypes were conceived elsewhere. Unearthed from first-century and early second-century graves were many glass urns, small flasks, and ribbed, decorated cups, together with glassware very similar to the products of the north Italian glass workshops.[2]

Glass imports from the Rhineland appeared as early as the second century. Arriving via the Danubian water route, the shipments started at the end of the second century, but increased in the third. In addition, the arrival of the eastern settlers[3] and the Syrian Cohors Milliaria Hemesenorum resulted in the appearance of Syrian glassware in Pannonia (Pl. LXIII).[4] Especially responsible for these imports were the Syrian merchants[5] who settled and built villas behind the portion of the *limes* protected by Syrian troops.[6] Many good examples of fine Syrian glassware have come from the burial places, sarcophagi, and family tombs of these merchants.

Some Pannonian glass workshops (Pl. LXIV) with half-finished or defective pieces and some wasted and unused raw materials have been discovered in Carnuntum,[11] Arrabona[7] Brigetio,[8] Aquincum,[5] and Gorsium.[4] After thorough analysis, researchers were able to determine the glass types that were of Pannonian origin. Let us consider chronologically some larger groups of Pannonian glassware. The so-called millefiori and miore glasses, the origin and style of which are Italian, were of the same quality in Pannonia as in Italia or in any of the other provinces of the Roman Empire. The single-handled or double-handled colored carafe, made in Italia at the beginning of the Roman Imperial period, was a form frequently employed in colorless glass in Pannonia as late as the second and third centuries.[9]

Extremely small, colored flasks from dark blue to yellow, from light blue to yellowish white, or white green, were very popular in all shades in Pannonia. That this type of glass was made in Pannonia is proved by some defective flasks discovered in Carnuntum. Another type of flask discovered had a long neck and flat projecting rim recessed toward the first third of the neck. Around the end of the first century and throughout the second century this style was very common in Pan-

381

nonia, emerging again in the late period of the Roman Empire. Several defective flasks stuck into sandiver (glass gall) have also been discovered in Pannonia. Moss green cups (Tröpfengläser, Tränengläser) made in the early Imperial period were introduced into Pannonia at the beginning of the first century and lasted, as a style, until the beginning of the second century. Greenish white glassware decorated with fused glass fiber on its surface emerged in the east during the middle Roman Empire. From the end of the first century to the end of the second, single-handled rectangular, block-shaped carafes were dominant. From the second century we have also some glass jugs with rectangular feet and geometric decorations.

From the Rhineland area came great quantities of ground (polished) glass vessels, usually made of clear white, transparent glass material with rich decoration. Their characteristic motifs were grapevine tendrils, grapes, palmettes, and circles. Also from the Rhineland came a technique called *Schlangenfadenauflage*, an imitation of which was seen on one of the footed chalices discovered in Carnuntum. Formed by glass fibers and attached to the vessel by means of a special sealing instrument were made the figures of aquatic birds and geese. Other footed chalices decorated in this manner have a lattice design. Although the technique of this decoration was connected with the *Schlangenfadenauflage* favored in the Rhineland, the pieces were very different in execution. It is possible that the technique used in the glass workshops in Carnuntum was influenced by craftsmen of the Near East or Israel who were active in Carnuntum from the end of the second century to the end of the third century A.D.

One of the most significant forms of Pannonian glassware was the globular flask (*Kugelflaschen*) (Pl. LXIV);[10] similar globular forms were developed in other provinces of the Roman Empire as well. On a dominant Pannonian flask, the neck had piles of glass fibers and the rim was twisted and unpolished. This type of neck formation appeared very often in the workshops of the Rhineland during the second and third centuries.[11]

Another Pannonian globular type of glass vessel had a long neck with an accentuated, polished rim; several malformed examples of this type have been discovered. The globular-bodied glasses made in Carnuntum, white with a slightly greenish tone, were extremely popular in the entire middle Danube area from the end of the third century to the beginning of the fourth.

A variety of the globular-shaped, long-necked flask had cornet-shaped necks and polished rims; they were mostly made of white glass. One of the most important centers of production was discovered in Carnuntum, where a great many remains of raw glass material, glass galls, and defective examples were found. The glass material used was not pure white but had greenish tone.

A typical product of the late Roman glass tradition, glassware decorated with blue dots (Pl. LXV), different in size and arrangement, has been found in large numbers in fourth-century graves. The basic color of these cups was white, bluish green, or moss green; their origin and provenance awaits future evaluation.

Another late Pannonian glass product was a moss green cup group with a hemispheric or cornet-shaped variety.

Remains of glass workshops have been uncovered in the following places: Carnuntum, Arrabona, Brigetio, Aquincum, Gorsium. Local provincial glass products formed a greater part of the Pannonian glass commerce than assumed at first. While earlier Pannonian research did not devote enough attention to local production, later research has proved that in spite of similarity of design and technique, the forms discovered in Pannonia were not all products from southern, eastern, or western glass-producing centers of the Roman Empire.

## NOTES

[1] É. B. BÓNIS, "Kaiserzeitliche Hügelgräber von Ivánc." *FolArch* 9 (1965): 79–80.
[2] L. BARKÓCZI, "Die datierten Glasfunde aus dem II. Jahrh. von Brigetio." *FolArch* 18 (1966): 167.
[3] A. RADNÓTI, *Intercisa II* (Budapest, 1957), pp. 141–163.
[4] E. B. THOMAS, *Römische Villen in Pannonien* (Budapest, 1964), p. 230.
[5] K. PÓCZY, *BudRég* 16 (1955): 41.
[6] M. KABA, *BudRég* (1958): 429.
[7] E. B. THOMAS, *Die Gläser des Espelmayrfeldes* (Linz, 1962), pp. 101–3.
[8] L. BARKÓCZI, Brigetio (*DissPann* 2:22) (Budapest, 1951), p. 8.
[9] A. BENKŐ, *Üvegcorpus—Glass corpus* (Budapest, 1962).
[10] A. S. BURGER, The Late Roman Cemetery at Ságvár. *ActaArchHung* 18 (1966): 99–180.
[11] E. B. THOMAS, Römisches Glas aus Carnuntum (Pannonien). Annal. du 4e congrès des Journées Internationales de Verre" (1967), pp.86–92.

# ARMOR

## EDIT B. THOMAS

The Roman soldier's armor was always practical and up-to-date. However, side by side with a practical character in close conformity with military science, the fine finish of the weapons was also an important quality.

Bravery in the Roman army had always been rewarded by the granting of badges of distinction and fine pieces of weaponry. Weapons were richly ornamented, although military and tactical usefulness were never forgotten. Studying these weapons has revealed new data about the occupation and Romanization of the province of Pannonia and the politics of late Roman times in the empire. In the context of the imperial spirit which had unified numerous peoples within the Roman Empire through the tools of a centrally controlled military propaganda, the study of symbolism of weapon ornaments has opened new fields for exploration.

We shall be dealing with only the most valuable ornamented weapons of Pannonia, not attempting to consider a full list of weapons. Beyond determining the period and the workshop, our aim is to describe fully the technology of the production and execution of each weapon considered.

The subject becomes particularly up-to-date because of the fact that in recent years a splendid dagger and exquisite shield-bosses have emerged from the Pannonian section of the Danube in the course of river-dredging activities.

## HELMETS

The armor of Roman Pannonia is quite prominent among that of the provinces of the Roman Empire. As far as helmets are concerned, the finds are very significant. The seven recovered helmets belong to two distinct types. Among the bronze helmets adorned with reliefs and the plate-covered, crested iron helmets we can find both simpler and more luxurious representatives of the types. They also embrace two different periods. The bronze helmets with reliefs illustrate one type of early imperial helmet fashion, whereas the iron ones represent a late Roman kind.

### Bronze Helmets with Reliefs[1]

*Bronze Helmet* (H.N.M. 2.1942.2) (Pls. LXVI and LXVII)

In 1942 at Szőny, when a housing development was being built, a blast furnace from the onetime interior of Brigetio's camp of the fourth century was uncovered. Among various bronze objects from the early Imperial Period, a wholly crushed bronze helmet, prepared for smelting, was lying next to the furnace.

Originally the helmet consisted of two parts, but the visor has never been re-covered. The helmet was made of 1-millimeter-thick bronze plate with an embossed pattern. On the inner side of the collar a dotted letter *R* in a reverse position is impressed (Pl. LXXIV. 2). On the surface can be observed traces of a white metal which must have originally covered the whole surface of the helmet. Only the deeper grooves of the pattern show the original color of the bronze underneath; the helmet seemed otherwise to have been made of silver. The full length of the Brigetio helmet is 29 centimeters.

### Bronze Helmet with Reliefs (H.N.M. 54.5.68) (Pl. LXVIII)

A relief-ornamented, rich bronze helmet, similar to the one described above, formed part of the Delhaes collection when it was purchased by the Hungarian National Museum.

This helmet must also have consisted of two parts, since in the part above the forehead there is a hook inside which served for attaching the visor. The helmet was made of an approximately 1-millimeter-thick bronze plate with relief orna-mentation. On the surface the white metal covering still exists, in many parts in great patches. On the flat surfaces, where it is not embossed, four-pronged stars are engraved. It seems that these were not covered with white metal, since they still show the reddish color of bronze. On the upper surface of the forepart, under the protruding top, there is a human face with softly rising lines. The helmet is 27.5 centimeters tall.

This extremely rare type of helmet formed part of a gala suit of armor. Such helmets were nearly always recovered from military camps and most of them were parts of a full set of equipment. According to inscriptions on the pieces, foot and mounted soldiers both used such helmets.

In ornamentation and execution these two helmets belong to the new Straubing finds and were probably contemporaneous with them. The Brigetio furnace, with its heterogeneous material, does not yield data which may determine the period, nor does the helmet deriving from a private collection of unknown origin furnish any evidence for dating. A comparison with the Straubing finds and Drexel's dating point to the second century as the period of ornamented armor. From the third-century stratum, as Krumbach also agrees, no similar pieces have emerged. We can say with certainty that both of our embossed, bronze-plated helmets date from the second century.

Recent studies on weaponry seem to suggest that both the Straubing finds and the Pannonian plated helmets, as well as the ornate shield-bosses made of bronze from the Rhineland and Pannonia, have an Oriental origin, though they were presumably produced in the middle Danube region. The weapons referred to above were all made of bronze covered with a white metal, and the engraved, embossed, and impressed patterns reveal the original color of bronze, giving a colorful appearance. It is particularly striking that the embossed technique of the ornamented armor and the surface covering of the objects with silver or white metal show a direct connection with cultic objects made of bronze and covered with white metal produced in honor of the Oriental deities, Mithras and Doli-chenus.

386

It is, therefore, very likely that this group of embossed ornamented armor displaying Oriental elements is connected with troops from *vexillationes* sent to, or soldiers recruited in, the East. The great quantity of this armor is well known in the Danube and Rhine areas. It is also striking that these objects had circulated in the greatest number in the area of the middle Danube. Their center or area of emergence in Pannonia coincides with Carnuntum, Brigetio, and the district of Aquincum. *Vexillationes* of the legions of the three camps in the second century had participated in wars in the East, and the three cities had received throughout the second century a considerable number of settlers from the East.

As can be seen in the forthcoming discussion, with regard to the bronze shield-bosses (whose motifs often are similar to those of the shields to be discussed below) and the remains of the ornamented armor in the Straubing hoard, there is no doubt that they are also of Oriental origin. Nor is an Oriental origin excluded if we suggest that an arms workshop had accompanied the troops which eventually settled near the flow of the middle Danube, since it continued to engage masters from the East.

## Crested Iron Helmets with Plate Coverings[2]

In 1909 excavations were conducted at Dunapentele, the Roman Intercisa, in the so-called Pozsgai vineyard. Inside the building which was uncovered, at point C of the room labeled 1, a group of fifteen to twenty iron helmets was found. On a few helmets even the crest and visor remained intact as recorded in Hekler's excavation report.

By studying the ground plan of the camp we can see that it is even possible the buildings had been used as warehouses. The great number of the helmets seems to prove that they did not get there at random but were deliberately collected and stored.

On the iron remains of the helmets a silver covering was found. It is possible that the helmets, originally all covered with silver, were divested of the valuable metal and stored in one place for smelting. If the silver covering had melted in some fire, then traces of silver would have remained among the iron pieces and would have been recovered during the excavation. Only four helmets could be restored completely but these preserved for us three different varieties of crests.

It is certain that the iron helmets had all been plated with silver because the rivets still contain bits of silver on them. As mentioned above, the helmets have been recovered in a mutilated shape. Nevertheless, recent research has solved certain problems, throwing light on their original condition and technology of production.

### *Helmet No. 1* (H.N.M. 61.13.195) (Pl. LXIX)

The two halves of this helmet were held together by a central longitudinal strip. The two segments were not of the same size; therefore, the halves were asymmetrical. On each side of the strip, two openings were made for the eyes; otherwise the segments remained unadorned. All around the rim there were little holes down below which served originally to fix the silver covering to the leather lining. Around the ears the helmet's segments had been cut out in arches. The middle strip on the helmet rose along the rim and on the surface of the segments are the marks of rivet holes 3 centimeters apart. It is possible that from these holes small buttonlike globes protruded, forming a sort of ornament on the crest. The iron plate to protect the neck must have also been covered with silver. The measurements of the helmet are as follows: outside circumference, 65 centimeters; width of strip, 9.3 centimeters; length of visor plate, 14 centimeters.

### Helmet No. 2 (H.N.M. 97.1909.249)

In shape this helmet is completely identical with the preceding one. The head of the helmet consists also of two segments. Besides the openings for the eyes, there is no other ornament.

### Helmet No. 3 (H.N.M. 56.40.11) (Pl. LXX)

On this helmet the flat strip rises to form a 2-centimeter-tall crest. The helmet has rivets with silver beads.

### Helmet No. 4 (H.N.M. 10.1951.2)

Helmets 1, 2, and 3 belong to the same type, there being only slight differences in the formation of the crest. Helmet No. 4 is identical in character with the preceding ones, but it differs in the arrangement of the crest. There is no iron strip on this helmet, but a crest rises where the two segments meet. On the forehead part we can also find openings for the eyes, and on the arched gaps for the ears two crescents are embossed. On this helmet, as on the former ones, we can find small holes to fix the helmet covering with the lining. The white-metal covering exists in traces around the gaps for the ears.

### Eskü Square Helmet No. 5 (H.N.M. 110, 1899) (Pls. LXXI–LXXIII)[3]

The so-called Eskü Square helmet emerged first 25 June 1898 on the left bank of the Danube when Elizabeth Bridge was being built and the foundation for the left-bank pillar had been made. According to official contemporary data the helmet was lying 3.5–3.8 meters deep under 0 point of the Danube, in a thick layer of gravel.

In 1932 and later in 1944 Lajos Nagy and Vilmos Bertalan made excavations and stated that between the Pest bridgehead of Elizabeth Bridge and the site of the City Church a camp was situated in Roman times. The excavations clarified two periods in the life of the camp. On the further side of the *limes*, it had served as bridgehead fortification.

The inside of this helmet was made of iron, and its outer surface was covered with gilt silver plate. In the embossed ornamentations there were patterns and colored paste beads. The edge of the helmet was bound by a hoop which was preserved almost fully intact and could easily be reconstructed. Ornamental designs framed by rows of pearls all contained the symbols of victory. We can see Victoria holding a wreath in her raised hands; in another frame Jupiter sits on a throne.

In another, lions facing each other defend a crater from both sides. The ornate strap on the right of the helmet, however, is defective. The ornamentation on the strap also represents Victoria and the lions defending the crater.

One needs say little else of the subject matter of the ornaments, since they speak for themselves — all referring to the symbolism of victory. The bottom band is divided into two parts. The top zone consists of small embossed crosses. Unfortunately, the bottom zone is badly damaged; originally there must have been a legend here. Before the helmet had been cleaned Hampel had seen the letter $R$ and somewhat further the letter $M$. Géza Nagy had deciphered the letter $O$ from the legend. After repeated cleanings, András Alföldi read the letters $ORAR$ consecutively and somewhat removed the letter $M$ — this has been photographed. However, today it is impossible to identify a single letter.

The approximately 5-millimeter-thick helmet was made of two hemispherical sections held together in the middle by a low, 1–2-millimeter-tall and 1-millimeter-wide band or crest.

Helmets numbered 1–5 were made of iron plates which, as we firmly believe, came from the same workshop. Earlier scholars, such as Hekler, Hampel, and Finály, seem to have been convinced that the Intercisa helmets must have originated with the sojourn of Cohors Milliaria Hemesenorum, and the symbols display Syrian or other Oriental elements. It becomes evident from the Intercisa monograph, however, that after Gallienus not a single inscription survived in Intercisa of Cohors Hemesenorum. This troop was no longer present in Intercisa in the early fourth century when the helmets were made.

The view which related Germanic elements to the origin of the helmets seems to have held its ground longest. It was believed that the Germanic influence manifested itself in the barbarically grandiose geometrical ornamentation and the large colored stone imitations. Recently Miriana Manojlović, in her review of the Berkasovo helmets, also insisted on Germanic influences in the ornamentation of helmets.

There is no doubt that in the ornamentation of these and similar helmets, Barbarian elements seem to preponderate. These elements, however, do not mean that the helmets derive from Barbarian workshops, but rather they reflect the fashion of ornamentation in late Roman court art, which was undoubtedly under Gothic–Germanic influences. The "barbarization" which characterized late Roman art at the end of the fourth century began under Diocletian and continued under Constantine the Great, when the Byzantine or Oriental style affected not only court clothing but court ceremony as well.

We believe that the helmets under discussion, which constitute a group with embossed patterns, precious metal covering, and inset paste stones, can be relegated to the circle of late Roman court art which was altogether under the influence of the late Roman court taste.

The workshop of armament factory which produced this group of helmets was not situated, we believe, very far from the place (Budapest, Intercisa) where they were found. A group of finds has for a long time been preserved in the Hungarian National Museum; the study of these may take us nearer to determining the origin of the helmets which belong to this group.

At one end of the Krászna gold bricks there was formed an allegorical figure framed with rows of pearls; it was a patron deity with a *corona muralis* stamped in. If we compare this stamp with the picture of Jupiter in a pearl frame on the Eskü Square helmet, it becomes evident that the two reflect a similar conception. On the Krászna gold bricks there are stamps under the figures with the letters *SIRM*. This is an obvious reference to Sirmium. The Sirmium mint had produced gold coins; that at Siscia, bronze ones. We believe it is justifiable to suppose that an imperial goldsmith workshop to deal in precious metals had been operating next to the mint at Sirmium. The Sirmium mint, as the coins indicate, had ceased production after Valentinian I. In determining the date of the Krászna gold bricks, help is provided by the stamps which preserve the half-length portraits of two adult men and a youth. According to Hampel's interpretation, the youth is Gratian (in 367, after he became Augustus) and the two adults are the emperors, Valens and Valentinian, who died in 375. According to these facts the Krászna gold bullions date back to 367 and 375.

In order to determine the date of production of the Eskü Square helmet, the following data are at our disposal. According to András Alföldi, the earliest occurrence of helmets of this type coincided with the third decade of the fourth century. Manojlović believes, according to the evidence of the legends on the Berkasovo helmets, that they were made under Licinius or Constantine. Together with the Berkasovo helmets, so-called propeller-shaped silver mountings had come to light; the same sort of mounting made of bronze emerged in greater numbers from fourth-century graves in the Transdanubian area, and some are also preserved in the Pécs, Keszthely, and Veszprém museums. The silver mountings of the Berkasovo helmets show a strong resemblance to the guards of fourth-century bulb-headed *fibulae*. As we have said in connection with the stamps above, the latest date for the Eskü Square helmets, considering the bead-framed stamps of the Krászna bricks, cannot be earlier than the sixth or seventh decade of the fourth century.

We consider our Pannonian helmets to be products of the middle Danube area, and we suppose that they were made in court workshops attached to the Sirmium mint for the purpose of producing ornamented armor. The decorations of the Eskü Square helmet — the polychrome stones and the embossed ornaments — reflect Barbarian taste, whereas the few pictorial decorations preserve something from the traditional imperial goldsmith's art. This type of decoration was characteristic of the late Roman court style.

## SHIELD-BOSSES

From the Roman period very few shields and shield-bosses have emerged.[4] Up to 1966 not a single ornamental shield-boss had existed in the archaeological collection of the Hungarian National Museum. However, in January and December of 1966 the collection of Pannonian armor was augmented with the acquisition of two richly chiselled, silver-plated, bronze shield-bosses.

Shields in the Roman period were made generally of layers of wood; they were often covered with leather, which was sometimes fastened and bordered with strips of leather. However, both modestly and richly ornamented shield-bosses had been made of metal. The boss (*umbo*) had a structural function. On the outer side of the shield, which faced the enemy, the boss had protruded spherically, and on the inner side, facing the warrior, there was the metal grip, the *ansa*, which accommodated his clasped hand. The structure of the shield-grip has been well preserved in unique wooden shields from the late second and the third century in Dura-Europos. Their colored decorations serve as important pieces to compare with our Pannonian shield-bosses.

One of the ornamented shield-bosses under discussion derives from Százhalombatta (Matrica) (Pls. LXXIV and LXXV); the second emerged from river dredging between Dunavecse and Harta when the gravel had been unloaded at Dunaföldvár. Along this stretch of the river are situated the remains of Roman camps at Matrica (Százhalombatta), Vetus Salina (Adony), Intercisa (Dunaújváros), and Annamatia (Kisapostag). It is not certain that the shield-bosses recently recovered got into the riverbed from any of these camps; rather we suppose that they were somehow connected with Roman boats sunk in the Danube. At Százhalombatta a Roman boat was in the riverbed, covered with a thick gravel-bank several meters wide; when dredging began thirty-five pieces of bronze vessels and the equipment of a boat's kitchen were recovered. The bronze finds from the boat seem to belong to the early Imperial period, the first and second centuries. The same dating is supported by ceramic pieces recovered from the boat.

The inventory number of the shield-boss which emerged from the Danube is H.N.M.66.1. Its full diameter is 20.4 centimeters. The diameter of the globe is 11.2 centimeters; its length, 6.3 centimeters; the width of the rim is 4.6 centimeters. The thickness of the bronze plate alternates between 0.4 and 0.6 centimeter. The patina is dark brown all over; the white metal covering has worn off at several points. The punched pattern is not precisely covered by the white metal covering, which is somewhat displaced. The shield-boss is richly decorated. On the upper field of the dome stands an eagle with outspread wings, clutching with its claws Jupiter's thunderbolt and holding in its beak the victor's wreath. In a broad, adorned field we can see animals chasing each other: a tiger, a stag, a lion, a deer, wild dogs, and a hare. On the rim alternating round decorations dominate; there are eyes at peacocks' feathers, and on the nails serving to fasten the shield-boss hang small wreaths. On the entire shield-boss the bronze-colored foundation alternates with the white metal covering.

It is worth observing that this shield-boss has the same characteristics as the shield-boss labeled Mainz A, as confirmed by Klumbach — namely, the white metal covering does not exactly coincide with the original punchings and small circles, but often diverges from them and definitely assumes an oblong character. In addition, the eagle in the upper round field of the Dunaföldvár shield-boss has wings outstretched in exactly the same way, as in Mainz A.

From Dunaföldvár (Pls. LXXVI and LXXVII) also derives a second richly ornamented shield-boss now in the Hungarian National Museum, its inventory number being 66.15.1. Its full diameter is 21.7 centimeters; that of the globe, 11 centimeters; its length is 6.7 centimeters; the width of the border is 5.5 centimeters; that of the bronze plate, between 0.9–1.1 millimeters.

The upper sectors of the hemispherical globe of the richly ornamented shield-boss are adorned with concentric circles; in two rings there are superimposed so-called bird's-eye pattern resembling the eye on a peacock's feather. The next ring is made up of flower-cups running all around. Then a field adorned with almond shapes follows the horizontal border of the globe. The border is filled with uniform foliage designs. However, when the design was planned, the place of the shield-nails had been left out of consideration. The design, the tendrils, and the peacocks' eyes preserved the original color of the bronze, while the white metal covering of the base is brightly shining even today.

Two inscriptions are still discernible on this richly ornamented shield-boss. According to one, the owner of the shield served in the *centuria* of Cassius Potens. The other names the second owner, Crescens Propincus, who belonged to the *centuria* of Antonius. On the inner side of the shield-boss there are the lightly incised letters *ROPINQUS*, from the name preserved on the outer side.

Although the broad border ornamentation with tendrils differs considerably from the geometrically proportioned border-edge ornamentations, this piece shows close resemblance to shield-bosses of Százhalombatta, Mainz A, Mainz B, and Halmágy.

By comparing the execution of various shield-bosses, we can come to the conclusion that the embossed remains of the Straubing ornamented shield and our ornamented shield-bosses trace back to the same origin. Among other marks, we can find in both groups ornamentation on embossed and smooth surfaces, other features being, for example, patterns made up of punched circles, where the circles themselves are connected with horizontal and perpendicular punched bars.

We are firmly convinced that the key to the definitive comparison can be furnished by the painted wooden shield of Dura-Europos. The painting of oval shields consisted exclusively of ornamental elements around the *umbo*. The shields were covered with Homeric scenes, the fight of the Amazons, and god figures. Only a circle-shaped field in the fess remained unpainted. The diameter of the unpainted surfaces varied between 18.5 and 22 centimeters. This measurement corresponds with that of the round bronze bosses which emerged at the excavations at Dura-Europos. It also corresponds with the measurements of the round shield-bosses under discussion.

On the oval and oblong shield we are particularly interested in the ornamentation which surrounded the bronze *umbones*. It would be impossible not to note a close parallel between the designs of the bronze shield-bosses we have discussed above and those around the bosses of the painted shields of Dura-Europos.

The central ornamentation of the so-called Homeric shields consists of closely connected plant elements. The wreath embracing the center is interspersed at times with ribbonlike ornaments. The ornamentations of Mainz B and of the shield-bosses of Dunaföldvár were conceived in the same undivided manner.

The *umbo* of the so-called Amazon-scene shield is divided into eight parts, garlanded by varied decorations, in exactly the same way as the decorations of the Százhalombatta shield *umbo*. Leaf, petal, and feather designs alternate. The center of the oblong Dura-Europos shield is also decorated with designs of petals, tendrils, leaves, and feathers.

The decorative motifs of the painted shields' *umbones*, as well as those of the Pannonians discussed above, and even partly those of the Mainz bronze *umbones*, show an indisputable similarity with each other.

Flower cups, tendrils, and flower designs decorate not only the Amazon shield and the painted *scutum* but also the Dunaföldvár, Mainz A, and Halmágy shield-bosses.

Nevertheless, we cannot believe that all these shields copy the same book of patterns. They may, however, trace to the same original workshop, a possibility supported not only by an identity of motifs but by a similar manner of execution. In the case of the bronze shield-bosses, let us take a simple motif, that of the punched circles, which are either put side by side or one above the other, or are joined by punched bars. The circles sometimes assume a slightly oblong shape, and this shape is highly characteristic of the same group.

The excavation report made on the Dura-Europos painted shields stated that they were all found together near a Roman camp, in the *canabae*, and may be considered to have been stored by their maker. The bronze *umbones*, as Hopkins remarks, had been fixed after the painting had been finished, on the spot left unpainted by the armor maker himself. Hopkins's remark that the artist who had made the shield was not a native of Dura, seems to be very significant because he did not paint in the characteristic local style. Indeed, although the shield decorations display a light, colorful manner, there is also a strict well-ordered composition. We believe that they follow a more disciplined imperial military style of decoration, although typical Oriental, Syrian, or Alexandrian marks are also present. This military style is evident not only in the subject matter but in the use of typical symbols of victory.

On the basis of similarities, and in many cases even identical features we may risk the supposition that our richly decorated Pannonian shield-bosses were produced in an imperial armament center either in Syria or near Alexandria and that this workshop had not excluded direct local artistic influences.

Concerning the purpose of the Pannonian bronze shield-bosses and similar pieces, we do not consider them as parts of exclusively ornamental armor, but more particularly as marks of distinction to reward official military accomplishments; they were fastened as ornaments on defensive shields used in battle.

The Pannonian shield-bosses and similar pieces in the group have been dated by various authors, on the basis of accompanying evidence, as within the period between Claudius and Severus. The Pannonian piece may be dated, with consideration of the circumstances found at the excavation, to the second century. Our shield-bosses of semicircular globes and flat rims must have been used by the Roman troops for approximately 150 to 170 years. Because of this time period the differences within the same type, which were discernible on the examples discussed above, are understandable. Nevertheless, agreement in motifs and execution is striking on certain pieces. These identical features permit us to conjecture that within the same group there must have been pieces which derived from the same origin at practically the same time.

# DAGGERS

In recent years various weapons from the Roman period have been identified, among them a group of daggers from the Imperial period in Pannonia. Richly ornamented daggers are among the finest pieces of weaponry from the Roman period.[5]

In the fall of 1967 a dagger was bought from workers engaged in river-dredging activities; it was rusty up to its blade. The length of this weapon, including its hilt, is 32.6 centimeters. The broadest part of the blade is 6.1 centimeters. On the shoulder and grasp of the dagger, holes mark the place of mountings. In the upper third of the grasp an iron plate protrudes on both sides in a semicircle. In the upper third of the blade, it narrows and widens further down to end in a point. On the blade's surface a middle rib protrudes. On both sides of the rib are channels, indented into the blade, which also end in a point. On the grasp and blade of the dagger, pieces of wood had been preserved by petrifaction.

The length of the deformed sheath is 28.4 centimeters; its widest point is 7.3 centimeters. The sheath also is made of iron. Its shape follows the dagger closely, though it ends not in a point but in a disc. A cover was fastened with six nails on both sides to the original wooden hilt. Whether the wooden hilt, which in front was covered by an encrusted iron piece, also had a leather covering at the back is impossible to state today.

The iron sheath was richly decorated with gold strips of 0.5–2.5 centimeter wide which were encrusted into the iron base. Rosettes, double rows of leaves, palmettes, gold incrustation, and red enamel inlay alternated in the decoration.

From the forty daggers which have come down to us from the first century, G. Ulbert selected a group which had seemed to him to consist of similar pieces. He came to the conclusion that the Vindonissa daggers and hilts dated by Exner to the middle of the first century were much thinner than the Roman dagger made in the early or middle decades of the century. As a result of lucky finds, he succeeded in dating them to the second half or end of the first century and in determining Vindonissa as the evident place of provenance.

At a later stage V. von Gonzenbach also selected a group of sword hilts, decorated with mythological and allegorical reliefs. He determined that they were used and produced during Tiberius's rule. It seems natural in connection with our new Pannonian finds of the Dunaföldvár daggers (Pls. LXXVIII and LXXIX) to differentiate a group of daggers with ornamental sheaths and suggest an exact dating. Such a group is characterized by a broad blade and sheath and by silver and gold filigree with frequent red, infrequent green or yellow, enamel encrustation. All the daggers and sheaths in this group were made of iron.

Within the type of the Dunaföldvár daggers two varieties can be distinguished: (a) dagger sheaths corresponding exactly with the Dunaföldvár one; (b) pieces agreeing with the style of decoration of the Dunaföldvár sheath, using variations of its patterns and other motifs too.

We are dealing only with Pannonian daggers belonging to the said category. A thorough discussion of the daggers found in the territory of the Roman Empire has been published elsewhere, so that here we are only summarizing results.

Besides the Dunaföldvár dagger another Pannonian dagger belongs to group (a); it emerged from the bed of the Kulpa River at Sisek, together with its sheath (both are preserved today in the Zagreb Museum). On this dagger the pattern was almost identical with that of the Dunaföldvár sheath, the decorations consisting of metal inlays and red enamel encrustation. To group (b) belongs the second dagger from the Kulpa which emerged at Sisek together with its sheath. Its pattern consists of metal encrustation and red enamel. Besides the encased rosettes and the palmettes running up to the point, the decorator introduced a square middle field.

Completing the chronology of G. Ulbert and V. von Gonzenbach, we can summarize what we know about the types of daggers in the first century. The ornamental daggers from the period of Augustus and Tiberius, the latest pieces coinciding with the time of Claudius's reign, were followed in the second third of the first century by daggers of the Dunaföldvár group with precious metal and encrustations of colored enamel. However, the daggers belonging to the group under discussion are mostly the types of the period between Claudius and Nero. Thus through a definitely characterized transitional period the daggers and sheaths belonging to the Vindonissa period dominate in the second half of the first century.

If we consider the daggers belonging to the Dunaföldvár group in the territory of the Roman Empire, it can be stated that some 75 percent of these emerged from rivers — from the Rhine at Mainz and Cologne, from the Danube at Dunaföldvár, from the Kulpa River at Sisek. The other 25 percent were found in Roman camps and fortifications.

The sites of the Pannonian finds furnish important data on the history of the Roman occupation of Pannonia. It becomes evident from the daggers which emerged from the rivers that Rome, before introducing inland trade and money circulation into Pannonia, had kept up lively traffic on the Danubian *limes*. For the transport of troops, waterways were given preference over inland routes, thus securing simultaneously the military supervision of the *limes* as well. The presence of daggers and blades in the waterways, particularly near the Danube, which was in direct contact with the Drava–Sava area, leads us to assume that the Romans had moved their troops around the Danubian *limes* of Pannonia during the reign of Claudius.

The question naturally arises whether these daggers had not been part of the armor of the officers of the Roman fleet. This pleasing theory, however, does not hold up because we know about numerous gravestones in the Rhineland where the same kinds of daggers appear as part of the equipment of foot soldiers, too.

We have every right to suppose that the daggers of the early Imperial period, with sheaths bearing polychrome decorations, had formed part of the armor of the officers of the legions. It is very likely that they were given as marks of distinction to their owners. In the decorations of the sheaths, wreaths made up of double rows of leaves seem to dominate, as the *corona triumphalis* had played an important role in Imperial military symbolism. The palmette as an ornament on the points of the blades was also the symbol of victory. Nor is the role of the *pelta apotropaicus* less noted among military symbols. Finally stylized *aediculae* on these daggers provided for the owners the protection of the fighting deities of the camp's sanctuary. Thus the decoration of the daggers retained the clear-cut military symbolism of the Roman Empire, which permits us to suppose that these daggers had been given as rewards after victorious battles to officers who deserved them.

Nevertheless, we do not possess sufficient data as yet to answer the question whether there existed a central military armaments workshop which produced the encrusted, enamel-colored daggers and sheaths. It seems more than likely, however, that we should look for such a workshop somewhere in North Africa, from where armaments were taken through Iberia to the provinces of the empire. We have no data to support the theory that armaments workshops existed near the Rhine, the Danube, or the Kulpa.

## NOTES

[1] E. B. THOMAS, *Helme, Schilde, Dolche: Studien über römisch-pannonische Waffenfunde* (Amsterdam, 1971).

[2] L. BARKÓCZI, "Római díszsisak Szőnyből — Römischer Paradehelm von Brigetio". *FolArch* 6 (1954): 45–48.

[3] L. LINDENSCHMIT, *Die Altenthümer unserer heidnischen Vorzeit*, vol. 5 (Mainz, 1900), p. 233, Table 41; J. HAMPEL, "Ein Helm von der pannonischen Reichsgrenze". *Zeitschrift für Historische Waffenkunde* 2 (1899): 192–201; J. HAMPEL, "Ókori sisak". *ArchÉrt* (1900): 361–74; G. NAGY, "Az eskütéri sisak" (The Helmet of Eskü-tér). *BudRég* 7 (1900): 69–83; F. DREXEL, "Römische Paraderüstung". *Strena Buliciana* (1924), pp. 55–72; A. ALFÖLDI, "Eine spätrömische Helmform und ihre Schicksale im germanisch–romanischen Mittelalter". *ActaA* 5 (1935): 99–144; L. NAGY, *Az eskütéri római erőd, Pest város őse* (Die römische Festung am Eskü-Platz, Vorgänger der Stadt Pest) (Budapest, 1946), p. 7; A. RADNÓTI, "Bronz Mithras-tábla Brigetióból — Le bas-relief mithriaque de bronze de Brigetio". *ArchÉrt* (1946–1948): 137–55; J. KEIM and H. KLUMBACH, *Der römische Schatzfund von Straubing*, vol. 1 (München, 1951), Table 15, Figs. 10, 11; Table 17, Fig. 1; Table 18, Figs. 10, 11; Table 29, Fig. 1; Table 23, Figs. 1, 2; Table 25, Fig. 1; *Intercisa* (Budapest, 1954), 1:8, Figs. 2, 3; L. BARKÓCZI, "Római díszsisak Szőnyből — Römischer Paradehelm von Brigetio". *FolArch* 6 (1954): 45–48.

[4] LINDENSCHMIT, Vol. 4, Table 52; V. HOFFILLER, "Oprema rimskoga vojnika u provo doba carstva". *Vjesnik Hrvatskog Arheološkog Društva* 12 (1912): 16–123, Figs. 48, 49; K. KÖRBER, "Einige römische Grabdenkmäler, die aus Mainz stammen oder im Altertumsmuseum daselbst aufbewahrt werden". *Mainzer Zeitschrift* 11 (1916): 87, Table 9, Fig. 7; K. EXNER, "Römische Dolchscheiden mit Tauschierung und Emailverzierung". *Germania* 24 (1940): 22–28, Figs. 1–6; K. MERKELIN, "Antiken im Hamburgischen Museum für Kunst und Gewerbe". *AA* (1928): 461, Fig. 172; W. DRACK, "Fragment einer römischen silbertauschierten Dolchscheide". *Zeitschr. f. Schweiz. Archäologie u. Kunstgesch.* 8 (1946): p. 113, Taf. 34; W. KRAMER, in *Cambodunum-Forschungen* 1953, vol. 1 (1957); L. BARKÓCZI, in *Intercisa* 2:501; G. WEBSTER, in *ArchJ* 1958 (1960): 115; J. YPEY, "Drei römische Dolche mit tauschierten Scheiden aus niederländischen Sammlungen". *Berichten v. d. Rijksdienst v. h. Oudheidkundig Bodemonderzoek* 10–11 (1960–1961); G. ULBERT, "Silbertauschierte Dolchscheiden aus Vindonissa". *Jahresber. Vindonissa 1961–1962* (1962), pp. 5–18; J. K. BOGAERS and J. YPEY, "Ein neuer römischer Dolch mit silbertauschierter und emailverzierter Scheide aus dem Legionslager Nijmegen". *Berichten v. d. Rijksdienst v. h. Oudh. Bodemond* 12–13 (1962–1963): 87 ff.; V. GONZENBACH, "Tiberische Gürtel und Schwertscheidenbeschläge mit figürlichen Reliefs", in *Helvetia Antiqua*. (Festschr. f. K. Vogt) (1966), pp. 184 f.; H. PETRIKOVITS, *Die römischen Streitkräfte am Niederrhein* (Düsseldorf, 1967), p. 24; G. ULBERT, *Römische Waffen des 1. Jahrhunderts n. Chr.* (Stuttgart, 1968), pp. 9 ff.

[5] DREXEL, p. 61; C. HOPKINS, F. E. BROWN and R. J. GETTENS, *Excavations at Dura–Europos, Preliminary Report of the Seventh and Eighth Seasons of Work, 1933–34 and 1934–35* (New Haven, 1939), p. 327, Tables XLII, XLV; J. KEIM and H. KLUMBACH, *Der römische Schatzfund von Straubing* (München, 1951), vol. 1. Table XXX; *Guide to the Antiquities of Roman Britain* (London, 1958), p. 67, Fig. 34; J. M. C. TOYNBEE, *Art in Britain under the Romans* (Oxford, 1964), p. 299, Table LXIXa; H. KLUMBACH, "Drei römische Schildbuckel aus Mainz". *JRGZ* 13 (1966): 165–89; PETRIKOVITS, p. 27; ULBERT, p. 15.

# PANNONIA – FIFTH TO NINTH CENTURIES

## ÁGNES SALAMON AND ÁGNES CS. SÓS

Until recently, Pannonia's late Roman period was considered to have ended in the last decades of the fourth, or at the beginning of the fifth century. New types of finds, which appeared at the turn of the fourth century and differed from the earlier fourth-century Roman material, were defined by archaeologists as those of the "Hun period".[1] Our recent historical studies developed the view that Pannonia's provincial status remained unchanged until the downfall of the West Roman Empire. Archaeological examinations have also proved with increasing conviction that the Barbarian peoples and parts of Barbarian groups recognized as *foederati* from the end of the fourth century altered Pannonia's ethnic composition and culture but that, apart from these changes, the local population and antique culture continued to survive even into the fifth century.[2]

Due also to its topographical position, Pannonia remained a territorial unit even after the decline of the West Roman Empire; during the flowering of the Ostgothic Empire, it belonged to the North Italian sphere of influence. It may be presumed that, after the death of Theoderich, Byzantium claimed this territory on the basis of the continuity of Roman law. This makes it clear that Emperor Justinian's grants of land and legal rights were settled on the Longobards in exchange for their military and political alliance. Analysis of archaeological evidence has also shown that the Danube *limes* remained the frontier even during the stay of the Ostrogoths and the Longobards. Traces of urban life have been found in Pannonia even in the sixth century.[3]

Changes occurred when the Avars occupied the whole of the Carpathian basin. Pannonia, as a territorial unit, appeared once again during Carolingian times and it is not coincidental that the centers of the Hungarian state subsequently developed in the areas of the one-time Roman cities.

In the following, we are going to trace the history of the peoples who settled in Pannonia based on our archaeological evidences.[4] Among these peoples or ethnic groups the Huns settled first in Pannonia with their own chieftains.

## THE HUNS

When Emperor Gratian (367- 383) settled the Hunno–Alanic– Gothic groups as *foederati* in Pannonia, the majority of the Huns still lived in the territories east of the Don River (Fig. 68).[5] The expansion of the united Huns first took place in the lower Danube district. Between 416 and 425, they bargained from here with

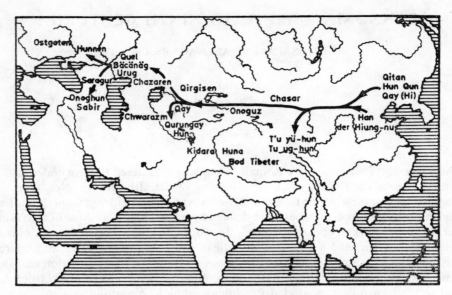

Fig. 68. Periphery of Iran and the western migration of the Huns
A.D. 310–374 (Historia Mundi V)

the Roman Empire; the negotiations ended with an official withdrawal of rights from the Huns who had settled in Pannonia at an earlier time. The Roman Empire gave them the status of *foederati* and in this period they moved away from the region of the lower Danube. Rua and Uptar became Hun chiefs after 427 and Attila took Uptar's place after the latter's death. In 427 Pannonia Secunda and Savia and, about 433, Pannonia Prima officially became part of the territory occupied by the Huns.[6]

At the time they settled in the Carpathian basin, the Huns lived in tribal confederation. They held together all the other peoples who lived in this area by their tribal confederation and the *foederati* rights. The "Hun Empire" reached its zenith during Attila's reign when he became the sole chief of the tribal union after Rua's death.[7] The West Roman Empire gave him the title of *Magister Militum* and the East Roman Empire gave him the title of *Basileus*. The earlier aspirations of the Huns to expand their territory were then fulfilled. Their warfare in the west made their impact felt as far as the Rhine River; because it was aimed mainly against the free Germanic peoples, they were not in opposition to the politics of the West Roman Empire.

The campaigns in the Balkans, however, created a strained relationship with Byzantium. To settle affairs, Priskos Rhetor, accompanied by a diplomatic mission entrusted with negotiations, arrived at Attila's court in 449.[8] It was probably due to these discussions that Attila again extended his wars waged against the west. His campaign against Gaul ended near Catalaunum with an undecided battle in 451. In 452, during his Italian campaign, after the successful sieges of Aquileia, Milan, and Pavia, Attila stopped before reaching Rome (Fig. 69). This was his

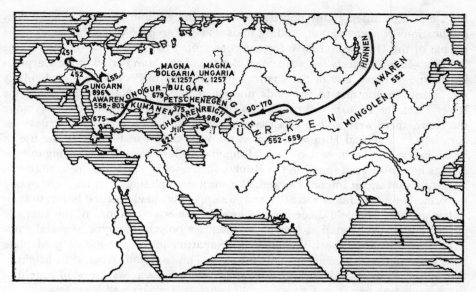

Fig. 69. Movement of North-European equestrian peoples (Historia Mundi V)

last great enterprise. After his death in 454, his sons could not retain the leading role which the Huns had hitherto maintained. The Huns' previous allies, led by Ardarich, king of the Gepids, defeated them in a battle near Nedao (this locality cannot be identified, it was perhaps in the Carpathian basin) and the majority of the Huns retreated to the area by the Black Sea.

Research concerned with Hun archaeological material dates back only to the past few decades. The greatest difficulty in selecting material was caused by the fact that the Hun Empire comprised a great variety of ethnic groups and therefore the finds were very heterogeneous in South Russia, too. In 1927 T. M. Minajeva pointed out that, in the Sarmatian finds near the Volga, the heritage of a new people from the east could be isolated.[9] According to her, their new burial rites and arms could be followed up to the Carpathian basin. Rather than calling this new material Hun, she identified it as that of a later Sarmatian group. In subsequent years, research has been considerably developed by A. Alföldi; W. Polaschek, and J. Werner. They defined the group of finds described by Minajeva as Hun, on the basis of those found in the Carpathian basin.[10]

In recent decades, the most comprehensive work has been J. Werner's two-volumed publication in 1956.[11] In it, the author coordinated archaeological observations and material with historical events and defined the geographical boundary of the fifth-century group of finds ascribed to the Huns between the areas of the Tienshan and Central Europe. Describing the equestrian nomadic culture of Attila's period, Werner demonstrated two different influences: one from the eastern steppes and another from the kingdom at Bosporus and the Sarmatian culture in the Kuban region. What interested Werner as well as the Hungarian researchers was the social organization of the "Hun Empire", its economic structure, and the causes of the flourishing and disintegration of tribal unity.[12] New publications concerning outstanding finds and the archaeological heritage of the people of the Hun period have further contributed to research.[13]

The so-called princely finds are characteristic of the Hun period in the Carpathian basin; of these, one in the territory of Pannonia in Pécsüszög is particularly significant. In contrast to the finds unearthed in Szeged-Nagyszéksós, near the River Tisza, the Pécsüszög finds did not originate from cremation burials. Among the latter, there were gold plates that were belt, armor (sword), and harness decorations.[14] A typical Hun weapon was the bow shown in Figure 70, the use of which had spread to the east as far as Mongolia. Such bone plates belonging to bow fittings have also been found in the Roman fortresses in Pannonia, at Vindobona, at Carnuntum and at Intercisa as well. The sheet-gold plating from the bow among the princely finds at Jakusowice (Poland) was probably made for the burial to symbolize might.[15] The gold diadem was a headdress worn by women of the nomadic chiefs. The diadem found at Csorna represents a polychrome type of metal work from South Russia. The little soldered compartments on a band of gold plate were decorated with almandine and cornelian. This technique reached its height in this period in the metal-working centers of South Russia; we know of nine diadems, including one from Csorna, found between Hungary and Kazahstan.[16]

Typical of Hun workmanship were the cauldrons which, according to the testimony of rock drawings in Southern Siberia in Kisil-kaja, were used for sacrifice. In the East, they were known in the Altai region and, in the West, in Moravia, Silesia, and France. The majority of these cauldrons have come to light in Hungary. Fragments of cauldrons emerged from the Roman fortresses of Intercisa and Brigetio. The most intact cauldrons were found in the valley of the River Kapos as well as at Bántapuszta and Törtel (Pl. CLII). The bronze cauldrons were cast and put together from two halves.[17]

Typical vessels from this period were the dark gray or black jugs ornamented with burnished pattern. One of the most beautiful examples was found in Murga (Fig. 73a; Pl. CLIII) together with almandine-decorated gold clasps.[18] One variety of this type of jug was found among late Roman ceramics; another one, wide-mouthed and small, was widespread mainly in the Sarmatian territory. The custom of skull-deformation came to Pannonia probably with the Huns in the fifth century. The deformation was begun in childhood. On the skulls, coming to light from the family burial place at Mözs, the manner of deformation can be clearly seen.[19]

## THE OSTROGOTHS

Apart from the Ostrogoths, who were settled by Emperor Gratian as a group of *foederati*, another important group of Ostrogoths arrived with the Huns in the Carpathian basin.[20] After the battle of Nedao, the Goths who had supported the Huns retreated to the territory of Pannonia and renewed their alliance with Emperor Martian. The retreat of the Huns forced those Germanic tribes which had settled in Pannonia in 455 into action (Fig. 71). These were the so-called Amal Goths who had three kings at this time: Valamer, Vidimer, and Thiudimer. Each

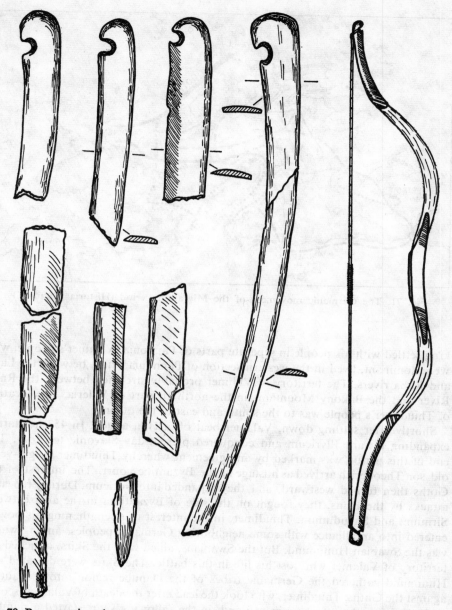

Fig. 70. Bow covering plates made of bone from Dunaújváros (Intercisa) and the reconstruction of a Hun bow

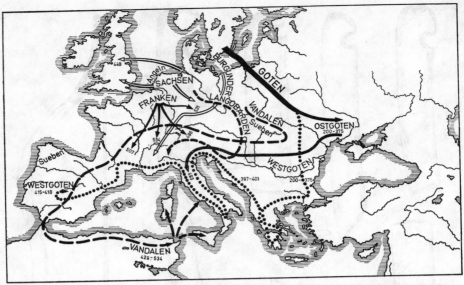

Fig. 71. The Germanic movement of the Migration Period (Historia Mundi V)

king settled with his people in separate parts of Pannonia. Valamer's people, who were dominant, lived in the western section of Pannonia Prima between the Lajta and Rába rivers. The territory of Vidimer probably stretched between the Rába River and the Bakony Mountains in the northern part of Valeria. The location of Thiudimer's people was to the south and east of the others.

Shortly after settling down, Valamer beat off a Hun attack. In 457 he started expanding toward Illyricum and conquered present-day Slavonic territory. The end of this period was marked by an agreement whereby Thiudimer's eight-year-old son Theoderich arrived as hostage in the Byzantine court. The interest of the Goths then turned westward, and they expanded into Noricum. During repeated attacks by the Huns, they fought on the side of Byzantium in the area between Sirmium and Singidunum. Thiudimer, in the interest of strengthening his power, entered into an alliance with some neighboring Germanic peoples, among whom was the Swabian Hunimund. But the Swabians, allied with the Skirs, attacked the territory of Valamer who lost his life in this battle. The Skirs were defeated and Hunimund gathered the Germanic tribes of the Danube region into a coalition against the Goths. Thiudimer, who took the lead after the death of Valamer, scored a victory over the attackers in 469 and, in the following year, started a punitive campaign against them. After these victories peaceful relationships were created for a time with their neighbors. However, on his return, Theoderich not only attacked the Sarmatians on the plains but occupied Singidunum as well. At this time, Vidimer's Goths left Pannonia. On his arrival in Italia, Vidimer died; subsequently, his son led his people to Gaul where they united with the Western Goths (Visigoths).

402

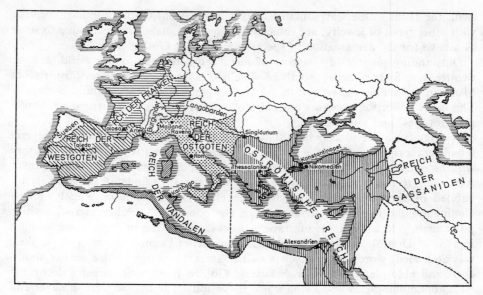

Fig. 72. Europe in the age of Theoderich the Great, A.D. 526
(Historia Mundi V)

After this event, Thiudimer continued his attacks against the East Roman Empire. Crossing the Sava River, he pressed on to Naissus where he died. The leadership was inherited by his son, Theoderich, in 471.

We have no records about the early period of King Theoderich's reign, but it seems possible that around this time the Goths expanded into the area of Lower Moesia. In 476 he sent troops to support Emperor Zeno; when he recaptured his throne from Basilius, he acknowledged the occupation of Moesia by the Goths. In 488 Theoderich with his Goths was directed to Italia against Odovacar with Byzantine approval. The final victory came with the occupation of Ravenna in 493. After this victory Theoderich as the personal representative of the East Roman Emperor and as the king of the Ostrogoths began his rule in Italia (Fig. 72). In 504, as a result of the Gepidic–Gothic warfare, Sirmium fell into Gothic hands and Pannonia Secunda and Savia also became part of the Roman Empire. Pannonia's provinces lying to the north of the Drava River belonged, in all probability, to Theoderich's sphere of interest, and it is possible that some Gothic tribes remained on these lands.[21]

Those finds, from which we conjecture that Ostrogoths appeared in Pannonia, can be traced back to the antique workshops on the Black Sea. The most characteristic feature of this metal work was the use of inlay technique: the decoration with almandine or garnet in cloissoné. In gold and silver jewels covered or decorated with inlay, filigree work, granulation, and decorations of niello technique frequently appeared.

The chronology of the East Gothic finds in Hungary is not final yet. No borderline has yet been defined between the earliest Goths who were settled as *foederati*

403

(with the Huns in the Carpathian basin) and the Amal Goths. Fibulae, together with other types of jewelry, are sufficiently typical in shape and technique to serve as a basis for the demarcation of successively settling Gothic tribes.

Outstanding records of this type of metal work, unique in the world, are the two treasure finds of Szilágysomlyó.[22] The first is housed in the Kunsthistorisches Museum in Vienna, the second in the Hungarian National Museum in Budapest. The latter collection contains ten pairs of golden fibulae, one large onyx fibula, gold torques, and gold cups.

Not far from the place where these treasures were found we know of several female graves (especially in the Tisza River region), where attires were held together with large silver-plate fibulae. Research has attributed them to the Ostrogoths.[23] According to our observation, there occur very few such finds on the territory inhabited by the Goths. In Pannonia (Répcelak, Domolospuszta, Regöly) jewels coming to light from three rich females' graves combine these characteristics which point mainly to eastern Mediterranean workshops.[24] One of these came to light in 1953 at Domolospuszta. The buried female from Domolospuszta (whose skull was deformed) wore silver bracelets on her arms, two large fibulae on her shoulders, and gold-plated beads on her neck. Golden polyhedric earrings decorated with almandine plates also belonged to these finds. The fibulae were of silver and had a gilded, chip-carved surface, and set with almandine plates. On the plate of the buckle, which was of similar style, birdlike figures were sitting. In the female's grave at Regöly (Pls. CLIV and CLV.1) were found a pair of fibulae covered with gold plates and her attire was decorated with numerous small pressed gold plates of different shapes.

The earliest relics of the runic inscription (Pl. CLVII. 1) used by the Goths came to light in Pannonia.[25]

## THE LONGOBARDS

A significant change occurred in the political situation of Pannonia after the death of Theoderich the Great in 526.

In 527 the Longobards under Vacho's leadership arrived in Pannonia[26] (Fig. 71) and the Gepidae occupied Sirmium in 536. The general political situation indicated that the Longobards appeared in the area of Pannonia Prima and Valeria in 539. After the death of Vacho, instead of his infant son, Auduin succeeded and became king of the Longobards after 547 or 548.

With the new dynasty, a new period started in Longobard politics. Emperor Justinian officially acknowledged the Longobard occupation in Pannonia and presented the Longobards with Savia and the eastern part of Noricum Mediterraneum as well. Although Pannonia's occupation by Byzantium is not documented, the acknowledgement and the territorial transfer prove that after the death of Theoderich, Byzantium renewed its earlier rights over this territory. To realize Justinian's ambition for the restoration of the empire, the acquisition of Sirmium was very significant. Thus, in the controversy between the Longobards and the Gepidae in 567, Byzantium helped the Gepidae. He obtained Sirmium in connection

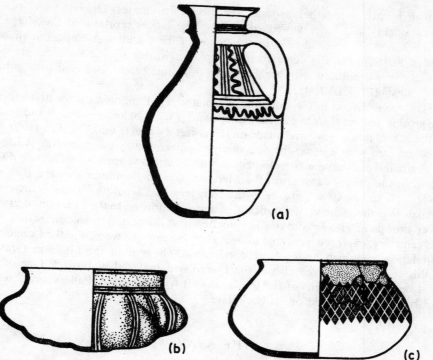

Fig. 73. (a) Jug of Murga; (b) Longobard vessel from the Vörs cemetery; (c) Longobard vessel from the Vörs cemetery

with this event. The king of the Longobards, Alboin, then entered into an alliance with Khagan Bajan who had previously been in contact with Byzantium and with his help scored a victory over the Gepidae.

According to an agreement with the Avars, the Longobards departed from Pannonia in 568 for northern Italia. Sources have revealed that several parts of the groups in Pannonia and also some Gepidae and Sarmatians attached themselves to the Longobards.

The Longobard finds of Pannonia[27] came to light primarily from burial grounds. After some research on this subject, new cemeteries were excavated but no publications appeared. Therefore, the present picture of the Longobard archaeological heritage — and more so that of Pannonia in the Longobard period — will be slightly modified.

So far, researchers have divided the Longobard finds from Pannonia into two parts: material produced in Pannonia and earlier finds brought from other locations. However, Longobard burial rites are connected with the earlier ones. The graves were in parallel lines and the dead were often placed in coffins. The typical jewels in females' graves were the fibulae (Pl. CLV.2). The fibulae were cast silver, often gilded, of medium size with chip-carving, and decorated with niello. They were found around the pelvis or between the legs. A small, S-shaped fibula held their attires together around the neck. In males' graves were placed spears, shields, and long, two-edged swords. The lances reproduced the shape of willow leaves. The shield-boss in the center of the shields was conical or slightly vaulted in shape. In the cemetery of Várpalota in one of the females' graves a necklace was decorated with amber, glass, millefiori beads, and also with golden bracteates[28] (Pl. CLVI. 1.2).

Clay vessels found in the Longobard cemeteries in Pannonia can be divided into two groups. In one, the pot was shaped by hand (Fig. 73 b) and was rough and small; in the other, it was gray and made on the potter's wheel (Fig. 73 c). The latter was often covered with a dense, netlike stamped pattern and its surface was polished; its characteristic shape was wide-bellied with a funnel-shaped neck. Bronze dishes found in the graves came from the Merovingians to Pannonia. (Such a dish was also found in one of the graves in the cemetery of Várpalota.[29]) The cemetery of Várpalota is therefore very significant because in addition to the Longobard graves, other groups of graves contained typical early Avar material. In connection with this cemetery, evidence has come to light showing that the women in the Longobard population remained under Avar domination. The type of the fibula in Grave 1 of this cemetery occurred also in the Longobard material in Italia.[30] One of the most beautiful specimens of the fibulae is that from Bezenye (Pallersdorf)[31] which had runic marks incised on its backplate.


## THE AVARS


The rule of the equestrian nomad Avars who arrived from Central Asia (Fig. 74) into the Carpathian basin, started in 568, or from the period when the Longobards left Pannonia, handing it over to their allies, the Avars (Fig. 69). The occupation progressed gradually, first in the area of the Tisza River, then in the Great Hungarian Plain which was Gepidic territory, and later in Pannonia.[32] In present Austria and in the eastern slopes of the Alps, there are very few Avar finds that can be dated earlier than the middle of the seventh century.[33] In the south, Sirmium represented an important base of Avaric power, and during the reign of Emperor Mauritius (582) they succeeded in occupying it.

Avaric power flourished during the reign of Khagan Bajan whose might was comparable to that of Attila. The period, which is called the first Khaganat in literature, was characterized by the establishment of Avaric power in the Carpathian basin and later by attacks against Byzantium. Owing to this, Byzantine sources contain a large amount of records relating to this period.[34] In the last third of the sixth and the first third of the seventh centuries, Thrace and Dalmatia suffered particularly from Avaric raids. At the beginning of the seventh century (601–602), a Byzantine counterattack ended in a battle near the Tisza River and the Avars suffered defeat. But neither this nor the birth of the so-called Samo state, which interrupted Avar rule over the Slavs living to the north of the Danube and over the Slavic Carantani from about 623 or 624 to about 660, broke their power.[35] The greatness of the Avar Empire declined with the fiasco at the gates of Constantinople in 626; after this event, the Avars were no longer a significant political fac-

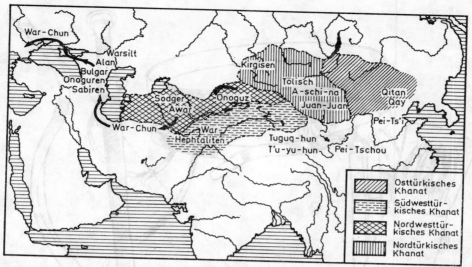

Fig. 74. The Turk Empire and the western migration of the Avars (A.D. 558, War-Chun)
(Historia Mundi V)

tor as far as Byzantium was concerned. Around 635, Kuvrat, ruler of the Onugor–
Bulgarians, shook off the Avar yoke and created the Onugor–Bulgarian Empire
(the so-called Great Bulgaria) near the Kuban River.[36] One of Kuvrat's sons settled his people on the territory of present-day Bulgaria in 679 or 680, blocking
Avaric expansion toward the east and southeast. Besides these events it is not
clear what kind of role the Islam conquest played, but without doubt it interfered with the west orientation of the Avars. The Avars subsequently turned west
after the dissolution of the Samo state and restored their empire in the north
and in the territory of present-day Austria or rather Slovenia. The Slav tribes
which lived in this area could only liberate themselves with foreign help around
the middle of the eighth century. In doing so, however, they came under Bavarian rule between 740 and 743.

The border between the Bavarians and the Avars during the reign of Tassilo,
a Bavarian prince, was the Enns River in the second half of the eighth century.
When the Franks annexed Bavaria in 788, they became immediate neighbors of
the Avars. Subsequently, one of Charlemagne's major political problems was the
settling of the Frankish–Avar border. Since negotiations with the Avars brought
no results, the Franks attacked them in 791. The Frankish victory in 803 ended
approximately 250 years of Avar domination.[37]

In evaluating the finds of the Avar period, researchers have considered the time
sequence of the finds and each subsequent group's connection with the ethnic
groups living under Avar domination. During this period and in this territory lived
the following ethnic groups: nomadic groups coming from Central Asia;[38] tribes
from the South Russian steppes, mainly Bulgarians;[39] a small group of people
belonging to the Ant tribe alliance, like the Slavs from the district of the Dnieper

407

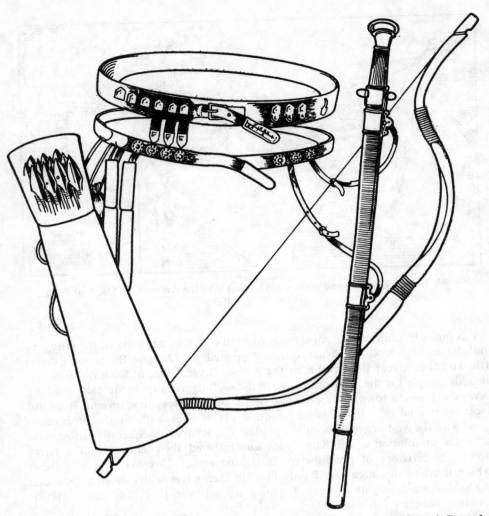

Fig. 75. Reconstruction of a richly ornamented Avar belt and armor belt; Szeged Csengele
(Gy. László, *ArchHung* 34 [1955])

River;[40] and former inhabitants of the Carpathian basin.[41] This latter group included surviving Sarmatians and Gepidae east of the Danube, the Romanized population of Pannonia (Pl. LVIII and CLXI. 2) and parts of Germanic tribes which had lived in this area before and together with the Longobards, as well as smaller Longobard groups who stayed behind after the Longobards left. On the basis of this consideration, it has been clearly proved that the archaeological material of the Carpathian basin dating between 568 and 800 may be considered as the so-called Avar period.

Recent research distinguishes three great chronological groups judging by archaeological finds from numerous cemeteries from the Avar period in the Carpa-

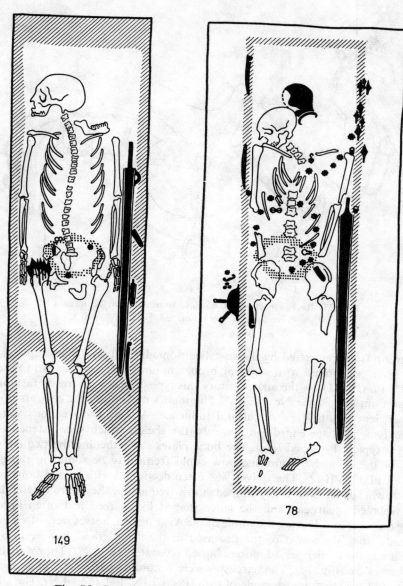

Fig. 76. Graves of armed men from Környe cemetery
(Á. Salamon and I. Erdélyi, *StudArch* 5 [1971])

thian basin.[42] The beginning of the earliest group is determined by the date of the
Avar conquest mentioned in the written sources. From the graves of this group
some Byzantine coins came to light. According to the material of this earliest group
the finds were in use until the middle of the seventh century. In this period, the
weight of Avaric occupation fell on the eastern half of Pannonia, on the southern
part of the Great Hungarian Plain, and on the area of the middle Tisza River.

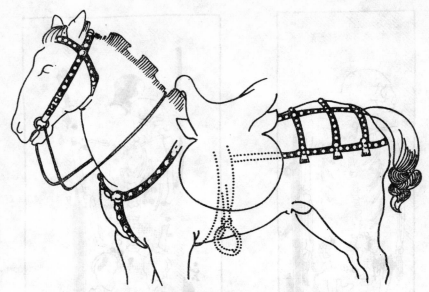

Fig. 77. Reconstruction of harness mounts from Kunágota
(Gy. László, *ArchHung* 34 [1955])

This group is characterized by the so-called nomadic belt with sidestraps (usually six or seven) decorated with silver or bronze mounts and strap-ends in repoussé (Fig. 75; Pl. CLXII). In the sixth century this type of belt was already fashionable in Byzantium and in the Near East.[43] The motifs decorating the mounts pointed to an ancient South Russian cultural influence. Among the weapons were the straight two- and one-edged long swords, the sheath which was fastened to the belt by a typical P-shaped loop. The bone plates strengthening the two ends and center of the nomadic composite bow could frequently be found in the graves (Fig. 76; Pl. CLVII. 2). The quiver was often decorated with carved bone plates. The arrowheads were diversely shaped: most frequently, they were trilateral. The narrow-bladed spearhead and the lance shaped like a reed leaf often appeared among the weapons. During the whole of the Avar period, horses were often buried. Generally, they were next to the deceased in the same grave, but we also find, especially in the earlier period, horses buried separately (Fig. 78). They were buried fully harnessed (Fig. 77). The harnesses were decorated with metal plate mounts; the bits and the stirrup were made of iron (Pls. CLIX. 1–2 and CLX). The stirrups changed with the period, the earliest type having circular or oblong-shaped loops. Characteristic among females' jewels were the gold and silver earrings with large globular pendants or with inverted pyramids, (CLXI. 1) the necklace made of so-called colored eye-beads and the bracelets which gradually widened toward their terminals (Pl. CLXVI. 2).

Around the middle of the seventh century, new types of objects appeared. Most characteristic belt ornaments were incised and of interlaced ribbon-like decorations. In the males' graves, decorated plates were frequently found on both sides

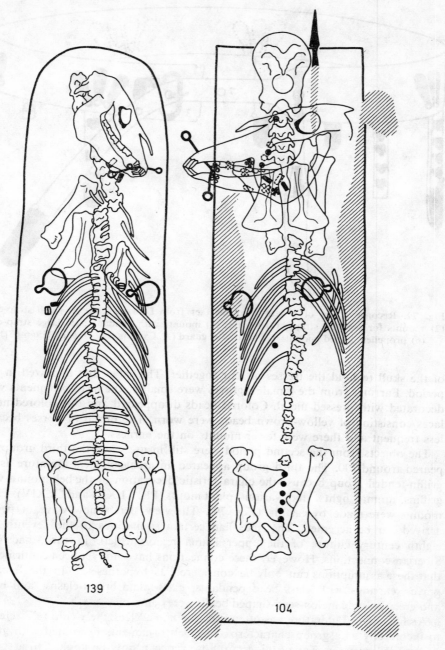

139

104

Fig. 78. Horse burials from Környe cemetery (Á. Salamon and I. Erdélyi, *Stud.Arch* 5 [1971])

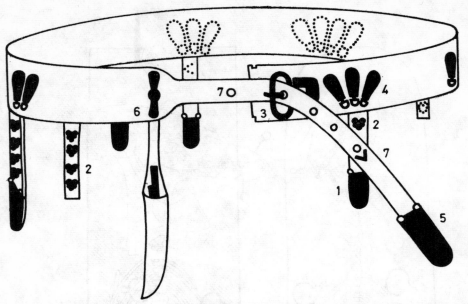

Fig. 79. Reconstruction of a late Avar belt set from Dunaszekcső: (1) small strap-end; (2) mounts for pendant straps; (3) buckle; (4) mounts with pendants; (5) large strap-end; (6) propeller-shaped mount; (7) belt-hole guard (Á. Cs. Sós, *FolArch* 18 [1966-67])

of the skull to hold the tresses of hair together. The saber first appeared in this period. Earrings from the females' graves were smaller; the head ornaments were decorated with pressed metal. Colored beads disappeared and one-colored necklaces consisting of yellow-brown beads were worn. The burial of horses became less frequent and there were fewer mounts on the harness.

The objects from the second period were still in use when the third group appeared around 700. The third group appeared in archaeological literature as the griffin-tendril group because the characteristic decoration of the belt mounts were griffins, animal fights, and various plant motifs (Pls. CLXIII and CLXIV). These mounts were made by casting (Fig. 79). The view that a new wave of people arrived can be accepted. The finds had certain similarities to the seventh and eighth century culture of the upper Kama region which, in turn, had West Siberian connections. However, these connections have not yet been confirmed so that these assumptions can only be considered as hypotheses.[44] In the females' graves, earrings with glass bead pendants, glass-inlaid breast-clasps, and black and green colored melon-seed-shaped beads were common.

Archaeological literature has often used the name "Keszthely culture"[45] referred to incorrectly as a group characterized by belt ornaments decorated with griffin and tendril patterns. Today this term, however, can only be applied to a special group of the finds of the Avar period concentrated in the southwestern region of the Lake Balaton. Earrings with large basket-pendants of filigree work and large dress pins and bracelets decorated with engravings were typical of this cul-

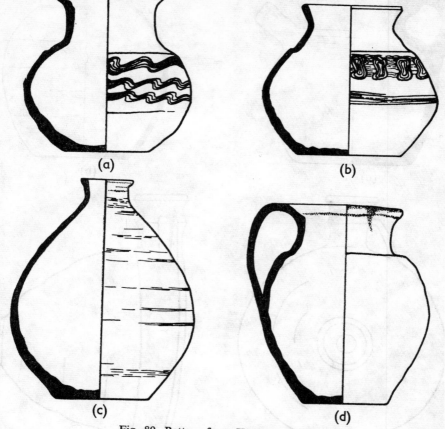

Fig. 80. Pottery from Környe cemetery

ture. Its roots, according to present research, must be found in late antique culture, this group being a later local flourishing (Pl. CLXV).

In addition to the above, pottery was often placed into the Avar graves.[46] They contained food meant for the other world or were leftovers from the funeral feast, as well as poultry or pork bones. Handmade vessels (Fig. 82b) and wheel-turned pottery were buried with the dead in all three periods. The earliest pottery made of gray, fine clay is represented by various types. The decoration of vessels was made up of straight and wavy lines incised with a comblike tool or they were stamp decorated (Figs. 80a, b and 81b). One group of the handmade vessels was influenced by Slavonic ceramics (Fig. 82b) and the other group imitated the metal vessels. Among wheel-made vessels, a separate group contained pilgrim flask-shaped ones (Fig. 81c), jugs with a spout (Fig. 81a) and bottles reminiscent of Roman shapes (Fig. 80c, d). The pear-shaped, thin-walled, yellow pottery appeared simultaneously with the griffin-tendril mounts, and their origin can be traced back to Central Asia (Fig. 82c–e). To late Avar pottery also belonged the usually brownish-gray pots

413

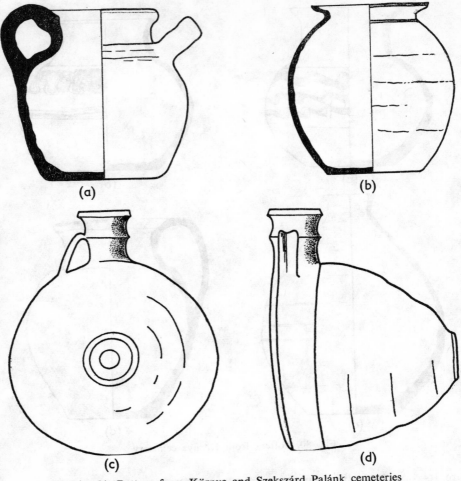

Fig. 81. Pottery from Környe and Szekszárd Palánk cemeteries

with incized decoration consisting of wavy and straight lines (Fig. 82a). On the latter, stamped marks on the bottom were frequent. This type of vessel was similar to the pottery of the so-called Burgwall period of the neighboring Slavonic territories.

Wooden buckets with metal fittings which were also found in the graves served a purpose similar to that of the clay vessels. The early ones were small and decorated with thin bronze bands, whereas the staves of the later buckets were held together by large iron hoops.

Different periods manifested themselves not only in types of objects, but, up to a point, in the burial rites too.[47] The east–west oriented graves in which the dead were placed with their heads to the west were characteristic of the early period. Similarly, in the early group were east–west or northeast–southwest oriented so-called pit-graves ("Nischengräber") especially in the Körös–Maros region. In the

414

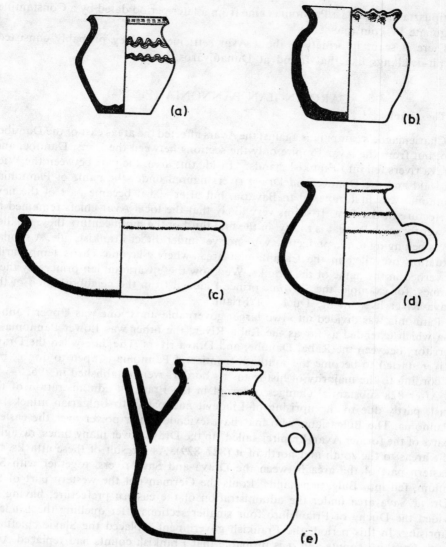

Fig. 82. (a, b) Pottery from Szekszárd Palánk Avar cemetery; (c, d) Yellow Avar pottery from Jánoshida; (e) Yellow Avar pottery from Bozómindszent

cemeteries of the second and third groups, a northwest–southeast orientation of the graves was dominant.

In the Avar period, special problems have been posed by the so-called chieftain burials which came to light east of Pannonia: Kunágota, Bócsa, Tépe, Kunbábony. Three finds were discovered along the old Danube *limes*, at Szentendre, Csepel and Dunaújváros. Of these, coins of Justinian I (518–527) and Phocas (602–610), Byzantine emperors, were found at Szentendre. The finds of Ozora-

Tótipuszta in southeast Pannonia came from a later period dated by a Constantine Pogonate IV coin (668–669).[48]

There is scant information about Avar settlements. They probably consisted of pit-dwellings like that found at Dunaújváros.[49]

## CAROLINGIAN PANNONIA (Fig. 84)

### The Franks

Charlemagne's campaigns against the Avars affected the areas east of the Danube, too, but from the Avar Empire only the sections between the Enns, Danube, and Drava rivers fell into Frankish hands.[50] Inside this area, the part between the Wienerwald and the Danube and Drava rivers figured under the name of Pannonia. Pannonia was first annexed to Bavaria, but after 803 it became part of the new prefecture in the East. It seems very likely that the local Avar chiefs remained in power for a while: that is to say, in the first third of the ninth century these smaller units belonging to the prefecture were not yet under direct Frankish rule. A similar situation prevailed in the Carantanian area, where Slavonic chiefs temporarily governed in the name of the Franks. We know these Carantanian princes by their names; for example, the Slavonic prince Liudevit from the neighborhhood of the Sava River lived in the Duchy of Friaul.

Pannonia was divided into two larger governable units: one was Upper Pannonia which extended as far as the Raba River; the other was Lower Pannonia, a territory between the Raba, Danube, and Drava rivers. The date when the Drava River started to become the southern frontier of Pannonia is open to doubt, but according to the majority of historians its borders were established in 796.

After 828 substantial changes occurred in the Frankish administration of the east, partly due to the uprising of Liudevit and partly to Bulgarian attacks on Pannonia. The Bulgarians, who may have extended their power over the eastern parts of the former Avar Empire, sailed up the Drava River many times, ravaging the areas to the south and north of it (827–829). As a result of these attacks, the eastern part of the area between the Drava and Sava rivers, together with Sirmium, fell into Bulgarian hands. Louis the German put the western part of the Drava–Sava area under the administration of the eastern prefecture, having divided the Duchy of Friaul into four smaller sections after quelling the Liudevit uprising. In this period, the Frankish government replaced the Slavic chieftains with Frankish counts and it is probable that Frankish counts had replaced Avar vassal-chiefs even at an earlier date. Of the latter, from Frankish sources many are known by their names. One of these was Theodor "Capcanus", who in 805 personally asked Charlemagne to protect his people against Slavic harassment and for whom the emperor marked out the territory between Savaria and Carnuntum to settle down. This location was in the upper territory of Pannonia, which was divided into a southern and a northern county in 828. Between 837 and 860, a certain comes Rikhari was the head of the southern county, while the lord of the northern county was Radpot, the praefect of Pannonia. The choice of the seat of the prefecture was evidently dictated by strategic reasons: it was near the Moravians, whose aspirations for independence and a flourishing Moravian Empire

416

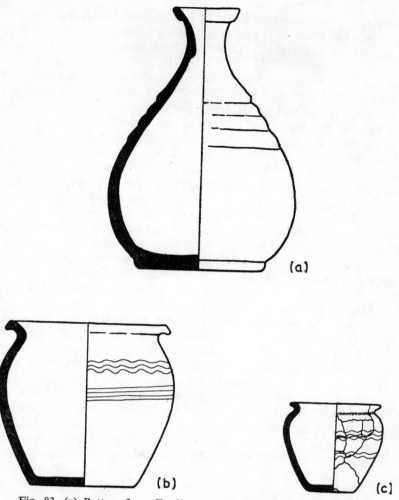

Fig. 83. (a) Pottery from Fenékpuszta; (b, c) Pottery from Zalavár

(in the second half of the ninth century) was menacing. The seat of the southern county was in all probability Savaria. We have no data about the territory extending to the east of the Raba River, although, according to some hypotheses, it was divided into two counties. The center of one was supposed to have been Mosaburg, which was situated in the present village of Zalavár on the lower Zala River.[51]

## The Slavs

The Frankophile Slavic chieftain, Pribina, banished from Nitra, settled at Zalavár around 840 and received a *beneficium* from the Franks. Later this became an *allodium* (in 847). The question of the so-called principality of Pribina has been much disputed.[52] The possibility that Pribina was head of the Slavic state in Pannonia has been raised. But this assumption contradicts the historical fact that it was the Frankish feudal state that exercised power over the people of Pannonia. Nor is the theory of a Pannonian–Slavic state supported by loan words in the Hungarian language.[53]

One of the possible reasons for Pribina to receive a fief in this area was due to numerous Slavs living in the region. These Slavs had settled in Pannonia partly during the Avar period and partly during the decades following the downfall of the Avar Empire. The presence of a Slavic population has been supported by geographical place names which the conquering Hungarians took over from the local Slavs.[54] The population of ninth-century Pannonia was strongly heterogeneous. In addition to the Slavic groups, there were surviving Avars, those surviving from the Avar period and including a small number of Bavarian settlers. A Bavarian ethnic group belonged primarily to the ruling stratum. It is possible that Pribina, and after his death around 860, his son Kocel, became Frankish officials.[55]

Pribina's castle, Mosaburg, was a type of Slavic *burgwall*, in the *mosa* land, as confirmed by archaeological excavations.[56] We know from a source originating from Salzburg around 871 that Mosaburg was an important religious center in the ninth century. This area belonged to the Church of Salzburg, while the section west of the Raba River was organized by the Church of Passau and the area between the Drava and Sava rivers belonged to the patriarchy of Aquileia. Recent research at Zalavár has proved that in Mosaburg and its neighborhood missionary work by the Frankish church had already started before Pribina's arrival.[57] It seems clear from various sources that the Church of Salzburg had consecrated numerous churches on Pannonian territory, but researchers have found it difficult to locate them. A few can be identified from a list of approximately thirty surviving names: Quinque Basilicae (ancient Sopianae, Pécs), Bettobia (Poetovio, Ptuj), Ortahu (Veszprém), and Durnauua (Somlóvásárhely).[58] The walls of the late Roman city (situated about nine kilometers from Zalavár) at Keszthely-Fenékpuszta were still standing in the ninth century. The cemetery outside these walls, also of the ninth century, confirmed that Fenékpuszta had survived during Carolingian times. It possibly defended the mouth of the Zala River, an important waterway leading to Mosaburg. Judged by the finds excavated near the walls of this late Roman city, we know that Christianity survived to a certain degree in this area.[59]

The work of the Frankish church was hindered during Kocel's rule because he supported the Slav bishop, Methodius. Pope Hadrian, at Kocel's specific request, had nominated Methodius as bishop of Pannonia with Sirmium as his seat. As Sirmium was in Bulgarian hands at that time, the actual seat of the archbishop could not have been in the town. The area of Methodius's activity was not Pannonia but the Moravian Empire.[60] After the death of Kocel around 874, Frankish counts again governed the area to the east of the Raba River. In 883 and in 884, Svatopluk, the ruler of Moravia, raided Pannonia and some sources report much devastation in this region. Whether these raids were coupled with occupation has not been sufficiently documented. On the basis of available information, we have to accept that the area of Pannonia known today as Transdanubia was under Frankish domination without interruption from the end of the Avar wars until the Hungarian conquest.[61]

418

Fig. 84. The Carolingian Empire (Kornemann, 1948)

Archaeological research has recovered only a few records which would advance a solution to the problems of the ninth century. The only large-scale, prolonged excavation was carried out in the center of Pribina around Mosaburg, where burial places and settlements and also the continuity of the system of fortifications can be traced to the sixteenth century. Archaeological material found in the course of the Zalavár excavations has shown many similarities to the material culture of the Moravian Empire (Pl. CLXVI. 1), simultaneously, the influence of the Carolingian culture and the cultural heritage of earlier local settlers can also be observed. Apart from the Slavic type pottery (Fig. 83b–c), reddish-yellow clay bottles with polished surfaces, which developed from antique and Avar traditions, were used (Fig. 83a). Frankish spurs, imported objects in Mosaburg, were placed in the graves of the leading men of Slavic origin to prove their owner's rank.[62]

The conquering Hungarians appeared in the Carpathian basin in 896 and gradually took possession of the whole area. In that year, Emperor Arnulf handed

Fig. 85. Find-places in the Carpathian basin

over the defense of Pannonia, including Mosaburg, to Brazlav, the chief of the Slavs in the area between the Drava and Sava rivers. Based on information of the *Annales Fuldenses* referring to the year 900, it is believed that this was the year when the Hungarians occupied the territory of Pannonia known today as Dunántúl (Transdanubian area), (Fig. 85; Pl. CLXVII).[63]

## NOTES

[1] This theory was mentioned by A. ALFÖLDI, *Der Untergang der Römerherrschaft in Pannonien* (Berlin, Leipzig, 1926), vol. 2; A. MÓCSY, R. E., Suppl. 9, *"Pannonia"*, pp. 579–82.

[2] Á. SALAMON and L. BARKÓCZI, "Bestattungen von Csákvár aus dem Ende des 4. und dem Anfang des 5. Jahrhunderts" *Alba Regia* 11 (1971): 35–80 and Á. SALAMON and L. BARKÓCZI, "Archäologische Angaben zur spätrömischen Geschichte des pannonischen Limes (Gräberfelder von Intercisa I)" *MittArchInst* 4 (1973): 73–95; L. BARKÓCZI and Á. SALAMON, "Das Gräberfeld von Szabadbattyán aus dem 5. Jahrhundert" (Manuscript). See also L. Barkóczi, pp. 116–20.

[3] L. BARKÓCZI and Á. SALAMON, "Remarks on the Sixth Century History of Pannonia". *ActaArchHung* 23 (1971): 139–43, and I. LENGYEL, "Chemico-Analytical Aspects of Human Bone Finds from the Sixth Century 'Pannonian' Cemeteries". *ActaArchHung* 23 (1971): 155–166. Research conducted for years in Pannonian cities have recently revealed evidences which support our views (Aquincum, Castra ad Herculem, Gorsium, Scarbantia, Sopianae). See also Notes 58 and 59.

[4] The above study confines itself only to the Pannonian phase of the history of peoples (Huns, Goths, Longobards, Avars, Slavs) which settled or were made to settle in the territory of Pannonia. Concerning researches of the years 1945–1969 see: I. BÓNA, "Ein Vierteljahrhundert der Völkerwanderungszeit-Forschung in Ungarn (1945–1969)." *ActaArchHung* 23 (1971): 265–336.

[5] The Huns first appeared in 375 on the Azov seacoast (Maeotia). L. DEGUIGNES, French sinologist, discovered as early as the eighteenth century, that relations existed between tribes near the Maeotis and the Hiung-nus mentioned in Chinese sources.—L. DEGUIGNES, *Histoire générale des Huns* (Paris, 1756–1758). These sources first mentioned this people in 318 B.C. The Hiung-nus, i.e., the Asian Huns had extended their power under their ruler, Hso-tung, in the west, as far as the Aral Sea and in the east, as far as Korea. In the north, their boundaries reached the Selenga, Yenisey, Ob, and the river-head of Irtis and, in the south, the Chinese border. In this huge empire numerous tribes had coexisted, yet the leading role was in the hands of the Turco-Huns. Their expansion filled the neighboring peoples with fear and started the first waves of migrations. However, the expansion of the Chinese Han dynasty terminated the domination of the Hiung-nus. The peoples that had been forced from their habitations allied themselves with Chinese forces and in 119 the Hiung-nus were driven to the Orchus region. Internal strife nurtured by these events undermined the very existence of their empire. Their leader who was banished during these revolts surrendered to the emperor of China and this last act brought about the end of the Eastern Hun Empire (54 B.C.). A Hiung-nus tribe migrated west and founded a new empire where present Kazakhstan is situated. However, this Hun Empire soon declined and the Hiung-nus were again obliged to move on. (For the history of Asian Huns, see L. LIGETI, *Az ázsiai hunok* (The Asian Huns) in *Attila és hunjai* (Attila and his Huns) ed. by GY. NÉMETH, (Budapest, 1940), and J. J. M. DE GROOT, *Die Hunnen der vorgeschichtlichen Zeit* (Berlin, Leipzig, 1921); for the Kazakhstan Empire, see LIGETI, p. 35). The latter tribe can be identified with a Hun group which appeared in the second century A.D. in the steppes west of the Caspian Sea. After the defeat of an Iranian tribe living here, the road was open to them toward South Russia and subsequently toward Europe. In the fourth century the Huns overran the Alan tribe and raided the eastern provinces of Byzantium. There were also raids into Asia Minor and Armenia. However, their advance only threatened the Eastern Empire when the Huns conquered the territory of the East Gothic people between the Don and the Dnyestr rivers (A.D. 370). See P. VÁCZY, *A hunok Európában* in NÉMETH, p. 66.

[6] Concerning the connection between the Huns and the Late Roman Empire, see L. VÁRADY, *Das letzte Jahrhundert Pannoniens* (Budapest, 1969).

[7] Concerning the personality of Attila and his military activity, see W. A. THOMPSON, *A History of Attila and the Huns* (Oxford, 1948) and F. ALTHEIM, *Attila und die Hunnen* (Baden-Baden, 1951).

[8] Priscos, Exc. de legat. (Ed. DE BOOR).

[9] T. M. MINAJEVA, *Pogrebeniya s Soscheniem blis gor. Pokrowska* (Saratow, 1927).

[10] A. ALFÖLDI, *Leletek a hunkorszakból és ethnikai szétválasztásuk—Funde aus der Hunnenzeit und ihre ethnische Sonderung* (Budapest, 1932) (*ArchHung* 9); E. POLASCHEK, "Wiener Grabfunde aus der Zeit des untergehenden römischen Limes". *WPZ* 19 (1932): 239–66; J. WERNER, "Bogenfragmente aus Carnuntum und von der unteren Wolga". *ESA* 7 (1932): 33–58.

[11] J. WERNER, *Beiträge zur Archäologie des Attila-Reiches* (München, 1956), vols. 1 and 2.

[12] J. HARMATTA, "Hun Society in the Age of Attila". *ActaArchHung* 2 (1952): 277–305; J. HARMATTA, "Préface", in *ArchHung* 32 (1953): 103–12.

[13] M. PÁRDUCZ, "Archäologische Beiträge zur Geschichte der Hunnenzeit in Ungarn". *ActaArchHung* 11 (1959): 309–98; M. PÁRDUCZ, *Die ethnischen Probleme der Hunnenzeit in Ungarn* (Budapest, 1963) (*StudArch* 1).

[14] "Princely finds" is an archaeological term, referring to such finds from graves where the dead are richly clothed, wearing sumptuous armor, and where there are vessels for food and drink. Such funerals were given only to the wealthiest members of society, i.e., to the political and economic top strata. N. FETTICH, *A Szeged–nagyszéksósi hun fejedelmi sírlelet—La trouvaille de tombe princière hunnique à Szeged–Nagyszéksósi* (Budapest, 1953) (*ArchHung* 32).

[15] J. WERNER, "Bogenfragmente aus Carnuntum und von der unteren Wolga". *ESA* 7 (1932): 33–58; Á. SALAMON, "Archäologische Angaben zur spätrömischen Geschichte des pannonischen Limes—Geweihmanufaktur in Intercisa". (Manuscript). For the golden bow of the Huns marking social rank, see J. HARMATTA, "The Golden Bow of the Huns". *ActaArchHung* 1

(1951): 107–51; Gy. László, "The Significance of the Hun Golden Bow". *ActaArchHung* 1 (1951): 91–106.

[16] J. Werner, *Beiträge*, pp. 61–68; M. A. Tichanova and I. T. Tcherniakov, "Nouvelle trouvaille d'une sépulture à diadème". *SowArch* (1970): 117–27.

[17] Werner, *Beiträge*, pp. 57–60; I. Kovrig, "Hunnischer Kessel aus der Umgebung von Várpalota". *FolArch* 23 (1972): 95–121 with appendix; K. Zimmer and M. Járó, "Spektographische Untersuchungen von hunnischen Kesseln", 122–25, and J. Imre, "Über die Herstellung des hunnischen Kessels aus der Umgebung von Várpalota", 126–27.

[18] Alföldi, pp. 42–58.

[19] Á. Salamon and I. Lengyel, (Manuscript).

[20] The Goths were an East Germanic tribe which presumably lived originally in present-day Sweden. In the first century A.D. they migrated to the lower Vistula region. Jordanes believes that a second wave of their migrations reached the Vistula estuary and during the second century they occupied land southwest from it down to the Black Sea. In 214 these Gothic tribes appeared near the western border of Dacia. During the third century, they divided into two groups: the West Gothic tribe settling in the west between the Dnyestr and Danube rivers and the East Gothic tribe between the Dnyestr and the Don. The East Gothic Empire flourished during the mid-fourth century. Their ruler at that time was Ermanarich. The raid of the Huns in 370 forced them to leave. L. Schmidt, *Die Ostgermanen* (München, 1941), pp. 195–337; N. Hachmann, *Die Goten und Skandinavien* (Berlin, 1970); J. Svannung, "Jordanes und die gotische Stammsage" and E. Lönroth, "Die Goten in der modernen kritischen Geschichtsauffassung", in *Studia Gotica*, Vorträge beim Gotensymposion Stockholm 1970, ed. U. E. Hageberg (Stockholm, 1972), 20–62.

[21] Barkóczi and Salamon, "Remarks on the Sixth Century History of 'Pannonia'," pp. 148–51.

[22] N. Fettich, *A szilágysomlyói második kincs— Der zweite Schatz von Szilágysomlyó* (Budapest, 1932) (*ArchHung* 8); J. Hampel, *Die Alterthümer des frühen Mittelalters in Ungarn* (Braunschweig, 1905), vol. 3, Tables 14–31.

[23] I. Kovrig, "A tiszalöki és a mádi lelet" (The Finds from Tiszalök and Mád). *ArchÉrt* 78 (1952): 113–18.

[24] J. Dombay, "Der gotische Grabfund von Domolospuszta". *Janus Pannonius Múzeum Évkönyve* (1956), pp. 104–30; Gy. Mészáros, "A regölyi korai népvándorláskori fejedelmi sír— Das Fürstengrab von Regöly aus der Frühvölkerwanderungszeit". *ArchÉrt* 97 (1970): 66–92; I. Lengyel, "A regölyi csontlelet vizsgálati eredményei" (The Result of an Analysis of the Bones from Regöly). *ArchÉrt* 97 (1970): 93.

[25] Á. Salamon, "Grave 5 from the cemetery at Hács-Béndekpuszta" and D. Székely, "A Lead Tablet with Inscriptions from Hács-Béndekpuszta" (Manuscript).

[26] The earliest data for the origin and history of the Longobards can be found in such ancient authors as Tacitus, Strabo, Ptolemaeus, etc. They confirm that they lived near the upper flow of the Elbe, moving southward at a later time. The first authentic data for the direction of their migration comes from the fifth century, when they settled in the northern section of present Lower Austria in 489 A. D. For the appearance of the Longobards in Pannonia we have two kinds of data: according to Paulus Diaconus and the Origo gentes Longobardorum the Longobards ruled. Pannonia for some forty-two years, while the Codex Gothanus attributes twenty-two years to the occupation of Pannonia by the Longobards. According to the first sources the Longobards lived in Pannonia from 526 onward; according to the Codex Gothanus, they lived in Pannonia from 546–547. The subject gets a detailed treatment by J. Werner, *Die Langobarden in Pannonien* (München, 1962), vols. 1 and 2. Recent research accepts the earlier date.

[27] The relics of the Longobards in Pannonia are summarized by Werner; those in Hungary are dealt with by I. Bóna, "Die Langobarden in Ungarn". *ActaArchHung* 7 (1956): 183–244; I. Bóna, "VI. századi germán temető Hegykőn— Germanisches Gräbfeld aus dem 6. Jahrhundert in Hegykő". *Soproni Szemle* 14 (1960): 223–41; 15 (1961): 131–40; K. Sági, "Das langobardische Gräberfeld von Vörs". *ActaArchHung* 16 (1964): 359–408; A. Kiss and J. Nemeskéri, "Das langobardische Gräberfeld von Mohács". *Janus Pannonius Múzeum Évkönyve, 1964* (1965), 95–127.

[28] BÓNA, *Die Langobarden*, Plate 30, Figs. 5.

[29] BÓNA, *Die Langobarden*, Plate 32, Figs. 5–9.

[30] BÓNA, *Die Langobarden*, Plate 35, Figs. 1–2.

[31] BÓNA, *Die Langobarden*, Plate 44, Figs. 1–2.

[32] As to the origin and earlier habitation of the Avars, opinions seem to differ. An identification of the Avars with the Jouan-jouans (first attempted by Deguignes) as with the Hephtalites (see J. MARQUART, "Über das Volkstum der Komanen". *Abhandl. d. K. Gesellsch. d. Wiss. zu Göttingen, Phil.-Hist. Kl.* 13 (1914): 71–75, has not been unanimously accepted. However, ethnic components in A. N. BERNSTAM's view are *Ocherk istorii gunnov* (Leningrad, 1951). Recent research has returned to an identification with the Hephtalites, see K. CZEGLÉDY, "Heftaliták, hunok, avarok, onogurok". *Magyar Nyelv* 50 (1954): 142–51, and H. W. HAUSSIG, "Indogermanische und altaische Nomadenvölker im Grenzgebiete Irans", in *Historia Mundi*, V. Frühes Mittelalter (Bern, 1956), 233–48. Concerning the problem of the route of the Avars see T. NAGY, "Studia avarica". Part 1, *Antiquitas Hungarica* 1 (1947): 56–63. Concerning the second period of the Avar-Byzantine relation, see T. NAGY, "Studia avarica". Part 2, *Antiquitas Hungarica* 2 (1948): 131–49.

[33] H. MITSCHA-MÄRCHEIM, "Awarisch–bairische Wechselbeziehungen im Spiegel der Bodenfunde". *ArchAu* 4 (1949): 125–31.

[34] Concerning the historical events of the Avar period, especially in regard to the Byzantine relations, see G. OSTROGORSKY, *Geschichte des byzantinischen Staates* (München, 1952), p. 67; A. KOLLAUTZ, "Abaria", in *Reallexikon der Byzantinistik* (Amsterdam, 1969); GY. MORAVCSIK, *Byzantinoturcica* (Budapest, 1942), vol. 1.

[35] Concerning the earlier references on the "Samo state", see G. LABUDA, *Pierwsze panstwo slowanskie. Panstwo Samena — Le premier État slave. 1. État de Samon* (Poznań, 1949).

[36] For a summary of the Bulgarian Empire see P. H. TRETIAKOV, S. A. NIKITIN and M. B. VALEV, *Istoriya Bolgari* (Moscow, 1954).

[37] Concerning the connections of the Avar Empire with the Bavarians and Franks, see J. DEÉR, "Karl der Grosse und der Untergang des Awarenreiches", in *Karl der Grosse* (Düsseldorf, 1965) vol. 1, 719–91.

[38] I. KOVRIG, "Contribution au problème de l'occupation de la Hongrie par les Avars". *ActaArchHung* 6 (1955): 163–84.

[39] D. CSALLÁNY, "Kora-avarkori sírleletek—Grabfunde der Frühawarenzeit". *FolArch* 1–2 (1939): 121–80; D. CSALLÁNY, "A bácsújfalusi avarkori hamvasztásos lelet. Adatok a kuturgur-bolgárok (hunok) temetési szokásához és régészeti hagyatékához—Trouvaille d'objects incinérés de l'époque avare à Bácsújfalu. Contributions à l'étude des rites funéraires et au legs archéologique de Koutourgours-Bulgares (Huns)". *ArchÉrt* 80 (1953): 133–40; N. FETTICH and A. MAROSI, *Dunapentelei avar sírleletek—Trouvailles avares de Dunapentele* (Budapest, 1936), 63 (*ArchHung* 18).

[40] Á. SÓS, "Das frühawarenzeitliche Gräberfeld von Oroszlány". *FolArch* 10 (1958): 105–24; Á. SÓS, "Vorläufige Mitteilungen über die Ausgrabungen in Pókaszepetk". *FolArch* 14 (1962): 67–82.

[41] D. CSALLÁNY, *Archäologische Denkmäler der Gepiden im Mitteldonaubecken* (Budapest, 1961) (*ArchHung* 38); D. SIMONYI, "Die Bulgaren des 5. Jahrhunderts im Karpatenbecken". *ActaArchHung* 10 (1959): 227–50; L. BARKÓCZI, "A Sixth-Century Cemetery from Keszthely-Fenékpuszta". *ActaArchHung* 20 (1968): 275–311; Á. SALAMON and I. ERDÉLYI, "Das völkerwanderungszeitliche Gräberfeld von Környe" with the collaboration of I. LENGYEL and T. TÓTH, (Budapest, 1971) *StudArch* 5; L. BARKÓCZI and Á. SALAMON, "Remarks on the Sixth-Century History of 'Pannonia' ". *ActaArchHung* 23 (1971): 139–53; I. BÓNA, "Beiträge zu den ethnischen Verhältnisse des 6–7. Jahrhunderts in Westungarn". *Alba Regia* 2–3 (1961): 49–68.

[42] D. CSALLÁNY, *Archäologische Denkmäler der Awarenzeit in Mitteleuropa* (Budapest, 1956); I. KOVRIG, *Das awarenzeitliche Gräberfeld von Alattyán* (Budapest, 1963) (*ArchHung* 40); Á. Cs. SÓS, "Zur Problematik der Awarenzeit in der neueren ungarischen archäologischen Forschung". *Berichte über d. II. Intern. Kongr. f. Slawische Arch.* (Berlin, 1973), 2:85–102; I. BÓNA, "Avar lovassír Iváncsáról—Grave of an Avar horseman at Iváncsa". *ArchÉrt* 87 (1970): 243–61.

423

[43] When examining the sixth-century finds in 'Pannonia', we have drawn the conclusion that the so-called nomadic belt usually with six or seven side-straps terminating in strap-ends could have appeared in the Carpathian basin and in Italia even before the Avaric period. See BARKÓCZI and SALAMON, "Remarks on the Sixth-Century History of 'Pannonia' ", 139–53.

[44] GY. LÁSZLÓ, "Les problèmes soulevés par le groupe à la ceinture ornée de griffon et de rinceaux de l'époque avare finissante". ActaArchHung 17 (1965): 73–75.

[45] The term "Keszthely culture" does not refer to sixth-century finds which follow late Roman traditions, earrings with small basket-pendants, pins, disc brooches as they were found in Fenékpuszta cemetery (near Horreum). See L. BARKÓCZI, "A Sixth-Century Cemetery" Today we call as "Keszthely culture" the jewels of silver and bronze which were made at Keszthely and its environments; these jewels are the following: earrings with large basket-pendants, pins of Keszthely types, and bracelets with incised and punched decorations. I. KOVRIG, "Megjegyzések a Keszthely-kultúra kérdéséhez — Contributions to the Keszthely Culture Problem". ArchÉrt 85 (1958): 66–74. The continuity of the Romanized population cannot be based merely on the significant finds which came to light in a few isolated localities in Pannonia. A. KISS, "Die Stellung der Keszthely-Kultur in der Frage der römischen Kontinuität Pannoniens". Janus Pannonius Múzeum Évkönyve (1967), pp. 49–59.

[46] D. BIALEKOVÁ, "Zur Frage der grauen Keramik aus Gräberfeldern der Awarenzeit im Karpatenbecken". Slovenská Archeológia 15 (1967): 5–65; É. GARAM, "A késő avarkori korongolt sárga kerámia" (The Late Avar Yellow Pottery Made on Potter's Wheel). ArchÉrt 96 (1969): 207–40. For Roman and local traditions, see Á. SALAMON and I. ERDÉLYI, Das völkerwanderungszeitliche Gräberfeld von Környe (Budapest, 1971), p. 59 (StudArch 5).

[47] For the different burial rites see KOVRIG, Das awarenzeitliche (ArchHung 40).

[48] Concerning the finds from Csepel and Szentendre, see GY. LÁSZLÓ, "Budapest a népvándorlás korában" (Budapest in the Migration Period), in Budapest története, vol. 1, pt. 2, p. 286; concerning the finds from Bócsa, Kunágota, Tépe, and Ozora, see GY. LÁSZLÓ, Études archéologiques sur l'histoire de la société des avars (Budapest, 1955) (ArchHung 34); E. H. TÓTH, "Preliminary Account of the Avar Princely Find at Kunbábony". Cumania (1972), pp. 143–60.

[49] I. BÓNA, "Abriss der Siedlungsgeschichte Ungarns im 5–7 Jahrhundert und die Awarensiedlung von Dunaújváros" AR 20 (1968): 605–18.

[50] Concerning the wars of the Franks and Avars and the Avars of Pannonia, see J. DEÉR, "Karl der Grosse und der Untergang des Awarenreiches", in Karl der Grosse (Düsseldorf, 1965).

[51] Concerning the administration of the Franks in the southeast part of their empire, see MITTERAUER, "Karolingische Markgrafen in Südosten". Archiv f. Österr. Geschichte 123 (1963); M. KOS, Conversio Begoariorum et Carantanorum (Ljubljana, 1936). The political situation of Pannonia judged by the archaeological finds is dealt with by TH. BOGYAY, "Mosaburc und Zalavári". Südostforschungen 14 (1955); G. FEHÉR, "A Dunántúl lakossága a honfoglalás korában" (Inhabitants of the Transdanubian Region during the Hungarian Conquest). ArchÉrt 83 (1956): 25–38; I. BÓNA, "Cundpald fecit". ActaArchHung 18 (1966): 279–325; Á. CS. SÓS, "Ausgrabungen in Zalavár". Cyrillo-Methodiana (Zur Frühgeschichte des Christentums bei den Slawen) (Köln, Graz, 1964); Á. CS. SÓS, "Die slawische Bevölkerung Westungarns im 9. Jahrhundert". Münchener Beiträge 22 (1973).

[52] KOS, Conversio.

[53] I. KNIEZSA, "A magyar állami és jogi terminológia eredete" (The Origin of the Hungarian Administrative and Legal Terminology). MTA II. Oszt. Közl. (1955).

[54] For the ninth-century population of Pannonia, see I. KNIEZSA, Ungarns Völkerschaften im 11. Jahrhundert (Budapest, 1938) (Ostmitteleuropäische Bibliothek 16).

[55] SÓS, "Die slawische".

[56] Á. CS. SÓS, Die Ausgrabungen Géza Fehérs in Zalavár (Budapest, 1963) (ArchHung 41); Á. CS. SÓS, "Über die Fragen des frühmittelalterlichen Kirchenbaues in Mosaburc-Zalavár", in Das östliche Mitteleurope in Geschichte und Gegenwart (Wiesbaden, 1966), p. 66.

[57] KOS, Conversio.

[58] TH. BOGYAY, "Die Kirchenorte der Conversio Bagoariorum et Carantanorum". Südost-

*Forschungen* 19 (1960): 52–77; Á. Cs. Sós, "Bericht über die Ergebnisse der Ausgrabungen von Zalavár-Récéskút in Jahren 1961–63". *ActaArchHung* 21 (1969): 51–103.

[59] Á. Cs. Sós, "Das frühmittelalterliche Gräberfeld von Keszthely-Fenékpuszta". *ActaArchHung* 13 (1961): 247–305; K. Sági, "Die zweite altchristliche Basilika von Fenékpuszta". *ActaAntHung* 9 (1963): 397–459.

[60] Concerning Method's activity in connection with Pannonia, see P. Váczy, *Die Anfänge der päpstlichen Politik bei den Slawen* (Leipzig, 1942) (*Archivum Europae Centro-Orientalis*).

[61] Gy. Török, *Sopronkőhida IX. századi temetője* (Ninth-Century Cemetery at Sopronkőhida) (Manuscript).

[62] Á. Cs. Sós, *Die Ausgrabungen Géza Fehérs* (*ArchHung* 41).

[63] Recent research holds that the conquest of Pannonia and its termination can be put to 901: Sz. Vajay, *Der Eintritt des ungarischen Stämmebundes in die europäische Geschichte* (862–933) (Mainz, 1968), pp. 32, 94.

# APPENDIX

# NEW METHODS AND RESULTS
## IN PALEOANTHROPOLOGY EMPLOYED
## IN PANNONIAN RESEARCH

IMRE LENGYEL

## INTRODUCTION

The sources from which our knowledge concerning human history is distilled are the remains of man's instinctive or self-conscious activity. From these records, archaeologists read and establish facts relating to the development of our culture and civilization. While these facts relate to the genius, spirit, and mind of our ancestors which gave birth to and which fertilized the characteristic cultures, events, and structures of society, they shed only a reflected light on man as a biological specimen. There are, it is true, portraits or even self-portrait-like representations which refer to the individual in the same way as monumental buildings, and monuments convey to us an idea of the physical capacity and aesthetic concepts of our ancestors. But about his health and sickness, his process of aging, and about him as an organism performing biological functions, we know precious little.

To become better informed — to know more about our ancestors as biological objects — we have to turn to the closest evidence of his biological existence: the skeletal remains. To use these remains, we need a method which would make an analogy between the bone remains of our ancestors and those of modern man, which would provide a common denominator for the comparison of men of various ages as biological objects. Such a comparative method "might put sinews on the dried bones and cover them with muscles, flesh, and skin" (Ezekiel 37:6). But, for the time being, the creators of the cultures of the dim hazy past are still resting mute and motionless in the depths of the earth: bones are turning into dust near the decaying objects which witness past activities. We believe, however, that once we become acquainted with the most modern methods of chemicoanalytical, serological, and histological processes and can evaluate their results, we will be able to extract some new information from the disintegrating bones of our ancestors.

Compelled by his curiosity and his desire for knowledge, and hoping to penetrate into the secrets of his own future by discovering the past of his own species, modern man turns with an eager interest toward these disintegrating bones. To satisfy his curiosity, he uses more and more the methods related to the natural sciences. By applying these methods to the special characteristics of the bones, he makes them accessible to the practice of anthropology.

Research oriented toward the recognition of our biological functions — the acquaintance with ourselves as biological phenomena — ranges from the examination of macroscopic events through microscopic dimensions to submicroscopic, molecular structures. As we know it today, a complicated molecular-biological

429

series of events lies in the background of our macroscopic biological functions. In the biomechanism of our organism the direction and intensity of the activity of the several cells as well as their morphological nature are determined by the chemical mechanism consisting of ribo- and desoxyribonucleic acids, intricate protein structures, and enzyme systems. Thus, there is a systematic relationship between the visible appearance of our organism and the chemical structures which are basic to its biomechanism. As the macroscopic biological function of the living organism and its internal structure may be determined and expressed on the molecular-biological level only, the research methods of historical anthropology may result in the complete biological reconstruction of the man of the past only when complemented by chemicoanalytical, serological, and histological research.

While historical anthropology attempts to reconstruct the image and bodily characteristics of our ancestors from the morphological appearance of the bones, modern laboratory methods use the internal analysis of these bones to reconstruct biological and disease patterns. Examination of the texture of the excavated bones through the metric method of physical anthropology creates a "snapshot" of individuals in different ages, while we may draw a synthesis of the biological phenomena of a smaller or a larger community from the results of biological reconstruction, aided by statistical methods.

We may look for causal similarities between the modification of biological qualities and a change in the direction of ecological development of some ethnic characteristics. We may read the effect of the change of life-conditions within the biological parameter of a given human group and vice-versa; from these effects we may speculate on essential changes in biological conditions. Some relationships known to modern medical science between the chemical and structural changes in the bone tissue and the physiological and pathological processes may be projected into the past. Various ethnic examples may be studied genetically and ecologically.

Modern research, as will be shown in the following chapter, may supplement the repertory of traditional anthropological methods, and anthropology may thereby bring alive the archaeological remains of an ancient necropolis.

## METHODOLOGY: PRELIMINARIES

The purpose of the following summary is to describe briefly the many-sided research which developed from the examination of bone tissue and the method which I am using in my own research.

The researches of others were directed mostly toward determining the historical age of the bone findings. My goal, however, lies beyond this task: it consists of "biological reconstruction". We may reach our goal only if we penetrate into the study of the biochemistry and histology of the bone tissue and, following this line, establish a method by which we may reconstruct the biology of men who vanished long ago.

I must say in advance that, in the selection of known literature in this field, I was influenced partially by methodological considerations and partially by factors which appeared as requisites for the chemical examination of bone tissue structures.

## Weight Loss by Cremation as a Result of the Evaporation of the Water Content of Excavated Bones

Van Bemelen (1897),[1] Oakley (1955),[2] and Heizer (1960)[3] looked for interrelations between the water content of the bones and their chronological age. However, they did not find a relationship between the loss of weight (loss of water content) of the bone samples as a result of the gradually rising temperature and their historical age. Their failure may be explained by the relationship between the characteristic liquid and dry material of the bone tissue.

The "free" liquid quantity which, in vivo, fills the interstitial space in the fresh autopsy material is considerable (8 to 10 percent). Its quantity depends on the structure of the bone tissue, more exactly on the quantitative ratio of the spongy and compact tissues. In the case of dried and cleaned bone material (without the soft tissues) or excavated bone material, besides the relationship of the spongy and the compact tissues, the porosity of the bones—i.e., its specific surface area and the moisture content of its surrounding—determines the "free" liquid quantity (2 to 4 percent). The "free" liquid content of the bone is lost as a result of drying and then is regained as a result of the wet surrounding (reversible liquid loss).

The water related to hydration is mainly bound to protein-structures of large molecular weights. This quantity of moisture cannot be extracted from the bone tissues at room temperature or in a dessicator; it will only evaporate during heat treatment. When by the action of heat (105°C) the protein molecules undergo structural changes they release chemically bound water. This amounts to 5 to 6 percent of the total weight in fresh bones. The water lost in this way cannot be reabsorbed (irreversible water loss). In excavated bones the amount of water bound by hydration depends primarily upon the changes in the water-binding properties of the protein, which in turn is related to the process of fossilization.

"Crystalline" water can be found in the apatitelike crystal lattice which forms the anorganic moiety of bone tissue. An increase in the temperature (around 450°C) brings about the disintegration of the crystal structure, releasing the "crystalline" water. In vivo, one sign of aging is that the hydration shell around the crystal structures becomes thinner, i.e., the amount of "crystalline" water diminishes. In fresh bones the amount of "crystalline" water varies between 3 and 5 percent, and there is a loose correlation with the biological age of the individual. In excavated bones the "crystalline" water content changes (but it does not necessarily diminish) only when the fossilizing process decomposes the crystal structure of the bone tissue.

## Interrelation between the Weight Loss by Cremation of Excavated Bones and the Quantity of their Organic Constituents

Van Bemelen (1897),[4] Barber (1939),[5] Bayle, Amy and Du Noyer (1939),[6] Jaffe and Sherwood (1951),[7] Cook and Heizer (1952),[8] and Oakley (1948)[9] looked for a relationship between the diminution of the organic material of the bones and their historical age. Although such a connection exists beyond doubt, rather few results

431

were obtained from their research. This may be explained by the following: First, the organic content of the fresh bones is influenced by various biological and other factors, so that not even the initial values are constant. Second, the diminution of the organic content of the bone tissue is not primarily the result of historical age. Third, the cremation method is by no means suitable for the determination of the quantity of the organic constituents of the bones. As a matter of fact, the loss of weight of the bone tissue is the result of complex processes: in the bone ashes which contain only inorganic materials, the bone tissue lacks both organic materials and liquid content. Besides, the carbon dioxyde burns out from the carbonate salts and the calcium and phosphorus compounds become oxydized in the cremation process.

## The Organic Content of the Bones

A. Examination of nitrogen-free organic materials (lipids and citrates) and $CO_3$ containing compounds: the decay of lipids was examined by Gangl (1936),[10] who searched for a dependency or relationship between these materials and the chronological age of the bones.

The fresh bone tissue contains lipids, measurable only as the order of magnitude permits. In excavated bones, the possible lipid content (which is in the range of one per thousand) is derived from the fat tissues of the corpse.

Under the influence of lipase-producing soil bacteria, the lipids undergo an oxydative disintegration process, after which their melting point and their acid and iodine number (index) change. The intensity and rate of this disintegration in the soil is controlled and determined by the partial pressure of oxygen, the moisture content and temperature of the soil, and the composition of microvegetation.

The citrate, being most important among the organic anions, plays a key role in the oxydative energy-producing process, and, as part of the tricarboxylic-acid cycle, it participates in the ossification process. About 70 per cent of the whole citrate content of the body is accumulated in the bone tissue. The changes in citrate content of the bone tissue in the context of time were examined by Thunberg (1947),[11] Pin (1950),[12] and Baud, Bodson, and Morgenthaler (1956).[13] The citrate content in fresh bone shows individual variations over a wide range. A relationship between the time element and the change of the citrate content was not detected, and no conclusions were drawn from other plausible signs either.

Carbon may appear in the bone tissue as a component of organic or inorganic materials. With acid treatment, the carbon may be liberated from its inorganic combinations of salts and carbonates. With cremation, the complete $CO_3$ content of the bone sample may be burned out. The difference between the results obtained by the two methods is more or less equal to the amount of the organic-bound carbon. Part of this carbon comes from citrate and lipid molecules. During the first 200 to 300 years after interment, the amount of both materials diminishes rapidly. The other part of the organic-bound carbon originates from proteins. The amount of this is in close relation with its $N_2$ content. Proteins are more resistant to the effect of decomposition than lipid molecules. The research based on the rate of decomposition and on the changes of the $CO_3/N_2$ ratio in the context of

432

historical times by Cook (1951), Heizer (1952),[14] and Oakley (1963)[15] are very encouraging.

Concerning $N_2$ containing organic materials (in fresh-oven-dried bones the complete amount totals between 76 and 92 percent of the total organic content), we deal mostly with collagen because of its resistance to decompositional processes.

Collagen is a protein of fibrous construction with a large molecular weight. Its nitrogen content measured in the dry ash-free material is 18.6 gram percentage (Eastoe and Eastoe, 1954).[16] Measurements of collagen, based on the nitrogen content of the bone tissue, may be made only on compact bone substance.

The nitrogen content of the spongy bone substance (even in the excavated bone material) is considerably influenced by the nitrogen content of the disintegrated soft-tissues, trapped in the cavities of the trabecular structure. Due to the lack of this consideration, the research of Sheurer-Kestner (1970)[17] and Koby (1938)[18] was indecisive. In fact, none of them distinguished between the spongy and compact bone substance. In addition Sheurer-Kestner prepared the bone substances with acidic decalcination, ignoring the consequent acidic hydrolysis of the proteins. Although the above considerations were ignored, both authors noticed connections between the age of the material and the decrease in the nitrogen content of the bone tissue.

Duerst (1926)[19] examined the nitrogen content of horse metacarpal bones which originated in various periods and went back 14,000 years. To eliminate the various disturbing factors, he worked out corrective mathematical methods. He used diagrams to demonstrate the drop in the nitrogen content as a function of time.

Although a statistical analysis between the age of a large number of cases shows a connection between the age of the bones and their nitrogen content, we do not believe that a world-wide, universal scale may be worked out which could determine the age of any bone finding on the basis of its nitrogen content. In agreement with Oakley (1963),[20] we may find the usual decrease in the nitrogen content of the bone tissue, together with other chemical components, useful as an auxiliary time index. In the evaluation of bone research in the laboratories, the application of collective data based on complex methods seems to lie in the future.

Noteworthy are the bone protein analyses of Abelson (1954, 1955)[21] and Cook (1951)[22] in which, based on the qualitative and quantitative examination of the amino acids (building stones of the bone proteins), conclusions were made about the chemical and time resistance of certain protein types and binding types (peptide bounds). They determined that the peptide bounds – as the remains of disintegrating proteins – may be found after several million years even in the skeletal tissue.

## Examination of Inorganic Materials in Excavated Bone Tissue

Inorganic materials impregnating the organic ground substance of the bone constitute an apatitelike (tricalciumphosphate-hydrate) crystal system. Calcium, phosphorus, carbon, oxygen, and hydrogen atoms constituting the above crystal structure are in a structural balance:

$$Ca_9 \cdot (PO_4)_6 \cdot H_2(OH)_2 + CaCO_3 \rightleftharpoons Ca_{10}(PO_4)_6(OH)_2 + CO_2 + H_2O$$

Within this system, in case of intact crystal structure in fresh bones, the calcium–phosphorus ratio is largely constant:

$$Ca:P = 1,667 \text{ (Dellamagne and Fabry, 1956)}.[23]$$

Many authors have dealt with the quantitative analysis of the inorganic components in excavated bones: Milne-Edwards (1860),[24] Middleton (1884),[25] Van Bemelen (1897),[26] Rivière (1905),[27] Barber (1939),[28] Bayle, Amy and Du Noyer (1939),[29] Tanabe (1944),[30] Pin (1950),[31] Macintosh (1953),[32] Neuman and Neuman (1957).[33] The results of their examinations were different due to the differences in their methods. However, in conclusion they basically agreed that no connection existed between the inorganic content of the bones and their historical age.

## Histological Examination of Excavated Bone Tissue

Several investigators have studied the microscopic changes which occur during the fossilization of the bone tissue. Several of them gave interpretations and searched for the time element involved.

Moodie (1926)[34] found recognizable oesateon structures in vertebrae much preceding man's appearance in geological periods.

Bayle, Amy and Du Noyer (1939)[35] performed parallel histological and bone chemical research, intending to determine the best method for investigating bones from different historical periods.

Graf (1949),[36] investigating bones from the Middle Ages, recognized the remains of various soft-tissues in bone remains.

According to Baud and Morgenthaler (1952)[37] the quality of bone remains depends on their historical age, the structure of the embedding earth, and local climate variations.

Ascenzi (1950)[38] demonstrated collagen fibers and other structural elements in the bone remains of the Neanderthal man with an electron-microscopic method.

Lengyel and Nemeskéri (1964a)[39] determined five phases in the decomposition of bone tissues on the basis of histological structure in relation to the chemical composition.

## Investigation of Bone Tissue's Fluoride Content

During the process of decomposition, the bone absorbs foreign materials from the soil, among them fluoride. The quantity of fluoride absorbed and built in the fossilizing bone material, depending on the fluoride content of the soil, increases with time. Recognition of this fact, which was the result of the works of Carnot (1893)[40] and Gassmann (1908)[41] attracted a whole series of scholars who then attempted to work out an exact method for determining the historical date of the bone findings.

The original method underwent various methodological changes in the hands of Willard, Winter (1933)[42] and Armstrong (1936),[43] was further modified by Balczó and Kaufmann (1951),[44] and was crystallized in its final form by Hoskins and Fryd (1955).[45] Oakley, who also played a key role in finalizing the method, worked out a solution for determining the age of excavated bones based on the fluor content (1948, 1955, 1963).[46]

The radiometric method, based on the measurements of the $C^{14}$ content of the bone tissue and using the most modern results and aids of isotope technique — similar to the argon-dating method — is not listed here since it cannot be ranged among the current laboratory techniques in a strict sense.

A question now arises: how is it possible that research was directed toward the chronological dating of bone findings, with paleophysiology and paleopathology mostly ignored, even though the idea of a complete biological reconstruction of our ancestors had already emerged at the end of the 1930s? (See Vallois, 1937; Angel, 1969; Nemeskéri and Acsádi, 1959; Howells, 1960; Nemeskéri and Harsányi, 1960.[47]) It is possible that the need for such an approach simply did not exist among archaeologists. Archaeologists seldom turned to classical anthropology for answers in reconstructing the complete biological appearance of our ancestors and hardly ever to biochemists or physiologists.

In my judgement, although bone research had been in process in various directions, there was no method which aimed beyond ascertaining chronology with physicochemical means which would have organized the examinations into a system, supplementing them with other needed examinations, putting them into the service of complete biological reconstruction.

## Theoretical Bases of the Complex Method

On the basis of these laboratory experiences, I was influenced by considerations described below in attempting to work out a new complex method. This method should establish a buttressing point for the anthropologist, the archaeologist, the historian, and the demographer by shedding light on genetical connections (in some respect), social structures, health condition, nutrition, and age divisions according to sex of our ancestors. I selected for my examinations a combination of chemicoanalytical, serological, and histological methods most suitable for my purposes, not based on the morphological attributes of the bones but on their material properties (chemical composition) and variations in their composition (see Lengyel and Nemeskéri, 1963; Lengyel, 1968[48]).

The results of chemicoanalytical, serological, and histological examinations performed on freshly excavated bones are reduced to a common denominator and then compared with one another.

The advantages of the complex methodology are as follows:

1. Since we base our conclusions on properties different from those on which current methods are based, the new technique makes it possible to control the reliability of data obtained by metrical analyses.

2. The new method compares changes in the chemical composition and the morphology of fossilized bones with the chemistry and morphology of bones obtained from the dissecting room and thereby facilitates inferences regarding the ecology of ancient populations.
3. Since few fragments of bones suffice for examinations, the new method yields reliable results even if the skeletal remains are scanty or poorly preserved.
4. Applied to the total skeletal material of a graveyard, the new method yields information that may be used for the biological reconstruction of the entire population or each separate individual.
5. By yielding information about the extent of decomposition suffered by the bones in the soil, the new method, in special cases, may permit conclusions as to the relative chronology of the series excavated.

This complex method may be divided into two theoretically related events. As a first step we are looking for an interrelationship between the individual's medical case history and the results of the bone examinations. We are basing our chemicoanalytical, serological, and histological method on fresh bone material. Applied to the spongiosa of lumbar vertebral bodies, the complex method yields nineteen final numerical results. These results will be compared to the individual's medical history in order to obtain a complete picture. (We understand that an individual medical history includes sectional findings which contain such items as the individual's sex, biological age, cause of death, physiological or pathologic condition of the organism, and blood group.) Analysis of more than a thousand fresh specimens and the literature regarding physiological and pathologic bio-mechanism have revealed the existence between the production of androgens and/or oestrogens and the citrate content of bones, the process of biological aging and the calcium, phosphorus, carbonate, and collagen content of the bones, the protein-polysaccharide complex content of the bones and a reliability of determining the individual's blood group, and some pathological processes upsetting the acid-basic and/or hormonal balance and the mineral and/or protein metabolism and the vitamin household of the organism, and changes in the normal chemical composition of the bones.

As a second step, based on excavated bones, we are looking for individual history through chemical analysis and on the basis of the above-mentioned inter-relationships.

When we take the cancellous part of a lumbar vertebral body from excavated skeleton, we determine the concentration of the same chemical components as in the case of fresh autopsy material, so that simultaneously nineteen numerical results are obtained. From these fixed results, we have to draw conclusions about a long-deceased individual's sex, biological age, blood group, and other matters. Actually, we have to compare the excavated bone findings to the known anamnesis and to the data of fresh bones in such a way that we should be able to determine the individual's history with the help of the previously specified correlations.

In our comparative system, the first unknown factor is the individual medical history of the excavated bones. The second, which basically influences the chemical composition of the excavated bones and consequently questions the credibility of our examinations, is decomposition.

We define decomposition as the summary of all those biological, chemical, and physical influences and impacts which act on the bones from the time of the individual's death until the beginning of our complex examinations. Furthermore, under the influence of decomposition, the original materials from which the bone is composed may change qualitatively, diminish quantitatively, or enlarge with material alien to the original bone components.

Anamnesis of a long-deceased individual whose excavated bones are examined cannot be ascertained without first eliminating the unknown factors of decomposition. It is justified to assume that anatomically identical bones, dating from the same historical age, interred in the same kind of soil, and excavated from the same depth have been exposed to roughly identical decomposing agents acting in the same manner and with the same intensity. Since factors of decomposition within a given graveyard or at least within certain soil units are practically uniform, the changes in the chemical composition of anatomically identical bones may be evaluated according to those statistical principles as in the case of fresh autopsy material. On this basis, examining the bone samples from several skeletons of certain necropolis, we consider the numerical and qualitative deviations of individual origin.

This theory is the basis of our complex examinations. Our complex chemico-analytical, serological, and histological method consists of the following manipulations, with both fresh and fossil bone samples:

A) *Determination of water content*:

1. Drying in weight constancy on 105°C. Loss of weight = with physically plus chemically bound water content;
2. Incineration (during which the microcrystalline structure disintegrates). Loss of weight = with the sum of the total water content, plus the burned-out organic materials, plus the carbonates escaping in the form of $CO_2$ gas.

B) *Determination of organic substances*:

I. Organic compounds containing nitrogen:

3. Total nitrogen content
4. Amount of nonprotein nitrogen
5. Amount of protein–polysaccharide complex
6. Water soluble organic fraction
7. Amount of bone collagen
8. Amount of resistant proteins (i.e., of proteinlike substances remaining after the extraction of collagen)

II. Organic compounds free from nitrogen:

9. Citrate content
10. Carbonate content
11. Phosphorus content in organic bounds

437

*C) Determination of inorganic substances:*

12. Calcium
13. Phosphorus content in inorganic bounds
14. Magnesium
15. Iron (ferrous and ferric iron together)

*D) Blood typing:*

16. Modified fluorescent antibody technique
17. Boyd–Candela's hemagglutination inhibition test

*E) Histological examinations:*

18. Staining methods:
    a) Haematoxylin–eosin staining
    b) Schmorl's method
    c) Azan staining
19. Histochemical procedures:
    a) Metachromatic staining reaction
    b) Peroxydase reaction.

## NOTES

[1] J. M. VAN BEMELEN, "Die Absorbtion. Anhäufung von Fluorcalcium, Kalk, Phosphaten in fossilen Knochen". *Zeitschr. f. anorg. u. allg. Chemie* 15 (1897): 90–122.

[2] K. P. OAKLEY, "Analytical Methods of Dating Bones". *Advancement of Science* 11 (1955): 3–8.

[3] R. F. HEIZER and S. F. COOK, *The Application of Quantitative Methods in Archaeology* (Chicago, 1960).

[4] VAN BEMELEN.

[5] H. BARBER, "Untersuchungen über die chemische Veränderung von Knochen bei der Fossilisation". *Paleobiologica* 7 (1939): 217–35.

[6] E. BAYLE, L. AMY and R. DU NOYER, "Contribution à l'étude des os en cours de fossilisation. Essai de détermination de leur âge". *Bull. Soc. Chim. France*, Ser. 5,6 (1939): 1011–24.

[7] E. B. JAFFE and A. M. SHERWOOD, *Physical and Chemical Comparison of Modern and Fossil Tooth and Bone Material*. Atomic Energy Commission, Technical Information Service (Oak Ridge, Tenn., 1951).

[8] S. F. COOK and R. F. HEIZER, "The Fossilization of Human Bone: Organic Components and Water". *Univ. Calif. Archeol. Survey Rept.* No. 17 (1952), pp. 1–24.

[9] K. P. OAKLEY, "Fluorine and the Relative Dating of Bones". *Advancement of Science* 4 (1948): 336–37.

[10] I. GANGL, "Altersbestimmung fossiler Knochenfunde auf chemischen Weg". *Ost. Chem. Z.* 39 (1936): 79–82.

[11] T. THUNBERG, "The Citric Acid Content of Older, Especially Mediaeval and Prehistoric Bone Material". *Acta Physiol. Scandinavica* 14 (1947): 244–47.

[12] P. PIN, "Contribution de la biochimie à l'étude des os préhistoriques". *Bull. et Mém., Soc. d'Anthrop. Paris*, Ser. 10, 1 (1950): 137–38.

[13] C. A. BAUD, P. BODSON and P. W. MORGENTHALER, "L'acide citrique et l'anhydride carbonique de l'os humain fossile". *Arch. Suisses Anthrop. Gén.* 21 (1956): 86–90.

[14] S. F. COOK and R. F. HEIZER, "The Fossilization of Human Bone: Organic Components and Water". *Univ. Calif. Archeol. Survey Rept.* No. 17 (1952), pp. 1–24.

[15] K. P. OAKLEY, "Dating Skeletal Material". *Science* 140 (1963): 356–488.

[16] J. E. EASTOE and B. EASTOE, "The Organic Constituents of Mammalian Compact Bone". *Biochem. J.* 57 (1954): 453–59.

[17] M. A. SCHEURER-KESTNER, "Sur la composition des ossements fossiles". *Bull. Soc. Chim. Paris*, Ser. 2, 13 (1970): 199–212.

[18] E. KOBY, "Une nouvelle station préhistorique (paléolithique, néolithique, âge du bronze): les cavernes de St. Brais (Jura Bernois)". *Verh. Naturforsch. Ges. Basel* 49 (1938): 138–96.

[19] J. U. DUERST, "Vergleichende Untersuchungsmethoden am Skelett bei Säugern". *Abderhaldens Handbuch d. Biol. Arbeitsmethoden*, Book 7, Part 1, fasc. 2 (1926), pp. 124–30.

[20] OAKLEY, "Dating Skeletal Material".

[21] P. H. ABELSON, "The Organic Constituents of Fossils". *Carnegie Inst. Washington, Yearbook*, 53 (1954): 97–101; 54 (1955): 107–9.

[22] S. F. COOK, "The Fossilization of Human Bone: Calcium, Phosphate, and Carbonate". *Univ. Calif. Publ. Amer. Archeol. and Ethnol.* 40 (1951): 263–80.

[23] M. J. DALLEMAGNE and C. FABRY, "Structure of Bone Salts". *Ciba Found. Symp.: Bone Structure and Metabolism* (London, 1956), pp. 14–32.

[24] A. MILNE-EDWARDS, "Études chimiques et physiologiques sur les os". *Ann. Sci. Nat.*, Ser. 4, 13 (1860): 113–92.

[25] J. MIDDLETON "On Fluorine in Bones: Its Sources and Its Application to the Determination of the Geological Age of Fossil Bones". *Proc. Geol. Soc. London* 4 (1844): 431–33.

[26] VAN BEMELEN.

[27] E. RIVIÈRE, "Sur l'utilité des recherches microscopique et de l'analyse chimique dans les études préhistoriques". *Bull. Soc. Préhist. Française* 2 (1905): 146–51.

[28] BARBER.

[29] BAYLE, AMY and DU NOYER.

[30] G. TANABE, "On the Calcium and Phosphorous Content of Human Bones from the Shell Mound of Homi". *Jour. Anthrop. Soc. Japan* 59 (1944): 1–5.

[31] PIN.

[32] N. W. G. MACINTOSH, "The Cohuna Cranium: Physiography and Chemical Analysis". *Oceania* 23 (1953): 277–96.

[33] W. F. NEUMAN and M. W. NEUMAN, "The Nature of the Mineral Phase of Bone". *Chem. Rev.* 53 (1957): 1–45.

[34] R. L. MOODIE, "Studies in Paleopathology. 13. The Elements of the Haversian System in Normal and Pathological Structures among Fossil Vertebrates". *Biologia Generalis* 2 (1926): 63–94.

[35] BAYLE, AMY and DU NOYER.

[36] W. GRAF, "Preserved Histological Structures in Egyptian Mummy Tissues and Ancient Swedish Skeletons". *Acta Anat.* 8 (1949): 236–50.

[37] C. A. BAUD and P. W. MORGENTHALER, "Recherches sur l'ultrastructure de l'os humain fossile". *Arch. Suisses Anthrop. Gén.* 17 (1952): 52–65.

[38] A. ASCENZI, "Some Histochemical Properties of the Organic Substance Neanderthalien Bone". *Amer. Jour. Phys. Anthrop.* N. S. 13 (1950): 557–66.

[39] I. LENGYEL and J. NEMESKÉRI, "Über die Blutgruppenbestimmung an Knochen mit Hilfe der Fluoreszenz-Antikörper-Methode". *Homo* 15 (1964): 65–72.

[40] A. CARNOT, "Recherches sur la composition générale et la teneur en fluor des os modernes et des os fossiles de différents âges". *Annales des Mines, Mémoires*, Ser. 9, 3 (1893): 155–95.

[41] T. GASSMANN, "Chemische Untersuchungen der Zähne. Teil I". *Zeitschr. f. Physiol. Chemie* 55 (1908): 455–65.

[42] H. H. WILLARD and O. B. WINTER, "Volumetric Method for Determination of Fluorine". *Ind. and Engin. Chem., Anal. Ed.* 5 (1933): 7–10.

[43] W. D. ARMSTRONG, "Microdetermination of Fluorine: Elimination of Effect of Chloride". *Ind. and Engin. Chem., Anal. Ed.* 8 (1936): 384–87.

[44] H. BALCZÓ and O. KAUFMANN, "Mikromassanalytische Bestimmung der Fluorins". *Mikrochemie* 38 (1951): 237–57.

439

[45] C. R. Hoskins and C. F. M. Fryd, "The Determination of Fluorine in Piltdown and Related Fossiles". *Jour. Applied Chem.* 5 (1955): 85–87.

[46] Oakley.

[47] H. V. Vallois, "La durée de la vie chez l'homme fossile". *Anthropologie* 47 (1937): 499–532; L. J. Angel, "Paleodemography and Evolution". *Amer. Journ. Phys. Anthrop.* 31 (1969): 343–53; J. Nemeskéri and Gy. Acsádi, "La paléodémographie, base nouvelle de l'analyse anthropologique", in *Selected Papers of the Fifth Internat. Congr. of Anthrop. and Ethnol. Sciences* (Philadelphia, 1959), pp. 692–97; W. W. Howells, "Estimating Population Numbers through Archaeological and Skeletal Remains", in Heizer and Cook, pp. 158–76; J. Nemeskéri and Gy. Harsányi, "Die Bedeutung paläopathologischer Untersuchungen für die historische Anthropologie". *Homo* 10 (1960): 203–26.

[48] I. Lengyel and J. Nemeskéri, "Application of Biochemical Methods to Biological Reconstruction". *Zeitschr. Morphol. Anthrop.* 54 (1963): 1–56; I. Lengyel, "Biochemical Aspects of Early Skeletons", in D. R. Brothwell, *The Skeletal Biology of Earlier Human Populations* (Oxford, 1968), 8:271–88.

# LABORATORY ANALYSES OF FOSSIL BONES

IMRE LENGYEL

Our complex method of examining osseous tissues was presented in the preceding chapter. As far as details are concerned, in view of the nature and arrangement of this book, we have to content ourselves with a discussion of only two of the procedures constituting the method.

It will be clear from the preceding discussion that the chemical composition of fossilized bones of dead persons admits of inferences to sex, biological age, A,B,O blood type, certain pathological processes, and, under favorable conditions, the chronological sequence of burials in a given graveyard. I intend to deal now with certain facts regarding the determination of sex and blood type.

A discussion of the determination of sex by way of the chemical evidence of fossilized bones seems to be worth-while because the results of laboratory analyses can be compared with those of morphological examinations, as with the archaeological interpretation of grave finds. Furthermore, examination of the entire skeletal material of a graveyard (which represents only a fraction of the population) may furnish valuable paleodemographic information.

The second subject that I propose to treat in more detail, namely, the determination of blood types, promises interesting results insofar as a comparison of archaeological profiles may provide new information about the ethnic relationships of earlier populations.

## DETERMINATION OF SEX OF FOSSIL BONES

As a preliminary step to sex determination by the chemical evidence of ancient human skeletal remains, we have to show that it is not only by way of manifest morphological characteristics but also through chemical properties that the dimorphism of the human skeleton manifests itself. According to literary data and our observations there is, from the time of sexual maturity to the time of climacteric, a significant difference between male and female human subjects as regards the amount of citrate contained in the cancellous substance of the bones.

Sherman, Mendel and Smith (1936)[1] were the first to study the distribution of citrate in the tissues as a product of endogenous metabolism. Krebs and Johnson, (1937)[2] elucidated the role played by citrate in the metabolism of carbohydrates, fats, and proteins or, more precisely, its role in the tricarboxylic-acid cycle. Dickens (1941)[3] established the fact that some 70 percent of the organism's total citrate is stored in the osseous tissues. It was demonstrated by Hennig and Theopold

441

(1951)[4] that citrate and calcium form in the bone a complex stereometric molecule of tricalcium dicitrate. While the amount of citrate in the cortical substance of the bone depends on the concentration of calcium (0.7 to 1.3 citrate molecule per 100 calcium atoms), the citrate level is lower in the spongiosa but does not depend as strictly on calcium contents (Kuyper, 1945).[5] Dixon and Perkins (1952)[6] proved the existence of intensive citrate metabolism in the bones and showed that the quantitative fluctuations in the urinary output of citrate are inversely related to its concentration in the bony tissues. Earlier, Shorr, Bernheim, and Taussky (1942)[7] observed the existence of correlations between changes in the excretion of citrate and the phases of the menstrual cycle. Subsequent investigations of the same authors made it clear that the urinary output of citrate is increased by estrogens and decreased by androgens. Having examined several cases, Thunberg (1947)[8] found that the concentration of citrate amounted to 0.71 gram percentage in the vertebral body of men, and to 1.11 gram percentage in that of women; he drew no conclusions from his observations. Our own observations confirm Thunberg's findings. The citrate content of the bones is thus a reliable index of hormonal conditions (estrogen–androgen production) at the time of death. Since, however, the amount of citrate in the cancellous substance of the bones depends also on the individual's biological age, the citrate content without the biological age is no key to sex.

No matter whether we are dealing with fresh or fossil bones, our method is as follows: First, the vertebral body is freed mechanically from all soft tissues and impurities of the soil; then it is pulverized and dried to constant weight at room temperature. The determination of citrate content is carried out by Taussky's method (1949), which was adapted to our purposes.[9] If pulverized bone is kept in a strongly acidic medium, e.g., in trichloroacetic acid, the citric acid contained in the bone-

TABLE 1

*Citrate Content in Recent Bones*

| Age | No. of events | | Mean values | | Upper end values | | Lower end values | | Difference between end values | |
|---|---|---|---|---|---|---|---|---|---|---|
| | ♂ | ♀ | ♂ | ♀ | ♂ | ♀ | ♂ | ♀ | ♂ | ♀ |
| 0–5 | 37 | 40 | 0.501 | 0.527 | 0.525 | 0.546 | 0.468 | 0.510 | 0.057 | 0.036 |
| 6–10 | 39 | 25 | 0.513 | 0.531 | 0.550 | 0.543 | 0.505 | 0.517 | 0.045 | 0.026 |
| 11–15 | 38 | 41 | 0.612 | 0.881 | 0.630 | 0.912 | 0.545 | 0.561 | 0.105 | 0.351 |
| 16–20 | 34 | 22 | 0.615 | 0.997 | 0.625 | 1.030 | 0.568 | 0.848 | 0.057 | 0.182 |
| 21–30 | 56 | 60 | 0.678 | 1.015 | 0.700 | 1.044 | 0.630 | 0.957 | 0.070 | 0.069 |
| 31–40 | 44 | 48 | 0.715 | 1.035 | 0.751 | 1.067 | 0.685 | 0.980 | 0.066 | 0.087 |
| 41–50 | 45 | 54 | 0.750 | 1.079 | 0.870 | 1.104 | 0.702 | 0.994 | 0.168 | 0.010 |
| 51–60 | 80 | 72 | 0.875 | 1.100 | 0.894 | 1.116 | 0.863 | 1.074 | 0.031 | 0.113 |
| 61–70 | 68 | 70 | 0.901 | 1.092 | 0.910 | 1.124 | 0.886 | 1.020 | 0.024 | 0.104 |
| 71–x | 47 | 35 | 0.976 | 1.020 | 1.007 | 1.002 | 1.897 | 0.991 | 0.110 | 0.011 |

Sum total:  488    467

N = 955

442

powder is transformed in the presence of bromide to pentabromacetate; the latter is then eluted with heptane and converted by means of sodium iodide to a complex iodine compound. The intensity of the color of this compound depends on the citrate level of the osseous tissue so that concentration can be determined by colorimetry.

During the first phase of our investigations we determined the citrate content of the vertebral bodies obtained from 955 freshly dissected cadavers. Sex and biological age of deceased individuals were known, and we studied the relationships between these particulars and the citrate contents of the vertebral bodies. Results are listed in Table 1.

Data shown in Table 1 will be understood better if we regard the citrate concentration in the vertebral bodies of females as 100 percent and refer the male values to this standard. The difference (to the advantage of females) amounts to the following:

| Percentage | Age class |
|---|---|
| 9.94 | 0–5 |
| 3.40 | 6–10 |
| 30.54 | 11–15 |
| 38.32 | 16–20 |
| 33.30 | 21–30 |
| 30.92 | 31–40 |
| 30.50 | 41–50 |
| 20.46 | 51–60 |
| 17.19 | 61–70 |
| 4.32 | 71–+ |

These figures illustrate that, in recent material, the concentration of citrate in the bone tissue admits of no inference to sex before puberty or after seventy years of age (Lengyel, 1969).[10]

During the second phase of our investigations we applied our findings derived from fresh bones to skeletal material from past ages: we tried to determine sex on the evidence of citrate concentration in disinterred vertebral bodies. Morphological sex determinations and analyses of grave furniture served as controls. In this phase we had to take two factors of uncertainty into account. First, we were ignorant of the biological age of the person to whom the examined sample had belonged. Even if the simultaneously performed morphological and chemical (calcium, phosphorus, carbonate, collagen) determinations yielded identical results as to age class, we had to infer the biological age of the members in the historical series from changes observed in the morphological characteristics of the skeletons. Second, we did not know to what extent the examined skeletal material had undergone decomposition. This factor could be eliminated by selecting a historical series which contained a statistically evaluable number of samples (N > 100) obtained from a graveyard that was used for a comparatively short time and in which the soil had an approximately homogeneous structure. The graveyard at Környe satisfied these requirements. One hundred and thirty samples were examined from the sixth-century cemetery, a burial ground that had been in use

TABLE 2

*Citrate Content in Group I*

| Age group | No. of events | | Mean values | | End values of the citrate content | | | |
|---|---|---|---|---|---|---|---|---|
| | | | | | upper | | lower | |
| | ♂ | ♀ | ♂ | ♀ | ♂ | ♀ | ♂ | ♀ |
| 0–5 | 3 | 11 | 0.300 | 0.320 | 0.315 | 0.340 | 0.280 | 0.300 |
| 6–10 | 4 | 4 | 0.310 | 0.325 | 0.315 | 0.580 | 0.300 | 0.300 |
| 11–15 | 3 | 4 | 0.340 | 0.592 | 0.345 | 0.680 | 0.335 | 0.330 |
| 16–20 | 6 | 1 | 0.405 | 0.675 | 0.450 | — | 0.360 | — |
| 21–30 | 5 | 7 | 0.420 | 0.670 | 0.460 | 0.685 | 0.400 | 0.650 |
| 31–40 | 5 | 12 | 0.435 | 0.670 | 0.500 | 0.680 | 0.405 | 0.640 |
| 41–50 | 8 | 5 | 0.560 | 0.685 | 0.600 | 0.690 | 0.500 | 0.645 |
| 51–60 | 2 | 3 | 0.645 | 0.700 | 0.655 | 0.705 | 0.635 | 0.695 |
| 61–+ | — | 3 | — | 0.705 | — | 0.710 | — | 0.700 |

for some 100–120 years. Archaeological evaluation of various grave findings was carried out by Salamon and Erdélyi (1971)[11] and the morphological analysis of the disinterred human skeletal remains by Tóth (see Salamon and Erdélyi, 1971).

The character of the grave furniture found in the cemetery of Környe shows that the persons interred there in the sixth century belonged to two categories that differed both in material culture and in social position. One category (group I) comprised (according to the attending archaeologist) poor persons from the lower social strata (Table 2), the other (group II) contained rich members of the higher classes (Table 3).

An extremely significant difference can be found regarding the proportion of sexes between the two groups (Table 4).

It can be seen that the citrate content of the vertebral bodies allows no distinction between the two sexes in the group Infans I. Sexual separation starts in the group Infans II and remains well distinguishable until senium. Chemical

TABLE 3

*Citrate Content in Group II*

| Age group | No. of events | | Mean values | | End values of the citrate content | | | |
|---|---|---|---|---|---|---|---|---|
| | | | | | upper | | lower | |
| | ♂ | ♀ | ♂ | ♀ | ♂ | ♀ | ♂ | ♀ |
| 0–5 | 1 | — | 0.290 | — | — | — | — | — |
| 6–10 | 1 | 2 | 0.310 | 0.310 | — | 0.300 | — | 0.320 |
| 11–15 | 5 | 2 | 0.320 | 0.320 | 0.345 | 0.315 | 0.310 | 0.325 |
| 16–20 | — | 1 | — | 0.605 | — | — | — | — |
| 21–30 | 8 | 2 | 0.410 | 0.625 | 0.440 | 0.630 | 0.390 | 0.580 |
| 31–40 | 6 | 2 | 0.420 | 0.630 | 0.460 | 0.640 | 0.390 | 0.620 |
| 41–50 | 10 | — | 0.430 | — | 0.460 | — | 0.400 | — |
| 51–60 | 2 | 1 | 0.580 | 0.685 | 0.600 | — | 0.560 | — |
| 61–x | 1 | — | 0.620 | — | — | — | — | — |

444

TABLE 4

*Proportion of Sexes*

|          | Men | Women | Total |
|----------|-----|-------|-------|
| Group I  | 36  | 50    | 86    |
| Group II | 34  | 10    | 44    |
| Total    | 70  | 60    | 130   |

$$\chi^2_{(1)} = 14.51883 \quad p < 0.01\%$$

properties no longer permit sexual distinction in old age. It is further clear that changes occurring in the citrate content of the male and female vertebrae in the course of life are of the same nature and intensity in both the historical and the recent materials.

Differences of citrate content between the two sexes in both series are best characterized by the values $b_{xy}$ (Table 3), with the exception of the group Infans I. There is a significant difference in all age classes between males and females as regards the regression value of citrate content. The practical value of our examinations is most fully shown by the control results, i.e., in the archaeological and anthropological findings. In 108 out of a total of 130 cases the sex of the interred person could be determined beyond any doubt. Morphological examination yielded the same result in 74 and the opposite result in 17 cases, while the condition of the bones made morphological examinations impossible in the rest (17 cases). The results of chemical analyses were in harmony with the archaeological findings in 86 cases, contradictory in 7, and gave no unequivocal results in 15 cases. It was possible to perform chemical sex determination in every one of the 130 cases.

In Table 5 the value of $\Gamma_{xy}$ and $b$ are positive in the given case. There is no significant difference between the two sexes in autopsy material as regards $b_{yx}$, whereas this value is significantly different between the two sexes in Környe group I and not significant in Környe group II. The value of $b_{xy}$ shows significant differences between males and females in the fresh autopsy material as also in Környe group I, but not so in Környe group II.

Sex differences between the samples are more clearly disclosed by the value $b_{xy}$; $t$-test shows regression values to be significantly different in men and women of all examined series (fresh and historical). Conditions of regression are different in the three series of samples also with regard to $b_{yx}$ values but not as explicit as when expressed by the values of $b_{xy}$.

Chemical analysis of the skeletal remains opened up in the graveyard of Környe justifies certain paleodemographic conclusions. First, it substantiates the archaeological finding that the population buried at Környe consisted of two culturally, socially, and probably ethnically different groups. Second, it shows that the proportion between the two sexes was different in the two groups (the difference being statistically significant according to Table 2). Third, the two groups differ as regards regression and correlation values ($b_{yx}$ and $b_{xy}$) of the males and females. Fourth, there is a further difference between the two groups in the average value

445

TABLE 5

*Correlation and Regression Values of Citrate Content*

| | | Recent material | | Környe group I | | Környe group II | |
|---|---|---|---|---|---|---|---|
| | | ♂ | ♀ | ♂ | ♀ | ♂ | ♀ |
| Biological | $n$ | 488 | 467 | 36 | 50 | 34 | 10 |
| age | $x$ | 38.02 | 37.38 | 26.11 | 26.60 | 32.90 | 27.00 |
| | $s$ | 21.1 | 20.6 | 16.1 | 18.6 | 14.8 | 15.5 |
| Citrate | $n$ | 488 | 467 | 36 | 50 | 34 | 10 |
| content | $y$ | 0.740 | 0.975 | 0.425 | 0.549 | 0.409 | 0.500 |
| | $s$ | 0.156 | 0.190 | 0.098 | 0.166 | 0.132 | 0.164 |
| Correlation | $r_{xy}$ | 0.958 | 0.807 | 0.977 | 0.887 | 0.585 | 0.253 |
| Regression | $b_{yx}$ | 0.00709 ±0.00013 | 0.00742 ±0.00033 | 0.00596 ±0.00023 | 0.0789 ±0.00059 | 0.00520 ±0.00127 | 0.00268 ±0.00362 |
| | $b_{xy}$ | 129.6 ± 2.3 | 87.5 ± 3.9 | 160.5 ± 6.08 | 99.3 ± 7.4 | 65.6 ±16.03 | 23.9 ±23.3 |

of citrate content in respect of males and of females per age class as applied to a single male or female member of that age class (Table 4). The period of life in which the important sexual-hormonal events (puberty and climacteric) occurred was different in both the male and the female prototypes of the two groups. The elucidation of factors lying in the background of these deviations may enrich the history of science with valuable information.

## DETERMINATION OF BLOOD TYPES IN FOSSIL BONES

We employed a modified version of the fluorescent antibody method for the purpose of determining blood types in cases of both fresh and fossil bones. Using frozen sections of soft tissues or bacteriological preparations, this method is suitable to demonstrate the antigen–antibody reaction taking place in the tissues (Pearse, 1960)[12] but cannot be used in its original form for the determination of the serological nature of fresh or fossil osseous tissues. The modifications made by us in view of adapting the method for our purposes are as follows: the bone tissue is subjected to a special preparative procedure, and the diagnostic serum is brought together (in an adequate proportion and under suitable control) with the fluorescent dye. The first modification enables us to section skeletal material without acid decalcification, the second modification increases the sensitivity and reliability of the method (Lengyel, 1974; Lengyel and Nemeskéri, 1964).[13]

In principle, the method is based on the fact that, if the reaction is positive, the antibodies, dyed with fluorescent stain and mounted on the section, are precipitated and indicate at the same time the part of the tissue in which the antigens are located. Exposed to ultraviolet light, fluorescent dye bound at the points of the antigen–antibody reactions emits a light visible in the microscope.

Determination of the blood type of persons whose fossil bones are unearthed in some ancient graveyard means the examination of a series of samples obtained from the population whose dead were buried in the given cemetery. The question which arises in connection with every series is invariably the same: is the distribution of blood groups in the examined series representative of the entirety of the population? In other words: is it justified to regard that fraction of the population which was laid to rest in a given graveyard as truly representing the entire population? Affirmative answer to this question is subject to certain strict reservations. If the absolute number of cases in a series is less than fifty, the results cannot be evaluated with regards to gene frequency ($p$, $q$, $r$).

It must be known (from the information provided by the archaeologist in charge of the excavations) whether the skeletal samples include all persons buried in the graveyard or only a fraction thereof. By possessing skeletal samples from all bodies of a completely excavated graveyard, we are in a position to examine that fraction of a population which was interred during the use of the cemetery. An additional factor of uncertainty has to be taken into account if the excavations do not extend to the entire graveyard or if the series does not include a sample from each body of a fully explored cemetery. Our examinations are restricted in such cases to an unknown proportion of a graveyard in which an unknown fraction of the population was buried. It would mean that one is trying to draw correct conclusions from a "wrongly planned" multistage sampling, a statistically and logically impossible procedure (Nemeskéri, 1962, 1970, 1972).[14]

Examination of a complete graveyard may yield results which reflect conditions of the entire local population even if they indicate only the distribution of blood types among the secretions that can be determined on the evidence of bone samples. If, however, a series of doubtful representative values is examined, the value of the results is further weakened by their including only the blood types of the secretory persons.

It should be borne in mind that the length of time during which a graveyard was in use is inversely related to the representative value of the fraction of the population buried there.

Although the extent to which a graveyard has been explored can be inferred with some approximation from a statistical analysis of the blood-type distribution it is safer to rely in respect of the gene frequency and the distribution of blood types A, B, and 0 on the information provided by the excavating archaeologist.

Considering that there exists no suitable method for the control of blood-type determinations of ancient skeletal material (the method of absorption being less reliable than that of antibody fluorescence), we had to elaborate a system of logical control based on statistical methods. Essentially, this control system is as follows.

Although, in theory, blood types may be distributed in any combination, familiarity with the laws governing the hereditary transmission of blood-type properties enables us to exclude some theoretical possibilities. A case in point, for instance, was when all individuals of a cemetery proved to possess B blood-type

447

properties — which seemed improbable from the beginning — and certain plant roots were found to have caused also the surrounding soil to develop an aspecific B-positive character (Lengyel and Nemeskéri, 1964).[15]

Gene frequency based on the observed number of different blood types, too, may reveal the reliability or falsity of our blood-type determination. Although the mathematical superiority of Bernstein's method (1924)[16] is beyond doubt, certain biological considerations induced us to use Fischer's technique (1950)[17] for the calculations of gene frequency.

If the distribution of the blood types A, B, and 0 is known, the law of Hardy–Weinberg makes it possible to calculate the probable frequency of group AB (Fraser, 1967).[18] Comparison of the calculated to the actually observed number of AB cases ($\chi^2$ test) yields two possibilities: that there is no statistically significant difference between the two values or that the difference is significant.

The absence of a significant difference means that one series of samples is in a state of genetic equilibrium, a fact that can be applied to the entire population from which the samples derive. By doing so we shall have increased the representative value of our results. Results of examinations carried out on serial bone samples excavated from ancient graveyards can be accepted only if there is no significant difference between the calculated number of AB groups and their actual occurrence as determined by the fluorescent antibody method.

If there is a significant difference between the two values either in a positive or a negative direction, the determination must have been erroneous.

If the numerical value of $\chi^2$ is not excessive there are two possibilities. If the deviation is negative, i.e., if the actual number of AB cases is less than expected, the examined population was not in a state of panmixia, that is to say, we are dealing with several noninterbreeding populations which differ in gene frequency. If, on the other hand, the deviation is positive, i.e., if the actual number of AB cases is more than calculated, it is probably due to heterosis of vitality in a population with high mortality rate. The values of p, q, and r may, therefore, be reliably used for further calculations.

Positive (not excessive) deviation may, however, be due to the examined series of samples being representative either because the entire graveyard has not been excavated or because the series does not contain all human skeletal remains of the completely excavated cemetery.

If the actual number of A, B, and O groups is known, a computation of the number of additional samples would be necessary to wipe out the difference between the expected and the observed AB cases, i.e., to deprive the value of $\chi^2$ of its significance, enabling us to determine the approximate size of the unexplored proportion of the graveyard and ascertain the full size of the original material.

The extent to which tombs have been destroyed can further be estimated if it is proved that although no tomb has been left unexplored and although the examined series contains samples from every tomb the values of AB are still significantly different.

Statistically significant $\chi^2$-values may indicate a biologically absurd situation as if the examined fraction of population had not been in a state of biological equilibrium. No such situation ever arises, so it is necessary to look for the cause of

TABLE 6
*Blood-Type Determination*

| Graveyard | Blood group | | | | | |
|---|---|---|---|---|---|---|
| | A | B | O | AB | Unknown | Total |
| Kajdacs<br>% | 11<br>42.31 | 7<br>26.92 | 5<br>19.23 | —<br>— | 3<br>11.54 | 26<br>100 |
| Kádárta<br>% | 1<br>50.00 | —<br>— | 1<br>50.00 | —<br>— | —<br>— | 2<br>100 |
| Kápolnásnyék<br>% | 2<br>50.00 | —<br>— | 1<br>25.00 | —<br>— | 1<br>25.00 | 4<br>100 |
| Mohács<br>% | 2<br>33.33 | 1<br>16.66 | 3<br>50.00 | —<br>— | —<br> | 6<br>99.99 |
| Rácalmás<br>% | 4<br>36.36 | 1<br>9.09 | 4<br>36.36 | —<br>— | 2<br>18.18 | 11<br>99.99 |
| Soponya<br>% | 1<br>50.00 | —<br>— | 1<br>50.00 | —<br>— | —<br>— | 2<br>100 |
| Szentendre<br>% | 32<br>41.60 | 13<br>16.90 | 18<br>13.40 | 4<br>5.20 | 10<br>12.90 | 77<br>100 |
| Testona<br>% | 11<br>39.29 | 6<br>21.43 | 5<br>17.86 | 2<br>7.14 | 4<br>14.28 | 28<br>100 |
| Várpalota<br>% | 6<br>50.00 | 2<br>16.66 | 3<br>25.00 | —<br>— | 1<br>8.33 | 12<br>99.99 |
| Vörs<br>% | 8<br>34.78 | 3<br>13.04 | 10<br>43.48 | —<br>— | 2<br>8.69 | 23<br>99.99 |
| Total<br>% | 78<br>40.84 | 33<br>17.28 | 51<br>26.70 | 6<br>3.14 | 23<br>12.04 | 191<br>100 |

such a result. Whether it is found or not, no conclusions may be drawn to the serological profile of an entire population if the value of $\chi^2$ is statistically significant.

We propose now to demonstrate the practical value of our method by presenting the results of blood-type determinations in a historical series of skeletal samples.

Migrating westward, Longobard tribes stayed in Pannonia (present-day Transdanubia in Hungary) from 526–546 to 568 (Werner, 1962).[19] Their route is distinctly marked by their graves. The short time of their stay explains the small number of tombs in their cemeteries. One of the largest Longobard graveyards in Pannonia is that at Szentendre (Pest County) where apparently two overlapping generations were laid to rest (Bóna, 1966).[20] Outside Pannonia, it is from the Longobard graveyard at Testona (Italy) that skeletal material has been placed at our disposal (Kiszely and Scaglioni, 1969).[21]

We have made blood-type determinations on 191 human skeletal remains of 10 Longobard graveyards from the time of the Great Migration (Table 6).

We assume that all examined samples stemmed from the skeletons of truly Longobard peoples; the combined values of the ten graveyards give the following serological gene frequency for the examined fraction of population:

$$p = 0.315$$
$$q = 0.151$$
$$r = \underline{0.534}$$
$$1.000$$

$$\chi^2_{(1)} = 4.171 \quad 5 > P > 1\% \text{ (significant)}.$$

The fact that, owing to a deficiency of AB in the examined series, the value of AB cases was smaller than expected, means that we classified several non-interbreeding subpopulations with different gene frequencies under the collective heading of "Longobards." In other words, our "Longobard" series was not in a state of panmixia.

If we reject the possibility that not all persons were Longobards who were interred with a Longobard type of grave furniture, there are two alternatives to explain the population-genetic phenomenon detected by our serological method; either it is not in all graveyards classified as Longobard that Longobards were interred, or all persons buried in the graveyards enumerated in the above list (except the cemetery at Testona) belonged to the so-called northern Longobard wave. This group included several genetically different, not interbreeding, elements, a theory supported by archaeological observations.

Which, then, is the serological profile that would distinguish the "true" Longobard from the other elements of our ethnically complex series? Careful analysis of the skeletons found in the Szentendre cemetery may enable us to solve this problem because the character and opulence of the grave finds as well as the nature of burial rites observable there offer the possibility of distinguishing the true Longobard remains from those of the local population.

Analysis based on this concept yields the following results (Tables 7 and 8).

TABLE 7

*Serological Profile of the Szentendre Cemetery*

| Blood type | True Longobards | | | Local population | | |
|---|---|---|---|---|---|---|
| | No. of events | % | Gene frequency | No. of events | % | Gene frequency |
| A | 20 | 57.14 | p = 0.494 | 12 | 28.57 | p = 0.225 |
| B | 3 | 8.57 | q = 0.105 | 10 | 23.81 | q = 0.192 |
| O | 5 | 14.28 | r = 0.401 | 13 | 30.95 | r = 0.583 |
| AB | 2 | 5.71 | 1.000 | 2 | 4.76 | 1.000 |
| ? | 5 | 14.28 | | 5 | 11.90 | |
| Sum total | 35 | 99.98 | | 42 | 99.99 | |

$$X^2_{(1)} = 0.21761 \qquad\qquad x^2_{(1)} = 1.29736$$

70 > P > 50% (not significant)      30 > P > 20% (not significant)

TABLE 8

*Distribution of Examined Prototypes*

|  | A | B | O | AB | Total |
|---|---|---|---|---|---|
| True Longobards | 20 | 3 | 5 | 2 | 30 |
| Local population | 12 | 10 | 13 | 2 | 37 |
| Total | 32 | 13 | 18 | 4 | 67 = N |

$\chi^2_{(3)} = 8.68324$; $5 > P > 1\%$ (significant).

This statistical analysis gives substance to the archaeological observations that the dead of two different ethnic groups were laid to rest at Szentendre. The two groups differed in both material culture and serological profile.

Having thus established the fact that the Longobard fraction of the Szentendre graveyard differs from the "local" fraction archaeologically and serologically alike, we can now make a blood-type comparison between the Longobard skeletons of Szentendre and the skeletal remains found in other Longobard graveyards (Table 9).

TABLE 9

*Phenotypes of True Longobards of Szentendre and Other Longobards*

|  | A | B | O | AB | Total |
|---|---|---|---|---|---|
| True Longobards of Szentendre | 20 | 3 | 5 | 2 | 30 |
| Other Longobards | 46 | 20 | 33 | 2 | 101 |
| Total | 66 | 23 | 38 | 4 | 131 = N |

$\chi^2_{(3)} = 5.40988$; $20 > P > 10\%$ (not significant).

TABLE 10

*Phenotypes of Local Population of Szentendre and Other Longobards*

|  | A | B | O | AB | Total |
|---|---|---|---|---|---|
| Local population of Szentendre | 12 | 10 | 13 | 2 | 37 |
| Other Longobards | 46 | 20 | 33 | 2 | 101 |
| Total | 58 | 30 | 46 | 4 | 138 = N |

$\chi^2_{(3)} = 1.98461$; $70 > P > 50\%$ (not significant).

451

Comparison between the "local" population of Szentendre and other Longobards yields the results found in Table 10.

Comparison between the entire material of Szentendre and that other Longobard graveyards gives the results following (Table 11):

TABLE 11

*Phenotypes of Entire Szentendre and Other Longobards*

|  | A | B | O | AB | Total |
|---|---|---|---|---|---|
| Szentendre complete | 32 | 13 | 18 | 4 | 67 |
| Other Longobards | 46 | 20 | 33 | 2 | 101 |
| Total | 78 | 33 | 51 | 6 | 168 = N |

$\chi^2_{(3/2)} = 2.26972$; $70 > P > 50\%$ (not significant).

It follows from the P-values of the three contingency tables that true Longobards were buried together with elements of the local populations in all graveyards classified as Longobard and that only the proportion of the two groups varies from graveyard to graveyard. Since, however, P-values show a closer connection between the "local" dead of Szentendre and those buried in the other Longobard graveyards than between the true Longobards of Szentendre and the skeletons of other Longobard graveyards, it seems justified to suppose that the bulk of our examined series consists of the skeletal remains of "local" people and not those of true Longobards.

Comparisons may also be carried out in order to determine similarities and dissimilarities between the serological properties of the Longobard series and that of the population of the historical period which followed chronologically, namely, the Avar period (Table 12).

TABLE 12

*Distribution of Phenotypes of All Longobards and All Avars Examined*

|  | A | B | O | AB | Total |
|---|---|---|---|---|---|
| Longobards (all together) | 78 | 33 | 51 | 6 | 168 |
| Avars (all together) | 128 | 135 | 106 | 61 | 430 |
| Total | 206 | 168 | 157 | 67 | 598 = N |

$\chi^2_{(3)} = 29.28465$; $P < 0.1\%$ (extremely significant).

It is evident, as regards the serological profile, that the respective population fractions of the two historically consecutive periods were significantly different.

Comparison between the respective serological profiles of "local" populations buried in two consecutive historical periods shows the following (Tables 13 and 14).

TABLE 13

*First Chronological Group in Alattyán*

|  | A | B | O | AB | Total |
|---|---|---|---|---|---|
| Local population of Szentendre | 12 | 10 | 13 | 2 | 37 |
| Local population of Alattyán, first chronological group in the Avar period | 13 | 30 | 13 | 7 | 63 |
| Total | 25 | 40 | 26 | 9 | 100 = N |

$\chi^2_{(3)} = 6.61058$; $10 > P > 5\%$ (undecided).

TABLE 14

*Distribution of Phenotypes in Szentendre and Early Hungarian Population*

|  | A | B | O | AB | Total |
|---|---|---|---|---|---|
| Local population of Szentendre | 12 | 10 | 13 | 2 | 37 |
| Early mediaeval Hungarian population with *A* predominance (local population?) | 71 | 59 | 62 | 23 | 215 |
| Total | 83 | 69 | 75 | 25 | 252 = N |

$\chi^2_{(3)} = 0.85437$; $90 > P > 70\%$ (not significant).

It seems probable from the last two contingency tables that populations classified as "local" actually represent a continuously resident aboriginal population which can be followed in our material from the sixth to the thirteenth centuries despite the diverse ethnic elements that were swept hither by the storms of history.

## NOTES

[1] C. C. SHERMAN, L. B. MENDEL and A. H. SMITH, "The Citric Acid Formed in Animal Metabolism". *J. Biol. Chem.* 113 (1936): 247–54.

[2] H. A. KREBS and W. JOHNSON, *Enzymetologia* 4 (1937): 132–34, in G. H. BOURNE, *The Biochemistry and Physiology of Bone* (New York, 1956), p. 309.

[3] F. Dickens, "The Citric Acid Content of Animal Tissues with Reference to Its Occurrence in Bone and Tumor". *Biochem. J.* 35 (1941): 1011–23.

[4] W. Hennig and W. Theopold, "Über Citronensäure-Calcium-Komplexverbindungen". *Ztschr. f. Kinderheilkunde* 69 (1950–1951): 55–61.

[5] A. C. Kuyper, "The Chemistry of Bone Formation I–II". *J. Biol. Chem.* 159 (1945): 411–24.

[6] T. F. Dixon and H. R. Perkins, "Citric Acid and Bone Metabolism". *Biochem. J.* 52 (1952): 260–65.

[7] E. Shorr, A. R. Bernheim and H. Taussky, "Relation of Urinary Citric Acid Excretion to Menstrual Cycle and Steroidal Reproductive Hormones". *Science* 95 (1942): 606–7.

[8] T. Thunberg, "The Citric Acid Content of Older, Especially Mediaeval and Prehistoric Bone Material". *Acta Physiol. Scandinavica* 14 (1947): 244–47.

[9] H. H. Taussky, "A Microcolorimetric Method for the Determination of Citric Acid". *J. Biol. Chem.* 18 (1949): 195–98.

[10] I. Lengyel, "Bestimmung der Geschlechtszugehörigkeit im Laboratorium". *Wiss. Ztschr. d. Humboldt-Univ. zu Berlin*, Math.-Nat. R. 18, 5 (1969): 977–79.

[11] Á. Salamon and I. Erdélyi, *Das völkerwanderungszeitliche Gräberfeld von Környe.* (Budapest, 1971).

[12] E. A. G. Pearse, *Histochemistry: Theoretical and Applied*, 2nd ed. (London, 1960), pp. 722–40.

[13] I. Lengyel and J. Nemeskéri, "Über die Blutgruppenbestimmung an Knochen mit Hilfe der Fluoreszenz-Antikörper-Methode". *Homo* 15 (1964): 65–72; I. Lengyel, *Paleoserology: Blood Typing with the Fluorescent Antibody Method* (Budapest, 1974); I. Lengyel and J. Nemeskéri, "A csontvázleletek dekompozíciójáról". *Anthrop. Közl.* 8 (1964): 69–82.

[14] J. Nemeskéri, "Problèmes de la reconstruction biologique en anthropologie historique". *VIᵉ Congrès Internat. d. Sciences Anthrop. et Ethnol. Paris* (1962), pp. 669–75; J. Nemeskéri, "A paleodemográfiai kutatások archeológiai és antropológiai feltételei". *Demográfia* 13 (1970): 32–72; J. Nemeskéri, "Die archäologischen und anthropologischen Voraussetzungen paläodemographischer Forschungen". *Praehist. Ztschr.* 47 (1972): 5–46.

[15] Lengyel and Nemeskéri, "A csontvázleletek", pp. 69–82.

[16] F. Bernstein, "Ergebnisse einer biostatischen zusammenfassenden Betrachtung über die erblichen Blutstrukturen des Menschen". *Klin. Wschr.* 3 (1924): 1495–1503.

[17] R. A. Fischer, *Statistical Methods for Research Workers*. 11th ed. (London, 1950).

[18] R. J. A. Fraser, *An Introduction to Medical Genetics* (London, 1967).

[19] J. Werner, *Die Langobarden in Pannonien. Beiträge zur Kenntnis der langobardischen Bodenfunde vor 568* (München, 1962).

[20] I. Bóna, *A Szentendrei Langobard temető* (The Longobard Cemetery from Szentendre)

[21] I. Kiszely and A. Scaglioni, "Lo sviluppo antropologico del sepolcreto Longobardo (barbaro) di Testona", in *Note antropologiche sul sepolcreto Longobardo di Testona* (Firenze, 1969).

# RECENT POTTERY ANALYSES

## GYÖRGY DUMA

A considerable part of archaeological finds, including the Pannonian ones, consists of utensils, pots, or building materials made out of burnt clay. It is understandable that in Hungary, as well as elsewhere, there is a growing demand to submit these important finds to modern scientific investigations, even though archaeologists themselves are still unclear how investigations of burnt clay products can further their work. Therefore, we feel that the results of our investigations of ancient pottery may give guidance for those concerned with archaeology.

## COMPLEX CHEMICAL AND MINERALOGICAL INVESTIGATION OF POTTERY

Natural raw materials which can be kneaded with water, called clay in common parlance, can be found almost everywhere, near the surface of the earth. The processing of clay was one of the most ancient occupations of mankind; experience has taught man that by burning this readily accessible and easily workable raw material a product can be obtained which effectively resists physical impact and chemical attack. For archaeological purposes clay products were tested almost exclusively on the basis of external formal signs. Because many questions have remained unanswered, newer experimental methods were sought which would classify pottery according to its basic material, determine similarities or dissimilarities independently of archaeological methods, illuminate the ceramic technologies under which pottery was produced, and reveal more about its actual function.

The scientific approach in the sphere of natural sciences led to the idea that the results of modern material-testing methods might be used for archaeological determinations as well. Generally applied, material-testing methods cannot produce satisfactory results in every case because of a number of special features relating to archaeological investigations.

Difficulties arise in the study of pottery because it is a handmade product. The original elutriation of clay and the addition of leaning agents to it change the natural ratio of the mineral components; molding and processing on a potter's wheel deform the texture. In addition, the chemical and physical changes which ensue as a result of baking considerably modify the mineral composition of clays. Last but not least, the original state of clay, its natural composition and macrostructure, is significantly transformed due to the physical and chemical effects of

455

being buried in the earth under natural conditions. Since it is generally not possible to discern which changes have taken place in the substance of pottery as a result of human work and which as a result of natural conditions, anthropogenic effects have also been used as distinctive features and characteristics when determining identities or dissimilarities.

Pottery was investigated for archaeological purposes even earlier than the beginning of the twentieth century. Since chemical studies have made many advances since then, it is understandable that chemical methods are applied in the investigation of ancient pottery. Since neither ceramic raw materials nor finished products can be characterized on the basis of their chemical components owing to their typical structure, determinations based exclusively on the results of chemical analyses can only rarely produce satisfactory results. In certain cases they have given us useful data for comparisons by means of spectroscopic analysis when a sufficient number of measurements were available to make statistical evaluation possible. Recently chemical components have been determined by micro sounding.

Because of the formation of ceramic raw materials, the scientific determination of both raw materials and products can be made on the basis of their mineralogical components. Methods generally applied for the determination of minerals and rocks are frequently applied to ceramic finds; mineralogical and petrographical examinations are also being used. Attempts have been made to draw conclusions from the minerals constituting the raw materials through calculations by using components established in chemical analysis. This kind of calculation, well known in silicate chemistry, may not be entirely successful in grouping or separating ceramic finds. In the case of potter's clay, the presence of clay minerals pertaining to the montmorillonite and mica group can hardly be characterized by chemical formulas. We have found that only a certain kind of evaluation, relying on determined mineralogical, petrographical, and chemical investigations of a particular type can help make determinations of pottery made out of potters' clay containing both montmorillonite and mica. There have been comparative examinations of pottery found in a number of cemeteries using this method which will be described in detail. For analytical purposes, average samples were taken, possibly from different parts of the same pot.

The comparative investigation of pottery bodies relies on quantitative chemical analysis. The aim of every chemical analysis is to determine the components of a material; hence, theoretically compounds are examined in which the components are uniformly distributed. Among silicates, it is mainly glasses and glazes that meet the requirement, whereas the composition of ceramic raw materials and products is very heterogeneous. With the exception of certain alkali silicates, silicates can be dissolved only by digestion, that is, by complete disintegration. Therefore, chemical analyses never throw light on the earlier relationship between the original components, i.e., on the original mineralogical structure. From this it follows that chemical composition in itself cannot be characteristic of raw and burnt products, that is, of pottery.

As far as clay is concerned, it is known that the transformation into pottery as a result of burning also brings about a loss of weight. In the case of quantitative chemical analysis, loss of weight is expressed by loss of ignition, which is an important and characteristic factor among the components of chemical analyses of raw materials. Baked clay, pottery, is only exposed to further loss if thermal transformation, which goes together with weight loss, has not come to an end during burning. If such conversion has taken place in the material of the pottery, similar processes can come about. It has been established that the reburning of the material of unglazed pots which were made by primitive technology and which have been lying in the earth generally brings about considerable loss of weight. In crockery that came to light in the course of excavations, the enrichment of organic and several inorganic materials could be observed frequently. Quite often a certain transformation or regeneration (rehydration) of clay minerals could also be established. Since these processes have considerable influence on the loss of ignition value in pottery, the loss of weight which takes place if pottery found at excavations is heated cannot be considered characteristic of either the original pottery or its one-time raw material.

It follows, then, that ceramic materials can be compared only on the basis of their constituent elements in the event of either full or zero loss of ignition. There is practically no loss of ignition in the case of dense products, whereas full loss of ignition can occur only in a state prior to burning. Loss of ignition determined in any interim condition renders uncertain the evaluation of comparative tests. Accordingly, the loss of ignition values of the investigated materials varies from 0.5 to 20 percent.

In view of the above, pottery brought to light in the course of excavations cannot be compared directly on the basis of a quantitative comparison of the chemical components. Hence, our calculations rely on the relative distribution of the selected components of the chemical analysis. When the selection is made, only such components can be taken into account which are characteristic of both the pottery and the processed raw material. Generally three components are selected: plotting applied in a ternary system was found very suitable for the comparison of the relative quantities of the components. In this system two components are independent variables, and the quantity of the third one is obtained if the sum of the two components is subtracted from the total amount. The number of independent variables is $n - 1$, which is similar to "n" component systems. Two dimensions suffice for the plotting of a ternary system; its form is an equilateral triangle. If the apexes of the triangle represent the single pure components (where their concentration is 100 percent), the sides of an equilateral triangle logically correspond to a binary system since the quantity of the third component is 0 percent along the lateral sides. Thus different concentrations of two components are placed on the sides of the triangle, whereas the mixtures of changing proportions of the three components lie inside the triangle. It is known that the sum of perpendiculars dropped on the sides from any point in an equilateral triangle is constant and equals the height of the triangle; in our case it equals the difference between 100 percent and 0 percent of the pure components. Each point inside the triangle expresses the percentage distribution of the three components. It also follows, therefore, that the points indicating the compounds of the unchanged proportion of the two components lie, in every case, along that straight line

457

which starts from the apex of the third component indicating 100 percent and tends toward 0 percent of the opposite side.

It should be emphasized that the present case does not refer to phase diagrams (ternary systems) customary in physical chemistry; the form of phase diagram was taken over in order to evaluate visually the relative quantity of the components selected from the chemical analysis.

In many cases it appeared that pottery could be characterized by the quantitative distribution of $Al_2O_3 - SiO_2 - K_2O$ components of the materials. This selection cannot be considered of general validity, for it can be characteristic only of certain types of clay. It can explain, however, the observation according to which hydromica and illite were the prevailing clay minerals in Pannonian potter's clay. In certain cases the total alkali ($K_2O + Na_2O$) content was taken into account. In this system one or more components in themselves can represent a relative value. In this way pottery derived from excavations was classified also in the system $CaO + (SiO_2/Al_2O_3) + (K_2O + Na_2O_3)$. If the ratio $Al_2O_3 : SiO_2$ is taken as one of the components (as above), one can in fact operate with the relative distribution of four components.

Those points that characterize one or another pottery and that lie either dispersed or dense at some areas inside the equilateral triangle representing the three components generally can be classified on the basis of the results of mineralogical and petrographical investigations (Figs. 86 and 87).

The mineralogical and petrographical study of pottery has been carried out by means of applying the thin-section method. This makes it possible to determine, as it is done in the case of rocks, the mineral composition of the sample, quantity, shape, and size of the components, as well as their spatial arrangement in relation to one another. Since petrographical methods have been adopted in this investigation, it seemed best to consider the pottery as rocks in spite of the anthropogenic effects; thus petrographical names have been used for the description of pottery.

The main aim of the mineralogical and petrographical study of ceramic finds is to obtain data on the mineral composition of the sample and, in this context, on the identity or differences of their texture. The classification and comparison on the basis of similar properties of data obtained by microscopic (or macroscopic) investigations cannot be confined to the statistical identification of the petrophysical properties of minerals. The raw material (clay, sand, etc.) of the investigated ceramic samples originates from the disintegration of a rock which came into being a long time ago. The chief aim and the most difficult task of sedimental petrography is to indicate or reconstruct the original location of these rocks. Therefore, one of the objectives of our work is to find an answer to the genetic question, depending upon the number and character of the available data.

In addition to the qualitative determination of mineral and detrital components, grain size distribution has also to be investigated in each case when pottery is tested for minerals. Typical groups can be established only if granulometric composition is taken into account. Generally, a number of larger mineralogical groups which are separable from each other can be established; it is possible, however, that differences between pottery bodies do not provide a reliable basis for this method and from a mineralogical aspect from one group. However, even

458

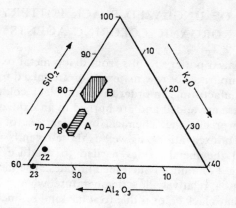

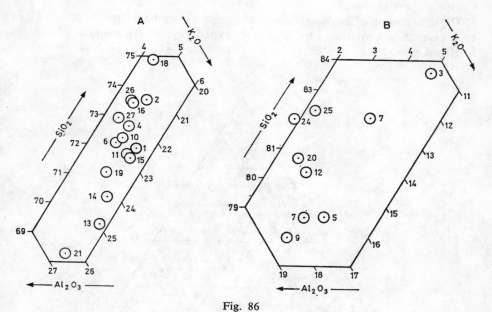

Fig. 86

the latter case, samples determined by chemical tests in the described ternary component graph can be classified definitely on the basis of identity and differences determinable from the results of mineralogical and petrographical investigations. There is no doubt that the standard of the investigation is considerably influenced by the number of samples: only in exceptional cases can reliable results be obtained from only a few samples.

# PRODUCTION OF UNGLAZED BLACK POTTERY CONTAINING ORGANIC COLORING AGENTS

The color of unglazed pottery results from dying metal oxides which either are present as natural impurities of raw materials or are mixed into the basic material in the course of manufacturing. In order to obtain black color, materials primarily containing manganese are added and are burned in an oxidizing atmosphere.

It is the more surprising that chemical investigations of unglazed black pots found in the course of excavations, as well as those recently made by folk potters, have not contained the metal oxides needed to produce black color or have contained only a slight amount of such metal oxides. The fact that these pots lose their black color at a relatively low temperature when burnt under oxidizing conditions, shows that black color is due to some kind of finely distributed organic coloring agent (carbon).

The distribution of a considerable carbon content in the pottery body, which in each case could be detected by laboratory experiments in the material of various unglazed pots, has shown characteristic differences between the pots.

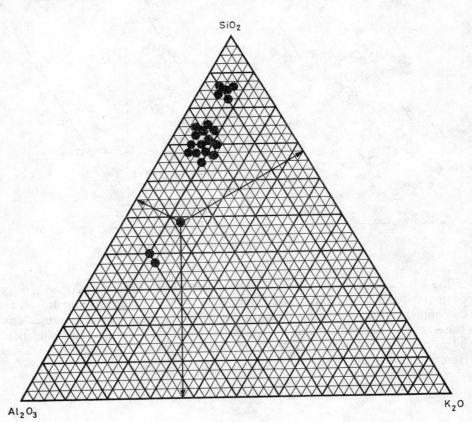

Fig. 87

There are pots in which the carbon content becomes enriched toward the surface layers, whereas in other cases equally large quantities of carbon can be found in the entire cross section of the pottery. On the basis of the distribution of carbon content which can be traced even with the naked eye on the fractures, unglazed pots can be classified into two distinct groups. The first group includes pots made out of clay which do not contain organic materials; in this case carbon has been incorporated in the course of firing. The second group consists of pots made out of basic material containing an organic additive (carbon).

## CARBON SEPARATION AND CARBON INCORPORATION OUT OF CARBON MONOXIDE AND GASES CONTAINING CARBON MONOXIDE

In the first group mentioned above, carbon becomes incorporated into the material of the pottery out of carbon monoxide and gases containing carbon monoxide; a similar phenomenon was observed at the end of the nineteenth century on the fire-clay lining of blast furnaces. In view of the industrial importance of the problem, a vast literature deals with questions concerning carbon separation. Numerous studies are concerned with the theoretical determination of the decomposition process of carbon monoxide and of gases containing carbon monoxide, and attempts have been made to support the expounded theories by experiments; in this context the role of materials acting as catalysts and factors influencing carbon separation have been investigated. Literature attributes a bursting effect to carbon in the presence of refractory materials; this assumption has been corroborated by a number of experiments.

It appeared in the course of our experiments that in spite of the carbon separation under suitable conditions, the carbon monoxide or gases containing carbon monoxide can be incorporated into pottery without impairing it. We have observed that the bursting effect is not caused by carbon, but by cementite, which acts as a catalyst in this process and results from the reduction of iron-containing compounds in the pottery (cementite also melts in silicate). The formation of cementite always precedes carbon separation. The fact that iron-cementite conversion brings about considerable volume expansion, causing bursting effect in the pottery, is also supported theoretically. Carbon, in itself, cannot produce a bursting effect, if only because it cannot fill the available space when separating out of carbon monoxide. When samples, under experimental conditions, have contained finely distributed iron contamination, the bursting action resulting from cementite formation has no visible effect. This explains that carbon, originating from carbon monoxide and gases containing carbon monoxide, can be incorporated without detrimental effects in the material of unglazed pots made out of washed clay; thus, unglazed black pots can be made by carbon incorporation.

461

# INVESTIGATION OF FAVORABLE CARBON INCORPORATION IN POTTERY

Our investigations also deal with cases in which a maximum degree of carbon separation in pottery is required. The experiments have been carried out partly under laboratory conditions in equipment specially designed for the purpose and partly in modern potters' kilns.

The laboratory experiments have been performed in pure carbon monoxide gas and gas mixtures corresponding to practical conditions, in purified gas originating from the dry distillation of various kinds of woods, and in gas products containing impurities (condensation products). The most favorable conditions of carbon separation have been determined in connection with various gas compositions, firing temperatures, and different ceramic raw materials. Carbon separation has been most favorable in gas flow nascent from the dry distillation of wood contaminated with volatile matter and condensation products. Carbon separation probably occurs within the same temperature range at which primitive unglazed pots were fired in ancient times. According to our experiments, carbon separation and carbon incorporation in the pottery took place in carbon monoxide gas at about 500° C, but characteristically changed in gas mixtures (Fig. 88). The most favorable carbon incorporation takes place in the phase of pottery body development; this can also be supported theoretically. Carbon separation in pottery depends on the iron content. It has been established that not the quantity of the iron content but its distribution and bond to minerals and silicates are decisive. Investigations were made to determine the relationship between the quantity of carbon separated in pottery and the formation of the black color, while another aim was to clear up not only the role of carbon but also the role played by iron compounds, primarily bivalent iron, in the development of black unglazed pottery surfaces.

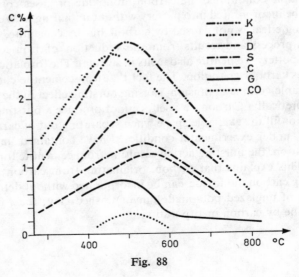

Fig. 88

Our experiments, in the course of which a number of unglazed pots were made, have revealed the production technology of primitive unglazed pots and have also provided a basis for the production of pottery in kilns by carbon separation in pottery.

## INVESTIGATION OF POTTERY MADE OUT OF CLAY CONTAINING ORGANIC MATTER

A separate group of unglazed black pots includes those whose color derives from the carbonization of the organic matter of processed clay and by preserving the carbon content in the course of burning. This requires a reducing atmosphere or at least a protective neutral atmosphere from the beginning of burning to its completion. Since primitive pots were generally made out of clay with a high iron content, the carbon incorporation probably occurred as a secondary process. Organic matter can also be preserved at low temperatures in a carbon dioxide atmosphere; however, as the temperature rises, the protective effect of carbon dioxide gradually decreases and subsequently exerts an oxidizing effect (Fig. 89).

The remaining organic matter of some pots in this group of ceramics has been thoroughly investigated. Among a very large group of unglazed, black, primitive pots there are some whose black layer contains finely distributed organic matter. The thermal decomposition of these pots could not have been complete in the course of burning. Nor did this matter decay while these pots were buried in the earth. Fluorescent analytical investigation has shown in the black pottery layers traces of the heme component (Figs. 90 and 91). A comparison technique gave evidence of the approximate quantity of the heme component. Factors influencing its traceability, as well as quantitative changes in the heme component under physical and chemical effects, have been investigated. Since the organic matter of the black pottery layers burns out in an oxidizing atmosphere at 400° C, the traceability of the heme component ceases at that point, too. In a carbon dioxide

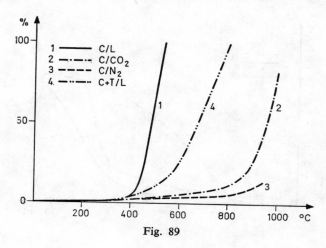

Fig. 89

463

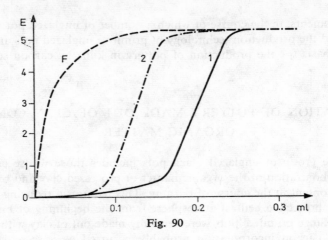

Fig. 90

atmosphere, however, a negative result ensues above 900° C. Quantity of the heme component considerably decreases the effect of acids which dissolve iron compounds. Enzymes which decompose hemoglobin had the same effect on the heme component in the black matter of the pots. The effect of decomposition factors on the heme component in the material of the pots was also investigated in autoclave. The heme component was less traceable in pottery processed in a high-pressure steam chamber. Bonds between pottery and heme components which strongly resist physical and chemical effects could develop in the black layers of the pottery only in the course of burning.

Up to now traces of the heme component have been found only in the material of pots excavated in the Mezőcsát cemetery, dating back to the Early Iron Age. However, it must have been customary to use blood elsewhere too, although it was hitherto unknown in the history of ceramics or in the history of religion. In order to prove the correctness of our experimental results it was deemed necessary to

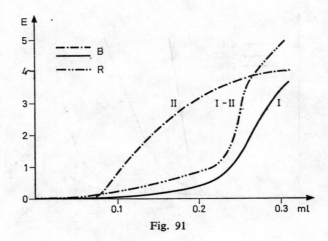

Fig. 91

repeat the test first made with ancient pottery with laboratory-made pottery consisting of clay mixed with blood. The results of the control tests were identical with those of the earlier tests in every respect.

## GLAZED POTTERY

Glazed pots which date back to Roman times and are to be found in the area bordered by the Carpathians can provide particularly interesting data for the archaeological investigation of ceramic finds. Although the chemical composition and structure of glazes are identical with the glasses, there are considerable differences between the two materials. Whereas glass is completely isotropic, the composition of glazes on the surface of pottery cannot be uniform because of the solvent effect of the boundary surface between glaze and pottery: this is why glazes are called vitreous materials. From this peculiar property of glaze it follows that methods applied for the investigation of glass cannot be used for the investigation of glazes.

Difficulties in the investigation of glazed surfaces frequently arise. The material of glazes, like that of glass, undergoes considerable changes as a result of corrosive effects while buried under ground; therefore glazes become unsuited for physical and chemical investigations. In addition, the pottery surface is generally covered only by a thin layer which can scarcely be separated from the basic material. Thus, quantitative chemical analysis can rarely be used in investigating the composition of glazes. Analysts have to content themselves with qualitative chemical investigations which may provide information on the lead content, the tin and antimony content, and on the possible presence of other components in the glazes under examination. In addition, vertical microscopic thin sections of the glazed surfaces can also provide useful information. Important data from the aspect of archaeological determinations can be obtained by intact mineral grains and crystallization on the boundary surface of glaze and pottery. We have found, moreover, that investigations of the melting conditions of glazes are of the utmost importance.

Owing to their structure, glasses and glazes have no melting point; these materials gradually melt with the rise of temperature. The melting conditions and changes of viscosity of glazes can be determined as a function of time and temperature, depending on their chemical composition. However, because viscosity measurements of glazes on pottery surfaces cannot be performed, a method had to be worked out for the characterization of glazes which would make possible the study of viscosity changes of burnt-in glaze surfaces.

One can speak of molten glazes if they are spread smoothly on the surface of pottery in kilns; the "smooth firing temperature" can be considered, practically speaking, to be the melting point of a glaze. Experiments have been carried out to determine the "melting point" of glazed surfaces reproduced as a function of time and temperature under laboratory conditions.

The melting conditions of glazed surfaces covering pottery have been examined by means of high-temperature microscope on glazed surfaces of $3 \times 4$ millimeters.

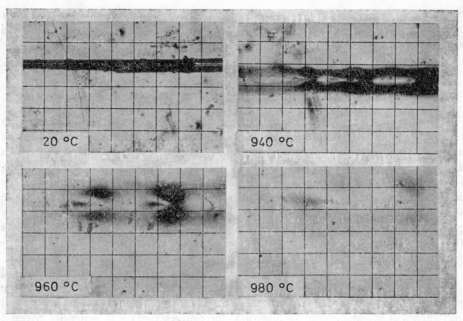

Fig. 92

The gradual fade-out of the horizontal groove (ditch) which takes place when the glaze surface melts and its viscosity decreases in line with the rise of temperature, all this could be observed easily by microscope (Fig. 92). This test can be carried out only on intact, noncorroded glaze surfaces. The determined "melting point" obtained by investigations of glazes melting with a smooth surface provide suitable data for comparisons. This method has been applied mainly for comparative investigations of medieval glazed pottery, and, to a lesser degree, of Roman pottery.

## LITERATURE

F. SAUTER and K. ROSSMANITH, "Chemische Untersuchung von Inkrustationen in hallstattlicher Keramik". *ArchAu* 40 (1966): 135–42; F. SAUTER and K. ROSSMANITH, "Chemische Untersuchungen von Inkrustationen in Mondsee-Keramik". *ArchAus* 43 (1967):1–5; H. W. CATTLING, E. E. RICHARDS and A. E. BLIN-STOYLE, "Correlations between Compositions and Provenance of Mycenaean and Minoan Pottery". *BSA* 58 (1963).

U. HOFFMANN and R. THEISEN, "Elektronenmikrosonde und antike Vasenmalerei". *Zeitschr. f. anorg. u. allg. Chemie* 341 (1965): 207–16; U. HOFFMANN and R. THEISEN, "Untersuchung antiker Vasenmalerei mit der Mikrosonde". *ArchAnzeiger* (1965): 164–67.

E. LENGYEL, "Adatok az Alföld rómaikori kerámiájához" (Data on Pottery of the Great Hungarian Plain Dating from the Roman Times). *Studies of the Arch. Institute of the Acad. of Sciences* 11 (1935): 226–28 (Petrographical investigation of the material of fired crockery and their place of occurrence); J. FRENCHEN, "Ausgrabungen in den Kirchen von Berberen und Doveren (Reg. Bez. Aachen): Ergebnisse der mineralogischen Untersuchungen". *Bonner Jb.* (1950): 192–228; J. FRENCHEN, "Der fränkische Friedhof in Rill bei Xanten: Ergebnisse der mineralogischen Untersuchungen". *Bonner Jb.* (1948), pp. 296–98; J. FRENCHEN, "Absatzgebiete frühgeschichtlicher Töpfereien nördlich der Alpen". *Antiquitas* 6 (1969): 143–46;

W. Hübener, "Die mineralogischen Untersuchungen". *Antiquitas* 6 (1969): 301–8; E. Schmidt "Ton und Magerung urgeschichtlicher Keramik von Schönberg, Gem. Erbringen, Ldkr. Freiburg". *Mitt. Bad. Landesver. Naturkunde u. Naturschutz* 9 (1966): 325–28; L. Grüniger, "Magerung und Technik der Keramik zweier prähistorischer Stationen im Schweizer Tafeljura". *Diss. Phil.-Naturwiss. Fak. d. Univ. Basel* (1965), pp. 1–15.

J. Derén, M. Dereniowa and L. Stoch, "Zastowanie metod fizykochemicznych i mineralogicznych do badan nad zabytkowa ceramika (na przykladzie dawnych kafli wawelskich) (Physico-Chemical and Mineralogical Methods Applied for the Investigation of Old Wawel Ceramics). *Oblike ze Sprawozdan z posiedzen komisji oddzialu PAN Krakowe* (1964), pp. 26–29; J. Derén, M. Piatkiewicz-Dereniowa and L. Stoch, "O surowcach i technologii wyrobu dawnych kafli wawelskich" (Raw Materials and Processing Technology of Old Wawel Glazed Tiles). *Kwartalnik Hist. Kultury Materialnej* (1966), pp. 21–29.

Gy. Duma, "Szénbeépüléssel készült mázatlan fekete kerámiák színének alakulása" (Color Change in Unglazed Black Ceramic Produced by Carbon Incorporation). *Építőanyag* 14 (1962): 463–70; Gy. Duma, "Der Brand von unglasierten schwarzen Tonwaren in Töpferbrennöfen". *Acta Ethnogr. Hung.* 12 (1963): 367–405; Gy. Duma, "Újabb vizsgálatok a kerámiai anyagokban történő szénkiválással kapcsolatban" (Newer Investigations on Carbon Decomposition in Pottery Clays). *Építőanyag* 13 (1961): 442–53; Gy. Duma and B. Galgóczi, "A kerámiai anyagokban történt szénkiválás és szénbeépülés elméleti alapjai" (Theoretical Principles of Carbon Deposition and Carbon Incorporation in the Material of Ceramics). *Bányászati és Kohászati Intézet Közleményei* 2 (1962): 205–15.

Gy. Duma and I. Lengyel, "Mezőcsát Pots Containing Red Blood Pigment (Haemoglobin)". *ActaArchHung* 22 (1970): 69–93.

Gy. Duma and I. Lengyel, "Őskori edények vértartalmú agyagokból" (Ancient Pots Produced Out of Clay Containing Blood). *Építőanyag* 19 (1957): 49–57; Gy. Duma and I. Lengyel, "Nachweis der hämkomponente des Blutfarbstoffes in keramischen Scherben antiker Gefässe". *Ber. d. Deutsch. Keram. Ges.* 45 (1968): 474–77; Gy. Duma and I. Lengyel, "Fluoreszenzanalytische Untersuchungen aus bluthaltigem Ton hergestellter vorzeitlicher Gefässe". *ArchAu* 45 (1969): 1–16.

G. E. Bourne, *An Introduction to Functional Histology* (London, 1963); T. Donáth and I. Lengyel, "Über Fluoreszenzprobleme der Erythrocyten". *Acta Histochemia* (1960): 260–71; S. Edelbach, *Handbuch der Biochemie* (Berlin, 1933); L. Pauling, *The Nature of the Chemical Bounds*, 2nd ed. (Ithaca, N. Y., 1940); R. Havemann, "Bestimmung des Haemoglobins mit dem lichtelektrischen Kolorimeter". *Klin. Wschr.* (1940): 503–11; H. Helmholtz, "Die theoretischen Grenzen für die Leistungsfähigkeit der Mikroskope". *Annalen d. Physik* (1874): 557–603; L. Klungsch and F. K. Stoa, "Spectrophotometrische Bestimmung der Sauerstoffsättigung des Haemoglobins". *Scandin. Journal of Klin. Labor. Invest.* (1945): 270–79; K. Hensberg and K. Lang, *Medizinische Chemie für den klinischen und theoretischen Gebrauch* (München, 1957); L. Magos and M. Szirtes, "Gleichzeitige Bestimmung von Methaemoglobin, Verdoglobin und Haemoglobin mit Pulfrich-Photometer". *Plasma* (1954): 551–65; M. Kiese, "Bestimmung des Gesamthämoglobins". *Archiv f. experim. Pathol. Pharmac.* (1947): 199–207.

Gy. Duma, "Wärmebehandlung von Majolikafliesen beim Restaurieren". *Deutsche Kunst und Denkmalpflege* (1971), pp. 75–78.

467

# PHOSPHATE TESTS USED IN PANNONIAN RESEARCH

## GYÖRGY DUMA

## SOIL INVESTIGATIONS

The chemical composition of the outer crust of the earth is on the average 0.12 percent phosphorus by weight. This element is a constituent of a number of rocks, minerals, and other inorganic and organic compounds. Minerals containing phosphorus can be found in almost every igneous rock; it is thus understandable that after weathering of the matrix rock, phosphorus becomes, secondarily, a constituent of various rocks of sedimentary origin. These are to be found in large quantities on and near the surface of the earth and consist mainly of loose clastic rocks. About 80 percent of clastic rocks of sedimentary origin are commonly called clay.

For the most part the phosphorus content of clay is of mineral origin and is bonded to various phosphates, phosphorites, and — under natural conditions sparingly digestible — to apatite minerals containing chlorine and fluorine to be found in the weathered remains of igneous rocks. It is known that clay layers near the surface of the soil contain large quantities of living organisms and substances of organic origin. Since phosphorus plays an important role in the tissue of both plant and animal organisms, it is understandable that, through biological accumulation, the phosphorus of the soil always exceeds by far the original phosphorus content of clays.

The living organisms of the soil decompose phosphorus minerals in clays and various phosphorus compounds; in this way phosphate ions permeate the soil, newer organic and inorganic phosphorus compounds can develop, and phosphate ions become incorporated into living organisms. In these processes phosphoric acid plays an important role; this is partly the result of the dissolving effect microorganisms exert on phosphorus minerals, particularly when organic substances decompose.

The decomposition of substances of plant and animal origin which have permeated the soil and accumulated there has been studied for some time under both aerobic and anaerobic conditions. In living organisms, a considerable amount of phosphorus is to be found in the phosphoric acid esters of more complicated carbon compounds containing alcoholic hydroxyl groups, primarily in the phosphoric acid esters of sugars and nucleotides. A large group of phosphorus compounds consists of phosphatides and phosphoproteids, the latter containing readily hydrolizable ester-bonded phosphoric acid. Phosphoric acid esters originate from muscular tissues, and fermenting liquids are to be found in every phase of the decomposition of organic substances. If the latter continue to decompose, transiently phosphoric acid ester of pyrroacemic acid, (enol) phosphoric acid — besides other products — develops.

The bonding of phosphate ions to the colloidal parts of the soil by way of absorption and ion exchange processes has been studied by many researchers, mainly in connection with clay minerals. It has been established that on the surface of kaolinite, mainly at the edges, $OH-PO_4$ ion exchange can also take place in case of suitable pH value. On the surfaces P–O bonds can develop of a structure similar to an Si–O tetrahedron; as a result of phosphorus intake, structural changes of kaolinite were observable. It is very probable that in addition to the above-mentioned ion exchange $OH-PO_4$, aluminum phosphates can develop too. It is difficult to distinguish among the roles that ion exchange, absorption, and chemical bonding play when phosphate ions become bonded to various clay minerals.

By the effect of phosphoric acid, however, new phosphates and minerals containing phosphorus can develop in the soil in which phosphorus is primarily bonded to calcium and, in a highly acidic soil, to iron and aluminum. In the newly developing group of minerals containing phosphorus, calcium phosphates are of the greatest importance. Of the latter, monocalcium and dicalcium phosphates readily dissolve in water, whereas tricalcium phosphates are practically insoluble. Since water-soluble phosphates also are transformed into tricalcium phosphate in time, calcium phosphates play a particularly important role in the binding of phosphate ions.

It has frequently been observed that phosphate ions become bonded in the soil within a very short time. It is well established that plants can use only a part of the entire phosphate content of the soil. This shows that very strong bonds develop between phosphate ions and the individual constituents of the soil. The relatively high stability of phosphates in the soil, as well as the fact that dissolved phosphates (phosphate ions) also form new, nearly insoluble phosphates, ensures that the phosphate will be inexhaustible and may even increase in certain cases. The phosphorus content that agricultural plants consume must be continuously regained for a regular cultivation of the soil; as a result, the unused phosphate content becomes gradually and permanently enriched. However, as long as human activity does not change the original condition of the soil or of vegetation in the soil, catabolic and anabolic processes are, with some exceptions, in equilibrium.

In larger areas the natural phosphate content of the soil develops relatively uniformly, according to the quality and character of the soil, the flora and fauna, climatic conditions, and a number of geographical and geological factors. Observations have revealed that the enrichment of the phosphate content of the soil in smaller areas is caused by local biological factors of generally anthropogenic origin and is attributable to the life processes of humans and animals. Human and animal organisms feed on nourishment containing phosphorus; as a result food scraps, excrement, and finally, the organisms' own decayed organic substances containing phosphorus get into the soil. The effect of phosphoric acid which develops as a result of the decomposition of these materials is that further phosphates are formed and that there is an increase, sometimes a very considerable increase, in the average phosphate content of the soil.

From the limited mobility and strong bonding of phosphate ions, it logically follows that changes and local enrichment of the phosphate content of the soil might be particularly suitable for archaeological determinations.

470

Tests for archaeological purposes, first made in 1931, have been increasingly applied for the investigation of cultural layers of diverse epochs. There is now a vast literature dealing with soil investigations for archaeological purposes. It appears from scholarly literature that this method lends itself particularly well to the task of locating traces of settlements and deserted villages, cemeteries, graves, old roads, and broken-up tumuli. It helps us to study the one-time population of former castles, to discover former residential areas, to distinguish between permanent fortresses and fortresses of refuge, to establish the ancient area of plowed fields within the greater area of land that is now cultivated, and to identify sites of early fireplaces, cesspits, and livestock pens. Further development of this method has led to the discovery of smaller details such as the placement of postholes and the original posture of corpses in graves without grave-spots.

While phosphate investigations for archaeological purposes have long been carried out in Hungary, the opportunity for large-scale research in the determination of the ancient organic content of pots has arisen only recently. In our investigations we have studied the decomposition of organic substances and the possibility of the bonding of phosphate ions and models under laboratory conditions and in various soils under natural conditions. We have found, in agreement with other scholars, that the natural phosphate content of soils as well as the possibility of phosphate-ion bonding depends upon the mineral composition and the granulometric distribution of the soil. The natural phosphate content is lower in sandy soils, and the highest in argillaceous-limy soils. In complete accordance with this fact, the mobility of phosphate ions is always greater in sandy soils than in argillaceous-limy soils. It appears that the migration and the penetration of phosphate ions could always more easily proceed straight downwards from the buried corpse than in any other direction. Both the model experiments and the tests of different kinds of soils in various graves have shown that phosphate ions of solutions bond in a very short time. The enrichment of the phosphate content of the soil could generally be established only immediately next to the buried corpse (Fig. 93).

The studies referred to above enable us to draw conclusions on the former presence of organic materials using as a basis the local enrichment of the natural phosphate content of the soil. For these investigations, therefore, a soil sample is also needed, a sample which has a natural phosphate content only (i.e. a phosphate content free from possible impurities). Because of the mixed quality of the soil filling the graves, it is best to take those soil samples on a level with the bottom of the grave, at a distance from the grave-spot.

Samples for archaeological investigations of the phosphorus content of the soil are generally taken along a straight line or at points determined by a grid. Our investigations rely on samples taken along a straight line. The frequency of sampling points was established from case to case according to necessity; in most cases samples were taken on a level with the bottom of the grave, and sometimes at different levels also. Since our aim was to clarify the extent to which the corpse influenced the phosphate content of the earth around it, the soil of the graves was nearly always investigated. It appeared that the former presence of organic materials, which supposedly decomposed without leaving a trace, could be positively detected by our investigations. In this way, the place of human or animal

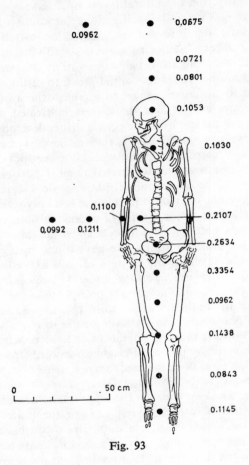

0.0962

0.0675

0.0721

0.0801

0.1053

0.1030

0.1100

0.2107

0.0992   0.1211

0.2634

0.3354

0.0962

0.1438

0.0843

0   50 cm

0.1145

Fig. 93

bodies could be determined precisely even if the remains of the bones were destroyed. Phosphate investigations could also detect traces of smaller amounts of organic compounds.

## INVESTIGATION OF EARTH IN POTS

Archaeological conclusions in the aforementioned studies and in our examinations were consistently based on changes in the natural phosphorus content of the soil. This was also true of investigation of earth in pots; the chemical analysis of part or all of the earth content of the pots has not provided any information on the function of the pots or the food they contained. Somewhat better results have been achieved by testing average samples of subsequent layers of earth inside the pots. In a number of cases phosphate determinations of soil samples taken from

472

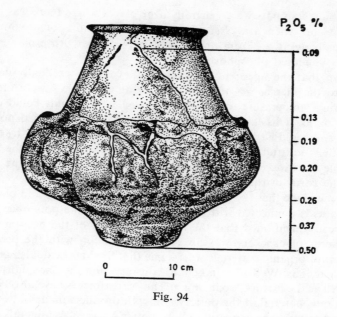

$P_2O_5$ %

- 0.09
- 0.13
- 0.19
- 0.20
- 0.26
- 0.37
- 0.50

0        10 cm

Fig. 94

the inner wall of the crockery or near it have proved most suitable. Conclusions about the pots must be made with comparative soil samples, one from the pot, and one identical to it but without contamination from the organic material presumed to have been in the pot; because the soil near to the graves is very heterogeneous, such soil samples are almost impossible to find. Therefore, the practical applicability of this method is very limited.

## INVESTIGATION OF UNGLAZED POTS

Clay used for the production of crockery is frequently procured from layers near the surface; the natural phosphate content of this kind of clay is not uniformly distributed. The phosphate content of the material from which pots have been made, however, is not unevenly distributed, because clays average out considerably in the course of being processed for use. This is even more true if the clay has been elutriated prior to processing. This has been corroborated by chemical tests both on pottery made in our own day and on ancient ceramic finds. Surprisingly, a large number of unglazed pots with enriched phosphate content have come to light in the course of excavations. In numerous cases values well above the average phosphate content of the pottery could be established, particularly in the lower parts of the pots; the phosphate content of many pots, including those found broken into fragments, increased from the brim toward the bottom (Fig. 94). Neither the original phosphorus content of the pots nor the fact that the pots had been lying underground for a long time provided a sufficient explanation for a systematic phosphate enrichment which could only have occurred before the pot was broken. Some pots which came to light showed the highest phosphate value

473

at the brim, but these pots were generally near places where there was high phosphate content and where traces survived of larger amounts of organic material. Animal bones lying at the brim or the proximity of the corpse may explain the origin of this increased phosphorus content.

It is known that the apparent porosity of the pottery gradually decreases in proportion to the increase of the temperature at which they were fired. At the estimated firing temperature of ancient unglazed pots, liquids could easily pass through the pores. This is why both the material of the unglazed pots and the surrounding soil could have been affected by the food in the pots or by the decomposition of the foodstuffs. The phosphate content of the soil in the immediate vicinity of the pots increased, sometimes considerably exceeding the phosphate content of the earth in the pots.

One must conclude that the phosphate enrichment of pottery and of soil in and around the pots is connected with the organic material the pots once contained. We have already mentioned that the local enrichment of the phosphate content of certain pieces of crockery was always in connection with the presence of a larger amount of organic material; we assume that fired clay can also permanently bond phosphate ions. We had to take into consideration the phosphorus intake of pottery as well as the factors producing it. Thus experiments have also been carried out with the raw material of the pottery and with products made out of these raw materials fired at different temperatures. In potter's clay, clay minerals which play an important role in the bonding of phosphate ions are represented by a number of mineral groups. Experiments have been carried out with relatively uniform raw materials containing no more than one clay-mineral group. Accordingly, in the first phase of our series of experiments, almost pure kaolinite and illite and bentonite of high montmorillonite content were used. Subsequently, experiments were carried out with potter's clay representing several characteristic types. The aim was to reproduce under laboratory conditions phosphorus enrichment in the material of pots; we used products made out of the above-mentioned raw materials and tested at different temperatures because we believed that phosphorus enrichment came about under natural conditions as a result of the decomposition of various organic materials and remained intact while the pots were buried in the ground.

Since phosphoric acid is an important factor in the development of minerals containing phosphorus and phosphates of the soil, it could rightly be assumed to play a decisive role in the processes which increase the phosphorus content of pottery in contact with organic substances. In order to study the bonding of phosphate ions in the pottery, we produced the phosphorus intake of ceramics directly out of the phosphoric acid solution. We also produced it from an organic compound from which phosphoric acid develops under experimental conditions. Theoretical considerations suggested adenosine triphosphate (ATP) for the latter purpose, since it occurs in relatively large quantities in various tissues.

The experiments have proved that a considerable phosphorus intake can occur both from phosphoric acid and from the aqueous solution of ATP. However, the extent of phosphorus intake was not proportionate to the decrease of porosity and the changes in surface which result from increased firing temperatures. It appeared that phosphorus intake can be plotted by a maximum curve whose peak values fall within the range of 600–800° Celsius, with certain variation from clay to clay. The degree of the phosphorus intake is much influenced by the mineral composition of the clay, firing time, temperature, and the concentration of phosphate ions in the solution. Our experiments have proved that pottery can permanently bond phosphate ions under determined conditions; therefore, from the phosphate enrichment of ancient pots, we can draw conclusions about their function and the food they contained.

## LITERATURE

O. ARRHENIUS, "Bodenanalyse in der Archaeologie". *Zeitschrift für Pflanzennährung und Bodenkunde* (1931): 427–39.

O. ARRHENIUS, "Markundersöking och Arkeologie". *Fornvänen* 30 (1935): 565–76.

O. ARRHENIUS, "Chemical Denudation in Sweden". *Tellus* 6 (1954): 326–41.

C. A. BLACK, "Phosphate Fixation by Kaolinite and other Clays as Affected by Phosphate Concentration and Time of Contact". *Soil Science Society of America* 7 (1942): 123–33.

R. D. BLECK, "Zur Durchführung der Phosphatmethode". *Ausgrabungen und Funde* 10 (1965): 213–18.

R. COLEMAN, "The Mechanism of Phosphate Fixation by Montmorillonitic and Kaolinitic Clays". *Soil Science Society of America* 9 (1944): 72–78.

S. F. COOK and R. F. HEIZER, "Studies on the Chemical Analysis of Archaeological Sites". *University of California Publications in Anthropology* (1965), 2:1–102.

I. W. CORNWALL, *Soils for the Archaeologist*. London, 1958.

GY. DUMA, "Égetett agyagok foszforfelvétele" (Phosphorus Intake of Burnt Clays). *Építőanyag* 20 (1968): 450–56.

GY. DUMA, "Methode zum Feststellen der Bestimmung von urzeitlichen Gefässen". *ActaArchHung* (1969): 359–72.

F. FELGENHAUER and F. SAUTER, "Phosphatanalytische Untersuchungen an paläolitischen Kulturgeschichten in Willendorf i.d. Wachau". *N. Ö. Archaeologica Austriaca* 12 (1959): 25–34.

E. FRAUENDORF and W. LORCH, "Einfache Bodenuntersuchung im Dienste der Vorzeitforschung". *Nachrichtenblatt für Deutsche Vorzeit* (1940), pp. 265–68.

H. GEIGER, "Bodenuntersuchung und Archäologie". *Die Naturwissenschaften* 24 (1936): 608.

P. GRIMM, "Phosphatuntersuchungen zur Besiedlung der Pfalz Tilleda". *Ausgrabungen und Funde* 7 (1962): 8–12.

U. W. GUYAN, "Beitrag zur topographischen Lokalisation einer Wüstung mit der Laktatmethode am Beispiel von Morgen". *Geographica Helvetica* 7 (1952): 1–9.

N. S. HALL and A. J. MACKENZIE, "Measurement of Radioactive Phosphorus". *Soil Science Society of America* 12 (1947): 100–6.

S. B. HENDRIKS and L. A. DEAN, "Basic Concepts of Soil and Fertilizer Studies with Radioactive Phosphorus". *Soil Science Society of America* 12 (1947): 98–100.

H. JAKOB, "Die Bedeutung der Phosphatmethode für die Urgeschichte und Bodenforschung". *Beiträge zur Frühgeschichte der Landwirtschaft* 2 (1955): 67–84.

H. JAKOB, "Über Ursachen anomaler Phosphatanreicherung auf Grabhügel". *Die Kunde* (1952), pp. 37–40.

H. JAKOB, "Wüstungsforschung und Phosphatmethode I.—II.". *Fränkische Blätter für Geschichtsforschung und Heimatpflege* (1951), pp. 97–104.

H. JAKOB, "Zur Gebrauchsbestimmung von Grabhügelkeramik". *Forschungen und Fortschritte* 30 (1954): 10–12.

K. JASMUND, *Die Silikatischen Tonminerale*. Weinheim, 1951, pp. 47–49.

I. LENGYEL and J. NEMESKÉRI, "A csontvázleletek dekompozíciójáról" (On the Decomposition of Skeletal Finds). *Anthropologiai Közlemények* 8 (1964): 69–82.

W. Lorch, "Burgenforschungen mit der Phosphatmethode". *Die Umschau in Wissenschaft und Technik* 47 (1943): 159.

W. Lorch, "Chemische Spuren im Boden als Zeichen früherer menschlicher Besiedlung". *Die Umschau in Wissenschaft und Technik* 45 (1941): 116–20.

W. Lorch, "Die Entnahme von Bodenproben und ihre Einsendung zur Untersuchung mittels der siedlungsgeschichtlichen Phosphatmethode". *Die Kunde* (1951), pp. 21–23.

W. Lorch, "Ergebnisse der Untersuchung würtenbergischer Burgberge mittels Phosphatmethode". *Die Naturwissenschaften* 32 (1944): 99–100.

W. Lorch, "Das Erkennen des Relickt-Charakters von Waldkapellen aus dem Phosphatgehalt des Bodens". *Zeitschrift für würtenbergische Landesgeschichte* 62 (1952): 246–53.

W. Lorch, *Methodische Untersuchungen zur Wüstungsforschung*. Jena, 1939.

W. Lorch, "Neue Methoden der Siedlungsgeschichte". *Geographische Zeitschrift* 35 (1939): 294–305.

W. Lorch, "Die Phosphatmethode im Dienst der Vorgeschichtsforschung". *German en Erbe* (1949) pp. 55–59, 90–95.

W. Lorch, "Die Siedlungsgeographische Phosphatmethode". *Die Naturwissenschaften* 28 (1940): 633–40.

C. D. McAuliffe, "Exchange Reactions between Phosphates and Soils: Hydroxylic Surfaces of Soil Minerals". *Soil Science Society of America* 12 (1947): 119–23.

F. Sauter, "Phosphatanalytische Untersuchungen an dem Spät Latène-Haus aus Oberbergen, p. B. Krems N.O.". *Archaeologica Austriaca* 12 (1959): 107–10.

F. Sauter, "Phosphatanalytische Untersuchungen von Erdproben aus einem avarischen Grab". *Archaeologica Austriaca* 20 (1967): 41–43.

E. Schmid, "Höhlenforschung und Sedimentanalyse". *Schriften des Institutes für Ur- und Frühgeschichte der Schweiz* (1958), pp. 21–36.

R. S. Solecki, "Notes on Soil Analysis and Archeology". *American Antiquity* (1951): 254–56.

W. Specht, "Chemische Abbaureaktionen bei der Leichenzersetzung III.". *Ergebnisse der Allgemeinen Pathologie* (1937), pp. 138–80.

P. R. Stout, "Alternations in Crystal Structure of Clay Minerals as a Result of Phosphate Fixation". *Soil Science Society of America* 4 (1939): 177–82.

K. Stoyo, "Die Anwendung der Phosphatmethode auf einem mittelalterlichem Friedhof". *Jahresbericht für mitteldeutsche Vorgeschichte* (1950), pp. 180–4.

K. Stoyo, "Die Phosphatmethode in ihrer Anwendung auf die Grabung bei Wahlitz (Kreis Burg)". *Beiträge zur Frühgeschichte der Landwirtschaft* 2 (1955): 87–91.

I. Utescher, *Das erdige phosphathaltige Sediment in der Ilsenhöhle von Ranin*. Berlin, 1948, pp. 1–15.

A. Uzsoki, *Előzetes jelentés a Mosonszentmiklós— Jánosházpusztai bronzkori temető ásatásának eredményeiről* (Preliminary Report on Excavations of Bronze Age Cemetery at Mosonszentmiklós—Jánosház-puszta). *Arrabona* (1959): 53–70.

K. Walcher, "Studium über die Leichenfäulnis, mit besonderer Berücksichtigung de Histologie derselben". *Wirchows Archiv*, no. 268 (1928), p. 17.

# CHRONOLOGICAL TABLE

## ALFONZ LENGYEL

| | |
|---|---|
| 4th century B.C. | — Celtic group reached the western part of Pannonia. |
| 4th century B.C. | — Boii nation was formed from Celtic tribes which invaded Italy at the beginning of the fourth century. After their defeat they left Italy and settled near the Danube River. During the fourth century they became the leading power in the northern part of Pannonia, while the Scordisci were in the south. |
| 3d century B.C. | — New Celtic group plundered the Balkans and entered Thrace and Greece. |
| End of 3d century B.C. | — Scordisci subdued the Pannonian tribes and expanded their territory to Southern Pannonia and Northern Serbia. |
| End of 2d century B.C. | — A new wave of Celts settled in the southern part of Slovakia and Northern Pannonia. |

*Note*:

- Posidonius stated that this area was ruled by the Boii.
- Strabo noted that the eastern border of the territory of the Boii reached the Tibiscus (Tisza) River.
- The capital of the Boiian settlement was around the present Bratislava (Pozsony).
- The Boii had territorial conflict with the Dacians (Strabo).

| | |
|---|---|
| 181 B.C. | — The Romans established Aquileia as a base for their eastward expansion. |
| 179 B.C. | — Scordisci occupied the whole Sava valley. |
| 168 B.C. | — The beginnings of minting of Celtic coins in Pannonia. |
| 156 B.C. | — Scordisci with the alliance of the Dalmatae came into conflict with the Romans. (Siege of Siscia.) |
| 141 B.C. | — Scordisci entered Macedonia. |
| 119 B.C. | — Scordisci fought again in the Dalmatian Confederation against the Romans. (Siege of Siscia.) |

*Note*:

- The territory between the Drava and Sava rivers was the strategically most important passage to Dalmatia and the East.

| | |
|---|---|
| 88 B.C. | — Scipio Asiagenus defeated Scordisci and liberated the Pannonian tribes. The Scordisci retreated to the eastern part. |
| 76 B.C. | — Cn. Scribonius Curio from the south reached the Danube. |
| c. 45 B.C. | — Norican expansion in Northwestern Pannonia. |
| Mid-1st century B.C. | — Under King Burebista, Dacians became a potentially strong nation. They engaged war against Boii, Taurisci, and Scordisci, and they completely destroyed all of them. |
| | — After the death of Burebista the Dacian state disintegrated, but was still considered an important enemy of the Romans. |
| 44 B.C. | — Boii were defeated by the Dacians. |

| | |
|---|---|
| 35–33 B.C. | — Octavian's Illyrian expedition. |
| 35 B.C. | — Octavian attacked the western part of the Balkans with two armies. (Iapodes and Siscia occupied.) |
| | — Soon the whole Sava-valley and the Pannonians came under Roman control. |
| 34 B.C. | — Conquest of Dalmatia began. |
| | — Emergence of Pannonia as a territorial unit from the Sava River to the Danube border (Boii, Taurisci, Scordisci, Breuci, Amantini and Andizetes dwelled in loose tribal relation there). |

*Note:*

| | |
|---|---|
| c. 20 B.C. | — Flourishing time of Gellérthegy-Tabán settlement. |
| | Other Pannonian tribes dwelled in Dalmatia, Herzegovina, and Bosnia. |
| 16 B.C. | — Pannonian and Norican tribes attacked the Romans on the Istria peninsula from Croatia and Bosnia. |
| 15 B.C. | — Tiberius (before he became emperor) defeated the Scordisci and made them Roman allies. |
| 14 B.C. | — First Dalmatian–Pannonian revolt. |
| 13–9 B.C. | — Tiberius quelled the revolt. Dalmatia submitted to the Romans. |
| 13 B.C. | — Augustus sent Agrippa against the Pannonians. |
| 12 B.C. | — New outbreak of the Pannonian revolution. |
| 10 B.C. | — Dacian invasion of southeastern Pannonia; Tiberius crushed the attack. |
| 9 B.C. | — Occupation of a part of the later Pannonia between the Drava and Sava rivers completed. |

*Note:*

In Monumentum Ancyranum is written the following: "The Emperor proclaims that the border of Illyricum was extended to the Danube."

| | |
|---|---|
| c. 8 B.C. | — Creation of Maroboduus' Marcomannian kingdom. |
| 8 B.C. | — M. Vinicius campaign to the eastern neighbors of the Marcomanns including the Dacians. |
| 6 A.D. | — Cn. Cornelius Lentulus reached with his army the Iron Gate of the Danube. |
| |     cf. Dio Cassius, LI. 23.2. and LIV. 36.2. |
| |     Tacitus: *Annales*, IV. 44. |
| | — Augustus tried to block the Marcomanns and moved with one army from the Rhein River and with another through Carnuntum. |

*Note:*

The campaign remained unfinished, due to the second Dalmatian and Pannonian revolt (Pannonians, the Amantini and Breuci under the leadership of Bato, and the Daesidiates under Pinnes).

| | |
|---|---|
| 6–9 A.D. | — Tiberius defeated the revolt. |
| |     cf. Dio Cassius, LV and LVI. |
| |     Velleius Paterculus: II, pp. 110–126. |

*Note:*

Velleius Paterculus was an eyewitness of the war; served as historian on Tiberius' staff.
— 200,000 infantry and 9,000 cavalry fought against the Romans.
— One rebellious army moved against Sirmium.
— The army defeated by local military governor.

|            |                                                                                            |
|------------|--------------------------------------------------------------------------------------------|
|            | — Sirmium liberated by Caecina Severus.                                                     |
|            | — Tiberius destroyed the harvest around Siscia and then invaded the city.                   |
| 8 A.D.     | — The Daesidiates incited the revolution again, despite the capture of their leader Pinnes. They moved against the Breucians and killed Bato. |
|            | — Tiberius finally defeated the Daesidiates in the Dalmatian mountain. (Pinnes was sent to exile in Ravenna.) |
|            | — Capitulation of the rebellious Pannonians.                                               |
|            | — Illyricum Superius became Dalmatia.                                                       |
|            | — Illyricum Inferius became Illyricum.                                                      |

*Note*:

— The Roman foreign policy shifted to this area. They interfered in the hostilities between the Marcomannian Maroboduus and the Germanic Ariminius. In the conspiracy against Maroboduus, Romans supported his enemy Cataulda. When the Marcomannians expelled Cataulda Drusus nominated Vannius from the tribe of Quadi as leader of the Marcomannians.

— Three legions were under Augustus in Pannonia:
in Emona, Legio XV Apollinaris
in Poetovio, Legio VIII Augusta
possibly in Siscia, Legio IX Hispana

| 14 A.D.       | — Augustus died (August 19).                                                             |
|---------------|------------------------------------------------------------------------------------------|
| 14–37 A.D.    | — Reign of Tiberius (A.D. 14–37).                                                         |
| 14 A.D.       | — Revolt of the legions in Pannonia was suppressed by Drusus.                            |
| c. 14 A.D.    | — Veteran settlements in western part: Emona, Scarbantia, and around the Lake Balaton.   |
|               | — Emona became *colonia*.                                                                |
| 17–20 A.D.    | — Drusus remained in Pannonia.                                                           |
| 19 A.D.       | — Break-up of Maroboduus kingdom, creation of Regnum Vannianum.                          |
| 41–45 A.D.    | — Reign of Claudius.                                                                      |
|               | — Construction of the *limes* began.                                                     |
| 43–45 A.D.    | — Legio IX Hispana left Pannonia permanently.                                            |
|               | — Legio VIII Augusta was replaced by Legio XIII Gemina in Poetovio.                      |
| 50 A.D.       | — Vannius was expelled and he and his followers settled in Pannonia.                     |
|               | — The successor of Vannius: Vangio and Sido became subjects of Rome.                     |
| c. 50 A.D.    | — Savaria became *colonia*.                                                              |
| 54–68 A.D.    | — Under Governor Tempius Flavius 50,000 Barbarians were settled in Pannonia.             |
| 62 A.D.       | — Legio XV Apollinaris left for the East.                                                |
| 62–68 A.D.    | — Legio X Gemina stayed in Carnuntum.                                                    |
| 69–73 A.D.    | — Reign of Vespasian (Vespasianus).                                                      |
|               | — At the beginning of the reign of Vespasian the former Illyrian Danubian province officially became Pannonia. |
|               | — Under Vespasian two legions stationed in Pannonia: Legio XIII Gemina in Poetovio and Legio XV Apollinaris in Carnuntum. |
| 69–79 A.D.    | — During the reign of Vespasian Pannonia became fully Romanized.                         |
|               | — The Flavians established *municipia* and *coloniae* in Pannonia:                       |
|               | Municipium Flavium Scarbantia                                                             |
|               | Municipium Flavium Latobicorum (Municipium Flavium Neviodunum) |
|               | Municipium Flavium Andautonia                                                             |

|  |  |
|---|---|
|  | Colonia Flavia Siscia |
|  | Colonia Flavia Sirmium. |
| 81–96 A.D. | — Extended work on the Danubian *limes* began. |
|  | — Due to the hostile attitude of the Dacians, Domitian established a permanent *limes*. |

*Note*:

— 81. At the lower Danube area the Roman army was attacked by Dacians.

— 85–86. The governor of Moesia became the victim of Dacian attack.

— 88. Domitian gained a victory and made peace with the Dacians.

At the same time the Suebian and Sarmatian tribes menaced Pannonia. The Sarmatians destroyed the Legio XXI Rapax.

| 89–93 A.D. | — Suebian–Sarmatian war in Pannonia. |
|---|---|
|  | — Domitian (Domicianus) spent 8 months in Pannonia (92 A.D.). |
|  | — The Sarmatian menace resulted in building an addition to Vetus Salina, the military camp of Lussonium (Dunakömlőd): Lugio (Dunaszekcső); Ad Militare (Kiskőszeg-Batina) on the Sarmatian front. |
|  | — On the German frontier were established the following camps for auxiliary units: Cannabianca (Klosterneuburg); Vindobona (Vienna); Ad Flexum (Magyaróvár); Arrabona (Győr); and Brigetio (Szőny). |
| 98–117 A.D. | — Trajan (Trajanus) continued the fortifications along the *limes* and particularly prepared Pannonia's resources for his move against Dacia. |
|  | — 100–102 and 105–106. Trajan led campaign against Dacia and gained a complete victory. |
|  | — According to the latest research, the Romans maintained routes between Pannonia and Dacia through the Sarmatian territory. (One route crossed Aquincum, another Lugio.) Regular postal service had been conducted also on those roads. |
| c. 106 A.D. | — Division of Pannonia: |
|  | Pannonia Superior (Carnuntum) |
|  | Pannonia Inferior (Aquincum) |
| 107 A.D. | — Sarmatians attacked Pannonia. (Aelius Hadrianus, the later emperor, defeated the Sarmatians.) |
| 117 A.D. | — Iazyges joined with the Roxolani and attacked Pannonia. |

*Note*:

— Under Trajan (Trajanus) the legions' assignments in Pannonia became final.

— Legio XIV Gemina, then Legio X Gemina in Vindobona.

— Legio XV Apollinaris, then Legio XIV Gemina in Carnuntum.

— Legio XI Claudia, then Legio I Adiutrix in Brigetio.

— Legio II Adiutrix in Aquincum.

| 117–138 A.D. | — Hadrianic era. |
|---|---|
| 117–124 A.D. | — Hadrian developed the Pannonian *limes* into its final form. |

*Note*:

A chain of watchtowers was set.

| 118 A.D. | — Hadrian defeated the Iazyges with Governor Marcius Turbo. |
|---|---|

*Note*:

Temporarily the military leadership of Pannonia and Dacia was unified during this war.
- Hadrian finalized the number of troops and their distribution in Pannonia.
- *Upper Pannonia:* Three legions (Vindobona, Carnuntum, and Brigetio).
- *Lower Pannonia:* One legion in Aquincum.

| | |
|---|---|
| 124 A.D. | — Visit of Hadrian (Hadrianus) in Pannonia. |
| | — Hadrian established *municipia* in Pannonia: Carnuntum, Aquincum, Mogentiana, Cibalae, Mursella, Sala, Halicanum. |
| 138 A.D. | — Campaign against the Quadi by the governor of Pannonia Superior, Titus Haterius Nepos. |
| 138–161 A.D. | — During the reign of Antoninus Pius peace was restored and maintained. |
| | — 140–144. On a coin of Pius was written: Rex Quadis Datus — A king was given to the Quadi. |
| 166–175 A.D. | — The Marcomannic war. |
| | — The great devastation of Pannonia during the reign of Marcus Aurelius. |
| | — 164–165. Eleven nations under the leadership of Quadi and Marcomanni attacked the *limes* at Upper Pannonia. |
| | — 166. Iallius Bassus, who got his military war training in the Parthian war, became the governor of Upper Pannonia. |
| | — 167. The Longobards and Obii established the first bridgehead between Brigetio and Arrabona. |

*Note*:

- Dio Cassius mentioned that the Marcomanni asked for peace after this defeat.
- 168. The praetorian Prefect Furius Victorinus suffered defea (The Barbarian tribes deeply penetrated into Pannonia.)

| | |
|---|---|
| 168 A.D. | — Plague was at peak in Italia. Aquileia besieged, Opitergium (Oderzo) burned. |
| 169 A.D. | — Lucius Verus died. |
| 172–173 A.D. | — The Roman army defeated the Marcomanni and Quadi. |
| 173 A.D. | — Battle shifted to the Iazyges front. |
| | — Marcus Aurelius moved his headquarters from Carnuntum to Sirmium. |
| 173–174 A.D. | — (Winter) Iazyges defeated, Banadaspos offered peace but the Romans refused. |
| | — The front moved against the Quadi. (Quadi and Iazyges became separated.) |
| 175 A.D. | — Roman offensive against the Iazyges. |
| | — Avidius Cassius, the Syrian governor, proclaimed himself emperor for the Romans. This fact forced the Romans to sign a peace treaty with the Iazyges. An area of 15 kilometers along the border was established as off limits (the first no man's-land idea). |
| 176 A.D. | — Marcus Aurelius defeated Avidius Cassius in Syria. |
| 177 A.D. | — The Sarmatian war broke out again. |
| 178 A.D. | — Marcus Aurelius defeated the Sarmatians (Sarmatia became a "client state"; the off-limits area was eliminated). He wrote his "Meditations" in the Garam valley. |

481

| | |
|---|---|
| | — 20,000 Romans stationed on the territory of Quadi and Marcomanni in Laugericium (Trenčin). |
| 180 A.D. | — The death of Marcus Aurelius ended the Roman occupation of Quadi and Marcomanni. (Client relationship reestablished.) |
| 180–190 A.D. | — Commodus rebuilt the *limes*. |
| 184–185 A.D. | — Watchtowers were built abong the Danube. |
| 192–193 A.D. | — Political confusion: P. Helvius Pertinax proclaimed as emperor (193) but he was soon assassinated. |
| 193–235 A.D. | — The Severan period. |

*Note*:

— After the Marcomannian war the Quadi and Iazyges were no longer Rome's main enemies. The new threat came from the new migrating people beyond the Carpathian basin.
— Great economic prosperity in Pannonia.

| | |
|---|---|
| 193 A.D. | — Septimius Severus, governor of Pannonia Superior, became emperor. |
| 194 A.D. | — Carnuntum and Aquincum became *coloniae*. During the reign of Septimius Severus Cibalae became *colonia*. |
| 196–203 A.D | — Sporadic fighting with Barbarians along the *limes* of the Danube. |
| | — S. Severus visited Pannonia in 202 A.D. |
| | — Upswing of the economic situation in Aquincum, Brigetio, Carnuntum, Siscia, and the military cities along the Danubian *limes*. |
| 212 A.D. | — Northern Dacia's *limes* was overrun by Carpi and Vandal tribes. The Pannonian *limes* also was attacked but the attack was repelled and the damage repaired. |
| 212–217 A.D. | — Reign of Caracalla. |
| | — During the reign of Caracalla Bassiana became a *colonia*. |
| 214 A.D. | — Brigetio was attached to Lower Pannonia. Bridgeheads on the left bank of the Danube between Carnuntum and Aquincum were reestablished or additionally built. (Dévény became the counterfortification of Carnuntum, Leányvár of Brigetio.) |
| | — Caracalla visited Pannonia. |
| 219 A.D. | — Elagabalus visited Pannonia. |

*Note*:

Archaeological findings demonstrate the richness of the Severan period.

| | |
|---|---|
| 236–238 A.D. | — Maximinus Thrax prepared in Sirmium his operation against the Barbarians. |
| 238 A.D. | — The Senate proclaimed D. Caelinus Balbinus and M. Clodius Pupienus as emperors. Maximinus moved against them, but his troops assassinated him. |
| 240 A.D. | — Fights along the *limes* during the reign of Gordian III. |
| 244 A.D. | — Goths put strong pressure on Pannonia and Dacia. |
| 247 A.D. | — A group of Dacian descent moved into the devastated area of Dacia (through the eastern Carpathian passage). |
| 249 A.D. | — Decius who was born in Sirmium became emperor. |
| 250–251 A.D. | — Decius fought against the Goths, but he was defeated and killed. |
| 250–253 A.D. | — Trebonianus Gallus (governor of Moesia) proclaimed himself emperor. |
| 258 A.D. | — Governor of Lower Pannonia Ingenuus defeated the Sarmatians and proclaimed himself emperor. Gallienus, the son of Valerian, came to Pannonia and defeated (killed) Ingenuus near Mursa. |

The Moesian army and Upper Pannonia backed Regalianus as counter-emperor. Lower Pannonia remained faithful to Gallienus.

| | |
|---|---|
| 260 A.D. | — Gallienus renewed the foederatus relationship with the Roxolani and forced the Sarmatians to leave the area of the Lower Pannonian *limes*. |
| 262 A.D. | — Gallienus married the daughter of the Marcomannian king and settled them in Pannonia Superior. |
| | — A mint in Siscia was established. |
| 268 A.D. | — Claudius II (Gothicus) fought with the Goths at the Lower Danube area. The Pannonian *limes* was quiet during his reign. |
| 269 A.D. | — Claudius II defeated the Alamanni and won a victory twice over the Goths. |
| 270 A.D. | — Aurelian (Aurelianus) defeated the Vandals, Gepids, and other German tribes along the Danubian border. (Vandals allied with the Suebi and Sarmatians. Then the Vandals and Gepids attacked Pannonia and destroyed the settlements). |
| 271 A.D. | — Aurelian abandoned Dacia. |
| 275 A.D. | — Probus, an Illyrian emperor who was born in Sirmium, defeated the Vandals at the Danube. |
| 282 A.D. | — Carus defeated the Sarmatians. |
| 284 A.D. | — Numerian (Numerianus) defeated the Quadi. |
| 284–305 A.D. | — Reign of Diocletian (Diocletianus). |
| 289–290 A.D. | — The first Sarmatian war of Diocletian. |
| 290–291 A.D. | — Diocletian stayed in Sirmium. War against the Goths and Vandals. |
| 292 A.D. | — Galerius I's war against the Sarmatians. |
| 293 A.D. | — Diocletian established the Tetrarchy. Sirmium became the emperor's seat. |
| 294 A.D. | — Galerius Caesar carried out Diocletian's military, financial, and economic reforms in Pannonia. |
| | — Pannonia and Noricum were administered with the capital in Sirmium. |
| | — Counterfortification of Bononia was rebuilt. |
| | — Modernization of the *limes* followed by the reorganization of the territories behind the *limes*. |
| 295 A.D. | — Carpi were settled in the area of Sopianae. |
| | — Pannonia divided into four parts: Pannonia Superior divided into: Pannonia Prima and Savia; Pannonia Inferior divided into: Pannonia Secunda and Valeria. |
| 297 A.D. | — Pannonian legions were sent against the Persians. |
| 299 A.D. | — Galerius led expedition against the Sarmatians. |
| 299–311 A.D. | — Galerius acquired the nickname Sarmaticus after he had gained a victory three more times over the Sarmatians. |
| 308 A.D. | — At the Imperial conference at Carnuntum in 308, the new Augustus, Licinius, received the leadership over Illyricum and Pannonia. |
| 313–314 A.D. | — Licinius and Constantine fought against the Sarmatians. (Both received the name Sarmaticus.) |

*Note*:

— October 314, near the Pannonian Cibalae (Vinkovce) the Gallii army under Constantine defeated Licinius. Licinius escaped to Sirmium, but Constantine pursued him.

| | |
|---|---|
| | — Constantine kept a permanent residence in Sirmium. |
| 320 A.D. | — Attacks again in the center and lower Danube area. |
| | — Establishment of the Sirmium mint. |
| 322 A.D. | — Constantine enlarged the Pannonian army and fought successfully against invaders. (He defeated and killed the Sarmatian king Rausimodus who attacked Campona south of Aquincum.) |
| | — Constantine returned to Sirmium. A series of gold coins were minted with inscription: *Sarmatia Devicta*. |
| 324 A.D. | — Sarmatians helped the army of Licinius who was in alliance with the Goths against Constantine. (Licinius was killed in this war.) |
| 324–337 A.D. | — Constantine became the sole ruler of the Roman Empire. |
| | — Constantine rebuilt the *limes* in the middle and lower Danube regions. (Major works were undertaken at the lower Danube area.) |
| | — Another defensive line was built (wall-system) in the present Great Hungarian Plain. This system, started at the Danube bend, returned to the Danube at Viminacium. |
| | — The Oltenian wall system was built. This system at Drobaeta (Turnu Severin) turned away from the Danubian *limes* and by creating an additional defensive line outside the said *limes*, joined the Danubian *limes* south of Braila. (Erection of fortresses, expanded town buildings, economic stabilization.) |
| 313–337 A.D. | — Sirmium became the center of serious Christian dogmatical conflicts. |
| 332 A.D. | — Goths attacked the Roxolani and the Sarmatians |

*Note*:

| | |
|---|---|
| | — Sarmatians with the help of the Romans repulsed the Goths. |
| 337 A.D. | — The empire (western part) was divided between Constantine II (337–340 A.D.) and Constans I (337–340), and Constantius II (337–350) ruled in the east. |
| 338–340 A.D. | — Constans defeated the Sarmatians three times. |
| 350 A.D. | — Ventramion proclaimed himself emperor in Sirmium but was defeated by Constantius II (351 A.D.). |
| 351 A.D. | — The usurper Magnius Magnentius took Sirmium but Constantius I defeated him at Mursa. |
| 357 A.D. | — Constantius II came to Sirmium to prepare for war against the joint forces of Quadi and Sarmatians. |
| 358 A.D. | — Victory of Constantius II over the Quadi and Sarmatians. |
| | — The Sarmatian king Zizais became the vassal of Rome. |
| | — The Quadi under Vitrodorus were defeated. |
| | — The Limigantes (Servi Sarmatae) invaded Moesia. |
| | — The Roman-vassal king Zizais and the Taifals forces defeated the former Sarmatian slaves. |
| | — The expelled Sarmatians returned. |
| 359 A.D. | — Constantius II proclaimed Arianism as state religion in Sirmium. |
| | — Constantius II visited Valeria. |

*Note*:

| | |
|---|---|
| | — Limigantes endangered the *limes* in Valeria. Constantius II moved against them. They sent a peace mission to Constantius, but this mission plotted against the Romans. The Roman army massacred them. |
| 361–363 A.D. | — Iulian became emperor and made his triumphal entry into Sirmium. |
| 364 A.D. | — Valentinian (364–367 A.D.) and Valens (364–378 A.D.) divided the empire between them. |

| | |
|---|---|
| 365 A.D. | — Valentinian and Valens met in Sirmium to prepare war against the Barbarians. |
| 366–370 A.D. | — Large-scale building program along the Danubian *limes* was financed by Valentinian. |
| 371–373 A.D. | — A chain of watchtowers completed the line of fortification. The concentration on fortification was between Aquincum and Brigetio. |
| 374 A.D. | — The Quadi objected to the Roman fortification on their territory. The king of Quadi was invited to Aquincum and assassinated. |
| 375 A.D. | — The expedition of Valentinian against the Quadi started out from Trier. |
| | — Merobaudes crossed the Danube at Carnuntum. |
| | — The emperor attacked the Quadi north of Aquincum at Nógrádverőce. Then he returned to Aquincum and soon established headquarters in Savaria. |
| | — The emperor then ordered the fortification between Arrabona and Brigetio and moved with his headquarters to Brigetio. |
| | — November 17. Emperor Valentinian died of stroke. |
| | — After the death of the emperor the expedition was abandoned. |
| 375–378 A.D. | — Emperor Valentinian II resided in Sirmium. |
| 378 A.D. | — Emperor Gratian (Gratianus) visited Pannonia. (Huns appeared.) Battle at Hadrianopolis. |
| | — Theodosius became Augustus of East and Illyricum. |
| 379 A.D. | — Theodosius was proclaimed emperor in Sirmium. |
| | — The regent of Vithimir's son Vitheric, the Ostrogoth Alatheus and the Alan Saphrac, who became the general of the Hunnish–Alan army, withdrew to the west. |
| | — This group attacked Valeria and Vitalianus failed to repulse them. (Amantius, the bishop of Iova, tried to convert them to Christianity.) |
| 379 A.D. | — Theodosius by regulating the unrest of the Foederati made extensive damages along Mursa and Stridon. |
| 387 A.D. | — The army of Maximus moved toward Pannonia Prima and Secunda. |
| | — In the battle of Siscia the Hunnish Foederati aided Theodosius, while in the siege of Poetovio the Eastern Goths and Alans fought within the loyal army units. |
| 395 A.D. | — The Foederati was again used to aid the Roman army, this time against usurper Eugenius. |
| | — The Foederati troops revolted. |
| | — The empire was again divided. Sirmium became part of Illyricum Occidentale. |
| 395–399 A.D. | — Under Arcadius and Honorius, the Romans had many difficulties with the Foederati. |
| | — Campaign from Aquincum. |
| 401 A.D. | — Invasion of Alaric passed through the western border of Pannonia. (Possibly they were helped by the Foederati of the Goths.) |
| 402 A.D. | — The people of Alaric settled in southern Pannonia. |
| | — Some original settlers started to move to Italia. |
| 405 A.D. | — Alaric invaded Italia. |
| 406 A.D. | — Radagaesus attacked Italia. |
| 410 A.D. | — Peace was restored in Pannonia. |
| 425–427 A.D. | — Rua requested to dissolve the Foederati status of the Hunnish–Alan–Gothic group. |
| 433 A.D. | — Rua made alliance with the Romans and the Great Huns entered Pannonia as Foederati. |

| | |
|---|---|
| | 416–425 A.D. The expansion of the United Huns started from the Don River. |
| | 427 A.D.      Rua and Uptar became leaders of the Huns. Pannonia Secunda and Savia under Hunnish rule. |
| | 433 A.D.      Pannonia Prima attached to the territory of the Huns. After the death of Rua, Attila became the leader of the Huns. The Western Roman Empire gave him the title of Magister Militum and the Eastern Roman Empire the title of Basileos. |
| | Despite this honor Attila attacked the Romans. |

451 A.D.      — Battle of Catalaunum (ended without a victory).

452 A.D.      — Attila stopped before Rome. (He took Milan, Aquileia, Pavia.)

454 A.D.      — Attila died.
         — After the death of Attila, the king of the Gepids, Ardarich defeated the Huns near Nedao.
         The Huns retreated to the area of the Black Sea.

455 A.D.      — Emperor Avitus visited Pannonia. (This proves that the Romans did not abandon Pannonia until the end of the Western Roman Empire.)
         — With the retreat of the Huns Germanic tribes had settled in Pannonia.
           — Amal Goths with the leadership of Valamer, Vidimer, and Thiudimer settled in Pannonia (Transdanubia).

457 A.D.      — Valamer conquered Slavonia.

*Note*:

     — The interest of the Goths turned westward and they expanded into Noricum.
     — Between Sirmium and Singidunum the Goths fought with the Byzantines against the Huns.
     — The Skirs were defeated and Hunimund gathered all Germanic tribes against the Goths.

469 A.D.      — Thiudimer united the Goths and was victorious over the Germans.

471 A.D.      — Thiudimer attacked the Eastern Roman Empire. He advanced up to Naissus, where he died. His son Theoderich (Theodoric) became the leader of the Goths (Ostrogoths).

476 A.D.      — Theodoric supported Emperor Zeno.

488 A.D.      — Gothic invasion started against Odoacer.

493 A.D.      — Theodoric occupied Ravenna (Eastern Gothic Empire).

504 A.D.      — Sirmium fell into Gothic hands from the Gepids. (Pannonia Secunda and Savia became part of the Eastern Gothic Empire. Soon north of the Drava River, Pannonia came under Gothic control.)

510 A.D.      — Pannonia was divided between Ostrogoths and Byzantines. (Sirmium remained under Theodoric's rule.)

526 A.D.      — Death of Theodoric.

527 A.D.      — The Longobards under Vacho's leadership arrived in Pannonia.

536 A.D.      — Gepids reoccupied Sirmium.

*Note*:

     — Iustinian in A.D. 535 liberated Sirmium from the Ostrogoths. The Gepids made a kingdom over the ruins of Sirmium.

539 A.D      — Longobards appeared in the area of Pannonia Prima and Secunda.

| | |
|---|---|
| 547 or 548 A.D. | — After the death of Vacho, Auduin became the leader of the Longobards. |
| 567 A.D. | — Iustinian aided the Gepids against the Longobards. |
| | — The king of the Longobards allied with the Bajans (Baians) and gained a victory over the Gepids. |
| 568 A.D. | — Longobards left Pannonia for Northern Italia. |
| | — From Inner Asia the Avars arrived in the Carpathian basin. First they took the Tisza River area, then Pannonia. |
| c. 580 A.D. | — Bishop of Scarbantia and other cities of Western Pannonia took part in the Council of Grado. |
| 582 A.D. | — The Avars occupied Sirmium. |
| 583 A.D. | — A great fire destroyed Sirmium. |
| 791–803 A.D. | — Charlemagne attacked the Avars and gained a victory. Enns–Danube–Drava region fell into Frankish hands. |
| 827–829 A.D. | — The eastern part of the area between the Drava and Sava rivers including Sirmium fell into Bulgarian hands. |
| 828 A.D. | — Carolingian administration on the territory occupied by the Franks of Pannonia. |
| c. 840 A.D. | — The Frankophile Slav Pribina from Nitra settled in Zalavár. |
| | — Pope Adrian appointed St. Methodius as Bishop of Pannonia with Sirmium as his seat. |
| 874 A.D. | — After the death of Kocel, the Frankish rule again extended to the east of the Raba River. |
| 883–884 A.D. | — Svatopluk, the Moravian emperor, raided Pannonia. |
| 896 A.D. | — The conquering Hungarians appeared in the Carpathian basin. |
| c. 900 A.D. | — The Hungarians occupied Pannonia. |

487

# INDEX

489

Andautonia 141, 143, 211, 239, 242, 255, 479
Andizetes 85, 88, 100, 141, 146, 242, 478
Andocs-Nagytoldipuszta 71
Animals 62, 282, 324, 330, 332
Annamatia 210, 215, 221, 391
Anne of Austria 19
Anonymus of Ravenna → Geographer
Anonymus (historian) 319
Anthropology 28—9, 429—54
Antiochia 267, 327, 333, 341
Antoninus Pius 93, 95, 100, 106, 129, 137, 341—2, 353, 392, 481
Ants 407
Apis 188
Apollo 166, 182, 184, 190, 258, 338, 349
Apollonius of Tyana 162
Appian 24, 57, 61, 88
Aqua Nigra 61
Aquae Balizae 141, 215
Aquae Iasae 165, 167, 264
Aqueducts 256, 262—4
Aquileia 24, 90, 97, 148, 167—8, 187, 193—4, 207, 211, 264, 267, 328, 330—2, 343, 360, 398, 418, 477, 481, 486
Aquincum 22, 33—4, 36—40, 44, 46, 50, 53—4, 90—2, 94—5, 100, 103—11, 115—6, 126-32, 134, 137-8, 141, 144—55, 162-73, 181, 185, 189—93, 196—8, 200, 209—15, 219—21, 225, 231—2, 239—40, 242—5, 247—8, 250—9, 275, 291, 299, 303, 305, 308, 312, 320, 325—34, 337, 352, 358, 360, 362—3, 365—7, 368, 374-75, 381—82, 387, 480—2, 484—5
Ara Augusti Provinciae 264
Arabia 167
Arcadius 118, 346, 485
Ardarich 399, 486
Arezzo 361
Argaragantes 112
Arianism 168, 196, 200—4, 351, 353, 484
Ariman 186
Ariminius 90, 479
Aristocracy 150, 161, 278, 324, 330, 332
Armor 385—96
Army 125—40
Arnulf 420
Árpád 320
Arrabo 61
Arrabona 47, 61, 95, 97, 116, 133, 165, 212, 214—5, 220—1, 329, 361, 366, 370—71, 372—5
Arts and crafts 349—55, 403
Arviates 85, 141, 146, 242,
Arvis 186
Asia, Central 413

Asia Minor 81, 152, 155, 162, 186, 189, 244, 327, 331, 358, 361, 366, 368
Askoi 73
Aszód 69
Athene → Minerva
Atrans 57—8, 211
Attila 18, 117—20, 319—20, 398—9, 406,486
Attis 171, 188—9
Auduin 404, 486
Aufidianus Rufus 139
Augustus 88—92, 100, 125, 131, 136, 333, 339 —41, 349, 351, 357, 361, 395, 478—9
Aunjetitz 79
Aurelia Marcellina 162
Aurelia Sabina 162
Aurelianus 105—6, 162, 192, 313, 316, 343, 345, 483
Aurelius 107
Aurelius Heraclitus 134
Aurelius Victor 26, 162
Aureolus 105
Aurochs 62
Austria 69, 78, 90
Auxilia Ursarensia 235
Avars 213, 246, 397, 405—16, 418, 452—3, 487
Avidius Cassius 98, 120, 481
Avitus 486
Azali 58, 85, 91, 99, 141—2, 170, 178
Azaum 51, 114, 220—1
Azilian 67

Baal 153, 165, 186, 189
Bacchus 166, 181
Bacuntius 61, 85
Baden culture 70—2
Bajan 405—6, 487
Bakony 402
Baláca → Nemesváros
Balaton 26, 34, 46, 58, 61—2, 65, 69—71, 74, 81, 91, 100, 108, 114, 120, 138, 143, 149— 50, 181, 197, 211—2, 246, 251, 262, 269, 275, 278, 282—5, 298, 300, 302, 313, 316, 318—9, 323, 325, 338, 351, 412, 479
Balatonaliga 365
Balatonalmádi 302
Balatonföldvár 298
Balatongyörök 297
Balbinus 104, 482
Balkans 22, 68—9, 71, 73, 81, 86, 111, 113, 116, 162, 170, 182, 207, 213, 331, 340, 344, 398, 477—8
Baltic Sea 90, 95, 207, 212
Baltis 186, 189
Banadaspes 97, 481
Banassac 362

Banat 94, 112, 116
Bándi, G. 75—7
Banner, J. 20—1, 71
Banoštor → Malata-Bononia
Bántapuszta 398
Baracs → Annamatia
Baracska → Iasulones
Baranya county 60, 72
Barátföldpuszta → Quadrata
Barbaricum 25, 50, 92, 95—6, 138, 154, 212—3, 246, 326, 329—32, 365
Barbius 187
Barkóczi, L. 6, 11, 20, 27—8, 34—5, 40—1, 46, 51, 85—124
Basante 61
Basilica 36, 41, 44—6, 156, 196—7, 245, 256, 261, 264, 267—8, 295—9, 319—20, 351
Basilius 403
Bassaeus Rufus 98
Bassiana 146, 152—3, 213, 243—4, 246, 258, 324, 482
Bastarnae 86, 110
Batavia 137, 166
Bathinus 61
Baths 37, 256, 266—8, 297, 326
Batina → Ad Militare
Bato 88—9, 478—9
Bauto 117
Bavaria 407, 416, 418
Békásmegyer 72, 357, 364
Bél, M. 33, 46
Béla (King) 18
Beleg 220
Belgites 85
Bella, L. 43
Bell-Beaker culture 73—4
Beneficiarii 47, 128—9, 262, 284, 332
Beneventum 165
Benjamin 193
Beograd → Singidunum
Beremend-Idamajor 318
Berkasovo 390
Bertalan, V. 37, 388
Bessarabia 111
Bezenye 406
Biatec 81, 338
Biblini Montes 60
Bicsérd 364—5, 370—1, 368
Bicske 68
Bihar 65
Bíró, E. 51
Biró-Sey K. 8, 11, 28, 337—48
Bitnitz, L. 43
Bivak cave 66
Black Sea 87, 399, 401, 486
Bócsa 415

Bodrogkeresztúr 69—70
Bogárzó 79
Bohemia 68, 338, 357
Boii 23, 81, 86—8, 90, 99—100, 141—2, 145—6, 149, 239, 242, 330, 332—3, 338—9, 477—8
Boleraz 70—1
Bolia 61
Bóna, I. 20, 73, 75—8
Bonfini 18, 33, 41
Bónis, É. B. 8—9, 11, 23, 27, 40, 357—80
Bononia 110—2, 132—3, 161, 212, 221, 483
Bonyhád 169, 198
Bosanska Gradiška → Servitium
Bosna 61
Bosnia 88, 478
Bosporus 399
Bosut 61—2
Botivo 262
Bozómindszent 415
Bölcske 74, 77
Börgönd 342
Braila 111, 484
Bratislava 23, 87, 338, 477
Brazda lui Novac 111
Brazlav 419—20
Brestanica 57
Breuci 85, 88—9, 141, 145—6, 243, 478
Bridges 210, 263, 267—8
Brigetio 28, 34—5, 44, 52, 58, 94, 97, 102—4, 106, 109, 112, 114—6, 126—8, 130—3, 141, 146—9, 152—3, 155, 164—5, 167—9, 172—3, 178, 185, 189, 191—2, 194—6, 198, 200, 209, 212—5, 219—21, 225, 244—5, 251—2, 258—9, 308, 310, 328, 337, 341, 343, 353, 358, 360, 363—7, 368, 372—5, 381—82, 385—7, 400, 480—2, 485
Britannia, Britain 81, 104, 109, 134, 225, 244, 340
British Museum 19
Brorup interstadial 66
Bronze age 22, 72—80
Bubastis 187, 266
Buda 18—9, 65, 70, 320, 325 → Aquincum
Budakalász 72
Budakeszi 376—7
Budalia 149, 251—2
Budapest 68—9, 73—4, 82, 132
Budapest → Aquincum
Bulgarians 407, 416, 418, 487
Buócz, T. 44
Burebista 87, 477
Burgenae 132—3, 221
Burgenland 17, 74, 85, 276
Burger, A. 27, 47, 51
Buri 99

491

492

495

Iran 186, 398
Irenaeus 26, 168, 194, 268
Iron age 80—2
Iron Gate 29, 478
Isaac 199
Isis, Iseum 44, 51, 167, 180, 187—8, 263, 265—6, 350, 353
Israel 382
Istria 88
Italia, Italici 58, 62, 86—7, 89—90, 96—7, 104, 106—7, 112—3, 116, 118—9, 125, 128, 130, 134—5, 137, 144—5, 163, 168, 170, 181—4, 192, 197, 207, 211, 213, 239, 242—3, 246, 275—6, 281, 284, 288—9, 292—4, 296, 300, 302, 306, 308—9, 326—9, 330—2, 340, 349—51, 353, 357, 359, 361—2, 366—7, 368, 381, 397, 402—3, 405—6, 485
Itinerarii 128, 149, 207—17
Itinerarium Antonini 17, 34—5, 37, 39—40, 42—4, 47, 60—1, 207, 212—7, 267
Iulia Mamaea 192—3
Iulius gentilicium 106—7
Iulius Ursus Servianus 126
Iulius Victorinus 173, 191
Iustinianus master 367
Iuvavum 247

Jakabffy, I. 11, 20
Jakusowice 400
Jankovich cave 66
Jánoshida 415
Járdányi-Paulovics, I. 35, 40—1, 44, 47
Jewelry 109, 328, 331, 403—4
Jews 26—7, 42, 47, 162, 187, 192—3, 199, 353
Jordanes 61
Joseph II 19
Juhász, L. 48
Julian Alps 58—60, 133
Julianus 115, 169, 173, 346, 484
Juno 43—4, 164, 167, 183—4, 190, 257, 265, 350
Jupiter 43—4, 163—4, 166—7, 180, 182—3, 189—90, 199, 257, 263, 265, 350, 390—1
Justunian 397, 404, 415, 486—7
Jutunghi 117

Kaba, M. 38—9
Kabhegy 313
Kabiri 190
Kádárta 447
Kafuan culture 65
Kajdacs 449
Kakasd 80
Kalenderberg 80
Kalicz, N. 69, 72

Kalocsa 68
Kama 412
Kamniške Planine 57—8
Kanzianberg 70
Kapella 58
Kápolnásnyék 447
Kapos 61, 81, 100, 339, 400
Kapospula 224, 245
Karavanka 58
Karst 60
Kassa 212
Kaszaháza 198
Katancsich, M. 40
Kazahstan 400
Kékkút 169, 196—8, 294—5, 298—9, 302, 319
Kendrisos 186
Keszthely 44, 46, 199, 325, 390
Keszthely culture 412, 416
Keszthely-Dobogó 172, 361, 370—1, 376—7
Keszthely-Fenékpuszta → Valcum
Keszthely-Újmajor 91, 172
Kéthely 313
Kézai, S. 18
Kilns 282, 357—61, 363—5, 368
Kisalföld 65
Kisapostag 75, 77
Kisárpás 224, 344, 350, 376—7
Kisdióspuszta 169
Kisdorog 201, 203
Kisil-kaja 398
Kiskevély cave 66
Kiskőrös 408
Kiskőszeg → Batina
Kismákfa 198—9
Kiss, Á. 312
Klosterneuburg → Cannabiaca
Klumbach, H. 27, 354, 359
Kocel 418, 487
Koller, J. 33, 48
Komin 105, 343
Koppány 182
Korabinsky, J. M. 46
Košice → Kassa
Koszider 77—9
Kölked → Altinum
Königshof 298
Környe 156, 224, 245, 409, 411, 413—4, 444—5
Körös 113, 414
Kralovánszky, A. 29
Krapina 66
Krászna 390
Kreisbach 168
Kretzoi, M. 65
Krka 61

Lussonium 189, 215, 221, 349, 480
Lužianky 69

Macedonia 23, 81, 89, 118, 127, 194, 338, 477
Macellum 255, 259, 265
Macrian 105
Macrinius Avitus Catonius Vindex 97, 133, 135, 138
Macrinus 154
Magdalenien 66
Magna Mater 188, 249, 259
Magnus Magnentius 484
Magyaróvár → Ad Flexum
Mahler, E. 41
Mainz 391—93, 395
Makó 72
Malata-Bononia → Bononia
Manaphus 167
Manojlović, M. 389—90
Mansiones 128, 149, 210, 213, 284, 288
Marble 62, 301—2, 330—1, 350—1
Marcellianus 115
Marcius Turbo 94, 480
Marcomannians, Marcomannic wars 58, 88—90, 93, 96—9, 102, 105, 110, 112, 126, 131—3, 135, 137—8, 147, 151, 154, 168, 221, 226, 230—1, 234, 236, 243, 248,2 257—60, 265, 269, 309, 318, 39, 337, 340—1, 478—9, 481—3
Marcus Antonius Victorianus 191, 340, 342
Marcus Aurelius 39, 95—101, 106—7, 109, 126, 130, 144, 152, 154, 191—2, 221—2, 243, 324, 332, 341, 481—2
Margitbánya 302
Maria Theresia 19
Mariaellend 209
Marinianis 58, 213
Marobuduus 90, 131, 478—9
Maros 78, 94, 110, 113, 213, 414
Marosi, A. 45
Mars 164, 166, 180, 184, 189, 328, 350, 376
Marsigli, J. 19, 29, 33, 35, 37, 40—1, 153
Marsonia 58, 216
Martian 398
Martinus (of Tours) 44, 197
Márton, L. 20
Marus 60
Mater Matuta 349
Matrica 30, 106, 196, 221, 306, 391—92
Matthias Corvinus 18, 33
Mauretania 95, 137
Mauritius 406
Maurus 106
Maxentius 110—1
Maximianus Herculius 46, 156
Maximinus master 367

Maximinus Thrax 103—4, 108, 138, 363, 482
Maximus Magnus 118, 485
Mecsek hills 138, 171, 181, 212
Mediolanum 128, 267, 344, 398, 486
Menas Ampulla 200
Meneandrus 267
Mercurius 165, 184, 190, 199
Méri, I. 21
Merobaudes 116, 485
Merovingians 406
Mesolithic 21, 67—8
Methodius 416, 487
Metubarbis insula 62, 85
Metulum 58, 88
Mezőcsát 80, 464
Mezőség 65, 73, 76—7
Mikoviny, S. 35
Milan → Mediolanum
Milanovce 212, 224
Milestones 104, 138, 208—9
Military → Army
Milles, J. 19, 33
Minajeva, T. M. 399
Mindel 65
Minerva 43—4, 153, 163—6, 180, 183—4,190, 261, 328, 350
Minicius Natalis Quadronius Verus 134
Mining 62, 66, 130, 268, 325
Minitra 164
Miskolc 212
Mithras, Mithreum 26, 40, 167—9, 190—2, 249, 256, 263, 330, 353, 364, 386
Mithridates 87
Mócsy, A. 20—4, 28—30, 34, 42, 51, 285
Modrič 168
Moesia 30, 88—90, 92—7, 102—3, 105, 111—2, 116, 127—8, 134—5, 211, 222, 251, 330, 333, 403, 480, 482—4
Mogentiana 46, 100, 146—8, 162, 212, 214, 242—3, 262, 326, 418, 481
Mohács 77, 449
Moldavia 101, 103, 111
Mommsen, Th. 41, 50
Mondsee 70
Money circulation 92, 332—4
Mongolia 400
Mons Aureus 60, 133, 324
Mons Cetius 57, 60
Mons Porphyreticus 60
Mons Scardus 86
Montans 362
Monumentum Ancyranum 90, 478
Mór valley 58
Morava river 60, 212
Moravia 21, 68—70, 400, 416—20, 487
Mórichida-Kisárpás → Mursella II

505

# PLATES

Plate I

1. Detail of a tombstone with hunting scene (Hungarian National Museum). Site: Császár;
2. Mucius Scaevola relief (Hungarian National Museum)

Plate II

1. Laurel-shaped stone blade from the Jankovich cave; 2. Neolithic clay vessel from the Zseliz
Culture. Site: Nagytétény; 3. Vessel from the Lengyel Culture

Plate III

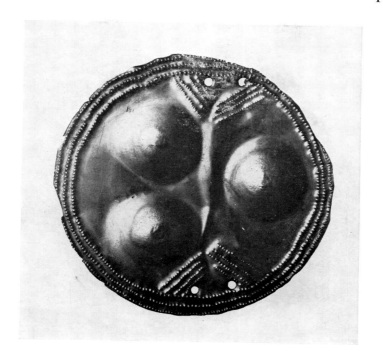

1. Gold disc of Csáford; 2. Two-parted bowl from the twenty-eighth grave of Alsónémedi cemetery

Plate IV

1. Bird-shaped vessel. Site: Zók; 2. A vessel belonging to the Bell-Beaker Culture; 3. Bronze
Age vessel. Site: Nyergesújfalu

Plate V

1. Chalk-incrusted pottery from South Pannonia; 2. Urn from the Vatya Culture

Plate VI

**1.** Black and red painted urn with animal figure from North Transdanubian Vaszar; 2. Bronze figure of a wild boar. Site: Báta

Plate VII

Milestones from the third century

Plate VIII

1. Wall of the corner tower of the Visegrád Roman Camp; 2. Building inscription from 372.
Site: *Burgus* Visegrád

Plate IX

1. Brick with stamp: Legio I adiutrix pia fidelis; 2. Valentinian brick with stamp: Terentianus Tribunus

Plate X

1. Excavation of gate-tower of fourth-century fortified settlement at Heténypuszta. Campaign 1970; 2. Detail of fourth-century *burgus* from Pilismarót

Plate XI

1. *Burgus* from Esztergom-Szentgyörgymező, fourth century; 2. Building inscription of a Commodean *burgus* from A.D. 184–185. Site: Intercisa

Plate XII

Detail of Roman Amber Route

Plate XIII

Statues of Capitolean Trias in Scarbantia

Plate XIV

1. Detail of forum in Gorsium with nympheums; 2. Peristyle yard of Aquincum Civilian City

Plate XV

Bath of a private house with "Wringlers" mosaics in Aquincum Civilian City

Plate XVI

1. Detail from laconicum of great public bath, Aquincum Civilian City; 2. Port with steps from Aquincum military amphitheater

Plate XVII

Wall and ceiling painting from Governor's Palace in Aquincum

Plate XVIII

Mosaic floor with the mythological scene Deianeira, Aquincum, *canabae*, "Hercules villa"

Plate XIX

Savaria, Iseum

Plate XX

Detail of a late-Roman fortified city wall in Scarbantia

Plate XXI

Detail of the *palatium* in Gorsium

Plate XXII

Gorsium, tombs from fourth-century cemetery

Plate XXIII

Gorsium, detail of fourth-century cemetery

Plate XXIV

Three-naved basilica of Fenékpuszta

Plate XXV

Silver lamp from the villa of Balácapuszta

Plate XXVI

Find-group of *lararium* of Nagydém

Plate XXVII

Head of *lar* from Nagydém

Plate XXVIII

Bronze statue of Apollo, *lararium* of Nagydém

Plate XXIX

Bronze statue of Concordia Domitilla from *lararium* of Tamási

Plate XXX

Bronze statue of Concordia Domitilla from *lararium* of Tamási, profile

Plate XXXI

Head of Concordia Domitilla from *lararium* of Tamási

Plate XXXII

Jupiter statue from *lararium* of Tamási

Plate XXXIII

Jupiter head from *lararium* of Tamási

Plate XXXIV

The Minerva of Tamási

Plate XXXV

The Minerva of Tamási, profile

Plate XXXVI

Commodus Hercules statue from Ajka

Plate XXXVII

Head of the Commodus Hercules statue, Ajka

Plate XXXVIII

The gravestone of Dexter and Julia Prisca from Ajka

Plate XXXIX

Typical Pannonian red sandstone column of Fonyód

Plate XL

Hypocaustum and wall heating tubes from Roman villa

Plate XLI

Wall painting with black background with figure of Child Dionysos from *tablinum* of Baláca-puszta villa

Plate XLII

Wall painting from red and black room of Balácapuszta villa, late first–early second centuries

Plate XLIII

Wall painting remains from *coenaculum* of Balácapuszta villa

Plate XLIV

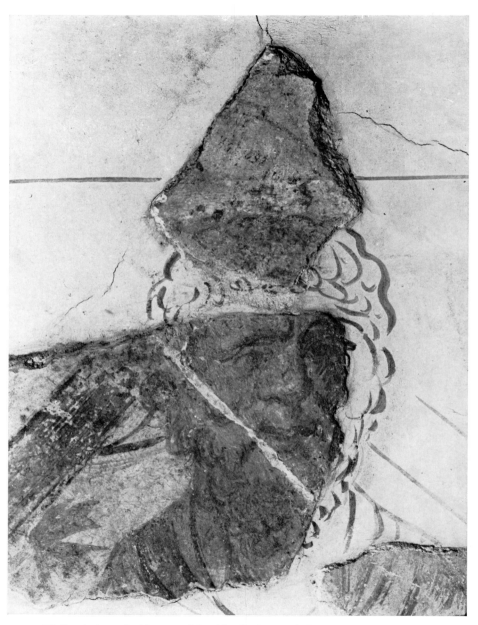

Wall painting of old man with white background from Balácapuszta villa,
second century

Plate XLV

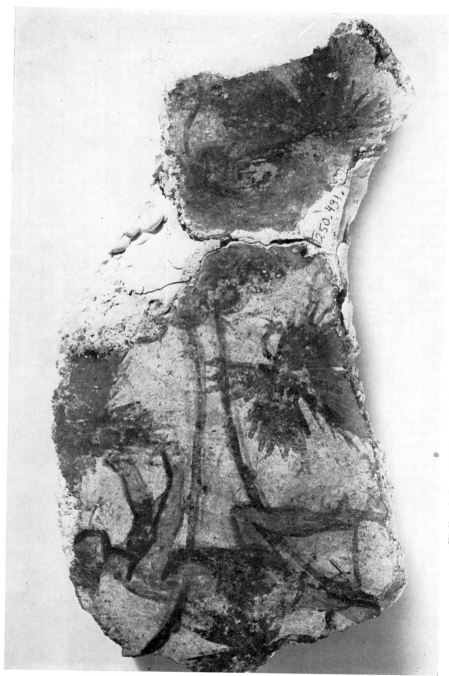

Wall painting with slave plucking grapes from Balácapuszta villa, second century

Plate XLVI

"Classical" stucco ledge from Balácapuszta villa, late first–early second centuries

Plate XLVII

Mosaic floor from Room 20 of Balácapuszta villa, late first –
early second centuries

Plate XLVIII

1. Scordiscus, tetradrachma; 2. Boii, stater; 3. Small silver coin from Réte; 4. Same, 17 gramme silver; 5. Biatec; 6. Nonnus; 7. Small silver coin from Tótfalu; 8.Tetradrachma from Noricum; 9. Tetradrachma with Apollo head; 10. Tetradrachma from Zichy hoard, Újfalu; 11. Tetradrachma, Mászlonypuszta find; 12. Tetradrachma of Regöly type; 13. Eraviscus denarius

Plate XLIX

1. Septimius Severus, bronze denarius; 2. Septimius Severus, subaeratus denarius; 3. Viminacium, large bronze coin of Gordianus III; 4. Alexander Severus bronze denarius counterfeit; 5. Contemporary bronze counterfeit of centenionalis; 6. Contemporary gold forgery of Constantius I; 7. Regalianus antoninianus; 8. Gallienus bronze antoninianus, Siscian coinage; 9. Constantius II centenionalis, Siscian coinage; 10. Valentinian centenionalis, Sirmian coinage

Plate L

Red-white painted vessel stamped with gem from Late Celtic period, Budapest-Tabán

Plate LI

Stamp of potter Resatus on bottom of a gray bowl

Plate LII

1. Red-striped vessels from Brigetio kiln; 2. Thin-walled cups from Italia

Plate LIII

*Terra sigillata* vessel from Rheinzabern, Gorsium

Plate LIV

1. "Raetian" cup; 2. "Motto" jar from Rhine region

Plate LV

Mold of handle of clay *patera*, Brigetio

Plate LVI

Brown glazed jug, Intercisa

Plate LVII

*Lucerna* imitating a gladiatorial helmet

Plate LVIII

Ring-shaped *lucerna* with several lights, Aquincum

Plate LIX

Crustulum form, Savaria

Plate LX

Clay turret, Aquincum

Plate LXI

Earthen gate model, Intercisa

Plate LXII

Cylindric two-handled bottle with engraved decoration, Intercisa; and one-handled bottle, Császár

Plate LXIII

Syrian-Oriental glassware

Plate LXIV

Globe-bodied Pannonian flask, third–fourth centuries, Intercisa

Plate LXV

Late-Roman glass with molten blue dots, fourth century, Intercisa

Plate LXVI

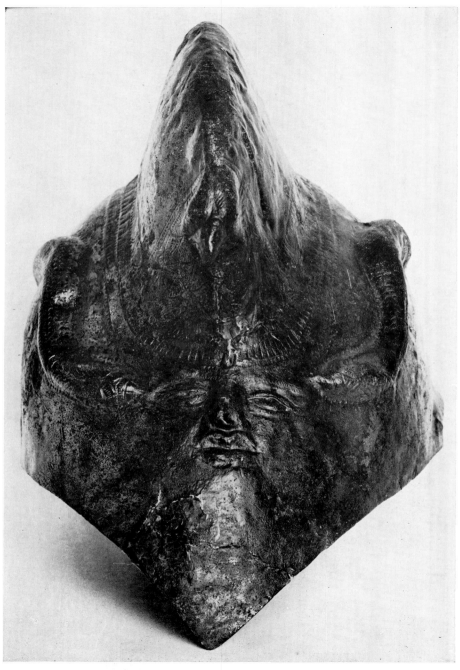

Bronze helmet with human face from Brigetio, second century

Plate LXVII

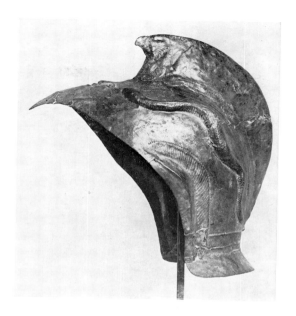

1. Side view of Brigetio bronze helmet; 2. Detail of eagle from Brigetio bronze helmet

Plate LXVIII

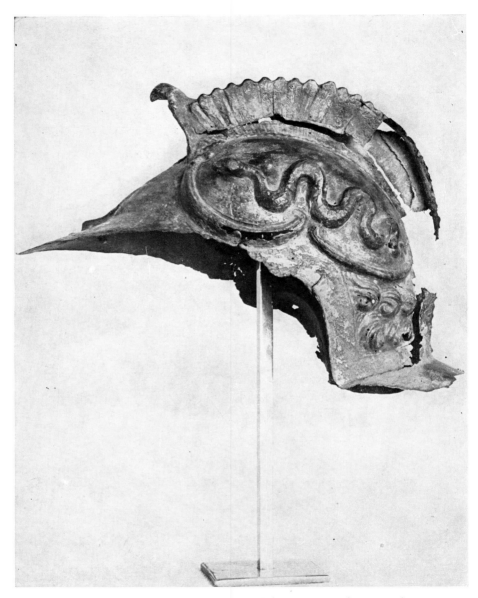

Bronze helmet with white metal covering and rich ornamentation, second century

Plate LXIX

Tall-crested iron helmet with remains of silver plate cover, fourth century, Intercisa

Plate LXX

Iron helmet, Intercisa, fourth century

Plate LXXI

Gilded bronze helmet from Eskü square, Budapest

Plate LXXII

Detail of iron helmet with gilded silver covering and relief ornamentation, precious stone imitations, second half of fourth century, Eskü square, Budapest

Plate LXXIII

Detail of helmet, Eskü square, Budapest

Plate LXXIV

1. Bronze shield-boss with white metal covering, wild dogs chasing a hare, incised scene, seco nd century, Százhalombatta; 2. Punched retrograde "R" from the Brigetio helmet

Plate LXXV

1. Badge with eagle, from shield-boss of Százhalombatta; 2. Bronze shield-boss with white metal covering, engraved plant ornaments, Százhalombatta

Plate LXXVI

Bronze shield-boss incised decoration, second century, Dunaföldvár

Plate LXXVII

Shield-boss with inscriptions

Plate LXXVIII

Iron dagger blade and sheath with gold, silver, and enamel inlay, early first century, Duna-földvár

Plate LXXIX

Detail of the dagger, Dunaföldvár

Plate LXXX

*Terra-cotta* votive in honor of Mater Matuta, Lussonium

Plate LXXXI

*Terra-cotta* votive in honor of Mater Matuta, Lussonium

Plate LXXXII

1. Bronze casket mount with personification of seasons, Kisárpás; 2. Bronze casket mount with a *thiasus* scene

Plate LXXXIII

Stone slab with figure of woman carrying casket, Intercisa

Plate LXXXIV

Isis on Sothis dog, detail of the frontal frieze of Isis shrine of Savaria

Plate LXXXV

Victory with palm, detail from frieze of Isis shrine in Savaria, first half of third century

Plate LXXXVI

Bronze jar with gold and silver inlay belonging to Isis Cult, Egyed

Plate LXXXVII

Bronze *patera* with gold and silver inlay belonging to Isis Cult, Egyed

Plate LXXXVIII

Stone carving with Silvanus figure from Aquincum

Plate LXXXIX

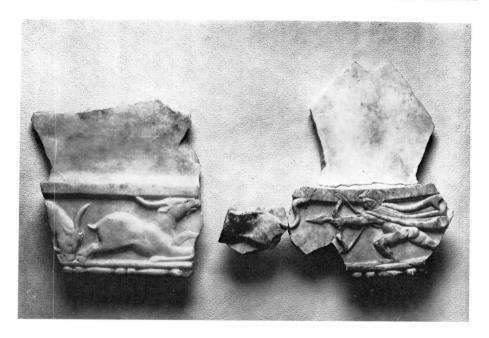

1. Figural rim ornamentation of Early Christian marble altar slab, Csopak; 2. Detail from Early Christian marble altar slab representing unicorn, Csopak

Plate XC

Gravestones with astral symbols of indigenous population of Pannonia  Szentendre and
Csákberény

Plate XCI

Gravestone of Boi woman called Belatusa, end of first century or early second, Bruck an der Leitha

Plate XCII

Gravestone of Veriuga, indigenous Pannonian woman vested in characteristic Pannonian clothes with "winged" fibulae, second century, Intercisa

Plate XCIII

Gravestone of M. Aur. Avitianus, miles of Legio Adiutrix, Tatabánya

Plate XCIV

Statue of Fortuna Nemesis from Governor's Palace of Aquincum, third century

Plate XCV

Chryselephantine Isis head from the *kithera* of Intercisa

Plate XCVI

Bronze gate-tower, once used as coach decoration, fourth century

Plate XCVII

Stone votive slab of the Jewish Cosmius, Intercisa

Plate XCVIII

Late Roman emperor bust, Sopianae

Plate XCIX

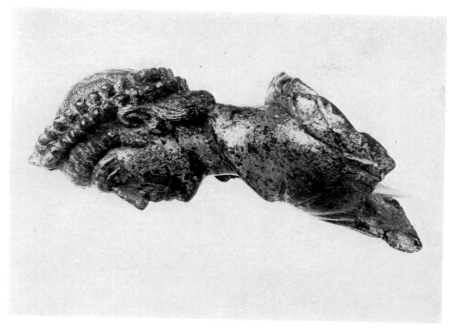

Back view and profile of a late-Roman emperor bust, Sopianae

Plate C

Silver trypos with Tryton and Nereid figures, Polgárdi

Silver trypos with Tryton and Nereid figures, Polgárdi

Plate CII

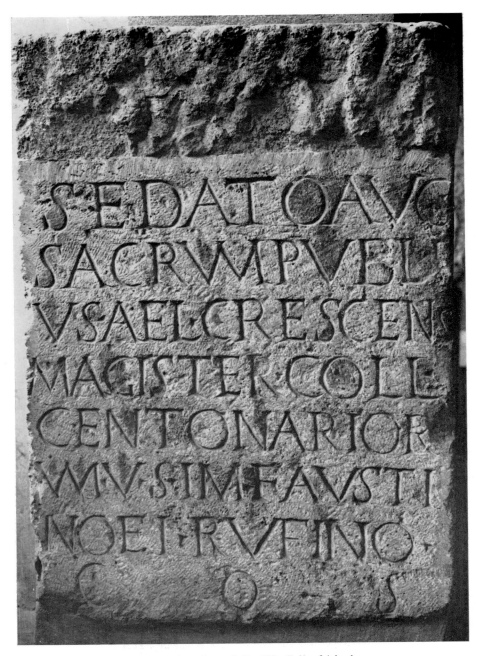

Sedatus altar from A.D. 210, Székesfehérvár

Plate CIII

Silvanus altar with Pan relief, Aquincum

Plate CIV

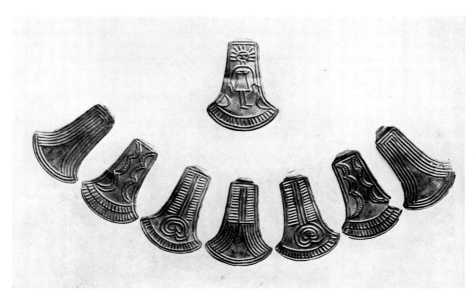

1. Silver radial crown of Sol and Luna priestesses, Szalacska; 2. Breast-plates of Sol and Luna priestesses, Szalacska

Plate CV

Satyr bust from coach mounts of Somodor

Plate CVI

Marble statue of Jupiter offering sacrifice, third century

Plate CVII

Minerva, marble torso from capitol of Savaria

Plate CVIII

Views of Jupiter pillar from Brigetio

Plate CIX

Views of Jupiter pillar from Brigetio

Plate CX

Inscription of Jupiter pillar, Brigetio

Plate CXI

Bronze Victory statuette, first century, Akasztó

Plate CXII

Votive slab of Priapos, god of fertility, second–third centuries, Savaria

Plate CXIII

Diana relief of marble, second–third centuries

Plate CXIV

Child Dionysos, ivory statue, third century, Savaria

Plate CXV

Bronze statuette of Liber Pater, onetime coach decoration, third century, Tata

Plate CXVI

Altar in honor of Liber and Libera, from A.D. 193 or 203, Brigetio

Plate CXVII

Relief of Liber Pater from altar in Brigetio

Plate CXVIII

Statue of Dionysos, Satyros, and Pan, top decoration of Roman carriage found at Somodor-puszta, early second century

Plate CXIX

Jupiter's bronze bust lamp, first century, Mór

Plate CXX

*Suovetaurilia.* Bronze relief, Savaria

Plate CXXI

Bronze bust of Juno in weight form

Plate CXXII

Crustulum mold, *terra-cotta* with emperor as *pontifex maximus*, Sirmium

Plate CXXIII

Silver plate bust of Trebonianus Gallus, Brigetio

Plate CXXIV

Bronze plaque for Jupiter Dolichenus, covered with white metal, third century, Lussonium

Plate CXXV

Reverse of plaque for Jupiter Dolichenus

Plate CXXVI

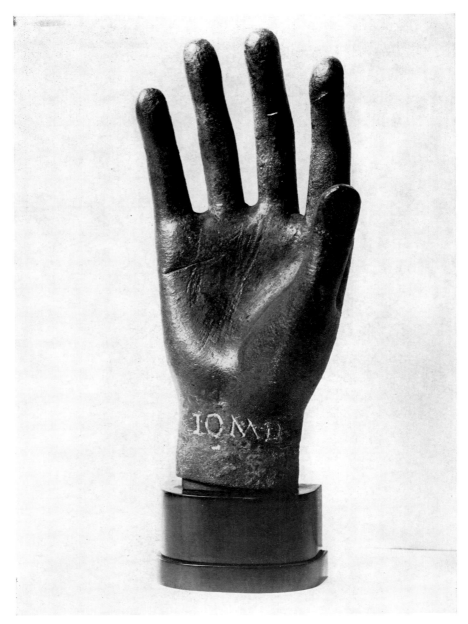

Bronze votive hand in honor of Jupiter Dolichenus

Plate CXXVII

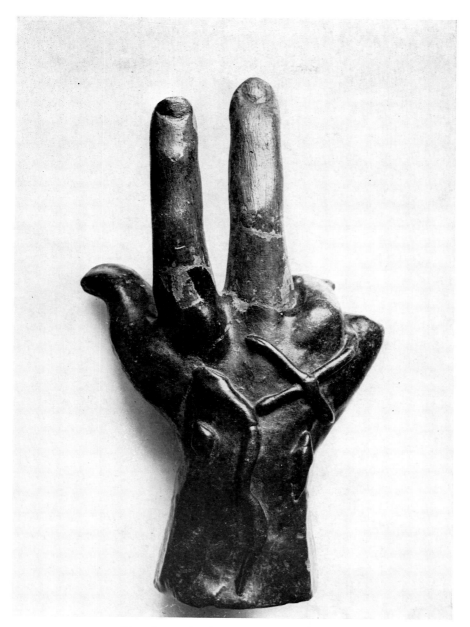

Glazed ceramic hand in honor of Sabasios

Plate CXXVIII

1. Bronze Osiris statuette; 2. Silver plate representing Sol in quadriga, Intercisa

Plate CXXIX

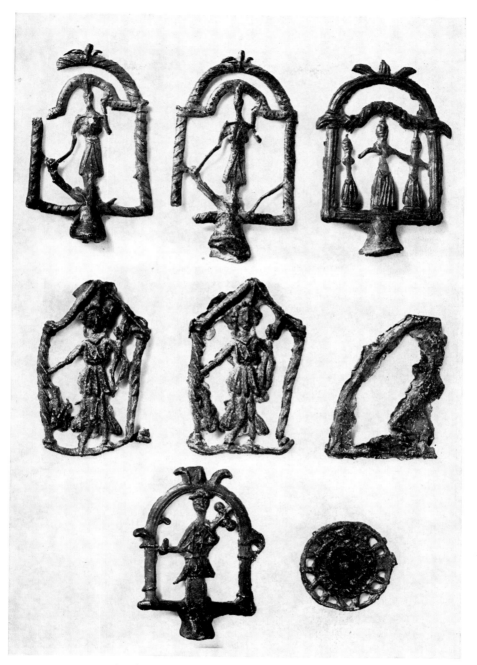

Plumb votive objects from Pogánytelek foundry

Plate CXXX

Stone figure of Attis from a tomb-building

Plate CXXXI

1. Bronze bull figure, Egyptian cult; 2. Plumb votive tablet for syncretic cult of Danubian cavalry god

Plate CXXXII

Votive marble figure in honor of Thracian cavalry god, fourth century

Plate CXXXIII

Votive table in honor of Mithras, bronze with white metal, third century, Brigetio

Plate CXXXIV

Gravestone of Jewish family with *Eis Theos* inscription and *menoras*, probably from Aquincum,
third century

Plate CXXXV

1. Detail of Jewish family gravestone, Aquincum; 2. Jewish lamp
with seven-branched *menora*, Savaria

Plate CXXXVI

1. Brick inscription fragment commemorating martyrdom, Brigetio;
2. Open-work brick Christogram, Kékkút

Plate CXXXVII

Early Christian lamp with Christ's initials

Plate CXXXVIII

Member of chain holding lamp with Christ's initials, Bonyhád

Plate CXXXIX

Early Christian silver *patera* with symbolic representations from the fifth century, Kismákfa

Plate CXL

Details of Christian sarcophagus, Szekszárd

Plate CXLI

*Vas diatretum* with Christian legend, Szekszárd

Plate CXLII

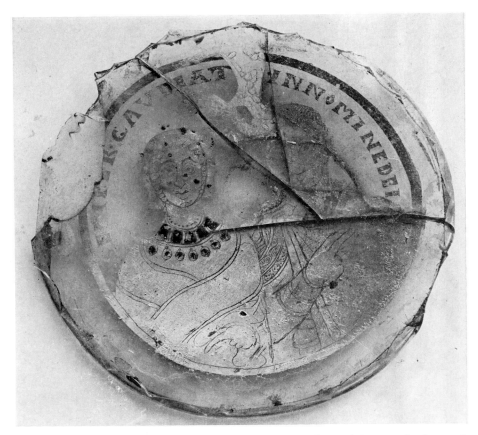

Fondo d'oro with "Semper gaudeatis in nomine Dei" inscription and the portrait of a couple, fourth century, Dunaszekcső

Plate CXLIII

Glass cup with molten blue dots with engraved text and ornament, Ságvár

Plate CXLIV

Bronze casket-mounts with personifications of cities

Plate CXLV

Bronze casket-mounts with Good Shepherd and Daniel among the lions, Ságvár

Plate CXLVI

Early Christian marble slab with Christ's initials, Savaria

Plate CXLVII

Wall painting in grave chamber with Christ's initials and figures of Peter and Paul, Sopianae

Plate CXLVIII

1. *Terra-cotta* Menas ampulla, Savaria; 2. NAMMIVS CUSTOR (sic) CYMETERI, gravestone inscription

Plate CXLIX

Brick with Christ's initials, Dombóvár

Plate CL

Incised brick with figure of Arius, Kisdorog

Plate CLI

1. Arius inscription, Kisdorog; 2. Anchor incised in brick, Kisdorog

Plate CLII

Hunnish kettle from Kaposvölgy

Plate CLIII

Jug of Murga

Plate CLIV

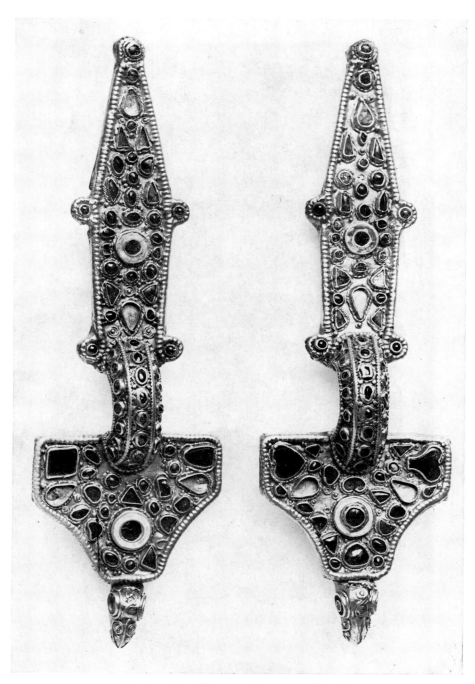

Jewelry of a rich woman's grave at Regöly

Plate CLV

1. Jewelry of a rich woman's grave at Regöly; 2. Longobard jewelry from Várpalota cemetery

Plate CLVI

1. Bracteates from Várpalota; 2. Enlarged bracteate

Plate CLVII

1. Lead plate with runic incisions from Béndekpuszta; 2. Bow-plate from Környe with runic inscription

Plate CLVIII

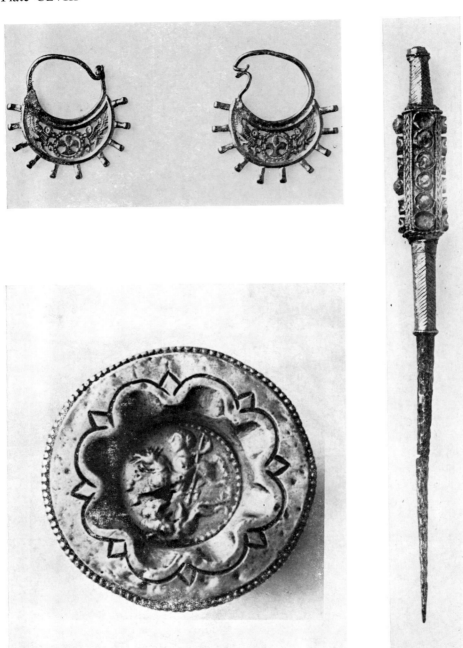

Finds from a rich woman's grave from Keszthely-Fenékpuszta, sixth century

Plate CLIX

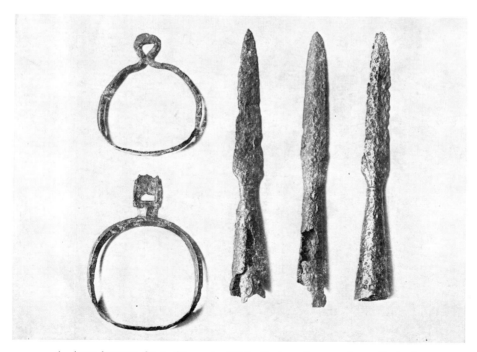

1. Avar harness from Grave A, Cikó; 2. Avar harness from Csolnok

Plate CLX

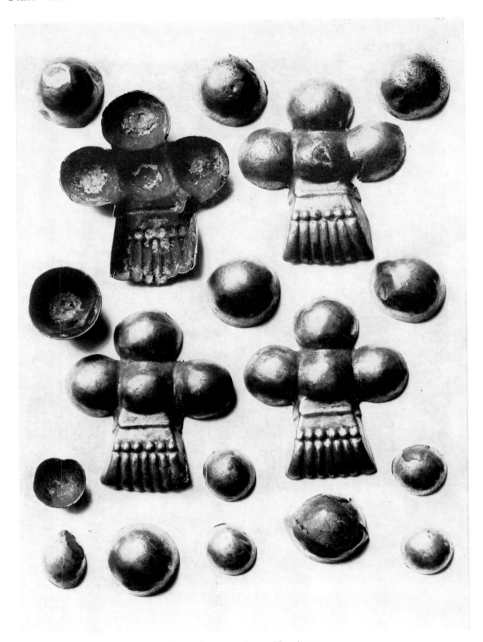

Avar harness from Kunágtoa

Plate CLXI

1. Jewelry from a rich Avar grave from Szentendre; 2. Jewelry from Keszthely

Plate CLXII

Avar gold belt-mounts from Kunágota

Plate CLXIII

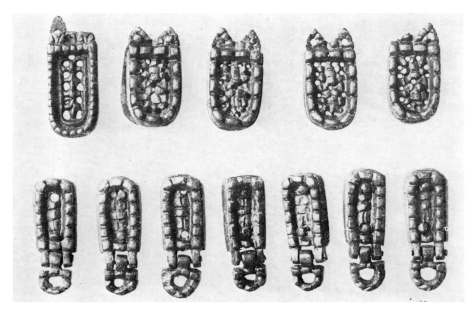

1, 2. Avar bronze belt-mounts from Dunaszekcső

Plate CLXIV

Avar bronze belt-mounts from Dunaszekcső

Plate CLXV

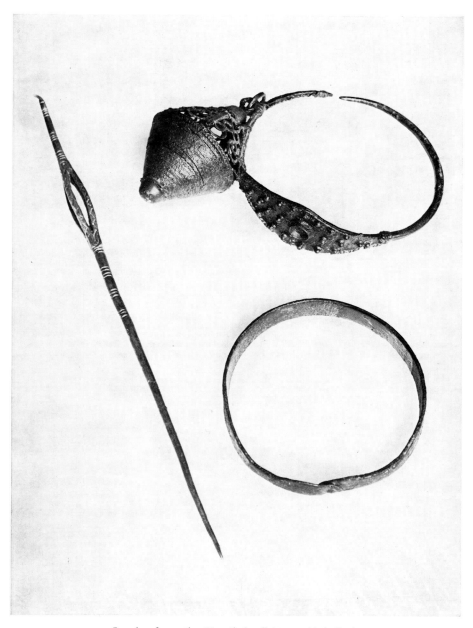

Jewelry from the Keszthely Culture, Alsópáhok

Plate CLXVI

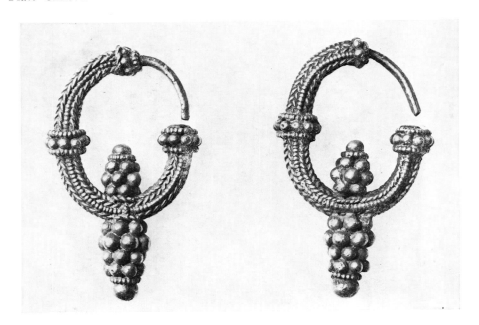

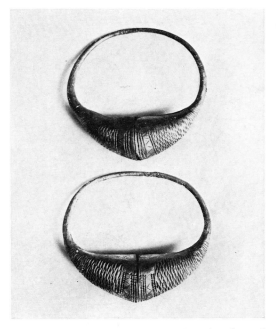

1. Gold earrings from Zalavár; 2. Silver bracelets from Szentendre

Plate CLXVII

Gilt silver purse-plate from the time of the Hungarian Conquest,
Szolnok-Strázsahalom